ENVIRONMENTAL REGULATORY CALCULATIONS HANDBOOK

ENVIRONMENTAL REGULATORY CALCULATIONS HANDBOOK

LEO STANDER
LOUIS THEODORE

A JOHN WILEY & SONS, INC., PUBLICATION

Library of Congress Cataloging-in-Publication Data:

Stander, Leo.
  Environmental regulatory calculations handbook/by Leo Stander and Louis Theodore.
       p. cm.
    Includes index.
    ISBN 978-0-471-67171-8 (cloth)
    1. Environmental law–United States.   2.   Environmental policy–United States.
    I.   Theodore, Louis.   II.   Title.
    KF3775.S73 2007
    344.7304'6–dc22

                                                                          2007040462

Printed in the United States of America

10 9 8 7 6 5 4 3 2 1

CONTENTS

4   CLEAN WATER ACT (CWA)   **135**

8 COMPREHENSIVE ENVIRONMENTAL RESPONSE, COMPENSATION AND LIABILITY ACT (CERCLA-SUPERFUND)   333

9 OCCUPATIONAL SAFETY AND HEALTH ACT (OSHA) 397

10   POLLUTION PREVENTION ACT (PPA)   453

PREFACE

In the past three and one half decades there has been an increased awareness of a wide range of environmental issues covering all sources: air, land, and water. More and more people are becoming aware of these environmental concerns, and it is important that professional people, many of whom do not possess an understanding of environmental regulatory problems, have the proper information available when involved with environmental regulations and calculations, i.e., all professionals should have a basic understanding of the technical and scientific terms related to these issues as well as the regulations involved. In addition to serving the needs of the professionals in industry, this *Handbook* also provides information of value to over 100,000 (in this country alone) regulatory officials.

This book is primarily intended for individuals with environmental regulatory responsibilities and concerns. It is presented in simple, understandable terms that provide the basic fundamentals of the many environmental regulatory topics that exist and may exist in the future. The authors' objective is to provide both background material and technical know-how on these issues.

This book is divided into ten Parts. Part I Provides an essay overview of the early history of environmental problems. Part II deals with the current regulatory framework. Part III–X constitutes the heart of the *Handbook*, including problems/solutions with sections on the following key laws and regulations:

Clean Air Act (3)
Clean Water Act (4)
Safe Drinking Water Act (5)
Resource Conservation and Recovery Act (6)
Toxic Substance Control Act (7)

Superfund and Comprehensive Environmental Response, Compensation and
   Liability Act (8)

Occupational Safety and Health Act (9)

Pollution Prevention Act (10)

In addition, the Introduction contains numerous regulatory references employed
in the preparation of this *Handbook*.

As is usually the case in preparing any text, the question of what to include and
what to omit has been particularly difficult. The *Handbook* provides the reader with
nearly 400 solved problems in the regulatory field. One of the key features of this
book is that the solutions to the problems are presented in a stand-alone manner.
Throughout the book, the problems are laid out in such a way as to develop the
reader's technical understanding of the regulatory subject in question. Each
problem contains a title, problem statement, reference to the key regulations
(where applicable) and data and solution, with the more difficult problems located
at or near the end of each problem set. Thus, this *Handbook* offers material not
only to individuals with limited technical background but also to those with extensive industrial experience.

The authors cannot claim sole authorship to all the problems and material in
this *Handbook*. The present book has evolved from a host of sources including: Code
of Federal Regulations; Federal Register; notes, homework problems, and exam problems prepared by L. Theodore for several chemical and environmental engineering
graduate and undergraduate courses; problems and solutions drawn (with permission)
from numerous Theodore Tutorials; Stander's personal notes; and problems and solutions developed by faculty participants during National Science Foundation (NSF)
Undergraduate Faculty Enhancement Program (UFEP) workshops.

The Appendix is another feature of this Regulatory *Handbook*. It contains three
Sections:

A. International regulations

B. ISO 14000

C. Miscellaneous topics

A short writing plus several problems (and solutions) are included in sections A and
B. The latter section treats the following topics:

C1. In state regulatory agency names

C2. Federation and preemption

C3. Hybrid systems

C4. Electromotive fields (EMFA)

C5. Life cycle analysis

C6. Environmental justice

One of the objectives of the aforementioned NSF workshops included the development of illustrative examples by the faculty. Some of the problems provided in

this *Handbook* were drawn, in part, from the original work of these faculty. We would like to acknowledge the following professors whose problems, in original or edited form, are included in this *Handbook*.

Prof. William Auberle; Civil and Environmental Engineering, Northern Arizona University

Dr. Howard Bein; Chemistry, U.S. Merchant Marine Academy,

Dr. Seymour Block; Chemical Engineering, University of Florida

Dr. Ihab Farag; Chemical Engineering, University of New Hampshire

Dr. Kumar Ganesan; Environmental Engineering, Montana Tech of the University of Montana

Dr. David James; Civil and Environmental Engineering, University of Nevada at Las Vegas

Dr. Christopher Koroneos; Chemical Engineering, Columbia University

Dr. SoonSik Lim; Chemical Engineering, Youngstown State University

Dr. Sean X. Liu; Civil and Environmental Engineering, University of California at Berkley

Dr. P.M. Lutchmansingh; Petroleum Engineering, Montana Tech of the University of Montana

Dr. Suwanchai Nitisoravut; Civil Engineering, University of North Carolina at Charlotte

Dr. Holly Peterson; Environmental Engineering, Montana Tech of the University of Montana

Dr. Lisa Reidl; Civil Engineering, University of Wisconsin at Platteville

Dr. Carol Reifschneider; Science and Math, Montana State University

Dr. Dennis Ryan; Chemistry, Hofstra University,

Dr. Dilip K. Singh; Chemical Engineering, Youngstown University

Dr. David Stevens; Civil and Environmental Engineering, Utah State University

Dr. Bruce Thomson; Civil Engineering, University of New Mexico

Dr. Frank Worley; Chemical Engineering, University of Houston

Dr. Ronald Wukash; Civil Engineering, Purdue University

Dr. Poa-Chiang (PC) Yuan; Civil Engineering, Jackson State University

During the preparation of this Handbook, the authors were ably assisted in many ways by a number of graduate students in Manhattan College's Chemical Engineering Master's Program. These students devoted time and energy researching and classroom testing various problems in the book. These students are acknowledged in the body of the book. Thanks are also due Stacey Shafer and Lucas Dorazio for their special contributions.

Reasonable care has been taken to assure the accuracy of the information contained in the *Handbook*. However, the authors and the publisher cannot be responsible for errors or omissions in the information presented or for any consequences arising from the use of this information.

LEO STANDER
LOUIS THEODORE

INTRODUCTION

Many environmental laws, in particular federal regulations as published in the code of Federal Regulations (CFR) were a generous resource for problem (and solutions). The entries in the Table below contain fields of information drawn from the Clean Air Act, Safe Drinking Water Act, etc. Many of these citations were employed in the development of problems illustrations examples contained in this calculations *Handbook*.

TABLE 1 Key Regulations

Regulation 2006

40 Code of Federal Regulations (CFR), Part xxx, 2006 Edition

Clean Air Act (CAA)
40 CFR Part 50, National Primary and Secondary Ambient Air Quality Standards
40 CFR Part 60, Standards of Performance for New Stationary Sources
40 CFR Part 61, National Emission Standards for Hazardous Air Pollutants
40 CFR Part 63, National Emission Standards for Hazardous Air Pollutants for Source
  Categories

SDWA (Safe Drinking Water Act)
70 FR 9071 February 24, 2005, "Contaminant Candidate List"

(Continued)

TABLE 1 *Continued*

Regulation 2006

40CFR141.11: MCL (maximum contaminant level) for inorganic chemicals
40CFR141.13: MCL for turbidity
40CFR141.23: Inorganic chemical sampling and analytical requirements
40CFR141.24: Organic chemicals other than total trihalomethanes, sampling and analytical requirements
40CFR141.40: Monitoring requirements for unregulated contaminants
40CFR141.40.j: Monitoring at the discretion of the state
40CFR141.50(b): MCLG (maximum contaminant level goal) for organic contaminants
40CFR141.51: MCLG for inorganic contaminants
40CFR141.61: MCL for organic chemicals
40CFR141.62: MCL for inorganic chemicals
40CFR141.80: General requirements for control of lead and copper
40CFR141.82: Description of corrosion control treatment requirement
40CFR141.87: Monitoring requirements for water quality parameters
40CFR142.62: Variances and exemptions from the maximum contaminant levels for organic and inorganic chemicals
40CFR143.3: Secondary maximum contaminant levels
40CFR148.10: Waste specific prohibitions-solvent wastes

RCRA (Resource Conservation and Recovery Act)
40CFR261.24: Toxicity characteristic (D waste) RCRA, 40 CFR302.4: designation of hazardous substances
40CFR261.31: Hazardous wastes from non-specific sources (or F waste) RCRA, 40CFR302.4: designation of hazardous substances
40CFR261.32: Hazardous wastes from specific sources (or K waste) RCRA, 40 CFR302.4: designation of hazardous substances
40CFR258.40: Design criteria for municipal solid waste landfill (MSWLF)
40CFR258-Appendix 1: Constituents for detection monitoring (for MSWLF)
40CFR258-Appendix 2: List of hazardous inorganic and organic constituents
40CFR261.33: Discarded commercial chemical products, off-specification species, container residues, and spill residues thereof. 261.33.e. (P waste)
40CFR261.33: Discarded commercial chemical products, off-specification species, container residues, and spill residues thereof. 261.33.f. (U waste)
40CFR261 Appendix VIII: Hazardous constituents, see also 40 CFR261.11
40CFR264 Appendix IX: Ground water monitoring list
40CFR266 Appendix IV: Reference air concentration
40CFR266 Appendix V: Risk specific doses
40CFR266 Appendix VII: Health based limits for exclusion of waste-derived residues
40CFR266 Appendix VIII: Organic compounds for which residues must be analysed
40 CFR Part 270, EPA Administered Hazardous Waste Permit Program
40 CFR Part 271, State Administered Hazardous Waste Program

(Continued)

TABLE 1 *Continued*

Regulation 2006

SUPERFUND & CERCLA (Comprehensive Environmental Response, Compensation and Liability Act)
40CFR302.4: Designation of hazardous substances
40CFR355-Appendix A: 40 CFR355-Appendix B: list of extremely hazardous substances and their threshold planning quantities
40CFR372.65(b): Chemicals and chemical categories to which this part applies (CAS number listing)
40CFR372.65: Chemicals and chemical categories to which this part applies (chemical categories in alphabetical order)
40CFR372.65(b): Chemicals and chemical categories to which this part applies (under diisocyanates)
40CFR372.65: Chemicals and chemical categories to which this part applies (under polycyclic aromatic compounds)

CWA (Clean Water Act)
40CFR1164.4: Designation of hazardous substances
40CFR117.3: Determination of reportable quantities
40CFR401.15: Toxic pollutants (identical to compounds in 40 CFR403 Appendix B)
40CFR413.02: Total toxic organics (TTOCs)
40CFR423-Appendix A: 126 priority pollutants

TSCA (Toxic Substances Control Act)
Chemical substance matrix by CAS registry number and trade name matrix in alphabetical order
40CFR712.30: Chemical lists and reporting periods
40CFR712.30d: Chemical listsed by reporting dates
40CFR712.30.c: Substances listed by categories
40CFR712.30.e: Aldehydes
40CFR712.30.e: Alkyl-, chloro-, and hydroxy-methyl diary ethers
40CFR712.30.e: Alkyl phosphates
40CFR712.30.e: Brominated flame retardants
40CFR712.30e: Chloroalkyl phosphates
40CFR712.30.e: Cyanoacrylates
40CFR712.30.e: IRIS chemicals
40CFR712.30.e: Isocyanates
40CFR712.30.e: Methyl ethylene glycol ethers and esters
40CFR712.30.e: OSHA chemicals in need of dermal absorption testing
40CFR712.30.e: Propylene glycol ethers and esters
40CFR712.30.e: Siloxanes
40CFR712.30.e: Substantially produced chemicals in need of subchronic tests
40CFR712.30.e: Sulphones
40CFR716.120.a-d: Substances and listed mixtures to which this subpart (specific chemical listing applies)

(Continued)

TABLE 1 *Continued*

Regulation 2006

40CFR716.120.a: List of substances
40CFR716.120.c: Substances listed by categories
40CFR716.120.c: Alkylepoxides, including all noncyclic aliphatic hydrocarbons with one or
more epoxy functional TSCA groups
40CFR716.120.c: Alkylphthalates - all alkyl esters of 1,2-benzenedicarboxylic acid (ortho-
phthalic acid)
40CFR716.120.c: Alkyltin compounds
40CFR716.120.c: Aniline and chloro-, bromo-, and/or nitroanilines
40CFR716.120.c: Aryl phosphates-phosphate esters of pherol or of alkyl-substituted phenols.
Triaryl and mixed alkyl and aryl esters are included but trialkyl esters are excluded
40CFR716.120.c: Asbestos-asbestiform varieties of chrycolite (serpentine); crocidolite
(riebeckite); amosite TSCA, (cummingtonite-grunerte); anthophyllite; tremolite; and actino lite
40CFR716.120.c: Bisazobiphenyl dyes derived from benzidine and its congeners, ortho-
toluidine (dimethylbenzidine)and dianisidine (dimethoxybenzidine)
40CFR716.120.c: Chlorinated benzenes, mono-, di-, tri- tetra-, and penta-
40CFR716.120.c: Chlorinated naphthalene, chlorinated derivatives of naphthalene (empirical
formula) $C_{10} H_x Cl_y$ where $x = y = 8$
40CFR716.120.c10: Chlorinated paraffins-chlorimated paraffin oils and chlorinated paraffin
waxes, with chlorine content of 35% through 70% by weight
40CFR716.120.c11: Ethyltoluenes-ethyltoluene (mixed isomers) and the ortho (1,2-), meta
(1,3-), and para (1,4-) isomers
40CFR716.120.c12: Fluoroalknes-general formula: Cn H2nxfx where $n - 2$ to and $X = 1$ to 6
40CFR716.120.c13: Glycidol (oxirancemethanol and its derivatieves)
40CFR716.120.c14: Halogenated alkyl esposcideshalogenated noncyclic aliphatic
hydrocarbons with one or more epoxy functional groups
40CFR716.120.c15: Phenylenediamines (benzenediamines) – all nitrogen unsubstituted
phenylenediamine and their salts with zero to two substitutents
40CFR716.120.d: Listed members of categories
40CFR716.120.d: Aldehydes
40CFR716.120.d: Alkyl-, chloro-, and hydroxy-methyl diary ethers
40CFR716.120.d: Alkyl phosphates
40CFR716.120.d: Brominated flame retardants
40CFR716.120.d: Chloralkyl phosphates
40CFR716.120.d: Cyanoacrylates
40CFR716.120.d: IRIS chemicals
40CFR716.120.d: Isocyanates
40CFR716.120.d: Methyl ethylene glycol ethers and esters
40CFR716.120.d: OSHA chemicals in need of dermal absorption testing
40CFR716.120.d: Propylene glycol ethers and esters
40CFR716.120.d: Siloxanes
40CFR716.120.d: Substantially produced chemical in need of subchronic test
40CFR716.120.d: Sulphones

(*Continued*)

TABLE 1 *Continued*

Regulation 2006

40CFR716.120.c: Bisazobiphenyl dyes derived from benzidine and its congeners, ortho-toluidine (dimethylbenzidine) and dianisidine (dimethoxybenzidine)

40CFR716.120.c: Chlorinated benzenes, mono-, di-, tri-, tetra-, and penta-

40CFA716-120.c: Chlorinated naphthalene, chlorinated derivatives of naphthalene (empirical formula) $C_{10} H_x Cl_y$ where $x = y = 8$

40CFR716.120.c: Chlorinated paraffins-chlorinated paraffin oils and chlorinated paraffin waxes, with chlorine content of 35% through 70% by weight

40CFR716.120.c: Ethyltoluenes-ethyltoluene (mixed isomers) and the ortho (1,2-), meta (1,3-), and para (1,4-) isomers

40CFR716.120.c: Fluoroalkenes-general formula: $C_n H_{2nx} F_x$ where $n = 2$ to 3 and $x = 1$ to 6

40CFR716.120.c: Glycidol (oxiranemethanol and its derivatives)

40CFR716.120.c: Halogenated alkyl eposcides halogenated noncyclic aliphatic hydrocarbons with one or more epoxy functional groups

40CFR716.120.c: Phenylenediamines (benzenediamines) – all nitrogen unsubstituted phenylenediamine and their salts with zero to two substitutents

CHAPTER 1

EARLY ENVIRONMENTAL HISTORY (1)

INTRODUCTION

BANG! The Big Bang. In 1948 physicist G. Gamow proposed the big bang theory of the origin of the universe. He believed that the universe was created in a gigantic explosion as all mass and energy were created in an instant of time. Estimates on the age of the universe at the present time range between 7 and 20 billion years, and with 13.5 billion years often mentioned as the age of the planet Earth.

The bang occurred in a split second and within a minute the universe was approximately a trillion miles wide and expanding at an unbelievable rate. Several minutes later all the matter known to humanity had been produced. The universe as it is known today was in place.

Gamow further believed that the various elements observed today were produced within the first few minutes after the big bang, when near infinitely high temperatures fused subatomic particles into the chemical elements that now comprise the universe. More recent studies suggest that hydrogen and helium would have been the primary products of the big bang, with heavier elements being produced later within stars. The extremely high density within the primeval atom caused the universe to expand rapidly. As it expanded, the hydrogen and helium cooled and condensed into stars and galaxies. This perhaps explains the expansion of the universe and the physical basis of Earth.

Environmental Regulatory Calculations Handbook, by Leo Stander and Louis Theodore
Copyright © 2008 John Wiley & Sons, Inc.

Flash forward to the present. More than any other time in history, the 21st century will be a turning point for human civilization. Human beings may be facing ecological disasters that could affect their ability to survive. These crises could force them to reexamine the value system that has governed their lives for the past two million years of existence (2). At some point during its journey human society lost its feeling of connectedness to nature, resulting in a "we can manage the world" attitude. This attitude might ultimately lead to the destruction of this country and the world.

How did it come to this? The answer lies in a knowledge of human history, a surprisingly brief chapter in the chronicle of the planet—how brief can be demonstrated by the use of a standard calendar to mark the passage of time on earth. The origin of the earth, estimated to be some several billion years ago, is placed at midnight January 1, 2007 and the present at midnight December 31, 2007. Each calendar day represents approximately 12 million years of actual history. Using this time scheme, dinosaurs arrived about December 10 and disappeared on Christmas day. The first humans can be placed at 11:45 P.M. on December 31. The recorded history of human achievement takes up only the last minutes of the world (2).

The remainder of this chapter describes the path that led to this dangerous predicament. However, this path is now leading a growing number of individuals to unite in a broad social movement called environmentalism: a movement that is building a potential road out of this predicament.

THE FIRST HUMANS

Environmental problems have bedeviled humanity since the first person discovered fire. The earliest humans appear to have inhabited a variety of locales within a tropical and semitropical belt stretching from Ethiopia to southern Africa about 1.9 million years ago. These first humans provided for themselves by a combination of gathering food and hunting animals. Humans, for the majority of their two million years' existence, lived in this manner. The steady development and dispersion of these early humans was largely due to an increase in their brain size. This led to the ability to think abstractly, which was vital in the development of technology and ability to speak. This in turn led to cooperation and more elaborate social organization (3). The ability to use and communicate the developed technology to overcome the hostile environment ultimately led to the expansion of these first human settlements.

With the use of primitive tools and skins of animals for clothes, the first humans moved outside Africa about one and a half million years ago. The migration led them into the frostfree zones of the Middle East, India, southern China and parts of Indonesia. The humans at this time could only adapt to those ecosystems found in the semitropical areas that contained a wide variety of vegetation and small, easily hunted animals to supplement their diet. Despite relatively easy access, Europe was not settled for a long period of time due to the deficient ecosystem, which was later overcome by an increase in technology. The first evidence of human settlement in Europe is dated to about 730,000 years ago. The settlement of America was almost the last stage in the movement of humans across the globe about 20,000 years ago. This was made possible by crossing to Alaska in the last glaciation when the reduced sea levels turned the Bering Strait into a land bridge. Once the first human settlers were able to move south through the passes, they found an enormously rich environment that supplied plenty of food. The human population multiplied rapidly and within a few thousand years had spread to the tip of South America.

By about 10,000 years ago humans had spread over every continent, living in small mobile groups. A minority of these groups lived in close harmony with the environment and did minimal damage. Evidence has been found where groups tried to conserve resources in an attempt to maintain subsistence for long periods of time. In some cases totemic restrictions on hunting a particular species at a certain time of the year or only in a certain area every few years helped to maintain population levels of certain animals (4). The Cree in Canada used a form of rotational hunting, only returning to an area after a considerable length of time, which allowed animal populations to recover. But the majority of these groups exploited the environment and the animals inhabiting it. In Colorado, bison were often hunted by stampeding them off a cliff, ending up with about 200 corpses, most of which could not be used. On Hawaii, within a thousand years of human settlement, thirty-nine species of land birds had become extinct (3). In Australia, over the last 100,000 years, 86 percent of the large animals have become extinct. The large numbers of species lost was largely due to the tendency for hunters to concentrate on one species to the exclusion of others. The main reason why these groups avoided further damage to nature was the fact that their numbers were so small that the pressure they exerted on the environment was limited.

THE DEVELOPMENT OF AGRICULTURE

A major shift in human evolution took place between 10,000 and 12,000 years ago. Humans learned how to domesticate animals and cultivate plants and in doing so made a transition from nomadic hunter gatherer to rooted agriculturalist. The global population at this time was about four million people, which was about the maximum that could readily be supported by a gathering and hunting way of life (3). The increasing difficulty in obtaining food is believed to be a major contributor to this sudden change. The farmer changed the landscape of the planet and was far more destructive then the hunter. While farming fostered the rise of cities and civilizations, it also led to practices that denuded the land of its nutrients and water-holding capacity. Great civilizations flourished and then disappeared as once-fertile land, after generations of over-farming and erosion, was transformed into barren wasteland.

The adoption of agriculture, combined with its two major consequences, settled communities and a steadily rising population, placed an increasing strain on the environment. The strain was localized at first, but as agriculture spread so did its effects. Agriculture involved removing the natural habitat to create an artificial habitat where humans could grow the plants and stock the animals they would need. The natural balance and inherent stability of the original ecosystem were thereby destroyed. Instead of a variety of plants and permanent natural ground cover, a small number of crops made only part-time use of the space available. The soil was exposed to the wind and rain to a far greater extent than before, particularly where fields were left barren for part of the year, leading to a higher rate of soil erosion than under natural ecosystems. Nutrient recycling processes were also disrupted and extra inputs in the form of manures and fertilizers were therefore required if soil fertility was to be maintained. The adoption of irrigation was even more disruptive since it created an environment that was even more artificial. Adding large amounts of water to a poor soil would allow the farmer to grow his preferred crop, but it would have catastrophic long term effects. The extra water would drain into the underlying water table, sometimes leading to rising water levels which caused the soil to become waterlogged. This additional water not only altered the mineral content of the soil but also increased the amount of salt and would eventually—especially in hot areas with high evaporation rates— produce a thick layer of salt on the surface that made agriculture impossible. The emergence of villages and towns meant that the demand for resources was now more concentrated.

These early societies were dependent on the production of a food surplus in order to feed and support the growing number of priests, rulers, bureaucrats, soldiers, and craftsmen. Forests suffered the most as the demand grew for wood to build houses, heat homes and cook. Local deforestation around settled areas added to the increase of soil erosion. Soil erosion then led to badly damaged landscape, declining crop yields and eventually an inability to grow a surplus of food. The first signs of wide-spread damage emerged in Mesopotamia, the area where the most extensive modi-fications to the natural environment were first made.

Both domestication of animals and the cultivation of plants had dramatic impacts on the environment. The nomadic hunters and gatherers were aware that they shared the earth with other living things. Animals and humans were able to live in the same area since the hunters and gatherers did not destroy the ecosystem to a great extent. The agriculturalist, on the other hand, deliberately transformed nature in an attempt to simplify the world's ecosystem. As an example, by ploughing and seeding a grassland, a farmer would eliminate a hundred species of native herbs and grasses, which would then be replaced with pure strands of wheat, corn, or alfalfa. This simplification reduced the stability of the ecosystem, making it uninhabitable for most animals.

COLONIZATION OF THE NEW WORLD

Only five hundred years, a mere second on the geological clock, have passed since Columbus' discovery opened a fresh and verdant new world to the Europeans: a land with few indications of human occupation except for a few thin plumes of smoke rising from cooking fires in small clearings in the woods. These clearings belonged to the Native Americans, which numbered about four million at this time. Over the centuries these people had created their own complex culture. Their means of sustaining themselves did not rely on scarring or subduing the earth, but on using what it offered. Native American society was not separate from nature but part of it. Geography, as well as history, began to change when Christopher Columbus anchored his little fleet off the island of San Salvador. Like most of those who followed, Columbus and his company risked the voyage to the New World for what they could take from it. They came for gold, a trade route to the spices of India and other riches of Asia: land, goods to sell, glory, adventure, religious and personal freedom, and in some cases to convert the heathen to Christianity (5). Although the explorers, adventurers, and settlers came to seize whatever riches and opportunities the land had to offer, it was what they brought with them, far from what they took, that changed the face of the continent forever. What they brought was Europe's two thousand or more years of western history, customs, prejudices and methodology. They brought European technology, philosophy, religion, aesthetics, a market economy and a talent for political organization. They brought European diseases that decimated the native people. They also brought with them European ideas of what the New World was and visions of what it should be.

In the beginning, the explorers and first settlers were faced by a dark forbidding line of forest behind which was a vast, unmapped continent, inhabited, they thought, by savages, and filled with ferocious wild beasts. Mere survival meant conquering the wilderness. The forest had to be cleared to make living space and to provide wood for shelters and fires (5). Behind the trees lurked the Indians, ready, the settlers suspected, to commit unspeakable atrocities. The forest was filled with wolves, bears, and panthers that would pounce on their children and domestic animals, or so they feared. The greater the destruction of the forest, the greater the safety for the tiny communities clinging to the edge of the hostile continent. Removing the trees also opened land for crops and cattle. Killing the wild animals not only filled the pot with meat but eliminated the deer and other grazing animals that stole the settlers' corn (5).

The European population quickly grew beyond the carrying capacity of the land. Cropland was frequently exhausted by permanent cultivation; cattle, swine, and sheep introduced by immigrants made far heavier demands on field and forest than wild animals. As each new field was harvested, the chemical, mineral, and biological nature of the soil itself was depleted. The Europeans also brought technology that contributed to the heavy impact they had on the land. Horses and oxen enabled the settlers to open and cultivate much broader acres. Plows could dig deeply into the soil, exposing far more loam. With draft animals, the Europeans could harvest heavier loads and transport them to markets. Sailing ships could then transport those loads along the coast or across the ocean.

Whereas the Native Americans would take from the land only what they could consume, the colonist and their successors sought to grow surplus that they could sell for cash or trade for manufactured goods and other commodities. The production of surplus led to the accumulation of capital and the creation of wealth, largely in the towns that served as marketplaces. That meant clearing more land, cutting more timber, planting more crops, and raising more cattle, all at a rate that could be sustained only at a cost of permanent damage to the land. The deforestation of New England and the disappearance of the beaver in the East are but two dramatic examples of how the demands of the market could deplete abundant resources in short order. By the time of the American Revolution, the wilderness along the eastern seaboard had been tamed. While some pockets of forest remained, the thirteen colonies were largely covered with farms, dotted with villages, and punctuated by a few substantial cities, notably Boston, New York, Philadelphia, and Charleston.

THE INDUSTRIAL REVOLUTION

Early in the nineteenth century, an awesome new force was gathering strength in Europe. The term "industrial revolution" was coined by the French as a metaphor of the affinity between technology and the great political revolution of modern times. Soon exported to the United States, the industrial revolution swept away any visions of America being an agrarian society. The steam engine, the railroad, the mechanical thresher, and hundreds of other ingenious artifacts that increased man's ability to transform the natural world and put it to use would soon be puffing and clattering and roaring in all corners of the land. The new machines swiftly accelerated the consumption of raw materials from the nation's farms, forests, and mines.

Lumbering became the nation's most important industry in the late eighteenth century. Wood was the most widely used raw material for heating, houses, barns, and shops; the same can be said for ships, furniture, railroad ties, factories and paper-making. The supply seemed inexhaustible since the forest still darkened huge parts of the country. The forest melted away before the axes of the advancing Americans. The settlers never thought of their axe work as deforestation, but as the progress of civilization. Soon after the tree cover was removed, the forest soil began to lose nutrients such as organic matter and materials. The soil began washing away, turning clear streams into slow, muddy ditches, filling lakes, and killing fish.

Meanwhile, the big cities and growing wealth of the East were creating a more rapidly expanding market for wheat, corn, beef, and other cash crops. New roads and canals, the steamboat and the locomotive, made domestic and foreign markets increasingly accessible to farms in the center of the continent. Eli Whitney's cotton gin, Cyrus McCormick's reaper, Benjamin Holt's combine and other ingenious inventions encouraged the development of a highly productive and efficient agriculture that sharply reduced the biological diversity of the land. Mining both preceded and quickly followed settlement of the interior, and left deep and permanent scars on the continent's land and waters. Gold in California, lead in Illinois, coal and oil in Pennsylvania, iron ore in Minnesota, and copper in Montana attracted fortune hunters and job seekers. Reports of a strike would draw thousands of prospectors and workers as well as those who lived off them. Mines were operated without care for the surrounding countryside. The picks and shovels, the hoses and dredges, and the smaller fires of the miners created the nation's first widespread pollution and environmental health problems. Mining left behind gutted mountains, dredged-out streams, despoiled vegetation, open pits, polluted creeks, barren hillsides and meadows, a littered landscape, and abandoned camps. Mining contributed to deforestation of the countryside. Woodlands were often cleared for mining operations; enormous amounts of timber were needed for the posts and beams that supported the mine shafts and fueled smelter operations (5).

Steam shovels came into use in the 1880s, enabling the coal operators of Pennsylvania and the iron ore producers of Minnesota to peel away the very crust of the earth to extract raw materials for industry and wealth for themselves. Spoils from the coal started to turn streams more acidic. The discovery of oil in Pennsylvania in 1859 brought drilling rigs that poked into the skyline: large areas of soil were soaked with black ooze (5). It was in the cities that environmental pollution and its effects were most pervasive. Garbage and filth of every kind were thrown into the streets, covering the surface, filling the gutters, obscuring the sewer culverts that sent forth perennial emanations. In the winter, the filth and garbage would accumulate in the streets to the depth of sometimes two or three feet. Most cities were nightmares of primitive sanitation and waste disposal systems. Privies for sewage and private wells for water were still widely used in metropolitan areas until the end of the nineteenth century.

Perhaps the national government could have done more to protect the land and its resources as well as public health. But, for most of the nineteenth century, the government was still a weak presence in most areas of the country. There was, moreover, no body of laws with which the government could assert its authority. Laissez-faire was the order of the day. By the end of the century there was a growing body of information about the harm being done and some new ideas on how to set things straight. Yet, there was no acceptable ethic that would impel people to treat the land, air, and water with wisdom and care. To a large extent the people did not know what they were doing (5).

As the nineteenth century was drawing to a close, three very special individuals made their entrance on the national stage. Gifford Pinchot, John Muir, and Theodore Roosevelt were to write the first pages of modern environmental history in the United States, which in turn led to the birth of the modern environmental movement early in the twentieth century. However, pollution and environmental degradation was a fact of life across most of America during the first half of the 20th century, phrases such as "the smell of money", "good, clean soot", "God bless it", "it's our life-blood", and "an index to local activity and enterprise" were used to describe air pollution.

The federal government ultimately entered into the environmental and conservation business in a significant fashion when Teddy Roosevelt's second cousin Franklin entered the White House in 1933. It was his political ideology as much as his love of nature that led Roosevelt to include major conservation projects in his New Deal reforms. The Civilian Conservation Corps, the Soil Conservation Service, and the Tennessee Valley Authority were among the many New Deal programs created to serve both the land and the people.

At this point in time, muscle and animal power were replaced by electricity, internal-combustion engines, and nuclear reactors. At the same time, industry was consuming natural resources at an incredible rate. All of these events began to escalate at a dangerous rate after World War II. Soon after, in the late summer of 1962, a marine biologist named Rachel Carson, author of *Silent Spring*, the best-selling book about ocean life, opened the eyes of the world to the dangers of attacking the environment. It was perhaps at this point that America began calling in earnest for reform of the destruction of nature and constraints on environmental degradation. Finally, in the 1970s, Congress began turning out environmental laws that addressed these issues. It all began in 1970 with the birth of the Environmental Protection Agency.

For additional literature regarding Early History and the Environmental Movement, the interested reader is referred to the book by Philip Shabecoff, titled *A Fierce Green Fire* (5). This outstanding book, as well as Ponting's A Green History of the World (3), is a "must" for anyone who works in, or has interests with the environment.

REFERENCES

1. Drawn (with permission) from M. K. Theodore and L. Theodore, "Major Environmental Issues Facing the 21st Century", contributing author (Chapter 1): A. Meier, Theodore Tutorials (originally published by Simon & Schuster), East Williston, NY, 1995.
2. Gorden, A. and Suzuki, D. *It's a Matter of Survival.* Cambridge, MA: Harvard University Press, 1991.
3. Ponting, C. *A Green History of the World.* New York: St. Martin's Press, 1991.
4. Goudie, A. *The Human Impact: Man's Role in Environmental Change.* Cambridge, MA: The MIT Press, 1981.
5. Shabecoff, P. *A Fierce Green Fire: The American Environmental Movement.* Harper Collins Canada Ltd, 1993.

CHAPTER 2

CURRENT ENVIRONMENTAL REGULATORY FRAMEWORK (1)

INTRODUCTION

It is now 1970, a cornerstone year for modern environmental policy. The National Environmental Policy Act (NEPA), enacted on January 1, 1970, was considered a "political anomaly" by some. NEPA was not based on specific legislation; instead it referred in a general manner to environmental and quality of life concerns. The Council for Environmental Quality (CEQ), established by NEPA, was one of the councils mandated to implement legislation. April 22, 1970 brought Earth Day, where thousands of demonstrators gathered all around the nation. NEPA and Earth Day were the beginning of a long, seemingly never ending debate over environmental issues.

The Nixon Administration at that time became preoccupied with not only trying to pass more extensive environmental legislation, but also implementing the laws. Nixon's White House Commission on Executive Reorganization proposed in the Reorganizational Plan # 3 of 1970 that a single, independent agency be established, separate from the CEQ. The plan was sent to Congress by President Nixon on July 9, 1970, and this new U.S. Environmental Protection Agency (EPA) began operation on December 2, 1970. The EPA was officially born.

In many ways, the EPA is the most far-reaching regulatory agency in the federal government because its authority is so broad. The EPA is charged to protect the nation's land, air, and water systems. Under a mandate of national environmental laws, the EPA strives to formulate and implement actions which lead to a

Environmental Regulatory Calculations Handbook, by Leo Stander and Louis Theodore
Copyright © 2008 John Wiley & Sons, Inc.

compatible balance between human activities and the ability of natural systems to support and nurture life (2).

The EPA works with the states and local governments to develop and implement comprehensive environmental programs. Federal laws such as the Clean Air Act, the Safe Drinking Water Act, the Resource Conservation and Recovery Act, and the Comprehensive Environmental Response, Compensation, and Liability Act, etc., all mandate involvement by state and local government in the details of implementation.

This chapter provides an overview of the eight key environmental protection laws and subsequent regulations that affect the environment in the US. Each of these laws will receive more detailed discussion in the next eight chapters.

THE REGULATORY SYSTEM

Over the past four decades environmental regulation has become a system in which laws, regulations, and guidelines have become interrelated. The history and development of this regulatory system has led to laws that focus principally on only one environmental medium, i.e., air, water, or land. Some environmental managers feel that more needs to be done to manage all of the media simultaneously. Hopefully, the environmental regulatory system will evolve into a truly integrated, multimedia management framework in the future.

Federal laws are the product of Congress. Regulations written to implement the law are promulgated by the Executive Branch of government, but until judicial decisions are made regarding the interpretations of the regulations, there may be uncertainty about what regulations mean in real situations. Until recently, environmental protection groups were most frequently the plaintiffs in cases brought to court seeking interpretation of the law. Today, industry has become more active in this role.

Enforcement approaches for environmental regulations are environmental management oriented in that they seek to remedy environmental harm, not simply a specific infraction of a given regulation. All laws in a legal system may be used in enforcement to prevent damage or threats of damage to the environment or human health and safety. Tax laws (e.g., tax incentives) and business regulatory laws (e.g., product claims, liability disclosure) are examples of laws not directly focused on environmental protection, but that may also be used to encourage compliance and discourage non-compliance with environmental regulations.

Common law also plays an important role in environmental management. Common law is the set of rules and principles relating to the government and security of persons and property. Common law authority is derived from the usages and customs that are recognized and enforced by the courts. In general, no infraction of the law is necessary when establishing a common law court action. A common law "civil wrong" (e.g., environmental pollution) that is brought to court is called a tort. Environmental torts may arise because of nuisance, trespass, or negligence.

Laws tend to be general and contain uncertainties relative to the implementation of principles and concepts they contain. Regulations derived from laws may be more specific, but are also frequently too broad to allow clear translation into environmental technology practice. Permits may be used to bridge this gap and prescribe specific technical requirements concerning the discharge of pollutants or other activities carried out by a facility that may impact the environment.

Most major federal environmental laws provide for citizen law suits. This empowers individuals to seek compliance or monetary penalties when these laws are violated and regulatory agencies do not take enforcement action against the violator.

LAWS AND REGULATIONS: THE DIFFERENCES

The following is a listing of some of the major differences between a Federal *law* and a Federal *regulation*, as briefly described in the previous section:

1. A law (or Act) is passed by both houses of Congress and signed by the President. A regulation is issued by a government agency such as the U.S. Environmental Protection Agency (EPA) or the Occupational Safety and Health Administration (OSHA).
2. Congress can pass a law on any subject it chooses. It is only limited by the restrictions in the Constitution. A law can be challenged in court only if it violates the constitution. It may not be challenged if it is merely unwise, unreasonable or even silly. If, for example, a law were passed that placed a tax on sneezes, it could not be challenged in court just because it was unenforceable. A regulation can be issued by an agency only if the agency is authorized to do so by the law passed by Congress. When congress passes a law, it usually assigns an administrative agency to implement that law. A law regarding radio stations, for example, may be assigned to the Federal Communications Commission (FCC). Sometimes a new agency is established to implement a law. This was the case with the Consumer Product Safety Commission (CPSC). OSHA is authorized by the Occupational Safety and Health Act to issue regulations that protect workers from exposure to the hazardous chemicals they use in manufacturing processes.
3. Laws include a Congressional mandate directing EPA to develop a comprehensive set of regulations. Regulations, or rulemakings, are issued by an agency, such as EPA, that translate the general mandate of a statute into a set of requirements for the Agency and the regulated community.
4. Regulations are developed by EPA in an open and public manner according to an established process. When a regulation is formally proposed, it is published in an official government document called the *Federal Register* to notify the public of EPA's intent to create new regulations or modify existing ones. EPA provides the public, which includes the potentially regulated community, with

an opportunity to submit comments. Following an established comment period, EPA may revise the proposed rule based on both an internal review process and public comments.

5. The final regulation is published, or promulgated, in the *Federal Register*. Included with the regulation is a discussion of the Agency's rationale for the regulatory approach, known as preamble language. Final regulations are compiled annually and incorporated in the Code of Federal Regulations (CFR) according to a highly structured format based on the topic of the regulation. This latter process is called **codification**, and each CFR title corresponds to a different regulatory authority. For example, EPA's regulations are in Title 40 of the CFR. The codified RCRA regulations can be found in Title 40 of the CFR, Parts 240–282. These regulations are often cited as 40 CFR, with the part listed afterward (e.g., 40 CFR Part 264), or the part and section (e.g., 40 CFR §264.10).

6. A regulation may be challenged in court on the basis that the issuing agency exceeded the mandate given to it by Congress. If the law requires the agency to consider costs versus benefits of their regulation, the regulation could be challenged in court on the basis that the cost/benefit analysis was not correctly or adequately done. If OSHA issues a regulation limiting a worker's exposure to a hazardous chemical to 1 part per million (ppm), OSHA could be called upon to prove in court that such a low limit was needed to prevent a worker from being harmed. Failure to prove this would mean that OSHA exceeded its mandate under the law, as OSHA is charged to develop standards only as stringent as those required to protect worker health and provide worker safety.

7. Laws are usually brief and general. Regulations are usually lengthy and detailed. The Hazardous Materials Transportation Act, for example, is approximately 20 pages long. It speaks in general terms about the need to protect the public from the dangers associated with transporting hazardous chemicals and identifies the Department of Transportation (DOT) as the agency responsible for issuing regulations implementing the law. The regulations issued by the DOT are several thousand pages long and are very detailed, down to the exact

size, shape, design and color of the warning placards that must be used on trucks carrying any of the thousands of regulated chemicals.

8. Generally, laws are passes infrequently. Often years pass between amendments to an existing law. A completely new law on a given subject already addressed by an existing law is unusual. Laws are published as a "Public Law #__-__" and are eventually codified into the United States Code.

9. Regulations are issued and amended frequently. Proposed and final new regulations and amendments to existing regulations are published daily in the Federal Register. Final regulations have the force of law when published.

THE ROLE OF THE STATES

The Resource Conservation and Recovery Act (RCRA), for example, like most federal environmental legislation, encourages states to develop and run their own hazardous waste programs as an alternative to EPA management. Thus, in a given state, the hazardous waste regulatory program may be run by the EPA or by a state agency. For a state to have jurisdiction over its hazardous waste program, it must receive approval from the EPA by showing that its program is at least as stringent as the EPA program.

States that are authorized to operate RCRA (or other) programs oversee the hazardous waste tracking system in their state, operate the permitting system for hazardous waste facilities, and act as the enforcement arm in cases where an individual or a company practices illegal hazardous waste management. If needed, the EPA steps in to assist the states in enforcing the law. The EPA can also act directly to enforce RCRA or other laws in states that do not yet have authorized programs. The EPA and the states currently act jointly to implement and enforce the regulations (3).

RESOURCE CONSERVATION AND RECOVERY ACT

Defining what constitutes a "hazardous waste" requires consideration of both legal and scientific factors. The basic definitions used in this Chapter are derived from: The Resource Conservation and Recovery Act (RCRA) of 1976, as amended in 1978, 1980, and 1986; the Hazardous and Solid Waste Amendments (HSWA) of 1984; and, the Comprehensive Environmental Response, Compensation and Liability Act (CERCLA) of 1980, as amended by the Superfund Amendments and Reauthorization Act (SARA) of 1986. Within these statutory authorities a distinction exists between a hazardous waste and a hazardous substance. The former is regulated under RCRA while the latter is regulated under the Superfund program (see Chapter 8).

Hazardous waste refers to " ... a solid waste, or combination of solid wastes, which because of its quantity, concentration, or physical, chemical or infectious characteristics may [pose a] substantial present or potential hazard to human health or the environment when improperly ... managed ... " [RCRA, Section 1004(5)]. Under RCRA regulations, a waste is considered hazardous if it is reactive, ignitable, corrosive, or toxic or if the waste is listed as a hazardous waste in Title 40 Parts 261.31–33 of the *Code of Federal Regulations* (4).

In addition to hazardous wastes defined under RCRA, there are "hazardous substances" defined by Superfund. Superfund's definition of a hazardous substance is broad and grows out of the lists of hazardous wastes or substances regulated under the Clean Water Act (CWA), the Clean Air Act (CAA), the Toxic Substances Control Act (TSCA), and RCRA. Essentially, Superfund considers a hazardous substance to be any hazardous substance or toxic pollutant identified under the CWA and applicable regulations, any hazardous air pollutant listed under the CAA and applicable regulations, any imminently hazardous chemical for which a civil action has been brought under TSCA, and any hazardous waste identified or listed under RCRA and applicable regulations.

The RCRA of 1976 completely replaced the previous language of the Solid Waste Disposal Act of 1965 to address the enormous growth in the production of waste. The objectives of this Act were to promote the protection of health and the environment and to conserve valuable materials and energy resources by (5,6):

1. Providing technical and financial assistance to state and local governments and interstate agencies for the development of solid waste management plans (including resource recovery and resource conservation systems) that promote improved solid waste management techniques (including more effective organizational arrangements), new and improved methods of collection, separation, and recovery of solid waste, and the environmentally safe disposal of nonrecoverable residues.

2. Providing training grants in occupations involving the design, operation, and maintenance of solid waste disposal systems.

3. Prohibiting future open dumping on the land and requiring the conversion of existing open dumps to facilities that do not pose danger to the environment or to health.

4. Regulating the treatment, storage, transportation, and disposal of hazardous waste that have adverse effects on health and the environment.

5. Providing for the promulgation of guidelines for solid waste collection, transport, separation, recovery, and disposal practices and systems.

6. Promoting a national research and development program for improved solid waste management and resource conservation techniques; more effective organization arrangements; and, new and improved methods of collection, separation, recovery, and recycling of solid wastes and environmentally safe disposal of nonrecoverable residues.

7. Promoting the demonstration, construction, and application of solid waste management, resource recovery and resource conservation systems that preserve and enhance the quality of air, water, and land resources.

8. Establishing a cooperative effort among federal, state, and local governments and private enterprises in order to recover valuable materials and energy from solid waste.

Structurewise, the RCRA is divided into eight subtitles. These subtitles are (A) General Provisions; (B) Office of Solid Waste; Authorities of the Administrator; (C) Hazardous Waste Management; (D) State or Regional Solid Waste Plans; (E) Duties of Secretary of Commerce in Resource and Recovery; (F) Federal Responsibilities; (G) Miscellaneous Provisions; and, (H) Research, Development, Demonstration, and Information. Subtitles C and D generate the framework for regulatory control programs for the management of hazardous and solid nonhazardous wastes, respectively. The hazardous waste program outlined under Subtitle C is the one most people associate with the RCRA (6).

MAJOR TOXIC CHEMICAL LAWS ADMINISTERED BY THE USEPA

People have long recognized that sulfuric acid, arsenic compounds, and other chemical substances can cause fires, explosions, or poisoning. More recently, researchers have determined that many chemical substances such as benzene and a number of chlorinated hydrocarbons may cause cancer, birth defects, and other long-term health effects. Today, the hazards of new kinds of substances, including genetically engineered microorganisms and nanoparticles are being evaluated. The EPA has a number of legislative tools to use in controlling the risks from toxic substances (Table 2.1).

The Federal Insecticide, Fungicide, and Rodenticide Act of 1972 (FIFRA) encompasses all pesticides used in the United States. When first enacted in 1947, FIFRA was administered by the U.S. Department of Agriculture and was intended to protect consumers against fraudulent pesticide products. When many pesticides were registered, their potential for causing health and environmental problems was unknown. In 1970, the EPA assumed responsibility for FIFRA, which was amended in 1972 to shift emphasis to health and environmental protections. Allowable levels of pesticides in food are specified under the authority of the Federal Food, Drug, and Cosmetic Act of 1954. Today, FIFRA contains registration and labeling requirements for pesticide products. The EPA must approve any use of a pesticide, and manufacturers must clearly state the conditions of that use on the pesticide label. Some pesticides are listed as hazardous wastes and are subject to RCRA rules when discarded.

The TSCA authorizes EPA to control the risks that may be posed by the thousands of commercial chemical substances and mixtures (chemicals) that are not regulated as either drugs, food additives, cosmetics, or pesticides. Under TSCA, the EPA can, among other things, regulate the manufacture and use of a chemical substance and require testing for cancer and other effects. TSCA regulates the production and distribution of new chemicals and governs the manufacture, processing, distribution, and use of existing chemicals. Among the chemicals controlled by TSCA regulations are PCBs, chloroflurocarbons, and asbestos. In specific cases, there is an interface with RCRA regulations. For example, PCB disposal is generally regulated by TSCA. However, hazardous wastes mixed with PCBs are regulated under RCRA. Under both TSCA and FIFRA, the EPA is responsible for regulating

TABLE 2.1 Major Toxic Chemical Laws Administered by the EPA

Statue	Provisions
Toxic Substances Control Act	Requires that the EPA be notified of any new chemical prior to its manufacture and authorizes EPA to regulate production, use, or disposal of a chemical.
Federal Insecticide, Fungicide, and Rodenticide Act	Authorizes the EPA to register all pesticides and specify the terms and conditions of their use, and remove unreasonably hazardous pesticides from the marketplace.
Federal Food, Drug, and Cosmetic Act	Authorizes the EPA in cooperation with the FDA to establish tolerance levels for pesticide residues on food and food producers.
Resource Conservation and Recovery Act	Authorizes the EPA to identify hazardous wastes and regulate their generation, transportation, treatment, storage, and disposal.
Comprehensive Environmental Response, Compensation, and Liability Act (CERCLA)	Requires the EPA to designate hazardous substances that can present substantial danger and authorizes the cleanup of sites contaminated with such substances.
Clean Air Act	Authorizes the EPA to set emission standards to limit the release of hazardous air pollutants.
Clean Water Act	Requires the EPA to establish a list of toxic water pollutants and set standards.
Safe Drinking Water Act	Requires the EPA to set drinking water standards to protect public health from hazardous substances.
Marine Protection, Research, and Sanctuaries Act	Regulates ocean dumping of toxic contaminants.
Asbestos School Hazard Act	Authorizes the EPA to provide loans and grants to schools with financial need for abatement of severe asbestos hazards.
Asbestos Hazard Emergency Response Act	Requires the EPA to establish a comprehensive regulatory framework for controlling asbestos hazards in schools.
Emergency Planning and Community Right-to-Know Act	Requires states to develop programs for responding to hazardous chemical releases and requires industries to report on the presence and release of certain hazardous substances.

certain biotechnology products, such as genetically engineered microorganisms designed to control pests or assist in industrial processes.

The CAA, in Section 112, listed 189 air pollutants. The CAA also requires emission standards for many types of air emission sources, including RCRA-regulated incinerators and industrial boilers or furnaces.

The CWA lists substances to be regulated by effluent limitations in 21 primary industries. The CWA substances are incorporated into both RCRA and CERCLA. In addition, the CWA regulates discharges from publicly owned treatment works (POTWs) to surface waters, and indirect discharges to municipal wastewater treatment systems (through the pretreatment program). Some hazardous wastewaters which would generally be considered RCRA regulated wastes are covered under the CWA because of the use of treatment tanks and a National Pollutant Discharge Elimination System (NPDES) permit to dispose of the wastewaters. Sludges from these tanks, however, are subject to RCRA regulations when they are removed.

The Safe Drinking Water Act (SDWA) regulates underground injection systems, including deep-well injection systems. Prior to underground injection, a permit must be obtained which imposes conditions that must be met to prevent the endangerment of underground sources of drinking water.

The Marine Protection, Research, and Sanctuaries Act of 1972 has regulated the transportation of any material for ocean disposal and prevents the disposal of any material in oceans which could affect the marine environment. Amendments enacted in 1988 were designed to end ocean disposal of sewage sludge, industrial waste and medical wastes (4).

LEGISLATIVE TOOLS FOR CONTROLLING WATER POLLUTION

Congress has provided the EPA and the states with three primary statutes to control and reduce water pollution: the Clean Water Act; the Safe Drinking Water Act; and the Marine Protection, Research, and Sanctuaries Act. Each statute provides a variety of tools that can be used to meet the challenges and complexities of reducing water pollution in the nation.

Clean Water Act

Under the Clean Water Act, the states adopt water quality standards for every stream within their respective borders. These standards include a designated use such as fishing or swimming, and prescribe criteria to protect that use. The criteria are pollutant-specific and represent the permissible levels of substances in the waters that would enable the use to be achieved. Water quality standards (WQSs) are the basis for nearly all water quality management decisions. Depending upon the standard adopted for a particular stream, controls may be needed to reduce the pollutant levels. Water quality standards are reviewed every three years and revised as needed.

National Pollutant Discharge Elimination System. Under the Clean Water Act, the discharge of pollutants into the waters of the United States is prohibited unless a permit is issued by the EPA or a state under the National Pollutant Discharge Elimination System (NPDES). These permits must be renewed at least once every five years. There are approximately 48,400 industrial and 15,300 municipal facilities that currently have NPDES permits. An NPDES permit contains effluent limitations and monitoring and reporting requirements. Effluent limitations are restrictions on the amount of specific pollutants that a facility can discharge into a stream, river, or harbor. Monitoring and reporting requirements are specific instructions on how sampling of the effluent should be performed to check whether the effluent limitations are being met. Instructions may include required sampling frequency (i.e., daily, weekly, or monthly) and the type of monitoring required. The permittee may be required to monitor the effluent on a daily, weekly, or monthly basis. The monitoring results are then regularly reported to the EPA and state authorities. When a discharger fails to comply with the effluent limitations or monitoring and reporting requirements, the EPA or the state may take enforcement action.

Congress recognized that it would be an overwhelming task for the EPA to establish effluent limitations for each individual industrial and municipal discharger. Therefore, Congress authorized the Agency to develop uniform effluent limitations for each category of point source such as steel mills, paper mills, and pesticide manufacturers. The EPA develops these effluent limitations on the basis of many factors, most notably efficient treatment technologies. Once the EPA proposes an effluent limit and public comments are received, the EPA or the states issue NPDES permits using the technology-based limits to all point sources within that industry category. Sewage treatment plants also are provided with effluent limitations based on technology performance.

Limitations that are more stringent than those based on technology are sometimes necessary to ensure that state-developed water quality standards are met. For example, several different facilities may be discharging into one stream, creating pollutant levels harmful to fish. In this case, the facilities on that stream must meet more stringent treatment requirements known as *water-quality-based limitations*. These limits are developed by determining the amounts of pollutants that the stream can safely absorb and calculating permit limits such that these amounts are not exceeded (7).

The CWA requires that Total Maximum Daily Loads (TMDLs) be developed only for waters affected by pollutants where implementation of the technology-based controls imposed upon point sources by the CWA and EPA regulations would not result in achievement of Water Quality Standard (WQS). At this time, most point sources have been issued NPDES permits with technology-based discharge limits. In addition, a substantial fraction of point sources also have more stringent water quality-based permit limits. But, because nonpoint sources are major contributors of pollutant loads to many waterbodies, even these more stringent limits on point sources have not resulted in attainment of WQS.

Strategies to address impaired waters must consist of a TMDL or another comprehensive strategy that includes a functional equivalent of a TMDL. In essence, TMDLs are "pollutant budgets" for a specific waterbody or segment, that if not exceeded, would result in attainment of WQS.

The EPA and the U.S. Army Corps of Engineers jointly implement a permit program regulating the discharge of dredged or fill material into waters of the United States, including wetlands. As part of this program, the EPA's principal responsibility as set forth in the Clean Water Act is to develop the substantive environmental criteria by which permit applications are evaluated. The EPA also reviews the permit applications and, if necessary, can veto permits that would result in significant environmental damage.

The National Estuary Program is also regulated under the Clean Water Act. States nominate and the EPA selects estuaries of national significance that are threatened by pollution, development, or overuse. The EPA and the involved state(s) form a management committee consisting of numerous work groups to assess the problems, identify management solutions, and develop and oversee implementation of plans for addressing the problems.

Safe Drinking Water Act

The EPA establishes standards for drinking water quality through the Safe Drinking Water Act. These standards represent the Maximum Contaminant Levels (MCL), and consist of numerical criteria for specified contaminants. Local water supply systems are required to monitor their drinking water periodically for contaminants with MCLs and for a broad range of other contaminants as specified by the EPA. Additionally, to protect underground sources of drinking water, EPA requires periodic monitoring of wells used for underground injection of hazardous waste, including monitoring of the ground water above the wells.

States have the primary responsibility for the enforcement of drinking water standards, monitoring, and reporting requirements. States also determine requirements for environmentally sound underground injection of wastes. The Safe Drinking Water Act also authorizes EPA to award grants to states for developing and implementing programs to protect drinking water at the tap and ground-water resources. These grant programs may be for supporting state public water supply, wellhead protection, and underground injection programs, including compliance and enforcement.

The Clean Water Act and the Safe Drinking Water Act place great reliance on state and local initiatives in addressing water problems. With the enactment of the 1986 Safe Drinking Water Act amendments and the 1987 Water Quality Act Amendments, significant additional responsibilities were assigned to the EPA and the states. Faced with many competing programs limited resources, the public sector will need to set priorities. With this in mind, the EPA is encouraging states to address their water quality problems by developing State Clean Water Strategies. These strategies are to set forth state priorities over a multiyear period. They will help target the most valuable and/or most threatened water resources for protection.

Success in the water programs is increasingly tied to state and local leadership and decision-making and to public support. The EPA works with state and local agencies, industry, environmentalists, and the public to develop environmental agenda in the following three areas:

1. *Protection of Drinking Water.* Although more Americans are receiving safer drinking water than ever before, there are still serious problems with contamination of drinking water supplies and of ground water that is or could be used for human consumption. Contaminated ground water has caused well closings. The extent and significance of contamination by toxics has not been fully assessed for most of the nation's rivers and lakes, which are often

used for drinking water supply. All of these issues are areas for continued work and improvement.

2. *Protection of Critical Aquatic Habitats.* Contamination or destruction of previously underprotected areas such as oceans, wetlands, and near coastal waters must be addressed.

3. *Protection of Surface-Water Resources.* The EPA and the states will need to establish a new phase of the federal–state partnership in ensuring continuing progress in addressing conventional sources of pollution (7).

Marine Protection, Research, and Sanctuaries Act (Title I)

EPA designates sites and times for ocean dumping. Actual dumping at these designated sites requires a permit. The EPA and the Corps of Engineers share this permitting authority, with the Corps responsible for the permitting of dredged material (subject to an EPA review role), and the EPA responsible for permitting all other types of materials. The Coast Guard monitors the activities and the EPA is responsible for assessing penalties for violations.

THE SUPERFUND AMENDMENTS AND REAUTHORIZATION ACT OF 1986

The 1986 amendments to the Comprehensive Environmental Response, Compensation, and Liability Act (CERCLA), known as the Superfund Amendments and Reauthorization Act (SARA), authorized $8.5 billion for both the emergency response and longer term (or remedial) cleanup programs. The Superfund amendments focused on:

1. *Permanent Remedies.* The EPA must implement permanent remedies to the maximum extent practicable. A range of treatment options are considered whenever practicable.

2. *Complying with Other Regulations.* Applicable or relevant and appropriate standards from other federal, state, or tribal environmental laws must be met at Superfund sites where remedial actions are taken. In addition, state standards that are more stringent than federal standards must be met in cleaning up sites.

3. *Alternative Treatment Technologies.* Cost-effective treatment and recycling must be considered as an alternative to the land disposal of wastes. Under RCRA, Congress banned land disposal of some wastes. Many Superfund site wastes, therefore, are banned from disposal on the land; alternative treatments are under development and will be used where possible.

4. *Public Involvement.* Citizens living near Superfund sites are involved in the site decision-making process. They are also able to apply for technical

assistance grants that further enhance their understanding of site conditions and activities.

5. *State Involvement.* States and tribes are encouraged to participate actively as partners with EPA in addressing Superfund sites. They assist in making the decisions at sites, can take responsibility in managing cleanups, and can play an important role in oversight of responsible parties.

6. *Enforcement Authorities.* Settlement policies were strengthened through Congressional approval and inclusion in SARA. Different settlement tools, such as *de minimis* settlements (settlements with minor contributors) are part of the Act.

7. *Federal Facility Compliance.* Congress emphasized that federal facilities "are subject to, and must comply with, this Act in the same manner and to the same extent ... as any non-government entity." Mandatory schedules have been established for federal facilities to assess their sites, and if listed on the National Priority List (NPL), to clean up such sites. EPA is assisting and overseeing federal agencies with these requirements.

These amendments also expanded research and development, especially in the area of alternative technologies. They also provided for more training for state and federal personnel in emergency preparedness, disaster response, and hazard mitigation.

Major Provisions of Title III of SARA (also known as Emergency Planning and Community Right-to-Know Act or EPCRA)

1. *Emergency Planning*: EPCRA establishes a broad-based framework at the state and local levels to receive chemical information and use that information in communities for chemical emergency planning.

2. *Emergency Release Notification*: EPCRA requires facilities to report certain releases of extremely hazardous chemicals and hazardous substances to their state and local emergency planning and response officials.

3. *Hazardous Chemical Inventory Reporting*: EPCRA requires facilities to maintain a material safety data sheet (MSDS) for any hazardous chemicals stored or used in the work place and to submit those sheets to state and local authorities. It also requires them to submit an annual inventory report for those same chemicals to local emergency planning and fire protection officials, as well as state officials.

4. *Toxic Release Inventory Reporting*: EPCRA requires facilities to annually report on routine emissions of certain toxic chemicals to the air, land, or water. Facilities must report if they are in Standard Industrial Classification (SIC) codes 20 through 39 (i.e., manufacturing facilities) with 10 or more employees and manufacture or process any of 650 listed chemical compounds in amount greater than specified threshold quantities. If the chemical compounds are considered persistent, bioaccumulative, or toxic, the thresholds are much lower. EPA is required to use these data to establish a national

chemical release inventory database, making the information available to the public through computers, via telecommunications, and by other means.

THE CLEAN AIR ACT

The Clean Air Act defines the national policy for air pollution abatement and control in the United States. It establishes goals for protecting health and natural resources and delineates what is expected of Federal, State, and local governments to achieve those goals. The Clean Air Act, which was initially enacted as the Air Pollution Control Act of 1955, has undergone several revisions over the years to meet the ever-changing needs and conditions of the nation's air quality. On November 15, 1990, the president signed the most recent amendments to the Clean Air Act, referred to as the 1990 Clean Air Act Amendments. Embodied in these amendments were several progressive and creative new themes deemed appropriate for effectively achieving the air quality goals and for reforming the air quality control regulatory process. Specifically the amendments:

1. Encouraged the use of market-based principles and other innovative approaches similar to performance-based standards and emission banking and trading.
2. Promoted the use of clean low-sulfur coal and natural gas, as well as innovative technologies to clean high-sulfur coal through the acid rain program.
3. Reduced energy waste and created enough of a market for clean fuels derived from grain and natural gas to cut dependency on oil imports by one million barrels per day.
4. Promoted energy conservation through an acid rain program that gave utilities flexibility to obtain needed emission reductions through programs that encouraged customers to conserve energy.

Several of the key provisions of the act are reviewed below.

Provisions for Attainment and Maintenance of National Ambient Air Quality Standards (NAAQS)

Although the Clean Air Act brought about significant improvements in the nation's air quality, the urban air pollution problems of ozone (smog), carbon monoxide (CO), and particulate matter persist. In 1995, approximately 70 million U.S. residents were living in counties with ozone levels exceeding the EPA's current ozone standard.

The Clean Air Act, as amended in 1990, established a more balanced strategy for the nation to address the problem of urban smog. Overall, the amendments revealed the Congress's high expectations of the states and the federal government. While it gave states more time to meet the air quality standard (up to 20 years for ozone in Los Angels), it also required states to make constant progress in reducing emissions. It required the federal government to reduce emissions from cars, trucks, and buses; from consumer products such as hair spray and window-washing compounds; and, from ships and barges during loading and unloading of petroleum products. The federal government also developed the technical guidance that states need to control stationary sources.

The Clean Air Act addresses the urban air pollution problems of ozone (smog), carbon monoxide (CO), and particulate matter (PM). Specifically, it clarifies how areas are designated and redesignated "attainment." It also allows the EPA to define the boundaries of "nonattainment" areas, i.e., geographical areas whose air quality does not meet federal ambient air quality standards designed to protect public health. The law also establishes provisions defining when and how the federal government can impose sanctions on areas of the country that have not met certain conditions.

For the pollutant ozone, the Clean Air Act established nonattainment area classifications ranked according to the severity of the area's air pollution problem. These classifications are marginal, moderate, serious, severe, and extreme. The EPA assigns each nonattainment areas one of these categories, thus triggering varying requirements the area must comply with in order to meet the ozone standard.

As mentioned, nonattainment areas have to implement different control measures, depending upon their classification. Marginal areas, for example, are the closest to meeting the standard. They are required to conduct an inventory of their ozone-causing emissions and institute a permit program. Nonattainment areas with more serious air quality problems must implement various control measures. The worse the air quality, the more controls these areas will have to implement.

The Clean Air Act also established similar programs for areas that do not meet the federal health standards for carbon monoxide and particulate matter. Areas exceeding the standards for these pollutants are divided into "moderate" and "serious" classifications. Depending upon the degree to which they exceed the carbon monoxide standard, areas are then required to implement programs such as introducing oxygenated fuels and/or enhanced emission inspection programs, among other measures. Depending upon their classification, areas exceeding the particulate matter standard have to implement reasonably available control measures (RACM) or best available control measures (BACM), among other requirements.

Provisions Relating to Mobile Sources

While motor vehicles built today emit fewer pollutants (60–80% less, depending on the pollutant) than those built in the 1960s, cars and trucks still account for almost half the emissions of the ozone precursors (volatile organic carbons (VOC) and nitrogen oxides (NO_x)), and up to 90% of the CO emissions in urban areas. The principal reason for this problem is the rapid growth in the number of vehicles on the roadways and the total miles driven. This growth has offset a large portion of the emission reductions gained from motor vehicle controls.

In view of the continuing growth in automobile emissions in urban areas combined with the serious air pollution problems in many urban areas, Congress made significant changes to the motor vehicle provisions on the Clean Air Act and established tighter pollution standards for emissions from automobiles and trucks. These standards were set so as to reduce tailpipe emissions of hydrocarbons, carbon monoxide, and nitrogen oxides on a phased-in basis beginning in model year 1994. Automobile manufacturers also were required to reduce vehicle emissions resulting from the evaporation of gasoline during refueling.

Fuel quality was also controlled. Scheduled reductions in gasoline volatility and sulfur content of diesel fuel, for example, were required. Programs requiring cleaner (so-called "reformulated") gasoline were initiated in 1995 for the nine cities with the worst ozone problems. Higher levels (2.7%) of alcohol-based oxygenated fuels were to be produced and sold in those areas that exceed the federal standard for carbon monoxide during the winter months.

The 1990 amendments to the Clean Air Act also established a clean fuel car pilot program in California, requiring the phase-in of tighter emission limits for 150,000 vehicles in model year 1996 and 300,000 by the model year 1999. These standards were to be met with any combination of vehicle technology and cleaner fuels. The standards became even more strict in 2001. Other states were able to "opt in" to this program, through incentives, not sales or production mandates.

Air Toxics

Toxic air pollutants are those pollutants which are hazardous to human health or the environment. These pollutants are typically carcinogens, mutagens, and reproductive toxins.

The toxic air pollution problem is widespread. Information generated in 1987 from the Superfund "Right to Know" rule (SARA Section 313) discussed earlier indicated that more than 2.7 billion pounds of toxic air pollutants were emitted annually in the United States. The EPA studies indicated that exposure to such quantities of toxic air pollutants may result in 1000–3000 cancer deaths each year.

Section 112 of the Clean Air Act includes a list of 189 substances which are identified as hazardous air pollutants. A list of categories of sources that emit these pollutants was prepared [The list of source categories included (1) major sources, or sources emitting 10 tons per year of any single hazardous air pollutants; and, (2) area sources, (smaller sources, such as dry cleaners and auto body refinishing)]. In turn, EPA promulgated emission standards, referred to as MACT or maximum achievable control technology standards, for each listed source category. These standards were based on the best demonstrated control technology or practices utilized by sources that make up each source category. Within 8 years of promulgation of a MACT standard, EPA must evaluate the level of risk that remains (residual risk) due to exposure to emissions from a source category, and determine if the residual risk is acceptable. If the residual risks are determined to be unacceptable, additional standards are required.

Acid Deposition Control

Acid rain occurs when sulfur dioxide and nitrogen oxide emissions are transformed in the atmosphere and return to the earth in rain, fog, or snow. Approximately 20 million tons of sulfur dioxide are emitted annually in the United States, mostly from the burning of fossil fuels by electric utilities. Acid rain damages lakes, harms forests and buildings, contributes to reduced visibility, and is suspected of damaging health.

It was hoped that the Clean Air Act would bring about a permanent 10 million-ton reduction in sulfur dioxide (SO_2) emissions from 1980 levels. To achieve this, the EPA allocated allowances in two phases, permitting utilities to emit one ton of sulfur dioxide. The first phase, which became effective January 1, 1995, required 110 power plants to reduce their emissions to a level equivalent to the product of an emissions rate of 2.5 lbs of SO_2/MM Btu × an average of their 1985–1987 fuel use. Emissions data indicate that 1995 SO_2 emissions at these units nationwide were reduced by almost 40% below the required level.

The second phase, which became effective January 1, 2000, required approximately 2,000 utilities to reduce their emissions to a level equivalent to the product of an emissions rate of 1.2 lbs of SO_2/MM Btu × the average of their 1985–1987 fuel use. In both phases, affected sources were required to install systems that continuously monitor emissions in order to track progress and assure compliance.

The Clean Air Act allowed utilities to trade allowances within their systems and/or buy or sell allowances to and from other affected sources. Each source must have had sufficient allowances to cover its annual emissions. If not, the source was subject to a $2,000/ton excess emissions fee and a requirement to offset the excess emissions in the following year.

The Clean Air Act also included specific requirements for reducing emissions of nitrogen oxides.

Operating Permits

The Act requires the implementation of an operating permits program modeled after the National Pollution Discharge Elimination System (NPDES) of the Clean Water Act. The purpose of the operating permits program is to ensure compliance with all applicable requirements of the Clean Air Act. Air pollution sources, subject to the program, must obtain an operating permit; states must develop and implement an operating permit program consistent with the Act's requirements; and, EPA must issue permit program regulations, review each state's proposed program, and oversee the state's effort to implement any approved program. The EPA must also develop and implement a federal permit program when a state fails to adopt and implement its own program.

In many ways this program is the most important procedural reform contained in the 1990 Amendments to the Clean Air Act. It enhanced air quality control in a variety of ways and updated the Clean Air Act, making it more consistent with other environmental statutes. The Clean Water Act, the Resource Conservation and Recovery Act and the Federal Insecticide, Fungicide, and Rodenticide Act all require permits.

Stratospheric Ozone Protection

The Clean Air Act requires the phase out of substances that deplete the ozone layer. The law required a complete phase-out of CFCs and halons, with stringent interim reductions on a schedule similar to that specified in the Montreal Protocol, including CFCs, halons, and carbon tetrachloride by 2000 and methyl chloroform by 2002. Class II chemicals (HCFCs) will be phased out by 2030.

The law required nonessential products releasing Class I chemicals to the banned. This ban went into effect for aerosols and noninsulating foams using Class II chemicals in 1994. Exemptions were included for flammability and safety.

Provisions Relating to Enforcement

The Clean Air Act contains provisions for a broad array of authorities to make the law readily enforceable. EPA has authorities to:

1. Issue administrative penalty orders up to $200,000 and field citations up to $5000.
2. Obtain civil judicial penalties.
3. Secure criminal penalties for knowing violations and for knowing and negligent endangerment.
4. Require sources to certify compliance.
5. Issue administrative subpoenas for compliance data.
6. Issue compliance orders with compliance schedules of up to one year.

Citizen suit provisions are also included to allow citizens to seek penalties against violators, with penalties going to a U.S. Treasury fund for use by the EPA for compliance and enforcement activities.

The following EPA actions represent recent regulations promulgated to implement the requirements of the Clean Air Act:

1. Clean Air Interstate Rule published on May 12, 2005 (70 FR 25161) amends requirements for State Implementation Plans (SIPs) and for the provisions for the Acid Rain Program.
2. Mercury Rules published on May 18, 2005 (70 FR 28605) amends New Source Performance Standards for electric utility steam generating units and some provisions of the Acid Rain Program.
3. Non-road Diesel Rule published on May 11, 2004 (69 FR 38957) amends provisions for mobile sources and for highway vehicles and engines.
4. Ozone Rules identified those areas that are designated as not attaining the ambient air quality standards for ozone.
5. Fine Particle Rules identified those areas that are designated as not attaining the ambient air quality standards for particulate matter.

Expanded details on these rules are provided in problem LCAA.2 in Chapter 3.

OCCUPATIONAL SAFETY AND HEALTH ACT (OSHA)

The *Occupational Safety and Health Act* (OSHAct) was enacted by Congress in 1970 and established the *Occupational Safety and Health Administration* (OSHA), which addressed safety in the workplace. At the same time the EPA was established. Both EPA and OSHA are mandated to reduce the exposure of hazardous substances over land, sea, and air. The OSHAct is limited to conditions that exist in the workplace, where its jurisdiction covers both safety and health. Frequently, both agencies regulate the same substances but in a different manner as they are overlapping environmental organizations.

Congress intended that OSHA be enforced through specific standards in an effort to achieve a safe and healthful working environment. A "general duty clause" was added to attempt to cover those obvious situations that were admitted by all concerned but for which no specific standard existed. The OSHA standards are an extensive compilation of regulations, some that apply to all employers (such as eye and face protection) and some that apply to workers who are engaged in a specific type of work (such as welding or crane operation). Employers are obligated to familiarize themselves with the standards and comply with them at all times.

Health issues, most importantly, contaminants in the workplace, have become OSHA's primary concern. Health hazards are complex and difficult to define. Because of this, OSHA has been slow to implement health standards. To be complete, each standard requires medical surveillance, record keeping, monitoring, and physical reviews. On the other side of the ledger, safety hazards are aspects of the work environment that are expected to cause death or serious physical harm immediately or before the imminence of such danger can be eliminated.

Probably one of the most important safety and health standards ever adopted is the OSHA hazard communication standard, more properly known as the "right to know" laws. The hazard communication standard requires employers to communicate information to the employees on hazardous chemicals that exist within the workplace. The program requires employers to craft a written hazard communication program, keep *material safety data sheets* (MSDSs) for all hazardous chemicals at the workplace and provide employees with training on those hazardous chemicals, and assure that proper warning labels are in place.

The *Hazardous Waste Operations and Emergency Response Regulation* enacted in 1989 by OSHA addressed the safety and health of employees involved in cleanup operations at uncontrolled hazardous waste sites being cleaned up under government mandate, and in certain hazardous waste treatment, storage, and disposal operations conducted under RCRA. The standard provides for employee protection during initial site characterization and analysis, monitoring activities, training and emergency response. Four major areas are under the scope of the regulation:

1. Cleanup operations at uncontrolled hazardous waste sites that have been identified for cleanup by a government health or environmental agency.

2. Routine operations at hazardous waste Transportation, Storage, and Disposal (TSD) facilities or those portions of any facility regulated by 40 CFR Parts 264 and 265.

3. Emergency response operations at sites where hazardous substances have or may be released.

4. Corrective action at RCRA sites.

The regulation addressed three specific populations of workers at the above operations. First, it regulates hazardous substance response operations under CERCLA, including initial investigations at CERCLA sites before the presence or absence of hazardous substance has been ascertained; corrective actions taken in cleanup operations under RCRA; and, those hazardous waste operations at sites that have been designated for cleanup by state or local government authorities. The second worker population to be covered involves those employees engaged in operations involving hazardous waste TSD facilities. The third employee population to be covered involves those employees engaged in emergency response operations for release or substantial threat of releases of hazardous substances, and post emergency response operations to such facilities (29 CFR, 1910.120(q) (9)).

USEPA'S RISK MANAGEMENT PROGRAM

Developed under the *Clean Air Act's* (CAA's) Section 112(r), the *Risk Management Program* (RMP) rule (40 CFR Part 68) is designed to reduce the risk of accidental releases of acutely toxic, flammable, and explosive substances. A list of the regulated substances (138 chemicals) along with their threshold quantities is provided in the Code of Federal Regulations at 40 CFR 68.130.

In the RMP rule, EPA requires a *Risk Management Plan* that summarizes how a facility is to comply with EPA's RMP requirements. It details methods and results of hazard assessment, accident prevention, and emergency response programs instituted at the facility. The hazard assessment shows the area surrounding the facility and the population potentially affected by accidental releases. EPA requirements include a three-tiered approach for affected facilities. A facility is affected if a process unit manufactures, processes, uses, stores, or otherwise handles any of the listed chemicals at or above the threshold quantities. The EPA approach is summarized in Table 2.2. For example, EPA defined Program 1 facilities as those processes that have not had an accidental release with offsite consequences in the five years prior to the submission date of the RMP and have no public receptors within the distance to a specified toxic or flammable endpoint associated with a worst-case release scenario. Program 1 facilities have to develop and submit a risk management plan and complete a registration that includes all processes that have a regulated substance present in more than a threshold quantity. They also have to: analyze the worst-case release scenario for the process or processes; document that the nearest public receptor is beyond the distance to a toxic or flammable endpoint; complete a five-year accident history for the process or processes; ensure that response

TABLE 2.2   RMP Approach

Program	Description
1	Facilities submit RMP, complete registration of processes, analyze worst-case release scenario, complete 5-year accident history, coordinate with local emergency planning and response agencies; and certify that the source's worst-case release would not reach the nearest public receptors.
2	Facilities submit RMP, complete registration of processes, develop and implement a management system; conduct a hazard assessment; implement certain prevention steps; develop and implement an emergency response program; and submit data on prevention program elements.
3	Facilities submit RMP, complete registration of processes, develop and implement a management system; conduct a hazard assessment; implement prevention requirements; develop and implement an emergency response program; and provide data on prevention program elements.

actions are coordinated with local emergency planning and response agencies; and certify that the source's worst-case release would not reach the nearest public receptors. Program 2 applies to facilities that are not Program 1 or Program 3 facilities. Program 2 facilities have to develop and submit the RMP as required for Program 1 facilities plus: develop and implement a management system; conduct a hazard assessment; implement certain prevention steps; develop and implement an emergency response program; and, submit data on prevention program elements for Program 2 processes. Program 3 applies to processes in Standard Industrial Classification (SIC) codes 2611 (pulp mills), 2812 (chloralkali), 2819 (industrial inorganics), 2821 (plastics and resins), 2865 (cyclic crudes), 2869 (industrial organics), 2873 (nitrogen fertilizers), 2879 (agricultural chemicals), and 2911 (petroleum refineries). These facilities belong to industrial categories identified by EPA as historically accounting for most industrial accidents resulting in off-site risk. Program 3 also applies to all processes subject to the OSHA Process Safety Management (PSM) standard (29 CFR 1910.119). Program 3 facilities have to develop and submit the RMP as required for Program 1 facilities plus: develop and implement a management system; conduct a hazard assessment; implement prevention requirements; develop and implement an emergency response program; and provide data on prevention program elements for the Program 3 processes.

THE POLLUTION PREVENTION ACT OF 1990

The Pollution Prevention Act, along with the Clean Air Act Amendments passed by Congress on the same day in November 1990, represents a clear breakthrough in this nation's understanding of environmental problems. The Pollution Prevention Act calls pollution prevention a "national objective" and establishes a hierarchy of environmental protection priorities as national policy.

Under the Pollution Prevention Act, it is the national policy of the United States that pollution should be prevented or reduced at the source whenever feasible; where pollution cannot be prevented, it should be recycled in an environmentally safe manner. In the absence of feasible prevention and recycling opportunities, pollution should be treated; and, disposal should be used only as a last resort.

Among other provisions, the Act directed the EPA to facilitate the adoption of source reduction techniques by businesses and federal agencies, to establish standard methods of measurement for source reduction, to review regulations to determine their effect on source reduction, and to investigate opportunities to use federal procurement to encourage source reduction. The Act initially authorized an $8 million state grant program to promote source reduction, with a 50% state match requirement (10).

The EPA's pollution prevention initiatives are characterized by its use of a wide range of tools, including market incentives, public education and information, small business grants, technical assistance, research and technology applications, as well as the more traditional regulations and enforcement. In addition, there are other significant behind-the-scenes achievements: identifying and dismantling barriers to pollution prevention; laying the groundwork for a systematic prevention focus; and, creating advocates for pollution prevention that serve as catalysts in a wide variety of institutions.

REFERENCES

1. G. Burke, B. Singh, and L. Theodore, "Handbook of Environmental Management and Technology" 2nd Edition, John Wiley & Sons, Hoboken, NJ, 2000.
2. U.S. EPA, *EPA J.* 14(2), March 1988.
3. Office of Solid Waste, *Solving the Hazardous Waste Problem*, EPA/530–SW–86–037.
4. U.S. EPA, Solid Waste and Emergency Response, *The Waste System*, November 1988.
5. P. N. Cheremisinoff, and F. Ellerbusch, Solid Waste Legislation, Resource Conservation & Recovery Act, A Special Report, Washington, D.C., 1979.
6. Bureau of National Affairs, Washington, D.C., Resource Conservation and Recovery Act of 1976, *International Environmental Reporter*, October 21, 1976.
7. U.S. EPA, *Environmental Progress and Challenges: EPA's Update*, EPA–230–07–88–033, August 1988.
8. The Clean Air Act Amendments of 1990 Summary Materials, November 15, 1990.
9. M. K. Theodore and L. Theodore, "Major Environmental Issues Facing the 21st Century", Theodore Tutorials (originally published by Simon & Schuster), East Williston, NY, 1995.
10. Office of Pollution Prevention, *Pollution Prevention News*, October 1991.

CHAPTER 3

CLEAN AIR ACT (CAA)

QUALITATIVE PROBLEMS (LCAA)

PROBLEMS LCAA.1–31

LCAA.1 EARLY LEGISLATION

Describe early air pollution legislation.

SOLUTION: Although air pollution has been around for a long time, it has only been during the past 50 years that the governments of this country and many others have taken active roles in controlling it. In the United States, the first efforts at regulating air quality were principally at the individual community level in cities such as Cincinnati, Ohio, Pittsburgh, Pennsylvania, and Los Angeles, California. State requirements for controlling air pollution on a statewide basis came later.

At the federal level, the Clean Air Act defines the national policy for air pollution abatement and control in the United States. It establishes goals of protecting health and natural resources and delineates what is expected of federal, state, and local governments to achieve those goals. From its modest beginnings as an effort to determine the causes of illnesses and deaths that occurred in such places as Donora, Pennsylvania, or in New York City to the comprehensive document that exists today, the evolution of the Clean Air Act is an example of the country's efforts to protect people and the environment.

Environmental Regulatory Calculations Handbook, by Leo Stander and Louis Theodore
Copyright © 2008 John Wiley & Sons, Inc.

The federal effort began with the first national air pollution control legislation, called the Air Pollution Control Act of 1955. This legislation was enacted after epidemiological studies began showing that the effects of air pollution had broader implications than previously believed and that injuries and even death were the result of high concentrations of air pollutants in the atmosphere. The Air Pollution Control Act provided the Public Health Service of the Department of Health, Education and Welfare with limited authority to conduct air pollution research and to provide technical assistance to state and local governments. Amendments to this legislation, enacted in 1960 and 1962, authorized special studies to evaluate the health effects of emissions from motor vehicles.

The Clean Air Act of 1963 encouraged state, regional, and local programs to abate and control air pollution and enabled the U.S. Public Health Service to intervene in interstate problem areas. It also encouraged additional research for pollution resulting from motor vehicles and the development of air quality criteria to guide states in establishing ambient air quality standards and emission standards for stationary sources. The Clean Air Act was amended in 1965 to authorize emission standards for motor vehicles.

In 1967, the Clean Air Act was again amended and was renamed the Air Quality Act of 1967. The 1967 amendments provided the basic framework for many of the current efforts to control air pollution. Specifically, states were required to develop and implement ambient air quality standards on a fixed time schedule after the Secretary of Health, Education, and Welfare issued air quality criteria and control techniques information.

Because of the slow process in developing ambient standards and the continuing deterioration of air quality across the country, the Clean Air Act of 1970 was enacted. These amendments required the newly formed Environmental Protection Agency (EPA) to adopt uniform national ambient air quality standards to protect health and welfare and required that these standards be attained and maintained across the country. Other aspects of the 1970 amendments included a significant federal enforcement authority; the development of state implementation plans to achieve the ambient air quality standards; more stringent motor vehicle emission standards; requirements for EPA to establish standards of performance for new and modified sources of air pollution; requirements for establishing national emission standards for hazardous air pollutants; and, provisions for citizen suits.

In 1977, the Clean Air Act was again amended to enable EPA to make some corrections in the country's efforts to achieve and maintain ambient standards. Additional time was allowed for achieving the National Ambient Air Quality Standards in the most polluted areas, and the concept of Prevention of Significant Deterioration was made part of the law. Additionally, EPA was given the flexibility to allow for some growth in dirty areas.

The 1990 Amendments to the Clean Air Act are the most recent changes in effect. Since 1977, the country had been unsuccessful in its efforts to achieve the ambient standards for many pollutants and EPA had been slow in identifying and requiring controls of hazardous air pollutants. The 1990 Amendments to the Clean Air Act, provided a listing of hazardous air pollutants (Title III) and required EPA to establish emission standards for certain industrial and commercial sources of these pollutants.

Under the 1990 Amendments to the Clean Air Act, EPA was authorized to require states to develop an operating permits program (Title V), to develop a program for reducing emissions of sulfur dioxide from power plants (Title IV), and to establish more stringent control requirements in those areas that continue to have polluted air (Title I). The EPA was also authorized to control and reduce those substances that are considered stratospheric ozone-depleting substances (Title VI) and to reduce emissions from motor vehicles (Title II).

The Clean Air Act (as amended in 1990) is composed of six subchapters, each of which establishes requirements for a portion of the nation's air pollution control effort. Subchapter I requires controls for stationary sources (including sources of toxic air pollutants as modified by Title III of the 1990 amendments to the Clean Air Act). Subchapter II establishes requirements for mobile sources. Subchapter III includes definitions and administrative provisions for development of regulations and for public involvement. The acid rain provisions are prescribed in Subchapter IV. Subchapter V (referred to as Title V) covers the operating permits program, and Subchapter VI provides requirements for protection of stratospheric ozone and implements the Montreal Protocol, an international agreement to reduce emissions and ultimately eliminate the use of chlorofluorocarbons and other ozone-depleting chemicals.

LCAA.2 RECENT KEY CLEAN AIR ACT (CAA) REGULATORY ACTIONS

Describe five recent Clean Air Act Regulations.

SOLUTION: The following five major rules were promulgated to achieve significant improvement in air quality, health, and quality of life.

1. *Clean Air Interstate Rule* (70 FR 25161, May 12, 2005)
    The Clean Air Interstate Rule provided states with a solution to the problem of power plant pollution that drifts from one state to another. The rule uses a cap and trade system to reduce the target pollutants by 70 percent.

2. *Mercury Rule* (70 FR 28605, May 18, 2005)
    EPA issued the Clean Air Mercury Rule on March 15, 2005. This rule builds on the Clean Air Interstate Rule (CAIR) to reduce mercury emissions from coal-fired power plants, the largest remaining domestic source of human-caused mercury emissions. Issuance of the Clean Air Mercury Rule marked the first time EPA regulated mercury emissions from utilities, and made the U.S. the first nation in the world to control emissions from this major source of mercury pollution.

3. *Nonroad Diesel Rule* (69 FR 38957, May 11, 2004)
    The Clean Air Nonroad Diesel Rule will change the way diesel engines function to remove emissions and the way diesel fuel is refined to remove sulfur. The Rule is one of EPA's *Clean Diesel Programs*, which were promulgated to produce significant improvements in air quality.

4. *Ozone Rules* (http://www.epa.gov/ozonedesignations/)
    The Clean Air Ozone Rules (dealing with 8-hour ground-level ozone designation

and implementation) designated those areas whose air did not meet the health-based standards for ground-level ozone. The ozone rules classified the seriousness of the problem and required states to submit plans for reducing the levels of ozone in areas where the ozone standards were not being met.

5. *Fine Particle Rules* (http://www.epa.gov/pmdesignations/)
   The Clean Air Fine Particles Rules designated those areas whose air does not meet the health-based standards for fine-particulate pollution. This rule required states to submit plans for reducing the levels of particulate pollution in areas where the fine-particle standards are not met.

See also *15th Anniversary of Clean Air Act Amendments of 1990* at http://www.epa.gov/air/cleanairact/

LCAA.3   SOURCES OF AIR POLLUTION

Discuss the sources of air pollution.

SOLUTION: Air pollution presents one of the greatest risks to human health and the environment in the United States. The long list of health problems caused or aggravated by air pollution includes lung diseases such as chronic bronchitis and pulmonary emphysema; cancer, particularly lung cancer; neural disorders, including brain damage; bronchial asthma and the common cold, which are most persistent with highly polluted air; and, eye irritations. Environmental problems range from damage to crops and vegetation to increased acidity of lakes that makes them uninhabitable for fish and other aquatic life.

A stationary source is any facility which directly emits or has the potential to emit air pollutants from a fixed emission point. These sources range in size from large processing plants to home heating units. Some of the common sources are power plants, steel mills, coke plants, painting and coating operations, industrial and commercial heaters and boilers, residential, commercial, medical, and municipal waste incinerators, dry cleaners, and printing establishments.

Stationary sources generate air pollutants as a result of the combustion of fuel and as byproduct of industrial processes. The combustion of coal, oil, natural gas, wood, and other fuels is a source of particulate matter, sulfur dioxide, nitrogen oxides, carbon monoxide, and hazardous air pollutants. Industrial processes that utilize petrochemicals are sources of volatile organic compounds and many hazardous air pollutants. Industrial processes that produce and utilize petrochemicals include oil refineries, chemical production facilities, painting and coating operations, dry cleaners, and gasoline marketing establishments.

LCAA.4   NANOTECHNOLOGY REGULATIONS: AIR

Comment on how and to what degree new legislation and rulemaking will be necessary for environmental control/concerns with nanotechnology from an "air" perspective.

SOLUTION: A likely source of regulation would fall under the provisions of the Clean Air Act (CAA) for particulate matter less than 2.5 μm (PM2.5 or $PM_{2.5}$). Additionally, an installation manufacturing nanomaterials may ultimately become subject as a "major source" to the CAA's Section 112 governing hazardous air pollutants (HAPs).

The reader is referred to the following two texts for additional information:

1. L. Theodore and R. Kunz, "Nanotechnology: Environmental Implications and Solutions," John Wiley & Sons, Hoboken, NJ, 2005.
2. L. Theodore, "Nanotechnology: Basic Calculations for Engineers and Scientists," John Wiley & Sons, Hoboken, NJ, 2006.

LCAA.5 ACRONYMS

Many acronyms are associated with the Clean Air Act and the corresponding amendments. Indicate what each of the following six acronyms stands for:

1. NAAQS
2. NSR
3. RFP
4. SIP
5. BACT
6. VOC

SOLUTION:

1. NAAQS = National Ambient Air Quality Standards
2. NSR = New Source Review
3. RFP = Reasonable Further Progress
4. SIP = State Implementation Plan
5. BACT = Best Available Control Technology
6. VOC = Volatile Organic Compounds

There are many other acronyms employed in the air pollution regulatory field.

LCAA.6 ENVIRONMENTAL LAW ENFORCEMENT

Environmental laws are administered by the EPA which control pollution to the outdoor environment by air pollutants. To administer these laws, EPA's responsibilities include which of the following (identify all of the correct answers):

1. Setting and enforcing standards.
2. Developing pollution control and measurement methods.
3. Working with and informing the public.
4. Getting more information on the air toxics.
5. Collecting information on short-term and lifetime toxicological studies.

SOLUTION: In carrying out law, EPA sets standards for environmental quality based on the intent of the law. By following specific criteria in developing standards under each environmental law, EPA has jurisdiction to control any release of air pollutants (e.g., air toxics).

EPA conducts programs to develop and test new ways to reduce the effects of pollutants in the environment. EPA's Pollution Prevention Office develops and implements programs to minimize pollution.

EPA programs provide information to the public and business about regulatory requirements, environmental programs, procedures to reduce exposure to pollutants, and the health effects of these pollutants. EPA provides information through published materials, training programs, and certification courses. Therefore the correct answers are 1, 2, and 3.

In order to assess the health effects due to exposure to air pollutants (including air toxics), toxicological studies are conducted using vertebrates such as mice, rats, and rodents. These studies include various pathways of exposure to toxicants and impact of the pollutants and toxicity tests such as short-term and long-term studies. The information collected in toxicological studies helps in predicting risks due to exposure to the hazardous chemicals (including air toxics). The EPA requires the producers of the air toxics to provide this information. Therefore 4 and 5 are not the correct answers.

LCAA.7 BUBBLE POLICY

Outline the key features of the "bubble" policy.

SOLUTION: The bubble policy allows an industrial facility to comply with some emission limits on a facility-wide basis (all sources in one "bubble"), rather than having to meet separate emission limits applied to each source operation, stack or vent, which perhaps number several hundred at a large manufacturing facility. The total allowable emissions remain the same, but the facility-wide approach enables the industry to reduce emissions more cost-effectively. For example, "over controlling" or even closing some source operations, while "under controlling" others, is likely to be preferred to uniform controls on all sources.

LCAA.8 NATIONAL AMBIENT AIR QUALITY STANDARD (NAAQS)

Discuss the present NAAQS.

SOLUTION: As described previously, a National Ambient Air Quality Standard (NAAQS) is the maximum level that will be permitted for a given pollutant. There are two kinds of such standards: *primary* and *secondary*. Primary standards are to be sufficiently stringent to protect the public health; secondary standards must protect the public welfare. EPA sets these standards after it issues a criteria document and a control technology document on the pollutant in question. Both the primary and secondary standards apply to all areas of the country. Ambient air quality standards are presented in Table 3.1.

TABLE 3.1   National Ambient Air Quality Standards

Pollutant	Primary Standards (Health Related)	Averaging Times	Secondary Standards (Welfare Related)
Carbon monoxide	9 ppm (10 mg/m^3)	8-hour[1]	None
	35 ppm (40 mg/m^3)	1-hour[1]	None
Lead	$1.5 \ \mu g/m^3$	Quarterly average	Same as Primary
Nitrogen dioxide	0.053 ppm ($100 \ \mu g/m^3$)	Annual (arithmetic mean)	Same as Primary
Particulate matter (PM_{10})	Revoked[2]	Annual[2] (arithmetic mean)	
	$150 \ \mu g/m^3$	24-hour[3]	
Particulate matter ($PM_{2.5}$)	$15.0 \ \mu g/m^3$	Annual[4] (arithmetic mean)	
	$35 \ \mu g/m^3$	24-hour[5]	
Ozone	0.08 ppm	8-hour[6]	Same as Primary
	0.12 ppm	1-hour[7] (applies only in limited areas)	Same as Primary
Sulfur oxides	0.03 ppm	Annual (arithmetic mean)	—
	0.14 ppm	24-hour[1]	—
	—	3-hour[1]	0.5 ppm ($1300 \ \mu g/m^3$)

[1]Not to be exceeded more than once per year.

[2]Due to a lack of evidence linking health problems to long-term exposure to coarse particle pollution, the agency revoked the annual PM_{10} standard in 2006 (effective December 17, 2006).

[3]Not to be exceeded more than once per year on average over 3 years.

[4]To attain this standard, the 3-year average of the weighted annual mean $PM_{2.5}$ concentrations from single or multiple community-oriented monitors must not exceed $15.0 \ \mu g/m^3$.

[5]To attain this standard, the 3-year average of the 98th percentile of 24-hour concentrations at each population-oriented monitor within an area must not exceed $35 \ \mu g/m^3$ (effective December 17, 2006).

[5]To attain this standard, the 3-year average of the fourth-highest daily maximum 8-hour average ozone concentrations measured at each monitor within an area over each year must not exceed 0.08 ppm.

[7](a) The standard is attained when the expected number of days per calendar year with maximum hourly average concentrations above 0.12 ppm is ≤ 1, as determined by Appendix H.

(b) As of June 15, 2005 EPA revoked the 1-hour ozone standard in all areas except the fourteen 8-hour ozone nonattainment Early Action Compact (EAC) areas.

A State Implementation Plan is a description of a state's strategies, efforts, and activities to achieve and maintain national ambient air quality standards. Similar plans are required for Indian country. These plans include the establishment and enforcement of emission limitations for existing stationary sources (commonly referred to as RACT), description of procedures and programs to monitor ambient air concentrations and to monitor emissions from stationary sources, regulation of modification and construction of stationary sources [new source review for attainment areas (PSD program) and nonattainment areas], and enforceable strategies for reducing emissions and impact from mobile sources. The requirements for such plans differ for those areas that have air quality cleaner than the limits established for the NAAQS (attainment areas) than for those areas with air quality poorer or dirtier than NAAQS (nonattainment areas). The requirements for such plans are published at 40 CFR Part 51. Such plans are adopted by states and submitted to EPA for review and approval. Following EPA approval, the SIP becomes enforceable by citizens and the U.S. Government. If a SIP does not adequately comply with federal requirements, EPA is required to promulgate it in those areas that lack approved SIPs. State plans as approved or Federal Implementation Plans as adopted are promulgated and published at 40 CFR Part 52.

LCAA.9   STANDARDS FOR NEW AND MODIFIED STATIONARY SOURCES (NSPS)

Discuss standards of performance for new and modified stationary sources.

SOLUTION: The NSPS program is described in the Clean Air Act in Section 111. These standards regulate emissions of criteria and other pollutants from new and modified sources in specific source categories. By focusing activities on new sources and modifications rather than existing sources of pollution, the NSPS prevents new air pollution problems and results in long-term improvements in air quality as the older existing plants are being replaced by new ones. The provisions of the NSPS are designed to ensure that new stationary sources are designed, built, equipped, operated, and maintained so as to keep emissions to a minimum. NSPS were promulgated for those categories of stationary sources that significantly cause or contribute to air pollution that could reasonably be anticipated to endanger public health or welfare. These standards are modified from time to time. The current standards can be found at 40 CFR Part 60.

A standard of performance is a standard for emissions of air pollutants that reflects the degree of emission limitation achievable through the application of the best system of emission reduction that (taking into account the cost of achieving such reduction and any non-air-quality health and environmental impact and energy requirements) the EPA determines has been adequately demonstrated. A list of some source categories for which performance standards have been established is provided in Table 3.2.

TABLE 3.2 New Source Performance Standards Source Categories

Municipal waste combustors	Phosphate fertilizer industry: Superphosphoric acid plants	Equipment leaks in synthetic organic chemical manufacturing industry
Municipal solid waste landfills	Phosphate fertilizer industry: Diammonium phosphate plants	
Sulfuric acid production units		Beverage can surface coating industry
Hospital/Medical/Infectious Waste Incinerators		
Fossil-fuel-fired steam generators	Phosphate fertilizer industry: Triple superphosphate plants	Bulk gasoline terminals
Electric utility steam-generating units	Phosphate fertilizer industry: Granular triple superphosphate storage facilities	New residential wood heaters
Industrial/commercial/institutional steam-generating units	Coal preparation plants	Rubber tire manufacturing industry
Small industrial/commercial steam-generating units	Ferroalloy production facilities	Polymer manufacturing industry
Incinerators	Steel plants: Electric arc furnaces	Flexible vinyl and urethane coating and printing
Portland cement plants	Steel plants: Electric arc furnaces and argon–oxygen decarburization vessels	Equipment leaks in petroleum refineries
Nitric acid plants	Kraft pulp mills	Synthetic fibre production facilities
Sulfuric acid plants	Glass manufacturing plants	Synthetic organic chemical manufacturing industry air oxidation unit processes
Hot asphalt facilities	Grain elevators	Petroleum dry cleaners

(Continued)

TABLE 3.2 *Continued*

Petroleum refineries	Surface coating of metal furniture	Equipment leaks from onshore natural gas processing plants
Storage vessels for petroleum liquids	Stationary gas turbines	Onshore natural gas processing plants; SO_2 emissions
Volatile organic liquid storage vessels	Lime manufacturing plants	Wool fiberglass insulation manufacturing plants
Secondary lead smelters	Lead-acid battery manufacturing plants	Petroleum refinery wastewater systems
Secondary brass and bronze production plants	Metallic mineral processing plants	Synthetic organic chemical manufacturing reactor processes
Primary emissions from basic oxygen process furnaces	Automobile and light-duty truck surface coating operations	Magnetic tape coating facilities
Secondary emissions from basic oxygen process steel-making facilities	Phosphate rock plants	Industrial surface coating: Surface coating of plastic parts for business machines
Sewage treatment plants	Ammonium sulfate manufacture	Calciners and dryers in mineral industries
Primary copper smelters	Graphic arts industry: Publication rotogravure printing	Polymeric coating of supporting substrates facilities
Primary zinc smelters	Pressure-sensitive tape and label surface coating operations	Municipal solid waste landfills
Primary lead smelters	Industrial surface coating: Large appliances	Commercial and industrial solid waste incineration units
Primary aluminum reduction plants	Metal coil surface coating	
Phosphate fertilizer industry: wet-process phosphoric acid plants	Asphalt processing and asphalt roofing manufacture	

LCAA.10   MAJOR STATIONARY SOURCES (40 CFR 52.21)

Define and list major stationary sources.

SOLUTION: Sources subject to PSD regulations (40CFR52.21) are major stationary sources and major modifications located in attainment or unclassified areas. A major stationary source is defined as any source listed in Table 3.3 with the potential to emit 100 tons/year (TPY) or more of any Clean Air Act pollutant. Sources not listed in Table 3.3 may also be classified as a major stationary source if the potential to emit exceeds 250 TPY. The "potential to emit" is defined as the maximum capacity to emit the pollutant under applicable emission standards and permit conditions (after application of any air pollution control equipment) excluding secondary emissions. A "major modification" is defined as any physical or operational change of a major stationary source producing a "significant net emissions increase" of any Clean Air Act pollutant (see Table 3.4).

TABLE 3.3   PSD Source Categories With 100 TPY Thresholds

Fossil fuel-fired steam-electric plants of more than 250 million Btu/hr heat input
Coal cleaning plants (with thermal dryers)
Kraft pulp mills
Portland cement plants
Primary zinc smelters
Iron and steel mill plants
Primary aluminum ore reduction plants
Primary copper smelters
Municipal incinerators capable of charging more than 250 tons of refuse/day
Hydrofluoric acid plants
Sulfuric acid plants
Nitric acid plants
Petroleum refineries
Lime plants
Phosphate rock processing plants
Coke oven batteries
Sulfur recovery plants
Carbon black plants (furnace process)
Primary lead smelters
Fuel conversion plants
Sintering plants
Secondary metal production plants
Chemical process plants
Fossil fuel boilers (or combinations thereof) totaling more than 250 million Btu/hr heat input
Petroleum storage and transfer units with a total storage capacity exceeding 300,000 barrels
Taconite ore processing plants
Glass fiber processing plants
Charcoal production plants

TABLE 3.4   Significant Increase Levels

Pollutant	Tons/Year
Carbon monoxide	100
Nitrogen oxides	40
Sulfur dioxide	40
Particulate matter (PM/PM_{10})	25/15
Ozone (VOC)	40 (of VOC)
Lead	0.6
Asbestos	0.007
Beryllium	0.0004
Mercury	0.1
Vinyl chloride	1
Fluorides	3
Sulfuric acid mist	7
Hydrogen sulfide	10
Total reduced sulfur compounds (including H_2S)	10
Benzene	Any
Arsenic	Any
Radionuclides	Any
Radon-222	Any
Polonium-210	Any
CFCs 11, 12, 122, 114, 115	Any
Halons 1211, 1301, 2402	Any

LCAA.11   SMALL SOURCES

Discuss applicable regulations for small sources.

SOLUTION: There is no short succinct answer to this question. All Clean Air Act regulations are different and have specific requirements that apply to certain types of sources.

A source is any facility which directly emits or has the potential to emit air pollutants. Stationary sources are comprised of process units that emit air pollutants. Depending on the regulatory requirements, these process units are referred to as emission units, affected facilities, emission points, or affected units. In other regulations, these emission units may be further categorized generally by size as significant, insignificant, or trivial.

Definitions for the term "major" when describing "stationary sources" vary within the Clean Air Act depending on whether the requirements are referring to criteria pollutants or hazardous air pollutants. In Section 302(j) of the Clean Air Act "major stationary source" and "major emitting facility" are defined as any stationary facility or source of air pollutants which directly emits, or has the potential emit, one hundred (100) tons per year or more of any air pollutant. In the case of sources located in nonattainment areas for ozone, particulate matter, or carbon monoxide, the term "major stationary source" refers to sources that emit smaller quantities of those pollutants. In Section 302(x), a "small source" means a source that emits less than 100 tons of regulated pollutants per year, or any class of persons that the

Administrator determines, through regulation, generally lack technical ability or knowledge regarding control of air pollution."

When discussing hazardous air pollutants, the term "major source" means any stationary source or group of stationary sources located within a contiguous area and under common control that emits or has the potential to emit considering controls, in the aggregate, 10 tons per year or more of any hazardous air pollutant or 25 tons or more of any combination of hazardous air pollutants. If a source is not a "major source" of hazardous air pollutants, it is considered an "area source." It should be noted, individual hazardous air pollutants are also criteria pollutants but individual criteria pollutants are not all hazardous. A single source could at the same time be considered a "small source" of an individual criteria pollutant such as particulate matter and a "major source" of hazardous air pollutants such as cadmium compounds. Likewise, another source could be identified as a "major source" of criteria pollutants and, at the same time, an "area source" of hazardous air pollutants.

These descriptors for sources of air pollutants help determine which requirements apply. In the case of State Implementation Plans, the various RACT and pre-construction requirements (including new source review and prevention of significant deterioration) apply to major sources of criteria pollutants. As previously indicated, major sources include one or more individual process units. Emissions of air pollutants from process units at the "major source" must meet the SIP requirements unless they are exempted as insignificant due to size or hours of operation. The construction of, or modification to, individual process units in a "major source" may be subject to new source review requirements. "Small sources" sometimes referred to as "minor sources" may be subject to different and in many cases less stringent emission limitations, permitting, and "monitoring, reporting, and record keeping" requirements.

In the case of sources of hazardous air pollutants, all sources, within specified source categories of hazardous air pollutants are subject to the prescribed national emission standard for hazardous air pollutants. The prescribed standards may have different requirements for sources of certain sizes and for the various emission units. In some cases, smaller sources (e.g., area sources) and some individual emission units are specifically exempted from the various requirements. In other cases, specific individual emission units located in smaller or "an area source" will be subjected to requirements.

In short, as the term "small" has different meaning in different regulations. Each individual regulation should be carefully evaluated to determine how the requirements apply to emission units in "smaller sources" of air pollution.

LCAA.12 TITLE V OPERATING PERMITS

Provide an overview of Title V Operating Permits.

SOLUTION: (Adapted from ERM, "Clean Air Act Primer," Exton, PA, 2004): One of the most substantial changes resulting from the CAA Amendments of 1990 was the institution of the Title V Operating Permit program (named after Title V of the

CAA). Title V is a comprehensive, facility-wide operating permit program affecting thousands of companies across the United States.

Prior to the Amendments of 1990, there were no federal programs in place that required facilities to obtain air quality operating permits. There were federal programs, such as the Prevention of Significant Deterioration (PSD) program, that required certain new and modified air emission units to obtain construction permits. Many states also had developed and implemented their own construction and operating permit programs for new and existing emissions units. However, for the first time under Title V, existing major sources and newly constructed major sources nationwide are required to obtain federally enforceable operating permits. Even those facilities and operations previously "grandfathered" from permitting requirements, or already subject to state operating permit programs, are required to obtain a Title V permit.

Title V was not intended to impose new emissions standards or operating requirements on existing facilities. Rather, Title V was meant to serve as the vehicle by which all requirements for a facility were rolled-up into one document to ensure better compliance with air quality regulations.

However, Title V has meant changes for affected facilities. One of the most significant impacts of Title V is that facilities are now required to demonstrate that they are operating in "continuous compliance" with each permit term and condition. For many facilities, this has meant new (and sometimes cumbersome and costly) monitoring, recordkeeping, and reporting requirements. Facilities unable to prove they are operating in compliance with regulations are considered to be out of compliance with the applicable standard and must report this information to state and federal agencies. This represents a dramatic shift from enforcement practices in the state air quality agencies prior to the Amendments of 1990. To bring an enforcement action under previous policies, most regulated authorities were required to demonstrate that a facility operated in non-compliance with an underlying standard. With Title V, it has become the facility's burden to prove routine compliance to prevent any enforcement action.

Most facilities subject to Title V submitted their applications for their first operating permits in the mid-1990s and received their Title V permits in the late 1990s– early 2000s. With a maximum five year term, many facilities have now applied for Title V permit renewals.

LCAA.13   HAZARDOUS AIR POLLUTANTS (HAPs)

Provide a list of regulated hazardous air pollutants (HAPs).

SOLUTION: The term *hazardous air pollutants* refers to those air pollutants whose presence in the atmosphere is known to result in a variety of adverse health effects, including cancer, reproductive effects, birth defects, and respiratory illness or a variety of adverse environmental effects. Other terms often used interchangeably with hazardous air pollutants include *toxic air pollutants*, *air toxics*, *noncriteria* *pollutants*, and *regulated substances*. The Clean Air Act, in Section 112(b), and the Federal regulations in 40 CFR Part 61 identify substances that are considered hazardous air pollutants. The EPA and state and local air pollution control agencies may add other substances to this list for their jurisdictions. These pollutants are all primary pollutants and may be present in the atmosphere as either particulate matter or gases. Some are pesticides, herbicides, of fungicides, and many are also volatile organic hydrocarbons. These substances are listed in Section 112(b) of the Clean Air Act.

TABLE 3.5   Hazardous Air Pollutants

Chemical Abstract Service (CAS) Number	Chemical Name
75070	Acetaldehyde
60355	Acetamide
75058	Acetonitrile
98862	Acetophenon
53963	2-Acetylaminofluorene
107028	Acrolein
79061	Acrylamide
79107	Acrylic acid
107131	Acrylonitrile
107051	Allyl chloride
92671	4-Amino biphenyl
62533	Aniline
90040	*o*-Anisidine
1332214	Asbestos
71432	Benzene (including benzene from gasoline)
92875	Benzidine
98077	Benzotrichloride
100447	Benzyl chloride
92524	Biphenyl
117817	bis(2-Ethylhexyl)phthalate
542881	bis(Chloromethyl)ether
75252	Bromoform
106990	1,3-Butadiene
156627	Calcium cyanamide
133062	Captan
63252	Carbaryl
75150	Carbon disulfide
56235	Carbon tetrachloride
463581	Carbonyl sulfide
120809	Catechol
133904	Chloramben
57749	Chlordane

(Continued)

TABLE 3.5 *Continued*

Chemical Abstract Service (CAS) Number	Chemical Name
7782505	Chlorine
79118	Chloroacetic acid
532274	2-Chloroacetophenone
108907	Chlorobenzene
510156	Chlorobenzilate
67663	Chloroform
107302	Chloromethyl methyl ether
126998	Chloroprene
1319773	Cresols/cresylic acid (isomers and mixture)
95487	*o*-Cresol
108394	*m*-Cresol
106445	*p*-Cresol
98828	Cumene
94757	2,4-D, salts and esters
3547044	DDE
334883	Diazomethane
132649	Dibenzofurans
96128	1,2-Dibromo-3-chloropropane
84742	Dibutylphthalate
106467	1,4-Dichlorobenzene (*p*)
91941	3,3-Dichlorobenzidene
111444	Dichloroethyl ether (bis(2-chloroethyl)ether)
542756	1,3-Dichloropropene
62737	Dichlorvos
111422	Diethanolamine
121697	N,N-Diethyl aniline (N,N-dimethyl aniline)
64675	Diethyl sulfate
119904	3,3-Dimethoxybenzidine
60117	Dimethyl aminoazobenzene
119937	3,3'-Dimethoxybenzidine
79447	Dimethyl carbamoyl chloride
68122	Dimethyl formamide
57147	1,1-Dimethyl hydrazine
131113	Dimethyl phthalate
77781	Dimethyl sulfate
534521	4,6-Dinitro-*o*-cresol, and salts
51285	2,4-Dinitrophenol
121142	2,4-Dinitrotoluene
123911	1,4-Dioxane (1,4-diethylene oxide)
122667	1,2-Diphenylhydrazine
106898	Epichlorohydrin (1-chloro-2,3-epoxypropane)
106887	1,2-Epoxybutane

(*Continued*)

TABLE 3.5 *Continued*

Chemical Abstract Service (CAS) Number	Chemical Name
140885	Ethyl acrylate
100414	Ethyl benzene
51796	Ethyl carbamate (urethane)
75003	Ethyl chloride (chloroethane)
106934	Ethylene dibromide (dibromoethane)
107062	Ethylene dichloride (1,2-dichloroethane)
107211	Ethylene glycol
151564	Ethylene imine (aziridine)
75218	Ethylene oxide
96457	Ethylene thiourea
75343	Ethylidene dichloride (1,1-dichloroethane)
50000	Formaldehyde
76448	Heptachlor
118741	Hexachlorobenzene
87683	Hexachlorobutadiene
77474	Hexachlorocyclopentadiene
67721	Hexachloroethane
822060	Hexamethylene-1,6-diisocyanate
680319	Hexamethylphosphoramide
110543	Hexane
302012	Hydrazine
7647010	Hydrochloric acid
7664393	Hydrogen fluoride (hydrofluoric acid)
123319	Hydroquindne
78591	Isophorone
58899	Lindane (all isomers)
108316	Maleic anhydride
67561	Methanol
72435	Methoxychlor
74839	Methyl bromide (bromomethane)
74873	Methyl chloride (chloromethane)
71556	Methyl chloroform (1,1,1-trichloroethane)
78933	Methyl ethyl ketone (2-butanone)
60344	Methyl hydrazine
74884	Methyl iodide (iodomethane)
108101	Methyl isobutyl ketone (hexone)
624839	Methyl isocyanate
80626	Methyl methacrylate
1634044	Methyl *tert*-butyl ether
101144	4,4-Methylene bis(2-chloroaniline)
75092	Methylene chloride (dichloromethane)
101688	Methylene diphenyl diisocyanate (MDI)

(Continued)

TABLE 3.5 *Continued*

Chemical Abstract Service (CAS) Number	Chemical Name
101779	4,4′-Methylenedianiline
91203	Naphthalene
98953	Nitrobenzene
92933	4-Nitrobiphenyl
100027	4-Nitrophenol
79469	2-Nitropropane
684935	N-Nitroso-N-methylurea
62759	N-Nitrosodimethylamine
59892	N-Nitrosomorpholine
56382	Parathion
82688	Pentachloronitrobenzene (quintobenzene)
87865	Pentachlorophenol
108952	Phenol
106503	p-Phenylenediamine
75445	Phosgene
7803512	Phosphine
7723140	Phosphorus
85449	Phthalic anhydride
1336363	Polychlorinated biphenyls (aroclors)
1120714	1,3-Propane sultone
57578	β-Propiolactone
123386	Propionaldehyde
114261	Propoxur (baygon)
78875	Propylene dichloride (1,2-dichloropropane)
75569	Propylene oxide
75558	1,2-Propylenimine (2-methyl aziridine)
91225	Quinoline
106514	Quinone
100425	Styrene
96093	Styrene oxide
1746016	2,3,7,8-Tetrachloridebenzo-p-dioxin
79345	1,1,2,2-Tetrachloroethane
127184	Tetrachloroethane (perchloroethylene)
7550450	Titanium tetrachloride
108883	Toluene
95807	2,4-Toluene diamine
584849	2,4-Toluene diisocyanate
95534	o-Toluidine
8001352	Toxaphene (chlorinated camphene)
120821	1,2,4-trichlorobenzene
79005	1,1,2-trichloroethane

(Continued)

TABLE 3.5 *Continued*

Chemical Abstract Service (CAS) Number	Chemical Name
79016	Trichloroethylene
95954	2,4,5-trichlorophenol
88062	2,4,6-trichlorophenol
121448	Triethylamine
1582098	Trifluraline
540841	2,2,4-trimethylpentane
108054	Vinyl acetate
593602	Vinyl bromide
75014	Vinyl chloride
75354	Vinylidene chloride (1,1-dichloroethylene)
1330207	Xylenes (isomers and mixture)
95476	*o*-Xylenes
108383	*m*-Xylenes
106423	*p*-Xylenes
0	Antimony compounds
0	Arsenic compounds (inorganic, including arsine)
0	Beryllium compounds
0	Cadmium compounds
0	Chromium compounds
0	Cobalt compounds
0	Coke oven emissions
0	Cyanide compounds[a]
0	Glycol ethers[b]
0	Lead compounds
0	Manganese compounds
0	Mercury compounds
0	Fine mineral fibers[c]
0	Nickel compounds
0	Polycyclic organic matter[d]
0	Radionuclides (including radon)[e]
0	Selenium compounds

Note: For all listings herein that contain the word *compounds* and for glycol ethers, the following applies: Unless otherwise specified, these listings are defined as including any unique chemical substance that contains the named chemical (i.e., antimony, arsenic, etc.) as part of that chemical's infrastructure.

[a]XCN, where X = H or any other group where a formal dissociation may occur. For example, KCN or Ca(CN)$_2$

[b]Includes mono- and diethers of ethylene glycol, diethylene glycol, and triethylene glycol R-(OCH$_2$CH$_2$)$_n$ –OR, where n = 1,2, or 3; R = alkyl or aryl groups; and R = R, H, or groups that, when removed, yield glycol ethers with the structure R-(OCH$_2$CH)$_n$-OH. Polymers are excluded from the glycol category.

[c]Includes mineral fiber emissions from facilities manufacturing or processing glass, rock, or slag fibers (or other mineral-derived fibers) of average diameter 1 μm or less.

[d]Includes organic compounds with more than one benzene ring, and which have a boiling point greater than or equal to 100°C.

[e]A type of atom that spontaneously undergoes radioactive decay.

LCAA.14   CONTINUOUS MONITORING

Discuss continuous emission monitoring.

SOLUTION: Continuous emissions monitoring is dependent on the source requirements (NSPS, NESHAP, NSR, SIP, or permit requirement). Interestingly, it does not have anything to do with the type of pollutant or the ambient standard.

Ambient monitoring is required for permitting a major new source or a significant modification of an existing source in an attainment area (NSR [new source review] or, if one might describe it, a PSD requirement). In these situations one would monitor the pollutants in the affected area using prescribed monitoring procedures (some of which may be continuous).

LCAA.15   OZONE DEPLETING SUBSTANCES (ODS) PROGRAM

Describe the Ozone Depleting Substances (ODS) Program.

SOLUTION: (Adapted from ERM, "Clean Air Primer," Exton, PA, 2004.) The ozone layer in the stratosphere (the upper atmosphere) protects the Earth's surface from excessive quantities of harmful ultraviolet (UVB) radiation. UVB radiation is linked to harmful health effects, such as various types of skin cancer and cataracts, and damage to certain crops. It has long been recognized that many classes of man-made compounds referred to as ozone depleting substances (ODS)—including many common refrigerants, foam blowing agents, and propellants—contribute to the depletion of stratospheric ozone.

ODS are widely used in many processes and products. Common types of ODS are chlorofluorocarbons (CFCs), hydrochlorofluorocarbon (HCFCs), and hydrofluorocarbons (HFCs). The use and release to the atmosphere of CFCs and HCFCs is of concern because these chemicals have relatively high ozone depleting potential (ODP). ODP is a measure of a chemical's effect on the Earth's ozone layer. HFCs and some other CFC alternative chemicals have low ODP, which means that they have little or no effect on stratospheric ozone. Table 3.6 lists common ODS covered by the Montreal Protocol.

In recognition of the potential harmful impacts of ODS, the Vienna Convention on Protection of the Ozone Layer was initiated in 1985, which eventually led to the 1987 Montreal Protocol on Substances that Deplete the Ozone Layer and its subsequent amendments. The Montreal Protocol was developed to control and reduce consumption of harmful ODS. The Montreal Protocol has now been ratified by 175 nations, including the United States.

To implement provisions of the Montreal Protocol in the United States, Congress passed Title VI of the Clean Air Act Amendments of 1990, known as "Protection

TABLE 3.6 Common Substances Covered by the Montreal Protocol

CLASS I-CFCs	CLASS II-HCFCs
CFC-11	HCFC-22
CFC-12	HCFC-123
CFC-13	HCFC-124
CFC-111	HCFC-141
CFC-112	HCFC-142
CFC-113	
CFC-114	
CFC-115	
Halons	Others
Halon-1211	Carbon tetrachloride
Halon-1301	Methyl chloroform
Halon-1202	Methyl bromide

of Stratospheric Ozone." The Stratospheric Ozone regulations outline a series of handling, management, training, and certification requirements for facilities that use equipment that contains regulated ODS compounds, including many CFCs, HCFCs, and their alternatives.

The Stratospheric Ozone regulations cover various types of ODS activities, including manufacturing of ODS, servicing of air conditioning systems in motor vehicles, so-called "nonessential uses" of banned ODS, and labeling of products using ODS. EPA has established the Significant New Alternatives Program (SNAP) to identify and encourage the use of alternatives to more harmful ODS.

The provisions that commonly affect industrial facilities most significantly are those related to handling and recycling of regulated ODS in air conditions systems, chillers, and other equipment, including provision requiring:

1. Certification of ODS technicians and reclaimers.
2. Certification of ODS recovery and recycling equipment.
3. Certification of persons servicing or disposing of air conditioning and refrigeration equipment.
4. Limitations on the sale of refrigerants to certified technicians and appliance manufacturers.
5. Service records and the repair of substantial leaks in air conditioning and refrigeration equipment with a charge of greater than 50 pounds.
6. Safe disposal requirements to ensure removal of refrigerant goods that may enter the waste stream with the charge to home refrigerators and room air conditioners.

LCAA.16 IDENTIFYING OZONE DEPLETING SUBSTANCES

What types of man-made substances are responsible for depleting the stratospheric ozone layer?

SOLUTION: The "halogen source gases" are the primary chemical group responsible for ozone depletion. These gases contain chlorine and bromine atoms, which are known to be harmful to the ozone layer. Chlorine containing chemical families includes chlorofluorocarbons (CFCs) and hydrochlorofluorocarbons (HCFCs), which are commonly used in refrigerant systems. Examples of bromine containing chemical family are the halons, which are commonly used in fire extinguishers. Chlorine and bromine source gases are provided in Figures 3.1 and 3.2, respectively.

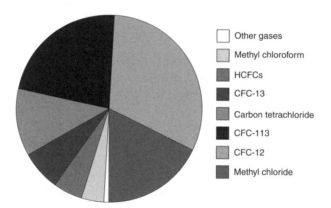

FIGURE 3.1   Chlorine source gases.

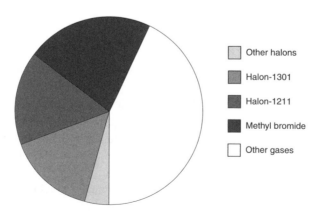

FIGURE 3.2   Bromine source gases.

LCAA.17 MECHANISM OF OZONE DESTRUCTION

Define ozone destruction and the role that bromine and chlorine play.

SOLUTION: *Ozone Destruction* refers to the conversion of ozone (O_3) to diatomic oxygen (O_2), which occurs in the stratosphere. Reactive gases containing *chlorine* (chlorine monoxide) and *bromine* (bromine monoxide) destroy stratospheric ozone in "catalytic" cycles made up two or more separate reactions. As a result, a single chlorine or bromine molecule can destroy many hundreds of ozone molecules before it reacts with another gas to break the cycle. Thus, small amounts of reactive gases can have a large effect on ozone depletion.

The reactive ozone depleting gases participate in three principal reaction cycles that destroy ozone.

Cycle 1

ClO	+	O	$\rightarrow$	Cl	+	O_2
Cl	+	O_3	$\rightarrow$	ClO	+	O_2
Net: O	+	O_3	$\rightarrow$	$2O_2$		

Cycle 2

ClO	+	ClO	$\rightarrow$	$(ClO)_2$	
$(ClO)_2$	+	sunlight	$\rightarrow$	ClOO	+ Cl
ClOO			$\rightarrow$	Cl	+ O_2
2(Cl	+	O_3	$\rightarrow$	ClO	+ O_2)
Net:		$2O_3$	$\rightarrow$	$3O_2$	

Cycle 3

ClO	+	BrO	$\rightarrow$	Cl	+ Br + O_2
ClO	+	BrO	$\rightarrow$	BrCl	+ O_2
BrCl	+	sunlight	$\rightarrow$	Cl	+ Br
Cl	+	O_3	$\rightarrow$	ClO	+ O_2
Br	+	O_3	$\rightarrow$	BrO	+ O_2
Net:		$2O_3$	$\rightarrow$	$3O_2$	

LCAA.18 ESTIMATION OF GASEOUS EMISSIONS

Emission estimates are required to develop emission control strategies, determine effects of sources and mitigation strategies, emission inventory, and permit applications. For today's environmental engineer, there are several methods available to estimate emissions. List five emission estimation methods and briefly describe each.

SOLUTION:

1. *Continuous Emission Monitoring (CEM).* This method refers to any monitoring effort that continuously measures (at short averaging times) emissions and records the data. This method is typically very expensive, but represents the most accurate (thus preferred) estimation method.

2. *Stack Test Data.* This method refers to a short-term scientific test to quantify an emission rate. The results are applicable only at the conditions existing at the time of the test. The results are typically reliable, but can be expensive.

3. *Material Balance.* This method uses a material balance to determine emission rates. The result is a reliable *average* emission rate. This method is best when a high percentage of material is lost to the atmosphere. This method may be inappropriate when a material is consumed or combined in a process.

4. *Emission Factors.* This method uses emission factors to estimate emission. Emission factors are ratios that relate emissions of a pollutant to an activity level that can be easily measured. Given an emission factor, simple multiplication can yield an estimate of emissions. The resulting estimate will be *typical* for industry, but not necessarily representative of a specific source.

5. *Engineering Estimate.* This method requires an engineer familiar with a specific process to generate an estimated emission rate based on experience and process information. This method is used when stack testing, material balance, or emission factors are not available.

LCAA.19   ETHYLENE OXIDE INFORMATION

An air pollutant of major concern to the government, industry and public is ethylene oxide. Provide key information on this chemical.

SOLUTION: The information below has been drawn, in part, from the 2000 J. Spero, B. Devito, and L. Theodore reference book "Regulatory Chemicals Handbook," published by Marcel Dekker (acquired by CRC Press/Taylor & Francis Group, Boca Raton, FL).

CAS/DOT #: 75-21-8/UN1040

Synonyms: Aethylenoxid, dihydrooxirene, dimethylene oxide, 1,2-epoxyethane, oxane.

Physical Properties: Colorless gas; often compressed, liquefied; soluble in water, alcohols and ether; MP ($-11.3°C$, $11.7°F$); BP ($10.7°C$, $51.3°F$); DN (0.8711 g/mL at $20°C$); VP (1093 mm Hg at $20°C$).

Chemical Properties: Stable; will polymerize; very flammable; reacts violently with acids, alcohols, alkali metals, ammonia, bases, oxidizers, active metal and their salts and water; FP($-4°F$); AT($804°F$); LFT (3%); UFL (100%).

Exposure Routes: Inhalation; ingestion; absorption; occupational exposure.

Human Health Risks: Acute Risks: irritation and tissue destruction in skin, eyes, mucous membranes and upper respiratory tract; burning sensation; frostbite; coughing; wheezing; laryngitis; shortness of breath; headache; nausea; vomiting; Chronic Risks: nerve and lung damage; dermatitis; probable carcinogen.

Hazard Risk: High fire hazard; decomposes on heating above 500°C; severe explosion hazard when exposed to flame or glycerol; gas/air mixtures explosive; decomposition emits acrid smoke and irritating fumes; gas heavier than air and may travel to ignition source, then flash back; NFPA Code: H 2; F 4; R 3.

Measurement Methods: Petroleum-based charcoal tube; dimethyl formamide; gas chromatograph with electron capture detection.

Major Uses: Used in the manufacture of acrylonitrate, nonionic surfactants and organic compounds; fumigant for foodstuffs and textiles; sterilization of surgical instruments; agricultural fungicide.

Storage: Keep in a tightly closed container away from heat, sparks, and open flame in a cool, dry area.

Fire Fighting: Use powder, alcohol-resistant foam, carbon dioxide or water spray; use water spray to cool exposed containers; if possible, allow fire to burn itself out.

Personal Protection: Wear full protective clothing and chemical-resistant gloves; wear chemical safety goggles; wear positive pressure self-contained breathing apparatus.

Spill Clean-up: Use water spray to cool and disperse vapors; dilute spills to form nonflammable mixtures; dilution required in enclosed areas such as sewers to eliminate flash potential; remove all ignition sources.

Health Symptoms: Inhalation (irritates eyes, skin, mucous membranes, and upper respiratory tract); skin absorption (allergic skin reaction, sever irritation, burning sensation).

General Comments: Oral rat LD_{50} (lethal dose, 50%) 72 mg/kg; First aid: immediately rinse eyes with large amounts of water; remove to fresh air if inhaled; wash out mouth with water if swallowed; provide respiratory support.

In addition to Spero et al., the reader may refer to following reference for specific details in this and other regulated chemicals.

1. R. Lewis, "Sax's Dangerous Properties of Industrial Materials," 9th edition, Van Nostrand, New York City, 1996.

2. "Suspect Chemicals Sourcebook," Roytech Publications, Bethesda, Maryland, 1996.

LCAA.20   BULK GASOLINE TERMINAL

Provide a method to determine whether a bulk gasoline terminal is an affected source.

SOLUTION: One of the methods used to determine whether a bulk gasoline terminal is an affected source under 40 CFR 63.420 (Subpart R—National Emission Standards for Gasoline Distribution Facilities Bulk Gasoline Terminals and Pipeline Breakout Stations) is to satisfy the following calculation.

An owner or operation of a bulk gasoline terminal is subject to the requirements of the regulations if the result, E_T, of the following equation is greater than 1 (unity):

$$E_T = C_F[0.59(T_F)(1 - C_E) + 0.17(T_E) + 0.08(T_{ES}) + 0.038(T_I)$$
$$+ 8.5 \times 10^{-6}(C) + KQ] + 0.04(OE) \tag{3.1}$$

where:

E_T = emissions screening factor for bulk gasoline terminals;

C_F = 0.161 for bulk gasoline terminals that do not handle any reformulated or oxygenated gasoline containing 7.6 percent by volume or greater methyl tert-butyl ether (MTBE), OR;

C_F = 1.0 for bulk gasoline terminals that handle reformulated or oxygenated gasoline containing 7.6 percent by volume or greater MTBE;

C_E = control efficiency limitation on potential to emit for the vapor processing system used to control emissions from fixed-roof gasoline storage vessels [value should be added in decimal form (percent divided by 100)]; *The term "C_E" represents the control efficiency of the control device used to process vapors from the fixed-roof tank. The value of C_E must be documented by the facility as meeting the definition of federally enforceable in subpart A of 40 CFR part 63 (General Provisions). If the facility is not controlling emissions from its fixed-roof tanks using a vapor control device, a value of zero will be entered for the term "C_E."*

T_F = total number of fixed-roof gasoline storage vessels without an internal floating roof;

T_E = total number of external floating roof gasoline storage vessels with only primary seals;

T_{ES} = total number of external floating roof gasoline storage vessels with primary and secondary seals;

T_I = total number of fixed-roof gasoline storage vessels with an internal floating roof;

C = number of valves, pumps, connectors, loading arm valves, and open-ended lines in gasoline service;

Q = gasoline throughput limitation on potential to emit in liters/day;

K = 4.52×10^{-6} for bulk gasoline terminals with uncontrolled loading racks (no vapor collection and processing systems), OR

$K = (4.5 \times 10^{-9})(EF + L)$ for bulk gasoline terminals with controlled loading racks (loading racks that have vapor collection and processing systems installed on the emission stream);

$EF =$ emission rate limitation on potential to emit for the gasoline cargo tank loading rack vapor processor outlet emissions (mg of total organic compounds per liter of gasoline loaded); *(EF ranges from 10 mg/l to 80 mg/l depending on agency regulations and age of facility—see Chapter 4.1.2.3 in "Gasoline Distribution Industry (Stage 1): Background Information for Proposed Standards" January 1994 EPA-453/R-94-002a).*

$OE =$ other HAP emissions screening factor for bulk gasoline terminals. The term OE equals the total HAP from other emission sources not specified in parameters in the equations for E_T; *(OE ranges from 0.1 to 3.5 tons/ year and should not be more than 5% of total facility HAP emissions— see 62FR9089 February 28, 1997).*

$L = 13$ mg/l for gasoline cargo tanks meeting the requirement to satisfy the test criteria for a vapor-tight gasoline tank truck in §60.501 of this chapter, OR

$L = 304$ mg/l for gasoline cargo tanks not meeting the requirement to satisfy the test criteria for a vapor-tight gasoline tank truck in §60.501 of this Chapter.

Different requirements apply if the results for E_T of the above calculations are less than 1.0 but greater than or equal to 0.50, and for those results that are less than 0.50.

LCAA.21   PIPELINE BREAKOUT STATION

Provide a method to determine whether a pipeline breakout station is an affected source.

SOLUTION: One of the methods used to determine whether a pipeline breakout station is an affected source under 40 CFR 63.420 (Subpart R—National Emission Standards for Gasoline Distribution Facilities Bulk Gasoline Terminals and Pipeline Breakout Stations) is to satisfy the following calculation.

An owner or operation of a pipeline breakout station is subject to the requirements of the regulations if the result, E_P, of the following equation is greater than 1 (unity):

$$E_P = C_F[6.7(T_F)(1 - C_E) + 0.21(T_E) + 0.093(T_{ES}) + 0.1(T_I)$$

$$+ 5.31 \times 10^{-6}(C)] + 0.04(OE) \tag{3.2}$$

where:

$E_P =$ emissions screening factor for pipeline breakout stations;

$C_F = 0.161$ for bulk gasoline terminals and pipeline breakout stations that do not handle any reformulated or oxygenated gasoline containing 7.6 percent by volume or greater methyl tert-butyl ether (MTBE), OR

C_F = 1.0 for bulk gasoline terminals and pipeline breakout stations that handle reformulated or oxygenated gasoline containing 7.6 percent by volume or greater MTBE;

C_E = control efficiency limitation on potential to emit for the vapor processing system used to control emissions from fixed-roof gasoline storage vessels [value should be added in decimal form (percent divided by 100)]; *The term "C_E" represents the control efficiency of the control device used to process vapors from the fixed-roof tank. The value of C_E must be documented by the facility as meeting the definition of federally enforceable in subpart A of 40 CFR part 63 (General Provisions). If the facility is not controlling emissions from its fixed-roof tanks using a vapor control device, a value of zero will be entered for the term "C_E".*

T_F = total number of fixed-roof gasoline storage vessels without an internal floating roof;

T_E = total number of external floating roof gasoline storage vessels with only primary seals;

T_{ES} = total number of external floating roof gasoline storage vessels with primary and secondary seals;

T_I = total number of fixed-roof gasoline storage vessels with an internal floating roof;

C = number of valves, pumps, connectors, loading arm valves, and open-ended lines in gasoline service;

OE = other HAP emissions screening factor for bulk gasoline terminals or pipeline breakout stations (tons per year). OE equals the total HAP from other emission sources not specified in parameters in the equations for E_T or E_P; *(OE ranges from 0.1 to 3.5 tons/year and should not be more than 5% of total facility HAP emissions—see 62FR9089 February 28, 1997).*

Different requirements apply if the results for E_P of the above calculations are less than 1.0 but greater than or equal to 0.50, and for those results that are less than 0.50.

LCAA.22 DEFINITIONS OF VOLATILE ORGANIC COMPOUNDS

Provide a definition of volatile organic compounds.

SOLUTION: Volatile organic substances have been defined as follows:

Organic materials mean chemical compounds of carbon excluding carbon monoxide, carbon dioxide, carbonic acid, metallic carbides, metallic carbonates and ammonium carbonates and having a vapor pressure of 0.02 pounds per square inch absolute (psia) or greater at standard conditions, including but not limited to petroleum fractions, petrochemicals and solvents.

A volatile organic compound (VOC) is also defined in 40 CFR Subpart A, General Provisions, §60.2, as any organic compound which participates in atmospheric photochemical reactions; or which is measured by a reference method, an equivalent method, or an alternative method; or which is determined by procedures specified under any subpart.

EPA [at 40CFR50.100(s)] defines *volatile organic compounds* as any compound of carbon, excluding carbon monoxide, carbon dioxide, carbonic acid, metallic carbides or carbonates, and ammonium carbonate, which participates in atmospheric photochemical reactions. This includes any such organic compound other than the following, which have been determined to have negligible photochemical reactivity: methane; ethane; methylene chloride (dichloromethane); 1,1,1-trichloroethane (methyl chloroform); 1,1,2-trichloro-1,2,2-trifluoroethane (CFC−113); trichlorofluoremethane (CFC−11); dichlorodifluoromethane (CFC−12); chlorodifluoromethane (HCFC−22); trifluoromethane (HFC−23); 1,2-dichloro 1,1,2,2-tetrafluoroethane (CFC−114); chloropentafluoroethane (CFC−115); 1,1,1-trifluoro 2,2-dichloroethane (HCFC−123); 1,1,1,2-tetrafluoroethane (HFC−134a); 1,1-dichloro 1-fluoroethane (HCFC−141b); 1-chloro 1,1-difluoroethane (HCFC−142b); 2-chloro-1,1,1,2-tetrafluoroethane (HCFC−124); pentafluoroethane (HFC−125); 1,1,2,2-tetrafluoroethane (HFC−134); 1,1,1-trifluoroethane (HFC−143a); 1,1-difluoroethane (HFC-152a); parachlorobenzotrifluoride (PCBTF); cyclic, branched, or linear completely methylated siloxanes; acetone; perchloroethylene (tetrachloroethylene); 3,3-dichloro-1,1,1,2,2-pentafluoropropane (HCFC−225ca); 1,3-dichloro-1,1,2,2,3-pentafluoropropane (HFC−225cb); 1,1,1,2,3,4,4, 5,5,5-decafluoropentane (HFC 43−10mee); difluoromethane (HFC−32); ethylfluoride (HFC−161); 1,1,1,3,3,3-hexafluoropropane (HFC−236fa); 1,1,2,2,3-pentafluoro propane (HFC−245ca); 1,1,2,3,3-pentafluoropropane (HFC−245ea); 1,1,1,2,3-pentafluoropropane (HFC−245eb); 1,1,1,3,3-pentafluoropropane (HFC−245fa); 1,1,1,2,3,3-hexafluoropropane (HFC−236ea); 1,1,1,3,3-pentafluorobutane (HFC−365mfc); chlorofluoromethane (HCFC−31); 1 chloro-1-fluoroethane (HCFC−151a); 1,2-dichloro-1,1,2-trifluoroethane (HCFC−123a); 1,1,1,2,2,3,3, 4,4-nonafluoro-4-methoxy-butane ($C_4F_9OCH_3$ or HFE−7100); 2-(difluoromethoxy methyl)-1,1,1,2,3,3,3-heptafluoropropane ($(CF_3)_2CFCF_2OCH_3$); 1-ethoxy-1,1,2,2, 3,3,4,4,4-nonafluorobutane ($C_4F_9OC_2H_5$ or HFE−7200); 2-(ethoxydifluoro methyl)-1,1,1,2,3,3,3-heptafluoropropane ($(CF_3)_2CFCF_2OC_2H_5$); methyl acetate, 1,1,1,2,2,3,3-heptafluoro-3-methoxy-propane (n-$C_3F_7OCH_3$, HFE−7000); 3-ethoxy-1,1,1,2,3,4,4,5,5,6,6,6-dodecafluoro-2-(trifluoromethyl) hexane (HFE−7500), 1,1,1,2,3,3,3-heptafluoropropane (HFC 227ea), and methyl formate ($HCOOCH_3$), and perfluorocarbon compounds which fall into these classes:

1. Cyclic, branched, or linear, completely fluorinated alkanes;
2. Cyclic, branched, or linear, completely fluorinated ethers with no unsaturations;
3. Cyclic, branched, or linear, completely fluorinated tertiary amines with no unsaturations; and,
4. Sulfur containing perfluorocarbons with no unsaturations and with sulfur bonds only to carbon and fluorine.

LCAA.23 INDUSTRIAL SURFACE COATING

Provide a general overview of industrial surface coating.

SOLUTION: *Process Description*: Surface coating is the application of decorative or protective materials in liquid or powder form to substrates. These coatings normally include general solvent type paints, varnishes, lacquers, and water thinned paints. After application of the coating by one of a variety of methods such as brushing, rolling, spraying, dipping, and flow coating, the surface is air and/or heat dried to remove volatile solvents from the coated surface. Powder type coatings can be applied to a hot surface or can be melted after application and caused to flow together. Other coatings can be polymerized after application by thermal curing with infrared or electron beam systems.

Coating Operations: There are both "toll" ("independent") and "captive" surface coating operations. Toll operations fill orders to various manufacturer specifications, and thus change coating and solvent conditions more frequently than do captive operations, which fabricate and coat products within a single facility and which may operate continuously with the same solvents. Toll and captive operations differ in emission control systems applicable to coating lines, because not all controls are technically feasible in toll situations.

Coating Formulations: Conventional coatings contain at least 30 volume percent solvents to permit easy handling and application. However, they typically contain 70 to 85 percent solvents by volume. These solvents may be of one component or of a mixture of volatile ethers, acetates, aromatics, cellosolves, aliphatic hydrocarbons and/or water. Coatings with 30 volume percent of solvent or less are called low solvent or "high solids" coatings.

Waterborne coatings, which have recently gained substantial use, are of several types: water emulsion, water soluble and colloidal dispersion, and electrocoat. Common ratios of water to solvent organics in emulsion and dispersion coatings are 80/20 and 70/30.

Depending on the product requirements and the material being coated, a surface may have one or more layers of coating applied. The first coat may be applied to cover surface imperfections or to assure adhesion of the coating. The intermediate coats usually provide the required color, texture or print, and a clear protective topcoat is often added. General coating types do not differ from those described, although the intended use and the material to be coated determine the composition and resins used in the coatings.

Coating Application Procedures: Conventional spray, which is air atomized and usually hand operated, is one of the most versatile coating methods. Colors can be changed easily, and a variety of sizes and shapes can be painted under many operating conditions. Conventional, catalyzed or waterborne coatings can be applied with little modification. The disadvantages are low efficiency from overspray and high energy requirements for the air compressor.

In hot airless spray, the paint is forced through an atomizing nozzle. Since, the volumetric flow is less, overspray is reduced. Less solvent is also required, thus reducing VOC emissions. Care must be taken for proper flow of the coating to avoid plugging and abrading of the nozzle orifice. Electrostatic spray is most efficient for low visocity paints. Charged paint particles are attracted to an oppositely charged surface. Spray guns, spinning discs or bell shaped atomizers can also be used to atomize the paint. Application efficiencies of 90 to 95 percent are possible, with good "wrap-around" and edge coating. Interiors and recessed surfaces are difficult to coat, however.

Roller coating is used to apply coatings and inks to flat surfaces. If the cylindrical rollers move in the same direction as the surface to be coated, the system is called a direct roll coater. If they rotate in the opposite directions, the system is a reverse roll coater. Coatings can be applied to any flat surface efficiently and uniformly and at high speeds. Printing and decorative graining are applied with direct rollers. Reverse rollers are used to apply fillers to porous or imperfect substrates, including papers and fabrics, to give a smooth uniform surface.

Knife coating is relatively inexpensive, but is not appropriate for coating unstable materials, such as some knit goods, or when a high degree of accuracy in the coating thickness is required.

Rotogravure printing is widely used in coating vinyl imitation leathers and wall-paper, and in the application of a transparent protective layer over the printed pattern. In rotogravure printing, the image area is recessed, or "intaglio", relative to the copper plated cylinder on which the image is engraved. The ink is picked up on the engraved area, and excess ink is scraped off the nonimage area with a "doctor blade". The image is transferred directly to the paper or other substrate, which is web fed, and the product is then dried.

Dip coating requires that the surface of the subject be immersed in a bath of paint. Dipping is effective for coating irregularly shaped or bulky items and for priming. All surfaces are covered, but coating thickness varies, edge blistering can occur, and a good appearance is not always achieved.

In flow coating, materials to be coated are conveyed through a flow of paint. Paint flow is directed without atomization toward the surface through multiple nozzles, then is caught in a trough and recycled. For flat surfaces, close control of film thickness can be maintained by passing the surface through a constantly flowing curtain of paint at a controlled rate.

Emissions: Essentially all of the VOC emitted from the surface coating industry is from the solvents which are used in the paint formulations used to thin paints at the coating facility or used for cleanup. All unrecovered solvent can be considered potential emissions. Monomers and low molecular weight organics can be emitted from those coatings that do not include solvents, but such emissions are essentially negligible. Additional details prove this in the next Problem.

Note that solvents used in coating processes are often classified according to their photochemical reactivity. Briefly, photochemical reactivity, sometimes shortened to "reactivity," is the tendency of an atmospheric system containing the organic compound in question and nitrogen oxides to undergo, under the influence of ultraviolet radiation (sunlight) and appropriate meteorological conditions, a series of chemical reactions that result in the various manifestations associated with photochemical air pollution. These include eye irritation, vegetation damage and visibility reduction.

LCAA.24 SURFACE COATING EMISSIONS

Qualitatively describe estimating procedures for emissions from surface coating.

SOLUTION: Emissions from surface coating for an uncontrolled facility can be estimated by assuming that all VOC in the coatings is emitted. Usually, the coating consumption volume will be known, and some information about the types of coatings and solvents will be available. The choice of a particular emission factor will depend on the coating data available.

All solvents separately purchased as solvent that are used in surface coating operations and are not recovered can subsequently be considered potential emissions. Such VOC emissions at a facility can result from onsite dilution of coatings with solvent, from "makeup solvents" required in flow coating and, in some instances, dip coating, and from the solvents used for cleanup. Makeup solvents are added to coatings to compensate for standing losses, concentration or amount, and thus bringing the coating back to working specifications. Solvent emissions should be added to VOC emissions from coatings to obtain total emissions from a coating facility.

Solvent density and solids content play a role in estimating emissions. Typical values are provided in Table 3.7.

TABLE 3.7 Typical Densities and Solids Contents of Coatings

Type of Coating	Density kg/liter	Density lb/gal	Solids (volume %)
Enamel, air dry	0.91	7.6	39.6
Enamel, baking	1.09	9.1	42.8
Acrylic enamel	1.07	8.9	30.3
Alkyd enamel	0.96	8.0	47.2
Primer surfacer	1.13	9.4	49.0
Primer, epoxy	1.26	10.5	57.2
Varnish, baking	0.79	6.6	35.3
Lacquer, spraying	0.95	7.9	26.1
Vinyl, roller coat	0.92	7.7	12.0
Polyurethane	1.10	9.2	31.7
Stain	0.88	7.3	21.6
Sealer	0.84	7.0	11.7
Magnet wire enamel	0.94	7.8	25.0
Paper coating	0.92	7.7	22.0
Fabric coating	0.92	7.7	22.0

LCAA.25 EMISSION FACTORS

Provide details on surface coating emission factors.

SOLUTION: Surface coating entails the deposition of a solid film on a surface through the application of a coating material such as paint, lacquer, or varnish. Surface coating operations are significant volatile organic compound (VOC) emission sources. Most coatings contain VOCs which evaporate during the coating application and curing processes, rather than becoming part of the dry film.

The EPA has issued Control Techniques Guidelines (CTGs) for many surface coating operations, including cans, metal coils, paper and paper products, fabrics, automobiles, light-duty trucks, metal furniture, large appliances, magnet wires, miscellaneous metal parts and products, graphic arts, and flatwood panelling. The emission limits recommended in these guidelines have been adopted by many state and local agencies. The EPA has also issued new source performance standards (NSPS) for many surface coating operations, including automobile, light-duty trucks, beverage cans, metal coils, large appliances, metal furniture, pressure sensitive tapes and labels, vinyl printing and topcoating, and publication rotogravure printing.

To assist states in defining reasonably available control technology (RACT), the EPA Office of Air Quality Planning and Standards (OAQPS) prepared a series of CTG documents described above. Individual stationary source categories are addressed by the documents. A list of the source categories are:

1. Surface coating of cans
2. Surface coating of metal coils
3. Surface coating of fabrics
4. Surface coating of paper products
5. Surface coating of automobiles and light-duty trucks
6. Surface coating of metal furniture
7. Surface coating of magnet wire
8. Surface coating of large appliances

For each source category, a CTG document describes the source, identifies the VOC emission points, discusses the applicable control methods, analyzes the costs required to implement the control methods, and recommends regulations for limiting VOC emissions from the source.

To comply with these regulations, a surface coating operator might elect to change to low VOC content coatings, to use add-on controls such as incineration or carbon adsorption, or to improve transfer efficiency. In cases where compliance is achieved by a change in coating alone, VOC emissions can be calculated from the VOC content of the coating as applied to the substrate. When add-on controls or transfer efficiency improvements are used, more complex calculations can be performed to determine the effectiveness of the control strategy. Some of these calculations are provided in this problem set.

Methods of calculating emission factors for coatings are normally in terms of the RACT units: lbs VOC/(gallon coating applied-H_2O). The purpose of excluding water is to avoid the problem of achieving compliance merely by diluting the paint. Another method of describing VOC emissions from a coating formulation is on the basis of lbs

VOC/gallon solids. If the surface area of the material being coated is known and if the dry coating thickness can be determined, the amount of VOCs emitted can easily be determined by finding the number of gallons of solids that would provide the necessary coverage. The analytical methods of measuring the volume of cured coating can be complex. As an alternative, the RACT limits were developed in terms of lbs VOC per gallon of uncured solvents and organic solvent less water.

A graphical procedure for these calculations is also given in the CTG "Control of Volatile Organic Emissions" from Existing Stationary Sources-Volume II," EPA-450/2-77-008, May 1977. The graphs can provide a quick way of determining emission levels and emission reductions. The graphs were developed from simple mathematical calculations and are not empirical in nature.

Many of the coating calculations can be solved by using the following expressions:

1. Emission factor, ef, in RACT units of lbs VOC/(gal − H_2O).

$$ef = vs\rho_v/(1 - ws) \tag{3.3}$$

where:

   ef = emission factor, lbs VOC/(gal − H_2O)

   v = % (by volume) organic volatiles in solvent/100

   w = % (by volume) H_2O in solvent/100

   s = % (by volume) solvent in the paint/100

   ρ_v = density of organic volatiles, lb/gal

2. Emission factor on a solids basis (lbs VOC/gal solids).

$$ef' = vs\rho_v/(1 - s) \tag{3.4}$$

where:

   ef' = emission factor, lbs VOC/gal solids

3. Percent emission reduction.

$$\% \text{ reduction} = [(ef'_{orig} - ef'_{rep})/ef'_{orig}]100 \tag{3.5}$$

where:

   ef'_{orig} = emission factor on a solids basis for original coating

   ef'_{rep} = emission factor on a solids basis for replacement coating

LCAA.26 FURNITURE MAXIMUM ACHIEVABLE CONTROL TECHNOLOGY (MACT)

Outline how to calculate the average volatile hazardous air pollutant (VHAP) content for all furniture finishing material.

SOLUTION: To calculate the average volatile hazardous air pollutant (VHAP) content, E, for all finishing materials used at a facility, the following equation is to be used. The value of E shall be no greater than 1.0:

$$E = (M_{c1}C_{c1} + M_{c2}C_{c2} + *** + M_{cn}C_{cn} + S_1W_1$$
$$+ S_2W_2 + *** + S_nW_n)/(M_{c1} + M_{c2} + *** + M_{cn}) \qquad (3.6)$$

where:

- M = the mass of solids in finishing material used monthly, kg solids/month (lb solids/month).
- C_c = the VHAP content of a finishing material (c), in kilograms of volatile hazardous air pollutants per kilogram of coating solids (kg VHAP/kg solids), as supplied; also given in pounds of volatile hazardous air pollutants per pound of coating solids (lb VHAP/lb solids).
- S = the VHAP content of a solvent, expressed as a weight fraction, added to finishing materials.
- W = the amount of solvent, in kilograms (pounds), added to finishing materials during the monthly averaging period.

Additional information is provided at (40 CFR 63.804(a)).

LCAA.27 WOOD FURNITURE MANUFACTURING OPERATIONS

Provide key calculation procedures for wood furniture manufacturing operations.

SOLUTION: Sources must keep a *Certified Product Data Sheet* (*CPDS*) for each coating or adhesive used for wood furniture manufacturing operations. They also must record how much of each material is used. The three (3) physical properties of a coating given in a CPDS—density, weight percent solids, and weight percent VHAP—plus each coating's monthly usage provide all necessary data to make the calculations required for this NESHAP.

1. *Individual Material Equations.*

C_n = NESHAP VHAP Content of an Individual Material

$$C_n = \frac{\text{Material}_n \text{ VHAP Weight, lbs.}}{\text{Material}_n \text{ VHAP Weight, lbs.}} = \frac{V_n}{S_n}$$

$$V_n = \text{Density}_n \text{ (lbs./gallon)} \times \frac{\text{Weight Percent VHAP}_n}{100} \times \text{Gallons}_n \qquad (3.7)$$

$$S_n = \text{Density}_n \text{ (lbs./gallon)} \times \frac{\text{Weight Percent Solids}_n}{100} \times \text{Gallons}_n$$

2. *As Applied Mixture Equation.*

$$C_{\text{Applied}} = \frac{V_{\text{Applied}}}{S_{\text{Applied}}} = \frac{V_1 + V_S}{S_1 + S_S} \qquad (3.8)$$

3. *Material Averaging Equations for Month x.*

$$C_{Average} = \frac{V_{Total}}{S_{Total}} = \frac{\text{Total VHAP Weight For Month}_x}{\text{Total Solids Weight For Month}_x}$$

$$V_{total} = V_1 + V_2 + V_3 + \cdots + V_n$$

$$S_{total} = S_1 + S_2 + S_3 + \cdots + S_n$$

(3.9)

LCAA.28   HALOGENATED SOLVENT CLEANING MACT

Outline how to calculate the emissions from solvent cleaning machines.

SOLUTION: Using the records of all solvent additions and deletions for the previous month as required under §63.464(a), solvent emissions (E_i) shall be determined using Equation (3.10) for cleaning machines with a solvent/air interface and Equation (3.11) for cleaning machines without a solvent/air interface:

$$E_i = \frac{SA_i - LSR_i - SSR_i}{AREA_i}$$

(3.10)

$$E_n = SA_i - LSR_i - SSR_i$$

(3.11)

where:

    E_i = the total halogenated HAP solvent emissions from the solvent cleaning machine during the most recent monthly reporting period i; kilograms of solvent per square meter of solvent/air interface area per month.

    E_n = the total halogenated HAP solvent emissions from the solvent cleaning machine during the most recent monthly reporting period i; kilograms of solvent per month.

    SA_i = the total amount of halogenated HAP liquid solvent added to the solvent cleaning machine during the most recent monthly reporting period i; kilograms of solvent per month.

    LSR_i = the total amount of halogenated HAP liquid solvent removed from the solvent cleaning machine during the most recent monthly reporting period i; kilograms of solvent per month.

    SSR_i = the total amount of halogenated HAP solvent removed from the solvent cleaning machine in solid waste during the most recent monthly reporting period i; kilograms of solvent per month.

    $AREA_i$ = the solvent/air interface area of the solvent cleaning machine; square meters.

LCAA.29   ACID RAIN PROVISIONS (CAA, TITLE IV)

Discuss acid rain regulations from an air regulatory perspective.

SOLUTION: Subchapter IV of the Clean Air Act-Acid Deposition Control contains comprehensive provisions to control the emissions that cause acid rain. The Act calls for reductions in sulfur dioxide emissions from the burning of fossil fuels, the principal cause of acid rain. It also mandates significant reductions in nitrogen oxide emissions.

The acid rain program that the EPA developed takes a flexible approach. It simply sets a national ceiling in sulfur dioxide emissions from electric power plants and allows affected utilities to determine the most cost-effective way to achieve compliance. It is estimated that this approach has resulted in at least a 20 percent cost saving over a traditional command-and-control program.

LCAA.30 SOLVENT SELECTION

List 10 factors that should be considered when choosing a solvent for a gas absorption column that is to be used as an air toxic control device.

SOLUTION: Solvent selection for use in an absorption column for gaseous pollutant removal should be based upon the following criteria:

1. *Solubility of the gas in the solvent*: High solubility is desirable as it reduces the amount of solvent needed. Generally, a polar gas will dissolve best in a polar solvent and a nonpolar gas will dissolve best in a nonpolar solvent.

2. *Vapor pressure*: A solvent with a low vapor pressure is preferred to minimize loss of solvent.

3. *Corrosivity*: Corrosive solvents may damage the equipment. A solvent with low corrosivity will extend equipment life.

4. *Cost*: In general, the less expensive, the better. However, an inexpensive solvent is not always the best choice if it is too costly to dispose of and/or recycle after it has been used.

5. *Viscosity*: Solvents with low viscosity offer benefits such as better absorption rates, better heat transfer properties, lower pressure drops, lower pumping costs, and improved flooding characteristics in absorption towers.

6. *Reactivity*: Solvents that react with the contaminant gas to produce an unreactive product are desirable since the scrubbing solution can be recirculated while maintaining high removal efficiencies. Reactive solvents should produce few unwanted side reactions with the gases that are to be absorbed.

7. *Low freezing point*: Resistance to freezing lessens the chance of solid formation and clogging of the column. See 2 on boiling point (vapor pressure).

8. *Availability*: If the solvent is "exotic," it generally has a higher cost and may not be readily available for long-term continuous use. Water is often the natural choice based on this criteria.

9. *Flammability*: Lower flammability or nonflammable solvents decrease safety problems.

10. *Toxicity*: A solvent with low toxicity is desirable.

LCAA.31 SELECTING A PLANT SITE

From an environmental perspective, list several guidelines that should be followed when selecting a site for a plant.

SOLUTION: The following guidelines should be followed when selecting a site for a plant:

1. *Topography.* A fairly level site is needed to contain spills and prevent spills from migrating and creating more of a hazard. Firm soil above water level is recommended.
2. *Utilities and water supply.* The water supply must be adequate for fire protection and cooling. The sources for electricity should be reliable to prevent unplanned shutdowns.
3. *Roadways.* Roadways should allow access to the site by emergency vehicles such as ambulances and fire engines in the event of an emergency.
4. *Neighboring communities and plants.* Population density and proximity to the plant should be considered for the initial site and in anticipation of a possible future expansion.
5. *Waste disposal.* Waste disposal systems containing flammable, corrosive, or toxic materials should be a minimum distance of 250 ft from plant equipment.
6. *Climate and natural hazards.* Lighting arrestors should be installed to reduce/eliminate ignition sources in flammable areas. Storm drainage systems should be maintained.
7. *Emergency services.* Emergency services should be readily available, well trained, and appropriately equipped.
8. Air and water quality standards. The location and operation of the facility should be consistent with efforts to maintain or protect air and water resources.

Key factors that should be considered in the layout of a new plant are:

1. Site selection
2. Water supply
3. Utilities
4. Offices and ancillary equipment
5. Storage and loading areas
6. Process equipment
7. Safety equipment
8. Access in and out of the plant
9. Emission reduction devices
10. Waste water treatment equipment

QUANTITATIVE PROBLEMS (TCAA)

PROBLEMS TCAA.1–40

TCAA.1 CALCULATIONS FOR STANDARD VOLUME (40 CFR §50.3)

Most atmospheric sampling techniques use a sampling train whereby air containing the pollutant of interest enters the train and passes through a sample collection device. The weight of the pollutant collected is compared to the volume of air drawn through the train to enable the calculation of the concentration of the pollutant in the ambient air. The concentration is typically expressed as a concentration corrected to EPA's standard temperature and pressure (STP). The EPA defines their standard (§40 CFR 50.3) conditions to be 298K and 760 mmHg (1 atm).

The equation used to correct sample volumes (V_s) to EPA standard volume (V_{std}) conditions is based on Charles' Law.

$$V_{std} = V_s \left(\frac{P_{atm}}{P_{std}}\right)\left(\frac{T_{std}}{T_{atm}}\right) \tag{3.12}$$

where:

V_{std} = corrected volume of gas sampled

V_s = sampled volume of gas

T_{std} = EPA standard temperature

P_{std} = EPA standard pressure

T_{atm} = sample standard temperature

P_{atm} = sample standard pressure

Determine the corrected sample volume if 10 ft^3 of air was sampled from air at an average temperature of 80°F and pressure of 14.3 psia.

SOLUTION: Convert ambient temperature and pressure to appropriate units.

$$T(K) = \frac{F - 32}{1.8} + 273 = \frac{80 - 32}{1.8} + 273 = 300 \, K$$

$$P(atm) = P_s\left(\frac{1 \, atm}{14.7 \, psi}\right) = 14.3 psi\left(\frac{1 \, atm}{14.7 \, psi}\right) = 0.97 \, atm$$

Calculate the corrected volume.

$$V_{std} = V_s \left(\frac{P_{atm}}{P_{std}}\right)\left(\frac{T_{std}}{T_{atm}}\right) = 10 \, ft^3\left(\frac{0.97 \, atm}{1 \, atm}\right)\left(\frac{273 \, K}{300 \, K}\right)$$

$$= 8.83 \, ft^3$$

TCAA.2   STACK VELOCITY

The exhaust gas flowrate from a facility is 1500 scfm. All of the gas is vented through a small stack which has an inlet area of 2.3 ft^2. The exhaust gas temperature is 350°F. What is the velocity of the gas through the stack inlet in ft/s. Assume standard conditions to be 70°F and 1 atm. Neglect the pressure drop across the stack.

SOLUTION: First note that the temperatures must be converted to an absolute scale. Thus,

$$T_a = 350 + 460 = 810°R,$$

$$T_s = 70 + 460 = 530°R$$

and

$$q_a = q_s \left(\frac{T_a}{T_s} \right)$$

$$= 1500 \left(\frac{810}{530} \right)$$

$$= 2292 \text{ acfm}$$

The average velocity may now be calculated:

$$v = \frac{q_a}{A}$$

$$= \frac{2292}{(2.3)(60)}$$

$$= 16.6 \text{ ft/s}$$

TCAA.3   CHECK FOR EMISSION STANDARDS COMPLIANCE

As a consulting engineer, you have been contracted to modify an existing control device used in fly ash removal. The standards for emissions have been changed to a total numbers basis. Determine if the unit will meet an emission standard of $10^{5.7}$ particles/acf. Data for the unit are given below.

Average particle size, $d_p = 10$ μm; assume constant
Particle specific gravity = 2.3
Inlet loading, IL = 3.0 gr/ft^3
Efficiency (mass basis), $E = 99.5\% = 0.995$

SOLUTION: The outlet loading, (OL) is given by

$$OL = (1 - E)(IL)$$
$$= (1.0 - 0.995)3.0$$
$$= 0.015 \ gr/ft^3$$

Assume a basis of 1.0 ft^3.

$$\text{Particle mass} = M = \rho_p V_p = \rho_p \frac{\pi d_p^3}{6}$$

$$= (7000 \ gr/lb) \left(\frac{\pi[(10 \ \mu m)(0.328 \times 10^{-5} \ ft/\mu m)]^3 (2.33)(62.4 \ lb/ft^3)}{6} \right)$$

$$= 1.881 \times 10^{-8} \ gr$$

Number of particles $= (0.015 \ gr)/(1.881 \times 10^{-8} \ gr/particle)$
$$= 0.8 \times 10^6$$
$$= 8.0 \times 10^5 \ \text{particles in 1 ft}^3$$

Allowable number of particles/ ft$^3 = 10^{5.7}$
$$= 5.01 \times 10^5$$

Therefore, the unit will not meet a numbers standard.

TCAA.4 CYCLONE SELECTION

Multiple-cyclone collectors (multiclones) are high-efficiency devices that consist of a number of small-diameter cyclones operating in parallel with a common gas inlet and outlet. The flow pattern differs from a conventional cyclone in that instead of bringing the gas in at the side to initiate the swirling action, the gas is brought in at the top of the collecting tube, and swirling action is then imparted by a stationary vane positioned in the path of the incoming gas. The diameters of the collecting tubes usually range from 6 inch to 24 inch with pressure drops in the 2- to 6-inch water guage inH$_2$O range. Properly designed units can be constructed and operated with a collection efficiency as high as 90% for particulates with diameters in the 5- to 10-μm range. The most serious problems encountered with these systems involve plugging and flow equalization.

Since the gas flow to a multiclone is axial (usually from the top), the cross-sectional area available for flow inlet conditions is given by the annular area between the outlet tubes and cyclone body. The outlet tube diameter (D$_O$) is usually one-half the body diameter (D$_B$).

A recently hired engineer has been assigned the job of selecting and specifying a cyclone unit to be used to reduce an inlet fly ash loading from 3.1 gr/ft^3 to an outlet value of 0.06 gr/ft^3. The flowrate from the coal-fired boiler is 100,000 acfm. Fractional efficiency data provided by a vendor are presented below (see Figure 3.3) for three different types of cyclones (multiclones).

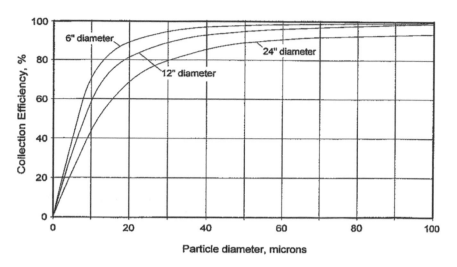

FIGURE 3.3   Fractional efficiency data.

Which type and how many cyclones are required to meet the above specifications? The optimum operating pressure drop is 3.0 inH$_2$O; at this condition, the average inlet velocity may be assumed to be 60 ft/s. Particle size distribution data is provided in Table 3.8.

SOLUTION: Calculate the required collection efficiency, E_R:

$$E_R = [(\text{IL} - \text{OL})/\text{IL}]100$$
$$= [(3.1 - 0.06)/3.1](100)$$
$$= 98\%$$
$$= 0.98$$

The average particle size associated with each size range is shown in Table 3.8.

For the 6-inch tubes, Table 3.9 provides the overall efficiency, E_6. Since $E_6 > E_R$, the 6-inch tubes will do the job.

TABLE 3.8   Particle Size Data

Particle Diameter Range (μm)	Average Particle Diameter (μm)	Weight Fraction (w_i)
5–35	20	0.05
35–50	42.5	0.05
50–70	60	0.10
70–110	90	0.20
110–150	130	0.20
150–200	175	0.20
200–400	300	0.10
400–700	550	0.10

TABLE 3.9   6-inch Tube Calculation

Average Particle Diameter (μm)	Weight Fraction w_i	Efficiency for 6-inch Tubes (%)	$E_i w_i$ for 6-inch Tubes (%)
20	0.05	89	4.45
42.5	0.05	97	4.85
60	0.10	98.5	9.85
90	0.20	99	19.8
130	0.20	100	20
175	0.20	100	20
300	0.10	100	10
550	0.10	100	10
			$E_6 = 98.95$

Table 3.10 is generated for the 12-inch tubes. Since the overall efficiency $E_{12} < E_R$, the 12-inch tubes will not do the job. Thus, it will be necessary to use the 6-inch tubes for a conservative design.

Since the outlet tube diameter is one-half the body diameter, the annular cross-sectional area (for axial flow) for each 6-inch (0.5-ft) tube will be

$$A = (\pi/4)(D_B^2 - D_O^2)$$

$$= 0.785(0.5^2 - 0.25^2)$$

$$= 0.147 \text{ ft}^2$$

Since the velocity in each tube is 60 ft/s, the number of tubes, n, is given by

$$\left(60 \frac{\text{ft}}{\text{s}}\right)\left(60 \frac{\text{s}}{\text{min}}\right)(0.147 \text{ ft}^2)n = 100,000 \frac{\text{ft}^3}{\text{min}}$$

Solving for n,

$$n = 190 \text{ tubes}$$

TABLE 3.10   12-inch Tube Calculation

Average Particle Diameter (μm)	Weight Fraction w_i	Efficiency for 12-inch Tubes (%)	$E_i w_i$ for 12-inch Tubes (%)
20	0.05	82	4.1
42.5	0.05	93.5	4.67
60	0.10	96	9.6
90	0.20	98	19.6
130	0.20	100	20
175	0.20	100	20
300	0.10	100	10
550	0.10	100	10
			$E_{12} = 97.97$

required in this multiple-cyclone unit. A 15×15, 14×14, or 12×16 design is recommended.

TCAA.5   ELECTROSTATIC PRECIPITATION (ESP) DESIGN PROCEDURE

Provide a design procedure for electrostatic precipitators, a control device employed for control of particulate air pollutants.

SOLUTION: No critically reviewed design procedure exists for ESPs. However, one suggested "general" design procedure (L. Theodore: personal notes, 1988) is provided below.

1. Determine or obtain a complete description of the process, including the volumetric flowrate, inlet loading, particle size distribution, maximum allowable discharge, and process conditions.
2. Calculate or set the overall collection efficiency.
3. Select a migration velocity (based on experience).
4. Calculate the ESP size (capture area).
5. Select the field height (experience).
6. Select the plate spacing (experience).
7. Select a gas throughput velocity (experience).
8. Calculate the number of gas passages in parallel.
9. Select (decide) on bus sections, fields, energizing sets, specific current, capacity of energizing set for each bus section, etc.
10. Design and select hoppers, rappers, etc.
11. Perform a capital cost analysis, including materials, erection, and startup costs.
12. Perform an operating cost analysis, including power, maintenance, inspection, capital and replacement, interest on capital, dust disposal, etc.
13. Conduct a perturbation study to optimize economics.

TCAA.6   FILTER BAG FABRIC SELECTION

It is proposed to install a pulse-jet fabric filter system to clean an airstream containing particulate pollutants. Select the most appropriate filter bag fabric considering regulatory requirements, performance, and cost. Pertinent design and operating data, as well as fabric information (see Table 3.11), are given below.

Volumetric flowrate of polluted airstream $= 10,000$ scfm ($60°F$, 1 atm)

Operating temperature $= 250°F$

Concentration of pollutants $= 4.00$ gr/ft^3

Average air-to-cloth ratio (ACR) $= v_f = 2.5$ cfm/ft^2 cloth

Regulatory collection efficiency requirement $= 99\%$

TABLE 3.11   Fabric Data

Filter Bag	A	B	C	D
Tensile strength	Excellent	Above average	Fair	Excellent
Recommended maximum temperature (°F)	260	275	260	220
Resistance factor	0.9	1.0	0.5	0.9
Cost per bag ($)	26	38	10	20
Standard size	8 in × 16 ft	10 in × 16 ft	1 ft × 16 ft	1 ft × 20 ft

Note: No bag has an advantage from the standpoint of durability under the operating conditions for which the bag was designed.

SOLUTION: A wide variety of woven and felted fabrics are used in fabric filters. Clean felted fabrics are more efficient dust collectors than woven fabrics, but woven materials are capable of giving equal filtration efficiency after a dust layer accumulates on the surface. When a new woven fabric is placed in service, visible penetration of dust may occur until buildup of the cake or dust layer. This normally takes from a few hours to a few days for industrial applications, depending on dust loadings and the nature of the particles.

When using woven fabrics, care must be exercised to prevent overcleaning so as not to completely dislodge the filter cake; otherwise, the efficiency will drop. Overcleaning of felted fabrics is generally impossible because they always retain substantial dust deposits within the fabric. Felted fabrics require more thorough cleaning methods than woven materials. If felted fabrics are used, filter cleaning is limited to the reverse-pulse method. When woven fabrics are employed, another cleaning technique may be used. Woven fabrics are available in a greater range of temperature and corrosion-resistant materials than felts and, therefore, cover a wider range of applications.

Bag D is eliminated since its recommended maximum temperature (220°F) is below the operating temperature of 250°F. Bag C is also eliminated since a pulse-jet fabric filter system requires the tensile strength of the bag to be at least above average.

Consider the economics for the two remaining choices. The cost per bag is $26.00 for A and $38.00 for B. Applying the Charles law once again, the gas flowrate and filtration velocity are

$$q_a = 10,000 \left(\frac{250 + 460}{60 + 460} \right)$$

$$= 13,654 \text{ acfm}$$

$$v_f = 2.5 \text{ cfm/ft}^2 \text{ cloth}$$

$$= 2.5 \text{ ft/ min}$$

The filtering (bag) area is then

$$A_c = q/v_f$$

$$= 13,654/2.5$$

$$= 5,462 \text{ ft}^2$$

For bag type A, the area and number, N, of bags are

$$A = \pi D h$$

$$= \pi \left(\frac{8}{12} \right)(16)$$

$$= 33.5 \text{ ft}^2$$

$$N = A_c/A$$

$$= 5{,}462/33.5$$

$$= 163$$

For bag type B:

$$A = \pi \left(\frac{10}{12} \right)(16)$$

$$= 41.9 \text{ ft}^2$$

$$N = 5462/41.9$$

$$= 130$$

The total cost (TC) for each bag is as follows:
For bag type A:

$$TC = N \text{ (cost per bag)}$$

$$= (163 \text{ bags})(26.00\$/\text{bag})$$

$$= \$4{,}238$$

For bag type B:

$$TC = (130)(38.00)$$

$$= \$4{,}940$$

Since the total cost for bag type A is less than bag type B, select bag A.

TCAA.7 COLLECTION EFFICIENCY FOR PARTICLES SMALLER THAN 1 MICRON

Explain why nearly all particle size collection efficiency curves for high-efficiency control devices take the form shown in Figure 3.4.

SOLUTION: As illustrated in Figure 3.4, the collection efficiency for particulate control increases with increasing particle size over nearly the entire particle size range. However, for particles less than approximately 1.0 μm, the trend reverses and efficiency increases with decreasing size. This phenomena is experienced by almost all high efficiency control devices, e.g., baghouses, venturi scrubbers,

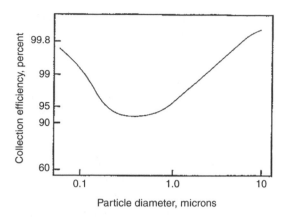

FIGURE 3.4   Effect of particle size on performance.

electrostatic precipitators, etc., and arises primarily because of molecular diffusion effects. The random, chaotic motion of submicron particles, similar to that predicted by the kinetic theory of gases, becomes more pronounced as the particle size decreases and approaches the molecular diameter of gases, resulting in higher efficiencies. This becomes an important consideration for systems requiring extremely high efficiencies, e.g., in excess of 99.5%.

TCAA.8   NANOPARTICLE BEHAVIOR

Refer to Figure 3.4 and Problem TCAA.7. Comment on the shape of the curve in particle sizes below 0.1 microns.

SOLUTION: The particle size at 0.1 μm (microns) is 100 nanometers (nm); at 0.01 μm it is 10 nm; and at 0.001 μm it is 1.0 nm. There are various statements in the literature regarding the size of nanoparticles. The most accepted value is the 3.0–75 nm range.

With respect to collection efficiency in Figure 3.4, the behavior of these nanoparticles in the above range appears to indicate that the curve continues to raise, approaching 100%. There is limited data available for size below 10 nm, but the authors believe the efficiency will continue to increase.

TCAA.9   DESIGN PROCEDURE FOR AN ABSORPTION COLUMN

A recently hired environmental engineer has been asked to provide a general design procedure for an absorption column, a control device employed for gaseous pollutant control. Fortunately, the young engineer just completed a course with Dr. L. Theodore—an internationally recognized authority in the air pollution control field. She decides to submit L. Theodore's personal notes (1984) from the course as a response.

SOLUTION: The general design procedure consists of a number of steps that have to be taken into consideration. These include:

1. Solvent selection.
2. Equilibrium data evaluation.
3. Estimation or specification of operating data (usually consisting of a mass and energy balance, where the energy balance decides whether the absorption process can be considered as isothermal or adiabatic).
4. Column selection (should the column selection not be obvious or specified, calculations must be carried out for the different types of columns and the final selection based on economic considerations).
5. Calculation of column diameter (for packed columns, this is usually based on flooding conditions, and for plate columns on the optimum gas velocity or the liquid-handling capacity of the plate).
6. Estimation of column height or the number of plates (for packed columns, the column height is obtained by multiplying the number of transfer units, obtained from a knowledge of equilibrium and operating data, by the height of a transfer unit; for plate columns, the number of theoretical plates is determined from the plot of equilibrium and operating lines. This number is then divided by the estimated overall plate efficiency to give the number of actual plates, which in turn allows the column height to be estimated from the plate spacing).
7. Determination of the pressure drop through the column (for packed columns, correlations dependent on packing type, column operating data, and physical properties of the constituents involved are available to estimate the pressure drop through the packing; for plate columns, the pressure drop per plate is obtained and multiplied by the number of plates).

TCAA.10   DESIGN OF A FIXED-BED ADSORBER

An engineering consultant has been requested to develop a design procedure for a fixed-bed adsorber, a control device employed for gaseous pollutant control.

SOLUTION: A rather simplified overall design procedure for a system adsorbing an organic that consists of two horizontal units (one on/one off) that are regenerated with steam is provided below (L. Theodore; personal notes, 1982).

1. Select adsorbent type and size.
2. Select cycle time; estimate regeneration time; set adsorption time equal to regeneration time; set cycle time equal to twice the regeneration time; generally, try to minimize regeneration time.
3. Set velocity; v is usually 80 ft/min but can increase to 100 ft/min.
4. Set the steam/solvent ratio.
5. Calculate (or obtain) working capacity (WC) for above.

6. Calculate the amount of solvent adsorbed (m_s) during one-half the cycle time (t_{ads}).

$$m_s = qc_i t_{ads}$$

where c_i = inlet solvent concentration, mass/volume
q = volume rate of flow, volume/time

7. Calculate the adsorbent required, m_{AC}:

$$m_{AC} = m_s/WC$$

8. Calculate the adsorbent volume requirement, V_{AC}:

$$V_{AC} = m_{AC}/\rho_B$$

9. Calculate the face area of the bed, A_{AC}:

$$A_{AC} = q/v$$

10. Calculate the bed height, Z.

$$Z = V_{AC}A_{AC}$$

11. Estimate the pressure drop from a graph or a suitable equation.
12. Set the L/D (length-to-diameter) ratio. Calculate L and D, noting that

$$A - LD$$

Constraints: $L < 30$ ft, $D < 10$ ft; L/D of 3 to 4 acceptable if $v < 30$ ft/min

13. Design (structurally) to handle if filled with water.
14. Consider designing vertically if $q < 2500$ actual cubic feet per minute (acfm). Consider designing horizontally if $q > 7500$ acfm.

For additional information, refer to L. Theodore, "Engineering Calculations: Adsorber Design Made Easy," CEP, p 16–17, March, 2005.

TCAA.11 CARBON MONOXIDE DESIGN VALUE CALCULATION

As a result of an EPA regulation, design values for ozone and carbon monoxide have received particular attention. Prior to this regulation, it sufficed to designate areas as either attainment or non-attainment, but now areas are further classified into different categories based upon the magnitude of the appropriate design value.

Carbon monoxide design values are discussed in terms of the 8-hour CO NAAQS, rather than the 1-hour NAAQS. For 8-hour CO, the maximum and second maximum non-overlapping 8-hour values at a site for the most recent 2 years of data are determined. Then, the higher of the second maximum is picked as the design value. For a given area, the highest design value is chosen as the area design value.

Determine the Site #1 design value, Site #2 design value, and area design value given the following data (all in ppm):

Site#1:	1st Max	2nd Max
1987	14.6	8.9

1988	13.9	10.9
Site#2:	1st Max	2nd Max
1987	12.2	11.1
1988	10.8	10.4

SOLUTION:

Site#1 Design Value $= 10.9$ ppm

Site#2 Design Value $= 11.1$ ppm

Area Design Value $= 11.1$ ppm

TCAA.12   CALCULATING PERCENT VOLATILE ORGANIC COMPOUNDS FROM LIQUID SAMPLES

As described earlier, volatile organic compounds (VOCs) are smog-forming chemicals found in many products such as paints, hair spray, chlorinated solvents, to metal-working fluids. The Clean Air Act regulates efforts to reduce emissions of VOCs and hazardous air pollutants. ASTM D2369 01el "Volatile Content of Coatings" procedure is used for the purpose of calculating the volatile organic content in solvent-reducible coatings. This information is useful to the producer, user, and to environmental interests for determining the volatiles emitted by coatings or fluids into the air.

The percent volatile matter, V, in liquid samples is calculated as follows:

$$\%V = 100 - \frac{w_2 - w_1}{S} \cdot 100 \qquad (3.13)$$

where:

$w_1 =$ weight of aluminum dish

$w_2 =$ weight of dish plus specimen after heating to 110°C for 60-min

$S =$ specimen weight

Then,

$\%\text{VOC} = \%V - \%$sample water content

Determine the %VOC for a 0.485 g sample using a 0.996 g aluminum dish. After heating, the combined weight of the dish and sample was 1.4021 g. The sample water content is 6.03%.

SOLUTION: Based on the problem statement,

$$w_2 = 1.4021g$$
$$w_1 = 0.9960g$$

Therefore, employing Equation (3.13),

$$\%V = 100 - \left(\frac{w_2 - w_1}{S}\right)(100)$$

$$= 100 - \left(\frac{1.4021 - 0.9960}{0.485}\right)(100)$$

$$= 16.27\%$$

$$\%\text{VOC} = \%V - \%\text{sample water content}$$

$$= 16.27\% - 6.03\%$$

$$= 10.24\%\text{VOC}$$

TCAA.13   NESHAP COMPLIANCE CALCULATION

The data tabulated below (see Table 3.12) are provided for the coatings and thinner solvent used by a facility.

Calculate:

1. Single coating compliant value.
2. As applied VHAP content; the as applied mixture consists of four gallons of basecoat and one gallon of solvent.
3. Average monthly VHAP content. Refer to Problem LCAA.27 and Equations 3.7–3.9 for details on the equations employed below.

SOLUTION:

1. On an individual basis, use the equation for NESHAP VHAP content and data in Table 3.12 to determine if each coating or thinner solvent complies.

$$C_n = \text{NESHAP VHAP Content} = \frac{V_n}{S_n} = \frac{\text{Material}_n\ \text{VHAP Weight, lb}}{\text{Material}_n\ \text{Solids Weight, lb}} \quad (3.7)$$

$$C_1 = \frac{V_1(\text{Lacquer})}{S_1(\text{Lacquer})} = \frac{7.24 \times (14.85/100) \times 553}{7.24 \times (16.85/100) \times 553} = \frac{594.6}{674.7}$$

$$= 0.8813\,\text{lb VHAP/lb Solids}$$

The calculated C_1 is less than 1.0.
Therefore, this lacquer is a compliant coating.
Note: Individual coating compliance can be calculated by dividing the coatings weight percent VHAP content by its weight percent solids content as is done below for other example coatings. This works for individual coating calculations because density and usage values cancel when using the full equation for single materials. Thinners, which contain no solids, must meet a maximum weight percent VHAP content limitation standard.

$$C_2(\text{Sealer}) = 30.48/30.47 = 1.0004;\ \text{Not Compliant}$$

$$C_3(\text{Stain}) = 4.42/3.25 = 1.3600;\ \text{Not Compliant}$$

$$C_4(\text{Basecoat}) = 24.51/24.51 = 1.0000; \text{ Compliant}$$
$$C_5(\text{Solvent}) = 8.03, \text{ which is less than 10\% VHAP; Compliant}$$

2. Determine compliance of the "as applied" material mixture consisting of four (4) gallons of basecoat and one (1) gallon of solvent.

   Compute the weight of VHAP in material mixed, and then total.

   $$V_1(\text{Basecoat}) = 8.16 \times 24.51/100 \times 4 = 8.00$$
   $$V_s(\text{Solvent}) = 6.67 \times 8.03/100 \times 1 = 0.54$$
   $$V_{\text{applied}} = 8.54 \text{ lb of VHAP}$$

   Compute the weight of solids in material mixed, and then total.

   $$S_1(\text{Basecoat}) = 8.16 \times 24.51/100 \times 4 = 8.00$$
   $$S_s(\text{Solvent}) = 6.67 \times 0/100 \times 1 = 0.00$$
   $$S_{\text{applied}} = 8.00 \text{ Pounds of VHAP}$$

   Compute the "as applied" VHAP content of the mixture.

   $$C_{\text{Applied}} = \frac{V_{\text{Applied}}}{S_{\text{Applied}}} = \frac{V_1 + V_s}{S_1 + S_s} = \frac{8.54}{8.00} = 1.07 \tag{3.8}$$

   This is a *non-compliant* mixture, demonstrating the misuse of a compliant cleaning solvent for thinning applications.

3. Calculate the average VHAP content of emissions from all finishing materials used at the facility for a month. An existing source must maintain a value of C_{Average} no greater than 1.0 to achieve compliance.

   Compute the weight of VHAP emitted from the materials used, and then total.

   $$V_1(\text{Lacquer}) = 7.24 \times 14.85/100 \times 553 = 594.6$$
   $$V_2(\text{Sealer}) = 7.68 \times 30.48/100 \times 325 = 760.8$$
   $$V_3(\text{Stain}) = 7.38 \times 4.42/100 \times 390 = 127.2$$

TABLE 3.12   Facility Data

	Lacquer	Sealer	Stain	Basecoat	Solvent
CPDS Data					
Density, lb/gallon	7.24	7.68	7.38	8.16	6.67
VHAP Content, Wt.%	14.85	30.48	4.42	24.51	8.03
Solids Content, Wt.%	16.85	30.47	3.25	24.51	0.00
Use Data					
Gallons used this month	553	325	390	228	55

$$V_4(\text{Basecoat}) = 8.16 \times 24.51/100 \times 228 = 456.0$$
$$V_5(\text{Solvent}) = 6.67 \times 8.03/100 \times 55 = 29.5$$
$$V_{\text{total}} = 1968.1 \text{ lb of VHAP}$$

Compute the weight of solids in the materials used, and then total.

$$S_1(\text{Lacquer}) = 7.24 \times 16.85/100 \times 553 = 674.7$$
$$S_2(\text{Sealer}) = 7.68 \times 30.47/100 \times 325 = 760.5$$
$$S_3(\text{Stain}) = 7.38 \times 3.25/100 \times 390 = 93.6$$
$$S_4(\text{Basecoat}) = 8.16 \times 24.51/100 \times 228 = 456.0$$
$$S_5(\text{Solvent}) = 6.67 \times 0/100 \times 55 = 0.0$$
$$S_{\text{total}} = 1984.8 \text{ lb of solids}$$

Compute the average emission for the month in NESHAP limitation units employing Equation (3.9)

$$C_{\text{Average}} = \frac{V_{\text{total}}}{S_{\text{total}}} = \frac{1968.1}{1984.8} = 0.9916 \frac{\text{lb VHAP}}{\text{lb Solids}} \qquad (3.9)$$

Note: This calculation illustrates how both complying and non-complying materials can be used while remaining in compliance via the averaging procedure.

TCAA.14   EMISSION FACTOR CALCULATION

Determine the emission factors ef and ef' for an organic solvent-borne coating which contains 40% organic solvent having a density of 7.36 lb/gal. Also determine the emission factors ef and ef' for a waterborne coating containing 65% solvent with 80% of the solvent being water. The density of the organic portion of the solvent is 7.36 lb/gal. Finally, calculate the percent reduction in volatile organic emissions achieved by switching from the solvent-borne to the waterborne coating. Employ Equations (3.3)–(3.5).

SOLUTION: Calculate the emission factor, ef, for the organic solvent-borne coating in lb/(gal-H_2O).

$$\begin{aligned} \text{ef} &= \text{vs}\rho_v/(1 - \text{ws}) \\ &= (1)(0.40)(7.36)/(1 - 0) \\ &= 2.94 \text{ lb/gal-}H_2O \end{aligned} \qquad (3.3)$$

Determine the emission factor, ef', for the organic solvent-borne coating in lb/gal solids.

$$ef' = vs\rho_v/(1 - S)$$
$$= (1)(0.40)(7.36)/(1 - 0.40)$$
$$= 4.91 \text{ lb/gal solids} \tag{3.4}$$

Calculate the emission factor, ef, for the waterborne coating in lb/(gal-H_2O).

$$ef = (0.2)(0.65)(7.36)/(1 - (0.80)(0.65))$$
$$= 0.957/(1 - 0.52)$$
$$= 1.99 \text{ lb/gal-}H_2O$$

Determine the emission factor, ef', for the waterborne coating in lb/gal solids.

$$ef' = (0.20)(0.65)(7.36)/(1 - 0.65)$$
$$= 0.957/0.35$$
$$= 2.73 \text{ lb/gal solids}$$

Finally calculate the percent reduction achieved in VOC emissions by switching from the solvent-borne to the waterborne coating.

$$\% \text{ reduction} = 100(ef'_{orig} - ef'_{repl})/ef'_{orig}$$
$$= 100(4.91 - 2.73)/4.91$$
$$= 44.4\% \tag{3.5}$$

The reader should again note that the same volume of coating solids must be deposited on an object to coat it to a desired film thickness regardless of the type of coating or volatile organic compound content of the coating used. Solids make the film. Volatiles (VOC, water, and non-photochemically reactive solvents) evaporate. For example, four gallons of a 25 volume percent solids coating must be used to get one gallon of coating solids. However, only two gallons of a 50 volume percent solids coating must be used to get one gallon of coating solids. This means that twice as much "work" can be done with a gallon of 50 vol% solids coating than with a gallon of 25 vol% solids coating, i.e., twice as many gallons of 25% solids coating are needed than gallons of 50% solids coating to do the same job. How do emissions from different coatings compare? A comparison of the percent difference in emissions between two coatings, or between a coating and an emission limit, must be performed on a solids basis. Suppose each gallon of the 25 vol% solids coating contains 5.5 lbs of VOCs. For each gallon of coating solids required, 22.0 lbs of VOCs are emitted. However, each gallon of the 50 vol% solids coating contains 3.7 (two thirds of 5.5) lbs of VOCs. So, for each gallon of coating solids, 7.4 lbs of VOCS are emitted. The emissions from the 50 vol% solids coating is approximately 66% less than those from the 25 vol% solids coating when providing an equal amount of solids to the process since $(22.0 - 7.4)/22.0 = 0.66$.

TCAA.15 BASIC CALCULATIONS FOR VOLATILE ORGANIC COMPOUND (VOC) COATINGS

Solve the following 6 problems that involve VOC coating emissions.

1. Determine the mass of VOC emitted per volume of solids for a solvent-borne coating. The following data are given:

   A. Coating Density $= 10.0$ lb/gal
   B. Total Volatiles $= 60$ percent by weight
   C. Water Content $= 0$
   D. Organic Volatiles Content $= 60$ percent by weight
   E. Nonvolatiles Content (Solids) $= 35$ percent by volume

2. Determine the mass of VOC emitted per volume of solids for a waterborne coating. The following data are given:

   A. Coating Density $= 9.0$ lb/gal
   B. Total Volatiles $= 70$ percent by weight
   C. Water Content $= 30$ percent by weight
   D. Organic Volatiles Content $= 70 - 30 = 40$ percent by weight
   E. Nonvolatiles Content (Solids) $= 19.6$ percent by volume

3. Determine the mass of VOC emitted per volume of solids for a coating that contains some negligibly photochemically reactive (NPR) solvents. The following data are given:

   A. Coating Density $= 11.0$ lb/gal
   B. Total Volatiles $= 80$ percent by weight
   C. NPR Solvent Content $= 40$ percent by weight
   D. Organic Volatiles Content $= 40$ percent by weight
   E. Nonvolatiles Content (Solids) $= 15$ percent by volume

4. Determine the mass of VOC emitted per volume of coating less water for a solvent-borne coating. The following data are given:

   A. Coating Density $= 10$ lb/gal
   B. Total Volatiles $= 60$ percent by weight
   C. Water Content $= 0$
   D. Organic Volatiles Content $= 60$ percent by weight

5. Determine the mass of VOC emitted per volume of coating less water for a water-borne coating. The following data are given:

   A. Coating Density $= 9.0$ lb/gal
   B. Total Volatiles $= 70$ percent by weight
   C. Water Content $= 30$ percent by weight
   D. Organic Volatiles Content $= 70 - 30 = 40$ percent by weight

6. Determine the mass of VOC emitted per gallon of coating less negligibly photo-chemically reactive material for a coating that contains some negligibly photochemically reactive material.
   The following data are given:

   A. Coating Density $= 10.5$ lb/gallon
   B. Total Volatiles $= 80$ percent by weight
   C. NPR Solvent Content $= 40$ percent by weight
   D. Organic Volatiles Content $= 40$ percent by weight
   E. NPR Solvent Density $= 11.0$ lb/gal

SOLUTION:

1. The mass of VOC emitted per volume of solids is:

$$\left(\frac{10.0\ \text{lb coating}}{\text{gal coating}}\right)\left(\frac{0.60\ \text{lb VOC}}{\text{lb coating}}\right)\left(\frac{1\ \text{gal coating}}{0.35\ \text{gal solids}}\right) = \frac{17.1\ \text{lb VOC}}{\text{gal solids}}$$

2. The mass of VOC emitted per volume of solids is:

$$\left(\frac{9.0\ \text{lb coating}}{\text{gal coating}}\right)\left(\frac{0.40\ \text{lb VOC}}{\text{lb coating}}\right)\left(\frac{1\ \text{gal coating}}{0.196\ \text{gal solids}}\right) = \frac{18.4\ \text{lb VOC}}{\text{gal solids}}$$

3. The mass of VOC emitted per volume of solids is:

$$\left(\frac{11.0\ \text{lb coating}}{\text{gal coating}}\right)\left(\frac{0.40\ \text{lb VOC}}{\text{lb coating}}\right)\left(\frac{1\ \text{gal coating}}{0.15\ \text{gal solids}}\right) = \frac{29.3\ \text{lb VOC}}{\text{gal solids}}$$

4. The mass of VOC per volume of coating less water is:

$$\left(\frac{10\ \text{lb coating}}{\text{gal coating}}\right)\left(\frac{0.60\ \text{lb VOC}}{\text{lb coating}}\right)\left(\frac{1\ \text{gal coating}}{(1-0)\ \text{gal coating less water}}\right)$$
$$= \frac{6\ \text{lb VOC}}{\text{gal coating less water}}$$

5. The mass of water in the coating is:

$$\left(\frac{9.0 \text{ lb coating}}{\text{gal coating}}\right)\left(\frac{0.3 \text{ lb water}}{\text{lb coating}}\right) = \frac{2.7 \text{ lb water}}{\text{gal coating}}$$

The volume of water in the coating is:

$$\left(\frac{2.7 \text{ lb}}{\text{gal coating}}\right)\left(\frac{1}{\dfrac{8.33 \text{ lb water}}{\text{gal water}}}\right) = \frac{0.32 \text{ gal water}}{\text{gal coating}}$$

The mass of VOC in the coating is:

$$\left(\frac{9.0 \text{ lb coating}}{\text{gal coating}}\right)\left(\frac{0.4 \text{ lb VOC}}{\text{lb coating}}\right) = \frac{3.6 \text{ lb VOC}}{\text{gal coating}}$$

The mass of VOC emitted per volume of coating less water is:

$$\frac{\dfrac{3.6 \text{ lb VOC}}{\text{gal coating}}}{\left(\dfrac{1 \text{ gal coating} - 0.32 \text{ gal water}}{\text{gal coating}}\right)} = \frac{5.3 \text{ lb VOC}}{\text{gal coating less water}}$$

6. The mass of VOC per volume of coating is:

$$\left(\frac{10.5 \text{ lb coating}}{\text{gal coating}}\right)\left(\frac{0.4 \text{ lb VOC}}{\text{lb coating}}\right) = \frac{4.2 \text{ lb VOC}}{\text{gal coating}}$$

The mass of NPR solvent in the coating is:

$$\left(\frac{10.5 \text{ lb coating}}{\text{gal coating}}\right)\left(\frac{0.4 \text{ lb NPR solvent}}{\text{lb coating}}\right) = \frac{4.2 \text{ lb NPR solvent}}{\text{gal coating}}$$

The volume of NPR solvent in the coating is:

$$\left(\frac{4.2 \text{ lb NPR solvent}}{\text{gal coating}}\right)\left(\frac{1}{\dfrac{11.0 \text{ lb NPR solvent}}{\text{gal NPR solvent}}}\right) = \frac{0.38 \text{ gal NPR solvent}}{\text{gal coating}}$$

The mass of VOC per gallon of coating less NPR solvent is:

$$\frac{\dfrac{4.2 \text{ lb VOC}}{\text{gal coating}}}{\left(\dfrac{1 \text{ gal coating} - 0.38 \text{ gal NPR solvent}}{1 \text{ gal coating}}\right)} = \frac{6.8 \text{ lb VOC}}{\text{gal coating less NPR solvent}}$$

TCAA.16 VOC TRANSFER EFFICIENCY

When spray guns are used to apply coatings, much of the coating material either bounces off the surface being coated or misses it altogether. Transfer efficiency (TE) is the ratio of the amount of coating solids deposited on the coated part to the amount of coating solids used. Regardless of the TE, all of the VOCs in the dispensed coating are emitted whether or not the coating actually reaches and adheres to the surface. Consequently, improved TE can reduce VOC emissions because less coating is used. EPA has defined baseline transfer efficiencies of 60 percent for RACT in metal furniture and appliance coating and 30 percent for RACT waterborne equivalence in the automobile industry (for both primer-surfacer and topcoat applications). If a base TE has not been documented by EPA, then the company must satisfactorily document their base TE prior to equivalency calculations/demonstrations. To obtain TE credits, a company must prove its baseline TE with documentation and document the new TE.

Solve the following transfer efficiency related problem. Determine the mass of VOC emitted per volume of solids applied given the following data:

$$\text{VOC content of coating} \quad = \frac{4.0 \text{ lb VOC}}{\text{gal solids}}$$

$$\text{Transfer efficiency} \qquad = 40 \text{ percent}$$

SOLUTION: For this case,

$$\left(\frac{3.0 \text{ lb VOC}}{\text{gal coating less water}}\right)\left(\frac{1 \text{ gal coating less water}}{0.55 \text{ gal solids}}\right)$$

$$\times \left(\frac{1 \text{ gal solids used}}{0.60 \text{ gal solids applied}}\right) = \frac{9.1 \text{ lb VOC}}{1 \text{ gal solids applied}}$$

Note: For a waterborne coating, care should be exercised in using pounds of VOC per gallon of coating less water and volume nonvolatiles content as a fraction of the total coating including water. These two items cannot simply be combined to get pounds of VOC per gallon of solids. The best method is to factor in transfer efficiency. Alternatively, the volume nonvolatiles content could be determined for the coating less water if the volume fraction water is known or calculated as follows:

Volume fraction nonvolatiles in coating less water

$$= \frac{\text{Volume fraction nonvolatiles in coating including water}}{1 - \text{volume fraction water}}$$

TCAA.17   VOC COATING COMPLIANCE DETERMINATIONS

Solve the following 6 coating compliance problems.

1. A coater is required to meet an emission limit of 3.5 pounds of VOC per gallon of coating less water. Does a coating with a density of 12 pounds per gallon that contains 25 weight percent VOC comply? The coating contains no water or negligibly photochemically reactive solvents.

2. A coater is required to meet an emission limit of 4.0 pounds of VOC per gallon of solids. Does a coating with a density of 10 pounds per gallon that contains 60 weight percent volatiles, 45 weight percent water, and 30 volume percent solids comply?

3. A coater is required to meet an emission limit of 10 pounds VOC per gallon of solids applied. Does the coating in Example 2 comply if it is applied at a transfer efficiency of 80 percent?

4. A metal furniture coater uses coating containing 0.40 kg VOC/liter of coating(less water and exempt solvents). The coating contains 55 volume percent solids. The transfer efficiency is 87 percent. Is the plant in compliance if the maximum allowable emissions are 1.0 kg VOC/liter solids applied?

5. A coater is required to meet an emission limit of 6 pounds of VOC per gallon of solids. What percent emission reduction is needed if the coater uses a coating with 22 pounds of VOC per gallon of solids?

6. A coater is required to meet an emission limit of 3.7 pounds of VOC per gallon of coating less water. What percent emission reduction is needed if the coater uses a solvent-borne coating with 5.0 pounds of VOC per gallon of coating less water and a volume solids content of 25 percent?

SOLUTION:

1. For part (1),

$$\left(\frac{12 \text{ lb coating}}{\text{gal coating}} \right) \left(\frac{0.25 \text{ lb VOC}}{\text{lb coating}} \right) \left(\frac{1 \text{ gal coating}}{(1-0) \text{ gal coating less water}} \right)$$

$$= \frac{3 \text{ lb VOC}}{\text{gal coating less water}}$$

Therefore, the coating complies with the regulation.

2. The weight percent organic volatiles is $60 - 45 = 15$.
   The VOC content of the coating is:

$$\left(\frac{10\,\text{lb coating}}{\text{gal coating}}\right)\left(\frac{0.15\,\text{lb VOC}}{\text{lb coating}}\right)\left(\frac{1\,\text{gal coating}}{0.30\,\text{gal solids}}\right) = \frac{5\,\text{lb VOC}}{\text{gal solids}}$$

   Therefore, the coating does not comply with the regulation.

3. From part (3),

$$\left(\frac{5\,\text{lb VOC}}{\text{gal solids}}\right)\left(\frac{1\,\text{gal solids used}}{0.80\,\text{gallon solids applied}}\right) = \frac{6.3\,\text{lb VOC}}{\text{gal solids applied}}$$

   Therefore, the coating complies with the regulation.

4. The solution is found by using the following basic equation:

$$\left(\frac{\text{mass of VOC used}}{\text{volume of coating solids used}}\right)\left(\frac{1}{\text{TE}}\right) = \frac{\text{mass of VOC used}}{\text{volume of coating solids applied}}$$

   Emissions are:

$$\left(\frac{0.40\,\text{kg VOC}}{1\,\text{liter of coating less water and exempt solvents}}\right)$$

$$\times \left(\frac{1\,\text{liter of coating less water and exempt solvents}}{0.55\,\text{liter solids}}\right)$$

$$\times \left(\frac{1\,\text{liter solids}}{0.87\,\text{liter solids applied}}\right) = \frac{0.84\,\text{kg VOC}}{1\,\text{liter solids applied}}$$

   Since 0.84 is less than 1.0, the coating operation is in compliance.

5. From part (5),

$$\left(\frac{22 - 6}{22}\right)(100) = \left(\frac{16}{22}\right)(100) = 73 \text{ percent emission reduction}$$

6. This calculation must be done on a solids basis. First, the emission limit must be converted to pounds of VOC per gallon of solids. To do this, an assumed VOC density of 7.36 pounds per gallon is used to calculate the volume solids content of the "presumptive" RACT coating.

$$\left(\frac{3.7\,\text{lb VOC}}{\text{gal coating less water}}\right)\left(\frac{1\,\text{gal VOC}}{7.36\,\text{lb VOC}}\right)(100)=50 \text{ volume percent VOC}$$

$$100-50=50 \text{ volume percent solids}$$

$$\left(\frac{3.7\,\text{lb VOC}}{\text{gal coating}}\right)\left(\frac{1\,\text{gal coating}}{0.50\,\text{gal solids}}\right)=\frac{7.4\,\text{lb VOC}}{\text{gal solids}}$$

Next, the VOC content of the coating used must also be calculated on a solids basis.

$$\left(\frac{5.0\,\text{lb VOC}}{\text{gal coating}}\right)\left(\frac{1\,\text{gal coating}}{0.25\,\text{gal solids}}\right)=\frac{20\,\text{lb VOC}}{\text{gal solids}}$$

Now the required percent reduction can be calculated.

$$\left(\frac{20-7.4}{20}\right)(100)=63 \text{ percent emission reduction}$$

Notes:

1. An erroneous result is obtained if this calculation is not performed on a solids basis. Using pounds of VOC per gallon of coating less water, the result would be:

$$\left(\frac{5.0-3.7}{5.0}\right)(100)=26\% \text{ emission reduction}$$

   This would not give equivalent emissions as it does not take into account that the "presumptive" RACT coating not only has a lower VOC content but also higher solids content as well.

2. An assumed VOC density of 7.36 pounds per gallon is used to calculate the volume solids content of the "presumptive" RACT coating because this same value was used to determine the "presumptive" recommended RACT emission limits from volume solids data.

3. The volume solids content of actual coatings should be determined directly from coating formulation data as described in the VOC data sheets. Occasionally, it may be useful to backcalculate volume solids from the VOC content and actual solvent or VOC density, but this must be performed with extreme caution. When an inspector gathers data on the actual coatings used at a facility, the volume solids content should be obtained from coating formulation data from the facility or the coating manufacturer. The volume solids content should not be backcalculated.

TCAA.18 VOC SURFACE COATING EQUIVALENCY DETERMINATION

Equivalency calculations are required when compliance decisions must be made for replacement coatings, bubbles, offsets, netting, etc. VOC equivalency calculations must be made on a solids basis. The amount of solids needed to coat a surface to a particular film thickness is the same regardless of the coating composition used. As discussed earlier, reducing the solids content of an organic solvent-borne coating increases the quantity of coating required and increases VOC emissions because more coating is used and the coating has a higher VOC content.

A surface coater uses 10 gallons per hour of coating that contains 5.5 lb VOC per gallon of coating and 25 volume percent solids. New regulations indicate that the coating formulation must meet an emission limit of 3.0 lb of VOC per gallon of coating (with a solvent density of 7.36 lb per gallon) or the coater must control VOC emissions to an equivalent level. Assuming that the production rate (solids usage rate), transfer efficiency, and film thickness stay constant, what are the coater's "allowable" hourly VOC emissions?

This problem introduces a method for determining hourly VOC mass emissions for offset calculations; however, RACT limitations should normally be based on either applicable coating formulations or control efficiency requirements. An hourly cap would normally only be used in addition to these RACT limitations.

SOLUTION: For the existing coating, the actual VOC emissions are:

$$\left(\frac{10\,\text{gal coating}}{\text{h}}\right)\left(\frac{5.5\,\text{lb VOC}}{\text{gal coating}}\right) = \frac{55\,\text{lb VOC}}{\text{h}}$$

The solids usage rate is:

$$\left(\frac{10\,\text{gal coating}}{\text{h}}\right)\left(\frac{0.25\,\text{gal solids}}{\text{gal coating}}\right) = \frac{2.5\,\text{gal solids}}{\text{h}}$$

For the complying coating, the VOC (solvent) volume fraction is:

$$\left(\frac{3.0\,\text{lb VOC}}{\text{gal coating}}\right)\left(\frac{1\,\text{gal VOC}}{7.36\,\text{lb VOC}}\right) = \frac{0.41\,\text{gal VOC}}{\text{gal coating}}$$

The complying coating solids volume fraction is:

$$1.0 - 0.41 = \frac{0.59\,\text{gal solids}}{\text{gal coating}}$$

Using the solids usage rate calculated above, the gallons of complying coating required are:

$$\left(\frac{2.5\,\text{gal solids}}{\text{h}}\right)\left(\frac{\text{gal coating}}{0.59\,\text{gal solids}}\right) = \frac{4.24\,\text{gal complying coating}}{\text{h}}$$

The emissions rate at the existing solids applied rate is:

$$\left(\frac{3.0\,\text{lb VOC}}{\text{gal coating}}\right)\left(\frac{4.24\,\text{gal coating}}{h}\right) = \frac{12.72\,\text{lb VOC}}{h}$$

TCAA.19 EQUIVALENCY CALCULATIONS FOR A CAN COATING OPERATION

RACT equivalence requirements for can coating operations were tabulated for a regulation calculation. An analysis of two coatings used in an actual plant is:

	Coating No. 1	Coating No. 2
1. Actual pounds of VOC per gallon of coating less water and exempt solvents as applied	5.42	1.09
2. Gallons of each coating applied per operation	110	240
3. Control efficiency, percent	0.81	—
4. Volume percent water and exempt solvents in coatings	—	41.3
5. Volume percent solids	26.4	50.0
6. Allowable emission limit, lb VOC/gal coating less water and exempt solvents	2.8	2.8

Calculate the following for the operation in question:

1. Gallons of solids applied.
2. Pounds of VOC per gallon of solids.
3. Pounds of VOC emitted.
4. Allowable VOC emissions.

SOLUTION:

1. The gallons of solids applied can be calculated as follows:

Gallons of solids applied = (gallons of coating used) × (volume percent solids) ÷ 100%

(Since the quantity of solids used in the equation appears as a percentage it is necessary to divide by 100%.)

Coating No. 1:

$$(110\,\text{gal coating applied})\left(\frac{26.4\,\text{gal solids}}{100\,\text{gal coating}}\right) = 29\,\text{gal solids applied}$$

Coating No. 2:

$$(240 \, \text{gal coating applied}) \left(\frac{50 \, \text{gal solids}}{100 \, \text{gal coating}} \right) = 120 \, \text{gal solids applied}$$

2. As noted earlier, the pounds of VOC per gallon of solids can be calculated as follows:

$$\frac{\text{lb VOC}}{\text{gal of solids}} = \frac{\left(\dfrac{\text{lb VOC/gal coating less water}}{\text{volume percent solids}} \right)}{100 - \text{volume percent water}}$$

Coating No. 1:

$$\left(\frac{5.42 \, \text{lb VOC}}{\text{gal coating less water}} \right) \left(\frac{100 \, \text{gal coating}}{26.4 \, \text{gal solids}} \right) \left(\frac{100 - 0 \, \text{gal coating less water}}{100 \, \text{gal coating}} \right)$$

$$= \frac{20.53 \, \text{lb VOC}}{\text{gal solids}}$$

Coating No. 2:

$$\left(\frac{1.09 \, \text{lb VOC}}{\text{gal coating less water}} \right) \left(\frac{100 \, \text{gal coating}}{50.0 \, \text{gal solids}} \right) \left(\frac{100 - 41.3 \, \text{gal coating less water}}{100 \, \text{gal coating}} \right)$$

$$= \frac{1.28 \, \text{lb VOC}}{\text{gal solids}}$$

3. The pounds of VOC emitted can be calculated as follows:

$$\left(\frac{\text{lb of VOC}}{\text{gal of solids}} \right) (\text{gal of solids})(1 - \text{overall control efficiency}^*)$$

where the overall control efficiency is equal to the fraction of total VOC used that is destroyed or recovered by the control system. Overall control efficiency = capture device efficiency × control device efficiency; fractional basis.
Coating No. 1:

$$\left(\frac{20.53 \, \text{lb VOC}}{\text{gal solids}} \right) (29.0 \, \text{gal solids})(1 - 0.81) = 113.1 \, \text{lb VOC}$$

Coating No. 2:

$$\left(\frac{1.28\,\text{lb VOC}}{\text{gal solids}}\right)(120\,\text{gal solids})(1-0) = 153.6\,\text{lb VOC}$$

4. The allowable VOC emissions can be calculated as described below.

As calculated in part (2), the gallons of solids applied for coatings No. 1 and No. 2 were 29 and 120, respectively. Assume that the coater will apply the same volume of solids with a RACT complying coating. Also, assume a VOC density of 7.36 lb VOC/gal.

Allowable pounds of VOC = gallons of complying coating applied × allowable emission limit,

$$\text{Allowable lb of VOC} = \left(\frac{\text{gal solids applied (per unit of time)}}{\text{volume fraction solids in complying coating}}\right)$$
$$\times \left(\frac{2.8\,\text{lb VOC}}{\text{gal coating less water and exempt solvents}}\right)$$

Volume fraction VOC:

$$\left(\frac{2.8\,\text{lb VOC}}{\text{gal coating}}\right)\left(\frac{1\,\text{gal VOC}}{7.36\,\text{lb VOC}}\right) = \frac{0.38\,\text{gal VOC}}{\text{gal coating}}$$

Volume fraction solids:

$$\text{Volume fraction solids} = 1 - 0.38\,\text{gal coating} = \frac{0.62\,\text{gal solids}}{\text{gal coating}}$$

Allowable pounds of VOC for Coating No. 1:

$$\text{Allowable lb of VOC(1)} = (29\,\text{gal solids})\left(\frac{1\,\text{gal coating}}{0.62\,\text{gal solids}}\right)\left(\frac{2.8\,\text{lb VOC}}{\text{gal coating}}\right)$$
$$= 131\,\text{lb VOC}$$

Allowable pounds of VOC for Coating No. 2:

$$\text{Allowable lb of VOC(2)} = (120\,\text{gal solids})\left(\frac{\text{gal coating}}{0.62\,\text{gal solids}}\right)\left(\frac{2.8\,\text{lb VOC}}{\text{gal coating}}\right)$$
$$= 542\,\text{lb VOC}$$

TCAA.20 COMPLIANCE DETERMINATION FOR AUTO PLANT PRIMER SURFACE (GUIDE COAT) OPERATION

An auto primer surface operation uses a coating that contains 3.58 lb VOC/gal of coating with a transfer efficiency of 50 percent. The RACT emission limit is 2.8 lb VOC/gal of coating less water at 30 percent transfer efficiency (waterborne equivalence). Is the operation in compliance?

The following information is provided.

1. The manufacturer's data show that the undiluted coating has 50.0 volume percent solids.
2. The plant adds 0.05 gal of thinner blend per gallon of undiluted coating:

   (a) 0.02 gallon of thinner No. 1/gallon undiluted coating (thinner density 7.36 lb/gal).
   (b) 0.02 gallon of thinner No. 2/gallon undiluted coating (thinner density 5.43 lb/gal).
   (c) 0.01 gallon of thinner No. 3/gallon undiluted coating (thinner density 9.52 lb/gal).

The density of the undiluted coating is 10.25 lb/gal.

Weight fraction of VOC in undiluted coating 0.333 lb VOC solvent/lb undiluted coating.

SOLUTION: First, verify the VOC content of the coating. In order to do this, the VOC content of the undiluted coating and the thinners must be calculated. The mass of VOC in the undiluted coating is:

$$\left(\frac{0.333 \text{ lb VOC}}{\text{lb undiluted coating}}\right)\left(\frac{10.25 \text{ lb undiluted coating}}{\text{gal undiluted coating}}\right) = \frac{3.41 \text{ lb VOC}}{\text{gal undiluted coating}}$$

The mass of thinner added per gallon of undiluted coating is:

$$\left(\frac{0.02 \text{ gal thinner No. 1}}{\text{gal undiluted coating}}\right)\left(\frac{7.36 \text{ lb thinner No. 1}}{\text{gal thinner No. 1}}\right) + \left(\frac{0.02 \text{ gal thinner No. 2}}{\text{gal undiluted coating}}\right)$$

$$\times \left(\frac{5.43 \text{ lb thinner No. 2}}{\text{gal thinner No. 2}}\right) + \left(\frac{0.01 \text{ gal thinner No. 3}}{\text{gal undiluted coating}}\right)\left(\frac{9.52 \text{ lb thinner No. 3}}{\text{gal thinner No. 3}}\right)$$

$$= 0.147 + 0.109 + 0.095 = \frac{0.351 \text{ lb thinner}}{\text{gal undiluted coating}}$$

The mass VOC per volume coating at application is:

$$\frac{3.41 \text{ lb VOC/gal undiluted coating} + 0.351 \text{ lb thinner/gal undiluted coating}}{1.05 \text{ gal coating/gal undiluted coating}}$$

$$= 3.58 \text{ lb VOC/gal coating; VOC content verified}$$

The undiluted coating has 50.0 volume percent solids. After the coating is diluted with 0.05 gallon of thinner per gallon of coating, the volume percent solids is:

$$\frac{0.50}{1 + 0.05} = \frac{0.48 \text{ gal solids}}{\text{volume coating}}$$

The equivalency calculations must be made on a solids basis. The formula for determining the maximum allowable emissions on a solids basis is:

$$\text{Allowable emissions} = \frac{\text{allowable mass of VOC per volume coating}}{(\text{baseline TE})(\text{baseline volume solids})}$$

As noted in earlier examples, an assumed VOC density of 7.36 gal is used to calculate the volume solids content of the "presumptive" RACT coating.

The volume of VOC in the "presumptive" RACT coating is:

$$\left(\frac{2.8 \text{ lb VOC}}{\text{gal coating}}\right)\left(\frac{1 \text{ gal VOC}}{7.36 \text{ lb VOC}}\right) = \frac{0.38 \text{ gal VOC}}{\text{gal coating}}$$

Therefore, the baseline volume of solids is:

$$1 - 0.38 = \frac{0.62 \text{ gal solids}}{\text{gal coating}}$$

Allowable emissions are:

$$\left(\frac{2.8 \text{ lbVOC}}{\text{gal coating}}\right)\left(\frac{1 \text{ gal coating}}{0.62 \text{ gal solids}}\right)\left(\frac{1 \text{ gal solids}}{0.30 \text{ gal solids applied}}\right) = \frac{15.1 \text{ lb VOC}}{\text{gal solids applied}}$$

The formula for actual emissions is:

$$\text{Actual emissions} = \frac{\text{actual mass VOC per volume coating}}{(\text{actual \% TE})\left(\dfrac{\text{actual volume solids}}{\text{gal coating}}\right)}$$

The actual mass of VOC per volume coating is 3.58 lb VOC/gallon coating. The actual transfer efficiency is 50 percent. The actual volume of solids in the sprayed coating is 48 percent. Actual emissions are therefore:

$$\left(\frac{3.58 \text{ lb VOC}}{\text{gal coating}}\right)\left(\frac{1 \text{ gal coating}}{0.48 \text{ gal solids}}\right)\left(\frac{1 \text{ gal solids}}{0.50 \text{ gal solids applied}}\right) = \frac{14.9 \text{ lb VOC}}{\text{gal solids applied}}$$

The actual emissions (14.9 lb VOC per gallon of solids applied) are less than the maximum allowable emissions (15.1 lb VOC per gallon of solids applied); therefore, the operation is in compliance.

TCAA.21   DETERMINING COMPLIANCE FOR A LARGE APPLIANCE COATING LINE USING SEVERAL TYPES OF SPRAY EQUIPMENT

A large appliance manufacturer has a coating operation that employs electrostatic spray coating equipment and manual spray coating equipment. The following data are available regarding the operation. Determine the compliance status. If this large appliance manufacturer is out of compliance, what percent reduction is required to achieve compliance?

	(A) Electrostatic coating	(B) Manual coating
Transfer efficiency, percent	90	40
Average volume percent of solids in coating	39	39
VOC content, lb VOC/gal coating less water	4.5	4.5
Gallons of coating used per day	30.4	47.1
Emission limit, lb/gallon less water	2.8	2.8
Baseline transfer efficiency for large appliances, percent	60	60

The baseline transfer efficiency is 60 percent for a large appliance coater. Table 3.13 provides a tabulation of the available data and calculated results. The actual calculations follow.

SOLUTION: Under the actual emissions category, the following calculations can be made.

For A and B, the mass of VOC per volume of solids is:

$$\left(\frac{4.5\,\text{lb VOC}}{\text{gal coating}}\right)\left(\frac{1\,\text{gal coating}}{0.39\,\text{gal solids}}\right) = \frac{11.5\,\text{lb VOC}}{\text{gal solids}}$$

For A, the mass of VOC per volume of solids applied is:

$$\left(\frac{4.5\,\text{lb VOC}}{\text{gal coating}}\right)\left(\frac{1\,\text{gal coating}}{0.39\,\text{gal solids}}\right)\left(\frac{1\,\text{gal solids}}{0.90\,\text{gal solids applied}}\right) = \frac{12.82\,\text{lb VOC}}{\text{gal solids applied}}$$

For B, the mass of VOC per volume of solids applied is:

$$\left(\frac{4.5\,\text{lb VOC}}{\text{gal coating}}\right)\left(\frac{1\,\text{gal coating}}{0.39\,\text{gal solids}}\right)\left(\frac{1\,\text{gal solids}}{0.40\,\text{gal solids applied}}\right) = \frac{28.85\,\text{lb VOC}}{\text{gal solids applied}}$$

TABLE 3.13 Large Appliance Multitransfer Efficiency Calculation

Spray Type	Gallons of Coating/Day	Solids, Vol. %	Lb VOC/Gallon Coating	Lb VOC/Gallon Solids	% TE	Lb VOC/Gallon Solids Applied	Gallons of Solids Applied/Day	Lb VOC/Day
				Actual Emissions				
A	30.4	39	4.5	11.5	90	12.8	10.7	136.8
B	47.1	39	4.5	11.5	40	28.8	7.3	212.0
Total								348.8
				Allowed Emissions				
A	28.8	62	2.8	4.5	60	7.5	10.7	80.6
B	19.6	62	2.8	4.5	60	7.5	7.3	54.9
Total								135.5

For A, the volume of solids applied per day is:

$$\left(\frac{30.4\,\text{gal coating}}{\text{day}}\right)\left(\frac{0.39\,\text{gal solids}}{\text{gal coating}}\right)\left(\frac{0.90\,\text{gal solids applied}}{\text{gal solids used}}\right)$$
$$=\frac{10.7\,\text{gal solids applied}}{\text{day}}$$

For B, the volume of solids applied per day is:

$$\left(\frac{47.1\,\text{gal coating}}{\text{day}}\right)\left(\frac{0.39\,\text{gal solids}}{\text{gal coating}}\right)\left(\frac{0.40\,\text{gal solids applied}}{\text{gal solids used}}\right)$$
$$=\frac{7.3\,\text{gal solids applied}}{\text{day}}$$

For A, the mass of VOC emissions per day is:

$$\left(\frac{4.5\,\text{lb VOC}}{\text{gal coating}}\right)\left(\frac{30.4\,\text{gal coating}}{\text{day}}\right)=\frac{136.8\,\text{lb VOC}}{\text{day}}$$

For B, the mass of VOC emissions per day is:

$$\left(\frac{4.5\,\text{lb VOC}}{\text{gal coating}}\right)\left(\frac{47.1\,\text{gal coating}}{\text{day}}\right)=\frac{212.0\,\text{lb VOC}}{\text{day}}$$

Under the allowed emissions category, the following calculations can be made. For A and B, the volume fraction of VOC in the baseline coating is:

$$\left(\frac{2.8\,\text{lb VOC}}{\text{gal coating}}\right)\left(\frac{1\,\text{gal VOC}}{7.36\,\text{lb VOC}}\right)=\frac{0.38\,\text{gal VOC}}{\text{gal coating}}$$

The volume fraction solids in the coating is:

$$1-\frac{0.38\,\text{gal VOC}}{\text{gal coating}}=\frac{0.62\,\text{gal solids}}{\text{gal coating}}$$

The baseline mass of VOC per volume solids is:

$$\left(\frac{2.8\,\text{lbVOC}}{\text{gal coating}}\right)\left(\frac{1\,\text{gal coating}}{0.62\,\text{gal solids}}\right)=\frac{4.5\,\text{lb VOC}}{\text{gal solids}}$$

For A and B, the maximum allowable emissions are:

$$\left(\frac{2.8\,\text{lb VOC}}{\text{gal coating}}\right)\left(\frac{1\,\text{gal coating}}{0.62\,\text{gal solids}}\right)\left(\frac{1\,\text{gal solids used}}{0.60\,\text{gal solids applied}}\right) = \frac{7.5\,\text{lb VOC}}{\text{gal solids applied}}$$

The volume of solids applied remains the same. Therefore, the gallons of complying coating used per day for A would be:

$$\left(\frac{10.7\,\text{gal solids applied}}{\text{day}}\right)\left(\frac{1\,\text{gal coating}}{0.62\,\text{gal solids}}\right)\left(\frac{1\,\text{gal solids used}}{0.6\,\text{gal solids applied}}\right)$$
$$= \frac{28.8\,\text{gal coating}}{\text{day}}$$

For B, the gallons of complying coating used per day would be:

$$\left(\frac{7.3\,\text{gal solids}}{\text{day}}\right)\left(\frac{1\,\text{gal coating}}{0.62\,\text{gal solids}}\right)\left(\frac{1\,\text{gal solids used}}{0.6\,\text{gal solids applied}}\right) = \frac{19.6\,\text{gal coating}}{\text{day}}$$

For A, the mass of VOC emissions allowed per day is:

$$\left(\frac{2.8\,\text{lb VOC}}{\text{gal coating}}\right)\left(\frac{28.8\,\text{gal coating}}{\text{day}}\right) = \frac{80.6\,\text{lb VOC}}{\text{day}}$$

For B, the mass of VOC emissions allowed per day is:

$$\left(\frac{2.8\,\text{lb VOC}}{\text{gal coating}}\right)\left(\frac{19.6\,\text{gal coating}}{\text{day}}\right) = \frac{54.9\,\text{lb VOC}}{\text{day}}$$

The total actual VOC emissions from A (136.8 lb) and B (212.0 lb) are 348.8 lb VOC per day. The total allowable VOC emissions (80.6 and 54.9) are 135.5 lb VOC per day. Therefore, the operation is out of compliance. To achieve compliance, the required reduction in emissions is:

$$\left(\frac{348.8 - 135.5}{348.8}\right)(100) = 61 \text{ percent}$$

TCAA.22 ESTIMATING OZONE EXCEEDANCES FOR A YEAR

In the following problem, the term "exceedances" equates to the 40 CFR §50.9 statement "days with maximum hourly average ozone concentrations above the level of the standard".

A site has the following data history for:

1978: 365 daily values; 3 days above the standard level.

1979: 285 daily values; 2 days above the standard level; 21 missing days satisfying the exclusion criterion.

1980: 287 daily values; 1 day above the standard level; 7 missing days satisfying the exclusion criterion.

During 1980, measurements were not taken during the months of January and February (a total of 60 days for a leap year) because the cold weather minimizes any chance of recording exceedances and the appropriate Regional Administrator had granted a monitoring waiver.

Determine the expected number of exceedances per year using the "exceedances formula" below:

$$e = v + \left(\frac{v}{n}\right) \cdot (N - n - z) \qquad (3.14)$$

N = the number of required monitoring days in the year

n = the number of valid daily maxima

v = the number of measured daily values above the level of the standard

z = the number of days assumed to be less than the standard level, and

e = the estimated number of exceedances for the year.

SOLUTION: When there are no missing daily values, the number of exceedances equals the number of days "above standard level". Thus, for 1978, there were no missing daily values and the number of exceedances is 3.

For 1979, the above equation is applied and the estimated number of exceedances is:

$$e = 2 + (2/285) \times (365 - 285 - 21) = 2 + 0.4 = 2.4$$

For 1980 the same estimation formula is used, but due to the monitoring waiver for January and February, the number of required monitoring days is 306 and therefore the estimated number of exceedances is:

$$1 + (1/287) \times (306 - 287 - 7) = 1 + (1/287) \times 12 = 1.0$$

Averaging these three numbers (3, 2.4 and 1.0) gives 2.1, which is the expected number of exceedances per year.

TCAA.23 CALCULATION OF 8-HOUR OZONE STANDARD

The eight-hour average ozone standard is one of the National Ambient Air Quality Standards designed by the EPA to maintain air pollution levels within healthy limits (40 CFR §50.10). This standard is set at 80 ppb, which means that there should be only an average of 80 ozone molecules per billion molecules over an eight-hour period. For any three-year period, the EPA will look at the four highest

8-hour average ozone pollution levels in each year, then drop the three highest levels from each year. They then average the forth-highest 8-hour averages from the three years to determine the three-year 8-hour average. If this average is higher than 80 ppb, they may designate the area to be in "non-attainment".

Given the following data, determine whether this area is in "non-attainment":

Year	S1	S2	S3	S4
1999	95	94	91	89
2000	98	95	89	84
2001	92	91	91	90

All units are ppb.

SOLUTION: Of the data collected, the three highest values are dropped for each year:

Year	S1	S2	S3	S4
1999	95	94	91	89
2000	98	95	89	84
2001	92	91	91	90

Then, the fourth highest level from each year is averaged for the three years:

$$\text{Average} = \frac{89 + 84 + 90}{3} = 87.6 \text{ ppb}$$

Thus, the three-year average 8-hour ozone value for this hypothetical case would be 87.6 ppb, which would be a violation of the 8-hour ozone standard.

TCAA.24 ESTIMATION OF CHLOROFORM EMISSION FROM A CHLOROMETHANE PROCESS

Emission factors are experimentally determined factors that define the quantity of pollutant generated per mass of a given source. These factors can be used to estimate pollutant generation for a given process.

Given the following data, determine the chloroform emission from this hypothetical chloromethane plant:

Annual methyl chloride production = 49,490,000 lb

Emission factor (EF) = 0.4496 lb chloroform/ton methyl chloride

SOLUTION: The facility estimate for chloroform emissions:

(49,490,000 lb methyl chloride)(1 ton/2000 lb)

   (0.4496 lb chloroform/ton methyl chloride) = 11,125 lbs chloroform

(11,125 lb chloroform)(1 ton/2000 lb) = 5.56 ton chloroform generated

TCAA.25   EMULSION ASPHALT

VOC regulations vary from state to state. The following is an excerpt from the regulatory provision provided 25 years ago by the state of Pennsylvania regarding VOC usage in cutback asphalt paving:

1. After April 30, 1980, and before May 1, 1982 no person may cause, allow, or permit the mixing, storage, use, or application of cutback asphalt for paving operations except when:

   a. long-life stockpile storage is necessary;

   b. the use or application between October 31 and April 30 is necessary; or

   c. the cutback asphalt is to be used solely as a penetrating prime coat, a dust palliative, a tack coat, a precoating of aggregate, or a protective coating for concrete.

2. After April 30, 1982, no persons may cause, allow, or permit the use or application of cutback asphalt for paving operations except when:

   a. long-life stockpile is necessary;

   b. the use or application between October 31 and April 30 is necessary; or

   c. the cutback asphalt is used solely as a tack coat, a penetrating prime coat, a dust palliative, or precoating of aggregate.

3. After April 30, 1982 emulsion asphalts may not contain more than the maximum percentage of solvent as shown in Table 3.14.

An automobile manufacturer that has a past history of applying sound pollution prevention procedures to its processes has proposed to construct a plant in a non-attainment area in Pennsylvania. The required offset is to be obtained by replacing emulsion asphalt by a water-borne mix. The area has traditionally used 200,000 lb/ month of medium setting emulsion asphalt. Calculate the maximum annual offset (20%) if this conversion is adopted. State of Pennsylvania Regulations for cutback asphalt are given above. Note that a 20% offset indicates that the ratio of actual emission reductions to new emissions is equal to or greater than 1.2 to 1.

SOLUTION: Calculate the average amount of medium setting emulsion asphalt used by the automobile manufacturer in lb/yr.

$$m = (200,000 \, lb/month)(12 \, months/yr)$$
$$= 2,400,000 \, lb/yr$$

TABLE 3.14 Maximum Solvent in Asphalt

Emulsion Grade	Type	% Solvent, Maximum
E-1	Rapid Setting	0
E-2	Rapid Setting (Anionic)	0
E-3	Rapid Setting (Cationic)	3
E-4	Medium Setting	12
E-5	Medium Setting	12
E-6	Slow Setting (Soft Residue)	0
E-8	Slow Setting (Hard Residue)	0
E-10	Medium Setting (High Float)	7
E-11	High Float	7
E-12	Medium Setting (Cationic)	8

Source: The provisions of section 129.64 amended April 21, 1981, effective June 20, 1981, 11 Pa. B. 2118.

Calculate the amount of medium setting solvent used per year.

$$m_s = 0.12\,m$$
$$= (0.12)(2,400,000)$$
$$= 288,000\,\text{lb solvent/yr}$$

Calculate the maximum annual offset assuming 75% of the solvent is vaporized.

$$\text{offset} = 0.75\,m_s/1.2$$
$$= (0.75)(288,000)/1.2$$
$$= 180,000\,\text{lb/yr}$$

Repeat this calculation assuming 100% of the solvent is vaporized.

$$\text{offset} = m_s/1.2$$
$$= 288,000/1.2$$
$$= 240,000\,\text{lb/yr}$$

Note that cutback asphalt contains VOCs but the water-borne mix has no VOCs. The above result indicates that if this conversion is adopted (using the 75% vaporization figure above) a new source could operate that emits a maximum of 180,000 lb/yr.

TCAA.26 DEGREASER EMISSIONS REDUCTIONS

Emissions from cold cleaners occur through:

1. waste solvent evaporation,
2. solvent carryout (evaporation from wet parts),

3. solvent bath evaporation,

4. spray evaporation, and

5. agitation.

Waste solvent loss, cold cleaning's greatest emission source, can be reduced through distillation and transport of waste solvent to special incineration plants. Draining cleaned parts for at least 15 seconds reduces carryout emissions. Bath evaporation can be controlled by using a cover regularly, by allowing an adequate freeboard height, and by avoiding excessive drafts in the workshop. If the solvent used is insoluble in, and heavier than, water, a layer of water two to four inches thick covering the halogenated solvent can also reduce bath evaporation. This is known as a "water cover." Spraying at low pressure also helps to reduce solvent loss from this part of the process. Agitation emissions can be controlled by using a cover, by agitating no longer than necessary, and by avoiding the use of highly volatile solvents. Emissions of low volatility solvents increase significantly with agitation. However, contrary to what one might expect, agitation causes only a small increase in emissions of high volatility solvents. Solvent type is the variable which most affects cold cleaner emission rates, particularly the volatility at operating temperatures.

As with cold cleaning, open top vapor degreasing emissions relate heavily to proper operating methods. Most emissions are due to diffusion and convection, which can be reduced by using an automated cover, using a manual cover regularly, spraying below the vapor level, optimizing work loads, or using a refrigerated freeboard chiller (for which a carbon adsorption unit would be substituted on larger units). Safety switches and thermostats that prevent emissions during malfunctions and abnormal operation also reduce diffusion and convection of the vaporized solvent. Additional emission sources are:

1. solvent carryout,

2. exhaust systems, and

3. waste solvent evaporation.

Carryout is directly affected by the size and shape of the workload, by racking of parts, and by cleaning and drying time. Exhaust emissions can be nearly eliminated by a carbon adsorber that collects the solvent vapors for reuse. Waste solvent evaporation is not so much a problem with open top vapor degreasers as it is with cold cleaners, because the halogenated solvents used are often distilled and recycled by solvent recovery systems.

Metal surfaces are often cleaned using organic solvents in an open top degreasing tank. One of the widely used solvents for such operations is 1,1,1-trichloroethane (TCE). TCE belongs to a group of highly stable chemicals known as ozone depleters. A figure depicting a typical degreasing operation is shown below in Figure 3.5. The emission factor for the process shown is estimated to be 0.6 lb/lb of TCE entering the degreaser. The solvent from the degreaser is sent to a solvent

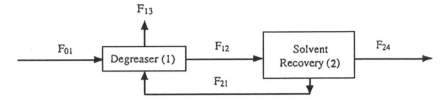

FIGURE 3.5   Schematic of a typical degreasing operation.

recovery unit where 80% of the solvent is recovered and 20% of the solvent is disposed with the sludge.

1. In order to ascertain the feasibility of the installation of a vapor recovery system (see Figure 3.6), determine the amount of TCE vented to the atmosphere per pound of fresh TCA used.
2. If the vapor recovery system is 90% efficient, determine the fraction of TCE lost to the atmosphere and the fraction going with the sludge.

SOLUTION: Write the mass balance equations around the two units for the process without the vapor recovery unit.

Assume a basis of 1 kg for F_{01}.

$$1 + F_{21} = F_{12} + F_{13}; \quad (1)$$

$$F_{12} = F_{21} + F_{24}; \quad (2)$$

Write the equation for the amount of TCE emissions in terms of the amount of TCE entering the degreaser.

$$F_{13} = 0.6(1 + F_{21}); \quad (3)$$

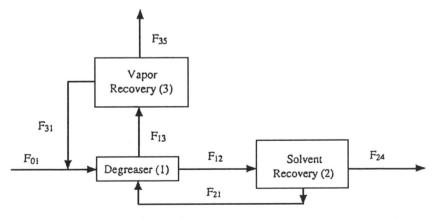

FIGURE 3.6   Schematic of degreasing operation with vapor recovery unit.

Write the equation for the amount of TCE recovered in terms of the amount entering solvent recovery.

$$F_{21} = 0.8F_{12}; \quad (4)$$

Calculate the amount of solvent recovered. There are four equations (1–4) and four unknowns (F_{21}, F_{12}, F_{13}, and F_{24}). First, rearrange equation (4) in terms of F_{12}.

$$F_{12} = 1.25F_{21}; \quad (5)$$

Substituting this result into equation (1) gives equation (6):

$$F_{21} = 1.25F_{21} + F_{13} - 1$$
$$0.25F_{21} + F_{13} - 1 = 0; \quad (6)$$

Simultaneously solving equations (3) and (6) gives the following result for F_{21}:

$$F_{21} = 0.471 \, \text{kg/kg fresh solvent}$$

Calculate the amount of TCE emissions per kg of fresh TCE. From the simultaneous solution above:

$$F_{13} = 0.882 \, \text{kg/kg fresh solvent}$$

Also calculate the amount of TCE in the sludge. First use the above equation to solve for F_{12}:

$$F_{12} = 1.25F_{21}$$
$$= (1.25)(0.471)$$
$$= 0.588$$

Now rearrange the above to solve for F_{24}:

$$F_{24} = F_{12} - F_{21}$$
$$= 0.588 - 0.471$$
$$= 0.118 \, \text{kg/kg fresh solvent}$$

Write the mass balance equations around the three units for the process with the vapor recovery unit.

$$1 + F_{21} + F_{31} = F_{12} + F_{13}; \quad (7)$$
$$F_{12} = F_{21} + F_{24}; \quad (8)$$
$$F_{13} = F_{31} + F_{35}; \quad (9)$$

Write the equation for the amount of TCE emissions in terms of the amount of TCE entering the degreaser.

$$F_{13} = 0.6(1 + F_{21} + F_{31}); \quad (10)$$

Write the equation for the amount of TCE recovered in terms of the amount entering solvent recovery.

$$F_{21} = 0.8F_{12}; \quad (11)$$

Write the equation for the amount of TCE lost to the atmosphere.

$$F_{35} = 0.1F_{13}; \quad (12)$$

Write the equation for the amount of TCE recovered in solvent recovery in terms of the amount of emissions from the degreaser.

$$F_{12} = 1.25F_{21}; \quad (13)$$

Therefore,

$$1 + F_{21} + F_{31} = 1.25F_{21} + F_{13}; \quad (14)$$

Arrange the above equations in terms of F_{31} and set them equal to each other:

$$0.25F_{21} + F_{13} - 1 = F_{13}/0.6 - 1 - F_{21}$$
$$1.25F_{21} = (2/3)F_{13}$$
$$F_{21} = (1.6/3)F_{13}$$

Write the equation for the amount of TCE recovered in vapor recovery in terms of the amount of emissions from the degreaser.

$$F_{13} = F_{31} + 0.1F_{13}$$
$$F_{31} = 0.9F_{13}$$

Calculate the amount of TCE emissions from the degreaser.

$$F_{13} = 0.6(1 + 1.6F_{13}/3 + 0.9F_{13})$$
$$F_{13} = 4.286 \text{ kg/kg fresh solvent}$$

Finally, calculate the amount of TCE lost to the atmosphere.

$$F_{35} = 0.1F_{13}$$
$$= (0.1)(4.286)$$
$$= 0.429 \text{ kg/kg fresh solvent}$$

Also calculate the amount of TCE lost in the sludge. First, solve for F_{21}:

$$F_{21} = 1.6F_{13}/3$$
$$= (1.6)(4.286)/3$$
$$= 2.286 \, kg/kg \text{ fresh solvent}$$

Now use the balance around the solvent recovery unit and the relation which gives the amount of solvent recovered:

$$F_{12} = F_{21} + F_{24}$$
$$F_{12} = 1.25F_{21}$$
$$F_{24} = F_{21}(1.25 - 1)$$
$$= (0.25)(2.286)$$
$$= 0.571 \, kg/kg \text{ fresh solvent}$$

For the system without vapor recovery, determine the amount of solvent lost to the atmosphere and the amount of fresh feed per solvent disposed of in the sludge.

$$F_{13}/F_{24} = 0.882/0.118$$
$$= 7.5 \, kg \text{ lost/kg waste solvent}$$
$$F_{01}/F_{24} = 1/0.118$$
$$= 8.5 \, kg \text{ fresh solvent/kg waste solvent}$$

For the system with vapor recovery, determine the amount of solvent lost to the atmosphere and the amount of fresh feed per solvent disposed of in the sludge.

$$F_{35}/F_{24} = 0.429/0.571$$
$$= 0.75 \, kg \text{ lost/kg waste solvent}$$
$$F_{01}/F_{24} = 1/0.571$$
$$= 1.75 \, kg \text{ fresh solvent/kg waste solvent}$$

Note that by employing the vapor recovery system, the amount of solvent lost due to evaporation and fresh solvent requirements are considerably reduced.

TCAA.27 BAKERY EMISSION RATES

As the manager of Butterbottom Bakeries in EI Paso, Texas, which is a small bakery producing 10,000 loaves of various types of bread per day, you have become aware of the new requirements of the 1990 Clean Air Act amendments for small businesses through your review of the law. Based on information you have found in U.S. EPA. 1992, *The Clean Air Act Amendments of 1990, A Guide for Small Business*, Office of Air and Radiation (ANR-443), 450-K-92-001, Washington, D.C., you know you are potentially a Major Source for VOC emissions as you are located in a "Serious" Ozone Non-Attainment Area, subjecting you to more stringent control requirements if you generate more than 50 T/yr VOCs. You are now going to take the next step and attempt to estimate the bakery's emission rates to see if it will be considered a Major Source. You have gathered together the following process information in order to carry out the necessary calculations:

Production rate = 10,000, 2 lb loaves/day   Process type = Sponge dough

Fermentation time = 6 h   Floor and Baking Time = 2 h

Oven exhaust rate = 450 scfm   Baker's Yeast in Bread = 2 wt%

In assembling information regarding responsibilities as a potential emission source, the bakery trade organization, the American Institute of Baking (AIB), was contacted and were very helpful in providing the following emission factor and emission estimate formula for bakeries:

EPA emission factors (lb/ton), bakeries, processing sponge dough, ethyl alcohol

Bread baking	7.55 lb/T product
Bakeries	3.53 lb/T product
Bakeries	2.0 lb/T product
Bakeries	14.07 lb/T product
Bakeries	5.95 lb/T product
Bakeries	5.9 lb/T product
Mean emission factor	6.5 lb/T product

AIB Ethanol emission factor formula

$$\text{Ethanol emission rate, lb/T bread} = [y\,(t)\,(0.444585)] + 0.40425 \qquad (3.15)$$

where y = wt% baker's yeast in bread; t − total fermentation, floor and baking time, h.

Now that all of the data needed for conducting emission rate estimates are compiled, proceed through the calculations and answer the following questions:

1. Estimate the annual ethanol emission rate from this process using EPA's mean emission factors for sponge dough bakeries.
2. Estimate the annual ethanol emission rate from this process using the AIB ethanol emission factor formula.
3. Is this bakery considered a Major Source based on the results of your calculations above? Discuss your answer in terms of what steps must be taken next to ensure compliance.

SOLUTION:

1. The annual ethanol emission rate from your process using the mean EPA emission factors for sponge dough bakeries is carried out by first calculating the total annual production rate of the facility in terms of T (T = 2000 lb)

bread product/yr, then multiplying this value by the mean emission factor of 6.5 lb ethanol/T product.

T product/yr $= (10,000, 2\,\text{lb loaves/day})(365\,\text{days/yr})/(2,000\,\text{lb/T})$

$\quad = (20,000\,\text{lb product/day})(365\,\text{days/yr})/(2,000\,\text{lb/T}) = 3,650\,\text{T product/yr}$

Ethanol emission rate $=$ (Ethanol emission factor)(Production rate)

$\quad = (6.5\,\text{lb ethanol/T product})(3,650\,\text{T product/yr}) = 23,725\,\text{lb ethanol/yr}$

$\quad = 11.9\,\text{T ethanol/yr}$

2. The annual ethanol emission rate from the process using the AIB ethanol emission factor formula is carried out by first plugging the appropriate input data from the problem statement into the AIB ethanol emission factor, also given in the problem statement, to yield an emission factor as follows:

Ethanol emission rate, $\text{lb/T bread} = [(y)(t)(0.444585)] + 0.40425$

where $y = \text{wt\% baker's yeast in bread} = 2\,\text{wt\%}$
$\quad\quad t = \text{total fermentation, floor and baking time} = 6 + 2 = 8\,\text{hr}$
Substituting into this AIB equation yields the following:

Ethanol emission rate, $\text{lb/T bread} = [(2)(8\,\text{hr})(0.444585)] + 0.40425$

$$= 7.5\,\text{lb ethanol/T bread}$$

Using the total bread production rate calculated from Part 2 above, the total annual ethanol emission rate is:

Ethanol emission rate $=$ (Ethanol AIB emission factor)(Production rate)

$$= (7.5\,\text{lb ethanol/T product})(3,650\ \text{T product/yr})$$
$$= 27,375\,\text{lb ethanol/yr}$$
$$= 13.7\,\text{T ethanol/yr}$$

3. Based on the results of both calculations above (both are below 50 T/yr), you are not considered a Major Source in the El Paso Serious Ozone Nonattainment area. You are therefore not going to be regulated as a Major Source, and have no requirements currently imposed upon your process by the Clean Air Act. You are off the hook! ... at least for now!!!

TCAA.28   DRY CLEANING SHOP

A dry cleaning shop in Los Angeles is trying to estimate if it will need to apply Reasonably Available Control Technology (RACT) to its operations. The Clean Air Act threshold for a major source in Los Angeles is 10 T (1 T = 2000 lb) of VOC emissions per year. The shop operates its cleaning facilities 8 hr/d, 6 d/ wk, 51 wk/yr. The shop's new vapor recovery equipment has a 250 gal reservoir and loses an average of 0.70 gal of solvent/hr during operations. The solvent is perchloroethylene, which has a liquid density of 1,483 kg/m^3 (12.37 lb/gal).

1. Does this shop need to apply RACT?
2. If, instead of Los Angeles, this shop were to be located in New York City, would it need to apply RACT? The VOC emission threshold for New York City is 25 T/yr.

SOLUTION: To calculate the shop's hourly mass emissions, the following expression is used:

$$mass\ rate = (volumetric\ solvent\ loss\ rate)\,(solvent\ density)$$
$$= (0.70\,gal/h)\,(12.37\,lb/gal) = 8.66\,lb/h$$

The mass emission rate in lb/d is then:

$$lb/d = (8.66\,lb/h)\,(8\,h/work\ day) = 69.3\,lb/d$$

The mass emission rate is then converted to annual emissions in lb/yr as:

$$lb/yr = (69.3\,lb/d)\,(6\,d/wk)\,(51\,wk/yr) = 21{,}206\,lb/yr$$

The emission rate is then converted to T/yr as:

$$T/yr = (21{,}206\,lb/yr)\,(1\,T/2000\,lb) = 10.6\,T/yr$$

The actual annual emission rate is then compared to the major source threshold for Los Angeles of 10 T/yr, i.e., 10.6 T/yr actual > 10 T/yr. Since the result marginally exceeds the 10 T/yr threshold, the shop must therefore apply RACT.

The major source threshold for New York City is 25 T/yr. Since this shop emits 10.6 T/yr, it is below the 25 T/yr threshold for New York City, and therefore does not need to apply RACT.

As a general rule of thumb, the release of 1 gal/hr of solvent with specific gravity 1.0 is equivalent to approximately 10 T/yr for a 52 week, 8 h work schedule, and 30 T/yr for a 24 hr work schedule. Hence, a source with apparently small hourly volatilization rates can have annual emissions that may be classified as a major source under the Clean Air Act. Careful measurements should be made if a facility's volatilization rates are suspected to be near the threshold to determine if it qualifies as a major source.

The reader should note that, in some respect, this is a trick question. According to CFR 51.100s, perchloroethylene (tetrachloroethylene) is not a VOC. However, there is a MACT for perchloroethylene dry cleaners, as noted in 40 CFR 63.320 (part 63, subpart M).

TCAA.29 BULK TERMINAL APPLICATION

A facility operates a such gasoline terminal under the following conditions;

1. Four gasoline storage tanks with a capacity of 150,000 bbl. It operates 24 hours per day and 7 days per week. There are no controls on the loading racks.

2. Throughput is 250,000 gallons/day.
3. The content of the additive MTBE exceeds 8.5%.
4. The facility has 10 pumps and 115 valves.
5. The four gasoline storage vessels consist of 1 fixed roof tank with no internal controls, 1 external floating roof gasoline storage facility with primary and secondary seals, and two fixed-roof gasoline storage tanks with an internal floating roof.
6. Total emissions of hazardous air pollutants from non-specified sources at the facility such as distillate fuel, additive tanks, wastewater storage/handling tanks, cleaning/degassing of tanks, subsurface recovery (and other remedial actions), service station tank bottoms storage, sample handling/laboratory activities, and pipeline transmix (interface) storage (i.e., transmix with no gasoline content) were estimated to be 3.5 tons per year. These emissions are less than 5 percent of the total emissions of hazardous air pollutants from the entire terminal.
7. Gasoline cargo tanks are in compliance with new source performance standards (see 40 CFR 60.501).

Is the facility subject to the requirements of 40 CFR Part 63 Subpart R? As noted in Problem LCAA.20, the facility is subject to the requirements of 40 CFR Part 63 Subpart R if the emission screening factor (E_T in the equation in the solution) for bulk gasoline terminals exceeds 1.0.

For this application,

$C_F = 1.0$ as the MTBE content exceeds 7.6 percent

$T_F = $ fixed roof tanks $ = 1.0$

$C_E = 0.0$ the fixed roof tank has no vapor controls in place

$T_E = $ external floating roof tanks $ = 0.0$

$T_{ES} = $ external floating roof gasoline storage vessels with primary and secondary seals $ = 1.0$

$T_I = $ fixed-roof gasoline storage vessels with an internal floating roof $ = 20$

$C = $ number of valves, pumps, connectors, loading arm valves, and open-ended lines in gasoline service $ = 125$

$Q = $ gasoline throughput in liters/day $ = 250,000$ gal/day $ = 962,500$ liters/day $ = 962,500$

$K = 4.52 \times 10^{-6}$ as there are no controls on the loading racks.

$OE = $ Total emissions of hazardous air pollutants from non-specified sources $ = 3.5$

SOLUTION: Substituting into Equation (3.1),

$$E_T = C_F[0.59\,(T_F)\,(1 - C_E) + 0.17(T_E) + 0.08(T_{ES}) + 0.038(T_I)$$

$$+ 8.5 \times 10^{-6}(C) + KQ] + 0.04(OE)$$

$$E_T = 1.0[0.59(1)\,(1 - 0.0) + 0.17(0) + 0.08(1.0) + 0.038(2)$$

$$+ 8.5 \times 10^{-6}(125) + (4.52 \times 10^{-6})\,(962,500)] + 0.04(3.5)$$

$$= 1.0[0.59 + 0 + 0.08 + 0.076 + 0.001 + 4.35] + 0.14 = 5.24$$

As this value exceeds 1.0, the bulk gasoline terminal is subject to the requirements of the regulation.

TCAA.30 PIPELINE BREAKOUT STATION APPLICATION

Is the facility described below subject to the requirements of 40 CFR Part 63 Subpart R?

A facility operates a pipeline break out station with the following parameters:

1. Four gasoline storage tanks with a capacity of 150,000 bbl. It operates 24 hours per day and 7 days per week.
2. Throughput is 250,000 gallons/day.
3. The content of the additive MTBE exceeds 8.5%.
4. The facility has 10 pumps and 115 valves.
5. The four gasoline storage vessels consist of 1 fixed roof tank with no internal controls, 1 external floating roof gasoline storage facility with primary and secondary seals, and two fixed-rood gasoline storage tanks with an internal floating roof.
6. Total emissions of hazardous air pollutants from non-specified sources at the facility such as distillate fuel, additive tanks, wastewater storage/ handling tanks, cleaning/degassing of tanks, subsurface recovery (and other remedial actions), service station tank bottoms storage, sample handling/laboratory activities, and pipeline transmix (interface) storage (i.e., transmix with no gasoline content) were estimated to be 3.5 tons per hazardous air pollutants from the entire terminal.

SOLUTION: The facility is subject to the requirements if E_P in the following equation exceeds 1.0. See problem LCAA.21 for additional details

$$E_P = C_F[6.7(T_F)(1 - C_E) + 0.21(T_E) + 0.093(T_{ES}) + 0.1(T_I)$$
$$+ 5.31 \times 10^{-6}(C)] + 0.04(OE) \tag{3.16}$$

The units in equation for E_P are dimensionless.

For this application:

$C_F = 1.0$ as MTBE content exceeds 7.6 percent

T_F = fixed roof tanks = 1

$C_E = 0.0$ as the fixed roof tank has no vapor controls in place

T_E = external floating roof tanks = 0

T_{ES} = external floating roof gasoline storage vessels with primary and secondary seals = 1

T_I = fixed-roof gasoline storage vessels with an internal floating roof = 2

C = number of valves, pumps, connectors, loading arm valves, and open-ended lines in gasoline service = 125

OE = Total emissions of hazardous air pollutants from non-specified sources = 3.5

Substituting above

$$E_P = 1.0[6.7(1)(1 - 0.0) + 0.21(0) + 0.093(1) + 0.1(2) + 5.31 \times 10^{-6}(125)]$$
$$+ 0.04(3.5) = 1.0[6.7 + 0.0 + 0.093 + 0.2 + 0.0007]$$
$$+ 0.14 = 6.99 + 0.14 = 7.13$$

As this value exceeds 1.0, the facility is subject to the requirements of the regulation.

TCAA.31 PLAN REVIEW OF A DIRECT FLAME AFTERBURNER

A regulatory agency engineer must review plans for a permit to construct a direct flame afterburner (Figure 3.7) serving a lithographer. Review is for the purpose of judging whether the proposed system, when operating as it is designed to operate, will meet emission standards. The permit application provides operating and design data. Agency experience has established design criteria that, if met in an operating system, typically ensure compliance with standards. Operating data from the permit application are provided below.

Application = lithography

Effluent exhaust volumetric flowrate = 7000 scfm (60°F, 1 atm)

Exhaust temperature = 300°F

Hydrocarbons in effluent air to afterburner (assume hydrocarbons to be toluene) = 30 lb/h

Afterburner entry temperature of effluent = 738°F

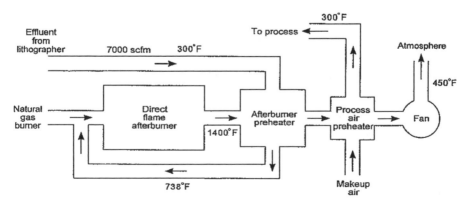

FIGURE 3.7 Direct flame afterburner system.

Afterburner heat loss = 10% in excess of calculated heat load

Afterburner dimensions = 4.2 ft in diameter (D), 14 ft in length (L)

Afterburner temperature = 1300–1500°F

Residence time = 0.3–0.5 s

Afterburner velocity = 20–40 ft/s

Standard Data

Gross heating value of natural gas = 1059 Btu/scf of natural gas

Combustion products per cubic foot of natural gas burned = 11.5 scf/scf natural gas

Available heat of natural gas at 1400°F = 600 Btu/scf of natural gas

Molecular weight of toluene = 92

Average heat capacity (C_{P1}) of effluent gases at 738°F (above 0°F) = 7.12 Btu/(lbmol °F)

Average heat capacity (C_{P2}) of effluent gases at 1400°F (above 0°F) = 7.38 Btu/(lbmol °F)

Volume of air required to combust natural gas = 10.33 scf air/scf natural gas

SOLUTION: The design temperature is already within agency criteria. To determine the fuel requirement for the afterburner, first calculate the total heat load (heating rate), $\dot{Q}$, required to raise 7000 scfm of the effluent stream from 738 to 1400°F in Btu/min:

$$\dot{n} = (7000 \text{ scfm})/(379 \text{ scf/lbmol})$$
$$= 18.47 \text{ lbmol/min}$$

Once again, note that at 60°F and 1 atm, 1.0 mole of an ideal gas occupies 379 ft³.

$$\dot{Q} = \dot{n}[C_{P2}(T_2 - T_b) - C_{P1}(T_1 - T_b)]$$
$$= 18.47[(7.38)(1400 - 0) - (7.12)(738 - 0)]$$
$$= 93,780 \text{ Btu/min}$$

Calculate the actual heat load required, accounting for a 10% heat loss, in Btu/min:

$$\text{Actual heat load} = (1.1)\dot{Q}$$
$$= (1.1)(93,780)$$
$$= 103,200 \text{ Btu/min}$$

Calculate the rate of natural gas required to supply the actual heat required to heat 7000 scfm of the effluent from 738 to 1400°F in scfm:

$$q_{NG} = \dot{Q}/HA$$
$$= (103,200)/(600)$$
$$= 172.0 \text{ scfm}$$

Determine the total volumetric flowrate through the afterburner, q_T, by first calculating the volumetric flowrate of the combustion products of the natural gas, q_1, in scfm:

$$q_1 = q_{NG}\left(11.5 \frac{\text{scf combustion products}}{\text{scf natural gas}}\right)$$
$$= (172.0)(11.5)$$
$$= 1978 \text{ scfm}$$

Also note that the volumetric flowrate of the effluent is 7000 scfm.

The volumetric flowrate of air required to combust the natural gas required, q_2, in scfm is

$$q_2 = q_{NG}\left(10.33 \frac{\text{scf air}}{\text{scf natural gas}}\right)$$
$$= (172.0)(10.33)$$
$$= 1776 \text{ scfm}$$

Calculate the total volumetric flowrate through the afterburner, q_T, in scfm. Since primary air is employed in the combustion of the natural gas, q_2 is not subtracted from q_T. Thus, the q_2 calculation is not required in this solution:

$$q_T = 7000 + q_1$$
$$= 7000 + 1978$$
$$= 8978 \text{ scfm}$$

Apply Charle's law to convert to acfm.
$$= (8978)(1400 + 460)/(60 + 460)$$
$$= 32,110 \text{ acfm}$$

To determine if the afterburner velocity meets the agency criteria, first calculate the cross-sectional area of the afterburner, S, in ft^2.

$$S = \pi D^2/4$$
$$= (\pi)(4.2)^2/4$$
$$= 13.85 \text{ ft}^2$$

The afterburner velocity, v, in ft/s is then

$$\begin{aligned} v &= q_{\mathrm{T}}/S \\ &= (32{,}110)/(13.85) \\ &= 2318 \ \ \text{ft/min} \\ &= 38.6 \text{ ft/s} \end{aligned}$$

Thus, the afterburner velocity is within the agency criterion.

The residence time t is then

$$\begin{aligned} t &= L/v \\ &= 14/38.6 \\ &= 0.363\,\text{s} \end{aligned}$$

This also meets the agency criterion.

TCAA.32 PLAN REVIEW OF A CATALYTIC AFTERBURNER

Plans have been submited for a catalytic afterburner. The installed afterburner is to incinerate a 3000 acfm contaminated gas stream discharged from a direct-fired paint baking oven at 350°F. The following summarizes the data taken from the plant:

Data Sheet

Exhaust flowrate from oven: 3000 acfm

Exhaust gas temperature from oven: 350°F (H at 350°F = 2222 Btu/lbmol)

Solvent emission to afterburner: 0.3 lb/min

Final temperature in afterburner: 1000°F (H at 1000°F = 6984 Btu/lbmol)

Gross heating value of natural gas: 1100 Btu/scf

Total heat requirement: 26,884 Btu/min

Natural gas requirement: 35.0 scfm

Furnace volume: 46.0 ft^3

Exhaust flow rate from afterburner at 1000°F: 6350 cfm

Gas velocity through catalytic bed: 8.6 ft/s

Number of type A 19 inch $\times$24 inch by 3.75 inch deep catalyst elements: 4

The following additional information and rules of thumb may be required to review the plans for the catalytic afterburner:

1. Heat will be recovered from the afterburner effluent, but that process will not be considered in this problem.

2. Catalytic afterburner operating temperatures of approximately 950°F have been found sufficient to control emissions from most process ovens.

3. Preheat burners are usually designed to increase the temperature of the contaminated gases to the required catalyst discharge gas temperature without regard to the heating value of the contaminants (especially if considerable concentration variation occurs).

4. A 10% heat loss is usually a reasonable estimate for an afterburner. This may be accounted for by dividing the calculated heat load by 0.9.

5. The properties of the contaminated effluent may usually be considered identical to those of air.

6. The natural gas is combusted using near stoichiometric (0% excess) external air.

7. The catalyst manufacturer's literature suggests a superficial gas velocity through the face surface of the catalyst element (in this case 19 inch × 24 inch) of 10 ft/s.

Three key regulatory questions are to be considered:

1. Is the operating temperature adequate for efficiency control?

2. Is the fuel requirement adequate to maintain the operating temperature?

3. Is the catalyst section properly sized?

SOLUTION: The plans indicate a combustion temperature of 1000°F; this is acceptable when compared to a 950°F rule-of-thumb temperature.

To determine if the fuel requirement is adequate to maintain the operating temperature, first calculate the lbmol/min of gas to be heated from 350 to 1000°F.

$$(3000 \text{ acfm})[(460 + 60)/(460 + 350)] = 1926 \text{ scfm}$$

$$1926/379 = 5.08 \text{ lbmol/min}$$

Also determine the heat requirement in Btu/min to raise the gas stream (air) temperature from 350 to 1000°F using the enthalpy (H) data provided.

$$H \text{ at } 350°F = 2222 \text{ Btu/lbmol}$$

$$H \text{ at } 1000°F = 6984 \text{ Btu/lbmol}$$
$$\dot{Q} = \dot{n}\Delta H$$
$$= (5.08)(6984 - 2222)$$
$$= 24,196 \text{ Btu/min}$$

The total heat requirement, $\dot{Q}_T$, in Btu/min is then (taking into account the heat loss)

$$\dot{Q}_T = \dot{Q}/0.9$$
$$= 24{,}196/0.9$$
$$= 26{,}884 \, \text{Btu/min}$$

The available heat of the natural gas in Btu/scf at 1000°F may be calculated from the following equation:

$$HA_{1000}/HV_G = (HA_{1000}/HV_G)_{reffuel}$$
$$HA_{1000} = 1100(745/1059)$$
$$= 774 \, \text{Btu/scf natural gas}$$

The natural gas (NG) requirement in scfm may be calculated from the last two results:

$$\text{Natural gas requirement} = \frac{\text{Total heat requirement}}{\text{Available heat in fuel}}$$
$$q_{NG} = \dot{Q}_T/HA$$
$$= 26{,}884/774$$
$$= 34.7 \, \text{scfm}$$

Thus, the natural gas requirement agrees with the given design data value.

To determine if the catalyst section is sized properly, first calculate the volume of flue products at 1000°F from the combustion of the natural gas:

$$q_c = (11.45)(35)[(460 + 1000)/(460 + 60)]$$
$$= 1125 \, \text{acfm}$$

Also calculate the volume of contaminated gases at 1000°F by applying Charles' law:

$$q = (3000)[(460 + 1000)/(460 + 350)]$$
$$= 5407 \, \text{acfm}$$

The total volumetric gas rate at 1000°F is then

$$q_T = q_c + q$$
$$= 1125 + 5407$$
$$= 6532 \, \text{acfm}$$

Calculate the number of $19 \times 24 \times 3.75$ inch catalyst elements, N, required:

$$N = (6532)(144)/[(19)(24)(10)(60)]$$
$$= 3.44$$

The catalyst section is sized properly. It is seen that four elements have been specified; this is a conservative design allowing a slightly slower gas flow through the elements. With four elements, the superficial velocity is reduced to 8.6 ft/s. This too is in agreement with the design specification.

TCAA.33   SIZING A PACKED TOWER WITH NO DATA

Qualitatively outline how one can size (diameter, height) a packed tower to achieve a given degree of separation without any information on the physical and chemical properties of a gas to be cleaned.

To calculate the height, one needs both the height of a gas transfer unit, H_{OG}, and the number of gas transfer units, N_{OG}. Since equilibrium data are not available, assume m (slope of equilibrium curve) approaches zero. This is not an unreasonable assumption for most solvents that preferentially absorb (or react with) the solute. For this condition:

$$N_{OG} = \ln(y_1/y_2) \tag{3.17}$$

where y_1 and y_2 represent inlet and outlet concentrations, respectively.

Since it is reasonable to assume the scrubbing medium to be water or a solvent that effectively has the physical and chemical properties of water, H_{OG} can be assigned values usually encountered for water systems. These are given in Table 3.15.

For plastic packing, the liquid and gas flowrates are both typically in the range of $1500-2000$ lb/(h·ft²) of cross-sectional area. (*Note:* The "flow rates" are more properly termed "fluxes"; a *flux* is a flow rate per unit cross-sectional area with units of lb/h·ft² in this problem. However, it is common practice to use the term *rate*.) For ceramic packing, the range of flowrates is $500-1000$ lb/(h·ft²). For

TABLE 3.15   Packing Data

Packing Diameter (in.)	Plastic Packing H_{OG} (ft)	Ceramic Packing H_{OG} (ft)
1.0	1.0	2.0
1.5	1.25	2.5
2.0	1.5	3.0
3.0	2.25	4.5
3.5	2.75	5.5

difficult-to-absorb gases, the gas flowrate is usually lower and the liquid flowrate higher. Superficial gas velocities (velocity of the gas if the column is empty) are in the 3–6 ft/s range. The height Z may be calculated from

$$z = (H_{OG})(N_{OG})(SF) \tag{3.18}$$

where SF is a safety factor—the value of which can range from 1.25 to 1.5. Pressure drops can vary from 0.15 to 0.40 inH$_2$O/ft packing. Packing size increases with increasing tower diameter.

The reader should note that the above design procedure and predictive equations provide only approximate answers. In many instances, it can be used to check plan review calculations that are often required in state permits. More rigorous approaches are available in the literature (see J. Reynolds, J. Jeris, and L. Theodore, *Handbook of Chemical and Environmental Engineering Calculations*, John Wiley & Sons, Hoboken, NJ, 2004).

SOLUTION: The reader is left the exercise of verifying Tables 3.16 and 3.17 below. A sample calculation is presented for the first chart.

Consider, as an example, a removal efficiency of 86.5% using 1.5-inch diameter plastic packing. Using a basis of 100 ppm for the inlet concentration, y_1, the outlet concentration, y_2, is obtained from

$$0.865 = \frac{100 - y_2}{100}$$

$$y_2 = 13.3 \text{ ppm}$$

The N_{OG} is then

$$N_{OG} = \ln(100/13.3)$$
$$= 2.02$$

TABLE 3.16 H_{OG} for Plastic Packing

| Removal Efficiency (%) | Plastic Packing Size (inch) | | | | |
	1.0	1.5	2.0	3.0	3.5
63.2	1.0	1.25	1.5	2.25	2.75
77.7	1.5	1.9	2.25	3.4	4.1
86.5	2.0	2.5	3.0	4.5	5.5
90	2.3	3.0	3.45	5.25	6.25
95	3.0	3.75	4.5	6.75	8.2
98	3.9	4.9	5.9	8.8	10.75
99	4.6	5.75	6.9	10.4	12.7
99.5	5.3	6.6	8.0	11.9	14.6
99.9	6.9	8.6	10.4	15.5	19.0
99.99	9.2	11.5	13.8	20.7	25.3

TABLE 3.17 H_{OG} for Ceramic Packing

Removal Efficiency (%)	Ceramic Packing Size (inch)				
	1.0	1.5	2.0	3.0	3.5
63.2	2.0	2.5	3.0	4.5	5.5
77.7	3.0	3.7	4.5	6.75	8.25
86.5	4.0	5.0	6.0	9.0	11.0
90	4.6	5.75	6.9	10.4	12.7
95	6.0	7.5	9.0	13.5	16.5
98	7.8	9.8	11.7	17.6	21.5
99	9.2	11.5	13.8	20.7	25.3
99.5	10.6	13.25	15.9	23.8	29.1
99.9	13.8	17.25	20.7	31.1	38.0
99.99	18.4	23.0	27.6	41.4	50.7

From Table 3.15 in the problem statement, for 1.5-inch diameter plastic packing, H_{OG} is 2.5 ft. The required packing height, Z(ft) is

$$Z = (2.5)(2.02)$$
$$= 5.0 \text{ ft; safety factor not included}$$

For ceramic packing, see Table 3.17.

An equation for estimating the cross-sectional area of the tower, S, in terms of the gas volumetric flow rate, q, in acfs (actual cubic feet per second) is

$$S(\text{ft}^2) = q(\text{acfs})/4$$

An equation to estimate the tower packing pressure drop, ΔP, in terms of Z is

$$\Delta P \text{ (in. } H_2O) = (0.2)Z \quad Z = \text{ft}$$

The following packing size(s) is (are) recommended:

$$\text{For } D \approx 3 \text{ ft, use 1-inch packing}$$
$$\text{For } D < 3 \text{ ft, use } <1 \text{ inch packing}$$
$$\text{For } D > 3 \text{ ft, use } >1 \text{ inch packing}$$

Additional details are available in the following references: L. Theodore, "Engineering Calculations: Sizing Packed-Tower Absorbers With No Data," CEP, New York City, May, 2005, and Reynolds et al., "Handbook of Chemical and Environmental Engineering Calculations," John Wiley & Sons, Hoboken, NJ, 2004.

TCAA.34 PACKED TOWER ABSORBER DESIGN WITH NO DATA

A 1600-acfm gas stream is to be treated in a packed tower containing ceramic packing. The gas stream contains 100 ppm of a toxic pollutant that is to be reduced to 1 ppm. Estimate the tower's cross-sectional area, diameter, height, pressure drop, and packing size. Use the procedure outlined in Problem TCAA.33.

SOLUTION: Key information from Problem TCAA.33 is provided below. Table 3.17 applies for ceramic packing.

The equation for the cross-sectional area of the tower, S, in terms of the gas volumetric flowrate, q, in acfs and the velocity v, in ft/s, is

$$S(\text{ft}^2) = q(\text{acfs})/4; \quad \text{v} = 4\,\text{ft/s} \tag{3.19}$$

An equation to estimate the tower packing pressure drop, ΔP, in terms of Z is

$$\Delta P\,(\text{in } H_2O) = (0.2)Z \quad Z = \text{ft} \tag{3.20}$$

The following packing size(s) is (are) recommended:

$$\begin{aligned}
&\text{For } D \approx 3 \text{ ft,} \quad \text{use 1-inch packing} \\
&\text{For } D < 3 \text{ ft,} \quad \text{use} <1\text{-inch packing} \\
&\text{For } D > 3 \text{ ft,} \quad \text{use} >1\text{-inch packing}
\end{aligned} \tag{3.21}$$

As a rule, recommended packing size increases with tower diameter.
    For the problem at hand,

$$S = 1600/4 = 400 \text{ ft}^2$$

The diameter D is

$$\begin{aligned}
D &= (4S/\pi)^{0.5} \\
&= [(4)(400)/\pi]^{0.5} \\
&= 22.6 \text{ ft}
\end{aligned}$$

For a tower this large, the 3.5-inch packing should be used.
    The removal efficiency (RE) is

$$RE = (100 - 1)/100 = 0.99 = 99\%$$

For 99% RE and a packing size of 3.5 inch, the required height is (see Table 3.17) 25.3 ft. Note that this is the packing height. The tower height can often be estimated by adding 10 ft to the packing height.

The pressure drop is

$$\Delta P = (0.2)(25.3) = 5.06 \, \text{inH}_2\text{O}$$

TCAA.35   CALCULATIONS ON AN ADSORPTION CANISTER

Small volatile organic compound (VOC) emission sources often use activated carbon, which is available in canisters or drums. An example of this is a modified form of Carbtrol model G-1, which is suitable for low air flow rates. The drum is not regenerated on site; it is returned to the manufacturer and a new drum is delivered.

A small pilot-scale reactor uses this modified model G-1 adsorber to capture methylene chloride (MEC) emissions in a 50 acfm nitrogen purge source. The following operating and design data are provided:

Volumetric flowrate of nitrogen purge = 50 acfm
Molecular weight of methylene chloride = 85
Operating temperature = 70°F
Operating pressure = 1.0 atm
Saturation capacity = 0.30 lb CH_2Cl_2/lb C
Working charge (operating capacity), WC = 0.2875 lb CH_2Cl_2/lb C
Methylene chloride concentration = 500 ppm
Weight of carbon in drum = 200 lb
Height of adsorbent in drum = 24 inch
Adsorption time = 6 h
Mass transfer zone (MTZ) = 2 inch

Based on the above data and information, estimate the number of purge stream batches that this G-1 model adsorber canister can treat to breakthrough.

SOLUTION: The drum capacity (DC) in pounds of methylene chloride is

$$
\begin{aligned}
\text{DC} &= (\text{WC}) \, (\text{carbon weight}) \\
&= (0.2875)(200) \\
&= 57.5 \, \text{lb MeCl}_2
\end{aligned}
$$

Determine the mole fraction of methylene chloride (MEC) in the purge stream:

$$
\begin{aligned}
y_{\text{MEC}} &= 500/10^6 \\
&= 0.0005
\end{aligned}
$$

The volumetric flowrate of MEC may now be calculated:

$$q_{MEC} = y_{MEC}q$$
$$= (0.0005)(50 \text{ acfm})(60 \text{ min/h})$$
$$= 1.5 \text{ acfh}$$

The density of the methylene chloride vapor in lb/ft^3 is

$$\rho_{MEC} = \frac{P(MW)}{RT}; \ R = 0.73 \text{ atm} \cdot \text{ft}^3/\text{lbmol} \cdot {}^{\circ}\text{R}$$
$$= \frac{(1)(85)}{(0.73)(70 + 460)} \tag{3.22}$$
$$= 0.220 \text{ lb/ft}^3$$

The weight W of methylene chloride emitted per batch is then

$$W = (q_{MEC}) \, (\rho_{MEC}) \, (6 \text{ h/batch})$$
$$= (1.5) \, (0.220) \, (6)$$
$$= 1.98 \text{ lb/batch}$$

The maximum number of batches per canister (MBC) is

$$\text{MBC} = \text{DC}/W$$
$$= 57.5/1.98$$
$$= 29$$

TCAA.36 ATMOSPHERIC DISCHARGE CALCULATION

Regulatory engineers have been requested to determine the minimum distance downstream from a toxic contaminated cement source emitting dust that will be free of cement deposit. The source is equipped with a cyclone. The cyclone is located 150 ft above ground level. Assume ambient conditions are at 60°F and 1 atm and neglect meteorological aspects. Additional data are given below.

$$\text{Particle size range of cement dust} = 2.5 - 50.0 \, \mu\text{m}$$
$$\text{Specific gravity of the cement dust} = 1.96$$
$$\text{Wind speed} = 3.0 \text{ miles/h}$$

Assume Stokes' Law applies for the calculation of the particle's terminal settling velocity.

SOLUTION: A particle diameter of 2.5 μm is used to calculate the minimum distance downstream free of dust since the smallest particle will travel the greatest horizontal distance. First calculate the particle density using the specific gravity given and determine the properties of the gas (assume air).

$$\rho_p = (SG)\,(\rho_{H_2O})$$

$$= (1.96)\,(62.4)$$

$$= 122.3\ \text{lb/ft}^3$$

$$\rho = P(MW)/RT \tag{3.23}$$

$$= (1)\,(29)/[(0.73)\,(60 + 460)]$$

$$= 0.0764\ \text{lb/ft}^3$$

Viscosity of air, μ, at $60°F = 1.22 \times 10^{-5}$ lb/(ft · s)
The terminal settling velocity in feet per second is given by Stoke's Law.

$$v = \frac{g d_p^2 \rho_p}{18\,\mu} \tag{3.24}$$

$$= \frac{(32.2)[(2.5)/(25,400)(12)]^2 (122.3)}{(18)(1.22 \times 10^{-5})}$$

$$= 1.21 \times 10^{-3}\ \text{ft/s}$$

The approximate time for descent is

$$t = h/v \tag{3.25}$$

$$= 150/1.21 \times 10^{-3}$$

$$= 1.24 \times 10^5\ \text{s}$$

Thus, the distance traveled, d, is

$$d = tu \tag{3.26}$$

$$= (1.24 \times 10^5)(3.0/3600)$$

$$= 103.3\ \text{miles}$$

TCAA.37 DISPERSION OF NANOPARTICLES

A plant manufacturing potentially toxic nanoparticles explodes one windy day. It disperses 100 tons of agglomated particles (specific gravity $= 0.8$) into the atmosphere (70°F, $\rho = 0.0752$ lb/ft^3). If the wind is blowing 20 miles/h from the west and the particles range in diameter of 2.1–1000 μm, calculate the distances from the plant where the particles will start to deposit and where they will cease to deposit. Assume the particles are blown vertically 400 ft in the air before they start to settle. Also, assuming even ground-level distribution through an average 100 ft wide path of settling, calculate the average height of the particles on the ground in the settling area. Assume the bulk density of the particles equal to half of the actual density. Also assume that Stokes' equation applies for smallest particle and the Intermediate Law equation applies for the settling velocity of the largest particle.

SOLUTION: The smallest particle will travel the greatest distance while the largest will travel the least distance. For the minimum distance, use the largest particle.

$$d_p = 1000 \, \mu m = 3.28 \times 10^{-3} \, ft$$

Use the intermediate range settling equation provided in Equation (3.24).

$$v = 0.153 \frac{g^{0.71} d_p^{1.14} \rho_p^{0.71}}{\mu^{0.43} \rho^{0.29}} \tag{3.27}$$

$$= 0.153 \frac{(32.2)^{0.71}(3.28 \times 10^{-3})^{1.14}[(0.8)(62.4)]^{0.71}}{(1.18 \times 10^{-5})^{0.43}(0.0752)^{0.29}}$$

$$= 11.9 \, ft/s$$

The descent time t is

$$t = H/v$$
$$= 400/11.9 = 33.6 \, s \tag{3.28}$$

The horizontal distance traveled, L_1, is

$$L_1 = (33.6)\left(\frac{20}{(60)\,(60)}\right)(5280) \tag{3.29}$$

$$= 986 \, ft$$

For the maximum distance, use the smallest particle:

$$d_p = 2.1 \, \mu m = 6.89 \times 10^{-6} \, ft$$

The velocity v is in the Stokes regime and is given by

$$v = \frac{d_p^2 g \rho_p}{18 \mu}$$

$$= \frac{(32.2)\ (6.89 \times 10^{-6})^2 (0.8)(62.4)}{(18)\ (1.18 \times 10^{-5})} \qquad (3.30)$$

$$= 3.59 \times 10^{-4}\ \text{ft/s}$$

The descent time t is

$$t = H/v = 400/3.59 \times 10^{-4}$$

$$= 1.11 \times 10^6\ \text{s} \qquad (3.31)$$

The horizontal distance traveled, L_2, is

$$L_2 = (t)(\text{wind velocity})$$

$$= (1.11 \times 10^6)\left(\frac{20}{3600}\right)(5280)$$

$$= 3.26 \times 10^7\ \text{ft}$$

To calculate the depth D, the volume of particles (actual), V_{act}, is first determined:

$$V_{\text{act}} = (m_p)(\rho_p); \quad m_p = \text{mass of operation}$$

$$= (100)\ (2000)/[(0.8)(62.4)]$$

$$= 4006\ \text{ft}^3$$

The bulk volume V_B is

$$V_B = V_{\text{act}}/\rho_B$$

$$= 4006/0.5 = 8012\ \text{ft}^3$$

The length of the drop area, L_D, is

$$L_D = L_2 - L_1$$

$$= 3.2 \times 10^7 - 994$$

$$= 3.2 \times 10^7\ \text{ft}$$

Since the width is 100 ft, the deposition area A is

$$A = (3.2 \times 10^7)(100) = 3.2 \times 10^9\ \text{ft}^2$$

Therefore,

$$V_B = AD; \quad D = \text{deposition height}$$

$$8012 = (3.2 \times 10^9)D \tag{3.32}$$

Therefore,

$$D = 2.5 \times 10^{-6} \, \text{ft}$$

$$= 0.76 \, \mu\text{m}$$

The deposition of these particles can be, at best, described as a "sprinkling."

TCAA.38 INSTANTANEOUS "PUFF" MODEL

A rather significant amount of data and information is available for sources that emit continuously to the atmosphere. Unfortunately, less is available on instantaneous or "puff" sources. Other than computer models that are not suitable for classroom and/or illustrative example calculations, only Turner's *Workbook of Atmospheric Dispersion Estimates*, USEPA Publication No. AP-26, Research Triangle Park, NC, 1970 provides an equation that may be used for estimation purposes. Cases of instantaneous releases, as from an explosion, or short-term releases on the order of seconds, are also often of practical concern.

To determine concentrations at any position downwind, one must consider both the time interval after the time of release and diffusion in the downwind direction, as well as lateral and vertical diffusion. Of considerable importance, but very difficult to determine, is the path or trajectory of the puff. This is most important if concentrations are to be determined at specific points. Determining the trajectory is of less importance if knowledge of the magnitude of the concentration for particular downwind distances or travel times is required without the need to know exactly at what points these concentrations occur.

An equation that may be used for estimates of concentration downwind from an instantaneous release from height, H, is

$$C(x,y,0,H) = \left[\frac{2m_T}{(2\pi)^{1.5}\sigma_x\sigma_y\sigma_z}\right]\left\{\exp\left[-0.5\left(\frac{x-ut}{\sigma_x}\right)^2\right]\right\}$$

$$\times \left\{\exp\left[-0.5\left(\frac{H}{\sigma_z}\right)^2\right]\right\}\left\{\exp\left[-0.5\left(\frac{y}{\sigma_y}\right)^2\right]\right\} \tag{3.33}$$

where m_T = total mass of the release
  u = wind speed
  t = time after the release
  x = distance in the x direction
  y = distance in the y direction
  $\sigma_x, \sigma_y, \sigma_z$ = dispersion efficients in the x, y and z direction, respectively.

The dispersion coefficients above are not necessarily those evaluated with respect to the dispersion of a continuous source at a fixed point in space. This equation can be simplified for centerline concentrations and ground-level emissions by setting $y = 0$ and $H = 0$, respectively. The dispersion coefficients in the above equation refer to dispersion statistics following the motion of the expanding puff. The σ_x is the standard deviation of the concentration distribution in the puff in the downwind direction, and t is the time after release. Note that there is essentially no dilution in the downwind direction by wind speed. The speed of the wind mainly serves to give the downwind position of the center of the puff, as shown by examination of the exponential term involving σ_x. In general, one should expect the σ_x value to be about the same as σ_y.

Unless another model is available for treating instantaneous sources, it is recommended that the above equation be employed. The use of appropriate values of σ for this equation is not clear cut. Consider the following problem.

A 20-m high tank in a plant containing a toxic gas suddenly explodes. The explosion causes an emission of 400 g/s for 3 min. A school is located 400 m west and 50 m south of the plant. If the wind velocity is 3.5 m/s from the east, how many seconds after the explosion will the concentration reach a maximum in the school?

SOLUTION: The time at which the maximum concentration will occur at the school is

$$t = (\text{downward distance/wind speed})$$
$$= (400\,\text{m}/3.5\,\text{m/s})$$
$$= 114\,\text{s}$$
$$\approx 2\,\text{min}$$

This problem will be revisited in Chapter 9, Problem TOSHA.15.

TCAA.39   GASOLINE SERVICE STATIONS APPLICATION

A gasoline service station has an average monthly throughput of approximately 30,000 gallons of gasoline. The station is equipped with a stage I submerged vapor balance system. There are no stage II controls. Based on the above information and the USEPA document AP-42 table given in Table 3.18, estimate the monthly emissions in lb/month.

SOLUTION:

1. Obtain the emission rate for balanced submerged filling in $\text{lb}/10^3$ gal.

$$\text{emission rate} = 0.3\,\text{lb}/10^3\,\text{gal}$$

TABLE 3.18 Hydrocarbon Emissions from Gasoline Service Station Operations

	Emission Rate	
Emission Source	lb/10^3 gal throughput	kg/10^3 liters throughput
Filling underground tank		
Submerged filling	7.3	0.88
Splash filling	11.5	1.38
Balance submerged filling	0.3	0.04
Underground tank breathing and emptying	1	0.12
Vehicle refueling operations		
Displacement losses (uncontrolled)	9	1.08
Displacement losses (controlled)	0.9	0.11
Spillage	0.7	0.084

Emissions include any vapor loss from the underground tank to the gas pump.

2. Calculate the emission rate for underground tank filling, E_1, in lb/month.

$$E_1 = (0.3 \, \text{lb}/1000 \, \text{gal}) \, (30{,}000 \, \text{gal/month})$$
$$= 9 \, \text{lb/month}$$

3. Calculate the emission rate from underground breathing and emptying, E_2.

$$E_2 = (1.0 \, \text{lb}/1000 \, \text{gal}) \, (30{,}000 \, \text{gal/month})$$
$$= 30 \, \text{lb/month}$$

4. Calculate the emission rare from vehicle refueling, E_3, with no stage II controls.

$$E_3 = (9.0/1000) \, (30{,}000)$$
$$= 270 \, \text{lb/month}$$

5. Calculate the emission rate from spillage, E_4.

$$E_4 = (0.7/1000) \, (30{,}000)$$
$$= 21 \, \text{lb/month}$$

6. Calculate the total emission rate.

$$E = E_1 + E_2 + E_3 + E_4$$
$$= 9 + 30 + 270 + 21$$
$$= 330 \, \text{lb/month}$$

TCAA.40 STRIPPING OF ETHYLENE OXIDE

Following the absorption of ethylene oxide by water from a process stream, a 600-lbmol/h water stream (prior to discharge) is steam-stripped of the ethylene oxide as part of the regeneration step. For a feed ethylene oxide concentration of 0.5 mol%, estimate (as part of a permit review) the actual amount of steam required for stripping ethylene oxide to a concentration of 0.03 mol%.

The liquid flowrate is 600 lbmol/h and 40% excess ethylene oxide free steam is required for the separation in a packed column with 1-in. Raschig rings.

The system is at 30 psia and uses saturated steam at system conditions. Assume Henry's law applies and $y = 20x$ as the ethylene oxide equilibrium data for the system.

SOLUTION: The reader is referred to the literature (J. Reynolds, J. Jeris, and L. Theodore "Handbook of Chemical and Environmental Engineering Calculations" John Wiley & Sons, Hoboken, NJ 2002 for technical details on the calculations that follow.

For $x_2 = 0.005$, $y_2 = 0.10$, the minimum liquid to gas ratio is

$$\frac{G_{min}}{L_{min}} = \frac{x_{a1} - x_{a2}}{y_{a1} - y_{a2}} \qquad (3.34)$$

$$= \frac{0.0003 - 0.005}{0 - 0.10}$$

$$= 0.047$$

For 40% excess

$$\frac{G}{L} = (0.047)(1.4)$$

$$= 0.0658$$

For stripping,

$$G_{min} = G_{actual}$$

Thus,

$$G_{actual} = (0.0658)(600\,\text{lbmol/h})(18\,\text{lb/lbmol})$$
$$= 710.64\,\text{lb/h of steam}$$

CHAPTER 4

CLEAN WATER ACT (CWA)

QUALITATIVE PROBLEMS (LCWA)

PROBLEMS LCWA.1–28

LCWA.1 CLEAN WATER ACT HISTORY

Provide the history of the Clean Water Act (CWA).

SOLUTION: Growing public awareness and concern for controlling water pollution led to enactment of the Federal Water Pollution Control Act Amendments of 1972. As amended in 1977, this law became commonly known as the Clean Water Act. The Act established the basic structure for regulating discharges of pollutants into the waters of the United States. It gave EPA the authority to implement pollution control programs such as setting wastewater standards for industry. The Clean Water Act also continued to set water quality standards for all contaminants in surface waters. The Act made it unlawful for any person to discharge any pollutant from a point source into navigable waters unless a permit was obtained under its provisions. It also funded the construction of sewage treatment plants under the construction grants program and recognized the need for planning to address the critical problems posed by nonpoint source pollution.

Subsequent enactments modified some of the earlier Clean Water Act provisions. Revisions in 1981 streamlined the municipal construction grants process, improving the capabilities of treatment plants built under the program. Changes in 1987 phased

Environmental Regulatory Calculations Handbook, by Leo Stander and Louis Theodore
Copyright © 2008 John Wiley & Sons, Inc.

out the construction grants program, replacing it with the State Water Pollution Control Revolving Fund, more commonly known as the Clean Water State Revolving Fund. This new funding strategy addressed water quality needs by building on EPA-State partnerships.

Over the years, many other laws have changed parts of the Clean Water Act. Title 1 of the Great Lakes Critical Programs Act of 1990, put into place parts of the Great Lakes Water Quality Agreement of 1978, signed by the U.S. and Canada, where the two nations agreed to reduce certain toxic pollutants in the Great Lakes. That law required EPA to establish water quality criteria for the Great Lakes, addressing 29 toxic pollutants with maximum levels that are safe for humans, wildlife, and aquatic life. It also required EPA to help the States implement the criteria on a specific schedule.

See also: The U.S. EPA. 2006. Laws and Regulations. http://www.epa.gov/region5/cwa.htm

LCWA.2   THE CLEAN WATER ACT

Describe the Clean Water Act (CWA).

SOLUTION: The CWA is the cornerstone of surface water quality protection in the United States. (The Act does not deal directly with ground water nor with water quality issues.) The statute employs a variety of regulatory tools to sharply reduce direct pollutant discharges into waterways, finance municipal wastewater treatment facilities, and manage polluted runoff. Those tools are employed to achieve the broader goal of restoring and maintaining the chemical, physical, and biological integrity of the nation's waters so that they can support "the protection and propagation of fish, shellfish, and wildlife and recreation in and on the water."

For many years following the passage of CWA in 1972, EPA, states, and Native American tribes focused mainly on the chemical aspects of the "integrity" goal. During the last decade, however, more attention has been given to physical and biological integrity. Also, in the early decades of the Act's implementation, efforts focused on regulating discharges from traditional "point source" facilities, such as municipal sewage plants and industrial facilities, with little attention paid to runoff from streets, construction sites, farms, and other "wet-weather" sources.

Starting in the late 1980s, efforts to address polluted runoff have increased significantly. For "nonpoint" runoff, voluntary programs, including cost-sharing with landowners are the key tool. For "wet weather point sources," like urban storm sewer systems and construction sites, a regulatory approach is being employed.

Evolution of CWA programs over the last decade has also included something of a shift from a program-by-program, source-by-source, pollutant-by-pollutant approach to more holistic watershed-based strategies. Under the watershed approach equal emphasis is placed on protecting healthy waters and restoring impaired ones. A full array of issues are addressed, not just those subject to CWA regulatory authority. Involvement of stakeholder groups in the development and implementation of strategies for achieving and maintaining state water quality and other environmental goals is another hallmark of this approach.

See also: U.S. EPA. 2006. Introduction to the Clean Water Act. http://www.epa.gov/watertrain/cwa/

LCWA.3 WATER QUALITY STANDARDS (WQS)

Detail Water Quality Standards (WQS).

SOLUTION: WQS are aimed at translating the broad goals of the CWA into waterbody-specific objectives. Ideally, WQS should be expressed in terms that allow quantifiable measurement. WQS, like the CWA overall, apply only to the waters of the United States. As defined in the CWA, "waters of the United States" apply only to surface waters–rivers, lakes, estuaries, coastal waters, and wetlands. Not all surface waters are legally "waters of the United States." Generally, however, those waters include the following:

1. All interstate waters
2. Intrastate waters used in interstate and/or foreign commerce
3. Tributaries of the above
4. Territorial seas at the cyclical high tide mark
5. Wetlands adjacent to all the above

The exact dividing line between "waters of the United States" according to the CWA and other waters can be hard to determine, especially with regard to smaller streams, ephemeral waterbodies, and wetlands not adjacent to other "waters of the United States." In fact, the delineation changes from time to time, as new court rulings are handed down, new regulations are issued, or the Act itself is modified.

Designated uses, water quality criteria, and an antidegradation policy constitute the three major components of Water Quality Standards Program.

The designated uses (DUs) of a waterbody are those uses that society, through various units of government, determines should be attained in the waterbody. The DUs are the goals set for the waterbody. In some cases, these uses have already been attained, but sometimes conditions in a waterbody do not support all the DUs.

Water quality criteria (WQC) are descriptions of the conditions in a waterbody necessary to support the DUs. These can be expressed as concentrations of pollutants, temperature, pH, turbidity units, toxicity units, or other quantitative measures. WQC can also be narrative statements such as "no toxic chemicals in toxic amounts."

Antidegradation policies are a component of state/tribal WQS that establish a set of rules that should be followed when addressing proposed activities that could lower the quality of high quality waters, i.e., those with conditions that exceed those necessary to meet the designated uses.

See also: U.S. EPA. 2006. Introduction to WQS. http://www.epa.gov/watertrain/cwa/cwa2.htm

LCWA.4   WATER QUALITY CRITERIA (WQC)

Detail Water Quality Criteria (WQC)

SOLUTION: WQC are levels of individual pollutants or water quality characteristics, or descriptions of conditions of a waterbody that, if met, will generally protect the designated use of the water. For a given DU, there are likely to be a number of criteria dealing with different types of conditions, as well as levels of specific chemicals. Since most waterbodies have multiple DUs, the number of WQC applicable to a given waterbody can be very substantial.

Water quality criteria must be scientifically consistent with the attainment of DUs. This means that only scientific considerations can be taken into account when determining what water quality conditions are consistently meeting a given DU. Economic and social impacts are not considered when developing WQC.

WQC can be divided up for descriptive purposes in many ways. For instance, numeric criteria (weekly average of 5 mg/L dissolved oxygen) can be contrasted with narrative criteria (no putrescent bottom deposits). Criteria can also be categorized according to what portion of the aquatic system they can be applied to: the water itself (water column), the bottom sediments, or the bodies of aquatic organisms (fish tissue). The duration of time to which they apply is another way of dividing WQC, with those dealing with short-term exposures (acute) being distinguished from those addressing long-term exposure (chronic).

Criteria can also be distinguished according to the types of organisms they are designed to protect. Aquatic life criteria are aimed at protecting entire communities of aquatic organisms, including a wide array of animals and various plants and microorganisms. These can be expressed as parameter specific (daily average of 30 μg/L of copper) or in terms of various "metrics" that directly measure numbers, weight, and diversity of plants and animals in a waterbody (community indices).

Human health criteria can apply to two exposure routes: (1) drinking water and (2) consuming aquatic foodstuffs.

Wildlife criteria, like human health/fish consumption criteria, deal with the effects of pollutants with high bioaccumulation factors. To date, EPA has issued and/or adopted fewer wildlife criteria than aquatic life or human health criteria species. Such criteria are designed to protect terrestrial animals that feed upon aquatic species. Examples are ospreys, herons and other wading birds, mink and otters.

Numeric criteria are usually parameter specific-they express conditions for specific measures, such as dissolved oxygen, temperature, turbidity, nitrogen, phosphorus, heavy metals such as mercury and cadmium, and synthetic organic chemicals like dioxin and PCBs. They do not consist merely of stated levels/concentrations, such as 15 μg/L or a pH above 5.0. They should also specify the span of time over which conditions must be met. This is the "duration" component of a WQC. Combining the concentration/magnitude and duration components of a WQC results in wording such as "the average 4-day concentration of pollutant X shall not exceed 50 μg/L."

A numeric WQC should also indicate how often it would be acceptable to go beyond specified concentration/duration combinations. This is often called the frequency or the recurrance interval component of the WQC. For instance, for protection of aquatic life, as a general rule, EPA recommends a recurrance interval of once in 3 years. The purpose of the recurrance interval is to recognize that aquatic ecosystem can recover from impacts of exposure to harmful conditions, but to make such conditions sufficiently rare as to keep the community of aquatic organism from being in a constant state of recovery.

Simply because one sample has exceeded the concentration component of a WQC does not necessarily mean the WQC has been violated and a designated use affected. This is true only in the case of "instantaneous criteria" levels that are never to be exceeded. But if there was a criterion of 50 mg/L of "X," for a 7-day average, then having one sample at a concentration above 50 mg/L would not "prove" that this criterion had actually been exceeded. Likewise, having just one or two samples below 50 mg/L is not a good basis for concluding a waterbody is indeed meeting WQS.

See also: U.S. EPA. Water Quality Criteria. http://www.epa.gov/watertrain/cwa/cwa10.htm

LCWA.5   TOTAL MAXIMUM DAILY LOADS (TMDLs)

Introduce the TMDL Strategy.

SOLUTION: If monitoring and assessment indicate that a waterbody or segment is impaired by one or more pollutants, and it is therefore placed on the 303(d) list, then the relevant entity (state, territory, or authorized tribe) is required to develop a strategy that would lead to attainment of the WQS.

The CWA requires that Total Maximum Daily Loads (TMDLs) be developed only for waters affected by pollutants where implementation of the technology-based controls imposed upon point sources by the CWA and EPA regulations would not result in achievement of WQS. At this point in the history of the CWA, most point sources have been issued NPDES permits with technology-based discharge limits. In addition, a substantial fraction of point sources also have more stringent water quality-based permit limits. But because nonpoint sources are major contributors of pollutant loads to many waterbodies, even these more stringent limits on point sources have not resulted in attainment of WQS.

Strategies to address impaired waters must consist of a TMDL or another comprehensive strategy that includes a functional equivalent of a TMDL. In essence, TMDLs are "pollutant budgets" for a specific waterbody or segment, that if not exceeded, would result in attainment of WQS.

TMDLs are required for "pollutants," but not for forms of "pollution." Pollutants include clean sediments, nutrients (nitrogen and phosphorus), pathogens, acids/bases, heat, metals, cyanide, and synthetic organic chemicals. As noted previously, pollution includes not only all pollutants but also flow alterations and physical habitat modifications.

At least one TMDL must be performed for every waterbody or segment impaired by one or more pollutants. TMDLs are performed pollutant by pollutant, although if a waterbody or segment were impaired by two or more pollutants, the TMDLs for each pollutant could be performed simultaneously.

EPA is encouraging states, tribes, and territories to do TMDLs on a "watershed basis" (e.g., to "bundle" TMDLs together) in order to realize program efficiencies and foster more holistic analysis. Ideally, TMDLs would be incorporated into comprehensive watershed strategies. Such strategies would address the protection of high quality waters (antidegradation) as well as restoration of impaired segments. They would also address the full array of activities affecting the waterbody. Finally, such strategies would be the product of collaborative efforts between a wide variety of stakeholders.

TMDLs must be submitted to EPA for review and approval/disapproval. If EPA ultimately decides that it cannot approve a TMDL that has been submitted, the Agency would need to develop and promulgate what it considers to be an acceptable TMDL. Doing so requires going through the formal federal rulemaking process.

LCWA.6 TMDL DETAILS

Provide details on the TMDL.

SOLUTION: The first element of a TMDL is "the allowable load," also referred to as pollutant "cap." It is basically a budget for a particular pollutant in a particular body of water, or an expression of the "carrying capacity." This is the loading rate that would be consistent with meeting the WQC for the pollutant in question. The cap is usually derived through use of mathematical models, probably computer based.

The CWA requires that all TMDLs include a safety factor as an extra measure of environmental protection, taking into account uncertainties associated with estimating the acceptable cap or load. This is referred to as the margin of safety (MOS).

Once the cap has been set (with the MOS factored in), the next step is to allocate that total pollutant load among various sources of the pollutant for which the TMDL has been done. This is in essence the "slicing of the pie."

TMDLs set loading caps for individual pollutants such as clean sediments, nitrogen, phosphorus, coliform bacteria, temperature, copper, mercury, and PCBs. Indicators of a group of forms of pollution can also be used, such as biochemical oxygen demand (BOD), which is often used when doing TMDLs for waterbodies with low dissolved oxygen. (Again, TMDLs are not required for non pollutant forms of "pollution," such as streamflow patterns and stream channel modification.) States, territories, and authorized tribes are free to develop TMDLs for such pollutants, as they see fit. The CWA and EPA regulations put no limits on these other government entities going beyond what the Act requires.

Though the CWA itself uses the term Total Maximum Daily Loads, EPA has determined that loadings rates (caps) can be expressed as weekly, monthly, or even yearly loads. Which time period to use depends on the type of pollutant for which the TMDL is being performed. Toxic chemicals that exhibit acute effects would probably call for daily or weekly loads, whereas nutrients and sediments could be expressed as monthly or yearly loading rates.

The CWA allows for seasonal TMDLs, i.e., it allows different rates of loading at different times of the year. For example, colder waters can absorb more oxygen-demanding substances than can warm water, so allowable loading could be higher in the winter than in the summer.

EPA regulations use the terms Wasteload Allocations (WLA) and Load Allocations (LA) to describe loading assigned to point and nonpoint sources, respectively.

Generally, point sources that are required to have individual NPDES permits are also required to be assigned individual WLAs. On the other hand, a group of sources covered under a "general" NPDES permit would be assigned one collective WLA.

Although load allocations should ideally be assigned to individual nonpoint sources, this is often not practical or even scientifically feasible; hence, loads can be assigned to categories of nonpoint sources (all soybean fields in the watershed, for example), or to geographic groupings of nonpoint sources (all in a particular subwatershed).

Even though the CWA provides no federal authority for requiring nonpoint sources to reduce their loadings of pollutants to the nation's waters, the Act does require states (and authorized territories and tribes) to develop TMDLs for waters where nonpoint sources are significant sources of pollutants. TMDLs do not create any new federal regulatory authority over any type of sources. Rather, with regard to nonpoint sources, TMDLs are simply a source of information that, for a given waterbody, should answer such questions as the following:

1. Are nonpoint sources a significant contributor of pollutants to this impaired waterbody?
2. What are the approximate total current loads of impairment–causing pollutants from all nonpoint sources in the watershed?
3. What fraction of total loads of the pollutant(s) of concern come from nonpoint sources vs. point sources?

4. What are the approximate loading from the major categories of nonpoint sources in the watershed?

5. How much do loads from nonpoint sources need to be reduced in order to achieve the water quality standards for the waterbody?

6. What kinds of management measures and practices would need to be applied to various types of nonpoint sources in order to achieve the needed load reductions?

LCWA.7   TMDL MISCONCEPTION

Explain TMDLs common misconception.

SOLUTION: A common misconception about TMDLs is that EPA has issued regulations specifying how the pollutant cap in a TMDL should be allocated among sources: equal reductions for all or equal loadings from each, for example. EPA has no such regulations. States, territories, and tribes are free to allocate to sources in any way they see fit, so long as the sum of all the allocations is no greater than the overall loading cap. However, when thinking about changing the share of allowed loads among sources, it is important to realize that in all but very small waterbody segments, load location matters. In many cases, the farther away from the zone of impact that a loading enters into the waterbody system, the less of an effect that load will have on the impaired zone. For example, studies of large watersheds, such as Long Island Sound (local to one of the authors), have indicated that one pound of pollutant (nitrogen in the case of the Sound) discharged close to the impaired zone has the same impact on that zone as 10 pounds discharged substantially farther away. Furthermore, even after accounting for locations-related relative impacts on a particular segment or zone, care must be taken to ensure that localized exceedences of WQS do not result from moving loads from one tributary/segment to another.

See also: U.S. EPA. 2006. TMDLs. http://www.epa.gov/watertrain/cwa/cwa29.htm

LCWA.8   REGULATORY EXPLANATIONS

Answer the following questions:

1. Identify key legislation directed at controlling wastewater discharges from municipalities. Briefly define the priorities of each legislation.

2. What is non-point-source (NPS) pollution? Why is NPS pollution difficult to deal with?

3. Highlight the difference between a "direct discharger" and an "indirect discharger."

4. What does the acronym NPDES stand for? What are the three major parameters regulated under the NPDES program for municipal wastewater discharges and what are the maximum concentrations allowed for each of these parameters in NPDES permits?

SOLUTION:

1. The *Clean Water Act* (Federal Water Pollution Act of 1972): created water quality standards for point discharges, established timelines and penalties, and regulated toxic discharges.

   The *Marine Protection, Research and Sanctuaries Act*: prohibits sewage discharge by ocean barge dumping.

   The *Water Quality Act of 1987*: strengthened federal water quality regulations, regulated toxic discharges, expanded treatment to nonpoint sources, and established new stormwater restrictions.

2. NPS pollution is described as pollution that enters water supplies over a large geographical area from a variety of human activities and surface runoff. NPS pollution is difficult to deal with because it is widespread and difficult to isolate. NPS pollution appears often in surges that are associated with periods of rainfall and snowmelt runoff.

3. A direct discharger is an industrial plant that discharges its effluent wastewater directly to a surrounding water source with no intermediate means of treatment. An indirect discharger is a plant that first discharges to a *publicly owned treatment works* (POTW) facility prior to release to the environment.

4. As noted several times earlier, NPDES stands for the *National Pollutant Discharge Elimination System*, a permit program established for each point source discharge in the United States under the Clean Water Act. The three major parameters regulated by the NPDES permit program for municipal wastewater treatment plants and their regulated maximum concentrations are shown Table 4.1.

TABLE 4.1 Effluent BOD$_5$ and TSS Concentrations for a Municipal Wastewater Treatment Plant as Defined by the NPDES Permit Program

	30-day Average Concentration	7-day Average Concentration
5-day BOD (BOD$_5$)[a]	30 mg/L	45 mg/L
TSS[b]	30 mg/L	45 mg/L
pH[c]	6 to 9	6 to 9

[a]See problem LCWA 10 and 11 for the definition of BOD.
[b]TSS = total suspended solids.
[c]pH = negative log$_{10}$ of hydrogen ion concentration in gmole/liter.

LCWA.9   WASTEWATER CHARACTERISTICS

Answer the following three questions:

1. List the physical characteristics important in establishing the quality of municipal wastewaters.
2. The characteristics of a municipal wastewater can change drastically over time. Why is this true?
3. List the three categories of industrial wastewater treatment. Provide examples from each category.

SOLUTION:

1. Numerous wastewater physical characteristics include:
   a. Temperature
   b. Color
   c. Odor
   d. Turbidity
   e. Suspended solids
   f. Volatile suspended solids
2. Municipal wastewater characteristics are dependent on a number of factors that include solids, wastewater volume, and chemical content. The factors can change rapidly in response to daily and seasonal variations, precipitation, industrial discharges, and public habits. Combinations of these factors can even result in significant hourly fluctuations in wastewater characteristics.
3. Wastewater treatment categories include:
   a. Physical treatment: clarification, flotation
   b. Chemical treatment: coagulation/precipitation/flocculation, neutralization
   c. Biological treatment: aerobic suspended growth processes, fixed film processes, aerated lagoons, anaerobic treatment processes

These are treated in separate problems later in this chapter.

LCWA.10   KEY REGULATORY WASTEWATER CHARACTERISTICS

List and describe some key wastewater characteristics that appear in the regulations.

SOLUTION: Key characteristics include:

Suspended Solid (SS): A measure of solids that are suspended (not dissolved) in the wastewater. Suspended solids will lower the amount of light and oxygen in a body of wastewater. Over time, suspended solids will settle from the wastewater, blanketing the floor with a layer of sediment called *sludge.*

Biochemical Oxygen Demand (BOD): Primarily the level of organic content in a wastewater measured by the demand for oxygen that can be consumed by living organisms in the wastewater. Wastewater with high BOD content is characterized by low oxygen content and high biological activity.

Floating Materials: Materials seen on the surface of wastewater that indicate the presence of insoluble fats, oils, greases, and other immiscible materials such as wood, paper, plastics, etc.

Color: Reveals the presence of dissolved material in wastewater. Color in wastewater may originate from dyes, decaying organics, etc.

The three major biodegradable organics in wastewater are composed principally of proteins, carbohydrates, and fats. If discharged untreated to the environment, their biological stabilization can lead to the depletion of natural oxygen resources, and to the development of septic conditions in rivers and other natural bodies of water. Other key characteristics include:

Pathogens: Pathogenic organisms that can transmit communicable diseases via wastewater. Typical infectious diseases reported are cholera, typhoid, paratyphoid fever, salmonellosis, and shigellosis. *E. coli* and fecal coli are indicators of pathogens.

Nutrients: Carbon is the major nutrient source while nitrogen and phosphorus are secondary. When discharged to the receiving water, these nutrients can lead to the growth of undesirable aquatic life. When discharged in excessive amounts on land, they can also lead to the pollution of groundwater.

Priority Pollutants: Organic and inorganic compounds designated on the basis of their known or suspected carcinogenicity, mutagenicity, or high acute toxicity. Many of these compounds are found in wastewater.

Heavy Metals: Heavy metals are usually added to wastewater from municipal, commercial and industrial activities, and may have to be removed if the wastewater is to be reused or discharged into a water body.

Additional details are provided below.

Floating solids and liquids oils, greases, and other materials that float on the surface, if not treated, cannot only make the river unsightly but also obstruct passage of light and clarity through the water.

Color contributed by textile and paper mills, tanneries, slaughterhouses, and other industries is an indicator of pollution. Compounds present in wastewaters absorb certain wavelength of light and reflect the remainder, a fact generally accepted that accounts for color development of streams. Color interferes with the transmission of sunlight into the stream and therefore lessens photosynthetic actions.

Regarding general definitions, municipal wastewater is composed of a mixture of dissolved and particulate organic and inorganic materials, and infections disease-causing bacteria. The total amount of each component accumulated in wastewater is referred to as the *mass loading* and is given units of pounds per day (lb/day). The concentration, given in pounds per gallon of water (lb/gal), of any individual component entering a wastewater treatment plant can change as a result of the activities that are producing this waste. The units used to express any concentration, e.g., lb/gal, can also be converted into other terms such as grams/liter, mg/L, ppm, or even μg/L, a term used frequently with very toxic substances. The concentration of each individual component, while in the treatment plant, is usually reduced significantly by the time it reaches the end of the plant prior to discharge.

Municipal wastewater normally contains approximately 99.9% water. The remaining materials (as described earlier) include suspended and dissolved organic and inorganic matter as well as microorganisms. These materials make up the physical, chemical, and biological qualities that are characteristic of municipal and industrial waters. Each of these three qualities are briefly described below.

Wastewater characteristics depend largely on the mass loading rates flowing from the various sources in the collection system. The flow in combined sewers is a composite of domestic and industrial wastewaters, infiltration into the sewer from cracks and leaks in the system, and flow from sanitary sewers. During wet weather, the addition of rainfall collected from the combined sewer system and the storm drain collection system (combined sewer overflow systems) can significantly change the characteristics of wastewater due to the increased flow carried by the sewer to the treatment plant. The peak flowrate can be two to three times the average dry (or sunny) weather flowrate and a portion of it may be bypassed to the receiving water. The mass loading rate into the plant also varies cyclically throughout the day with the usual peak occurring in the afternoon and the low in the early morning hours. The impact of flowrate is an important determining factor in the design and operation of wastewater treatment plant facilities. The records kept by the treatment plant should include the minimum, average, and maximum flow values (gallons per unit time) on a hourly, daily, weekly, and monthly basis for both wet and dry weather conditions. A moving 7-day daily average flow, and mass loading rate entering the plant and at various locations throughout the plant can then be computed from this record. This intricate form of record keeping of all factors affecting a wastewater treatment plant must be considered to assess the wastewater flow and variations of wastewater strength in order to operate a facility correctly. The parameters used to indicate the total mass loading in the wastewater entering the treatment plant are the measurements of total suspended solids (TSS), and total dissolved solids (TDS). The parameters used to indicate the organic and inorganic chemical concentration in the wastewater are the measurements of the biochemical oxygen demand (BOD) and the chemical oxygen demand (COD). Both BOD and COD are discussed in more detail later in this section. Additionally, the total nutrients (carbon, nitrogen, and phosphorus), any toxic chemicals, and trace metals are also characterized for the plant design so that the treatment processes can successfully treat the varying waste loads.

A wastewater plant can handle a large volume of flow if its units are properly designed. Unfortunately, most wastewater plants are already in operation when a request comes to accept the flow of waste from some new industrial concern or expanding community.

Finally, other harmful constituents in industrial waste can cause problems. Some problem areas and corresponding effects are:

1. Toxic metal ions that cause toxicity to biological oxidation.
2. Feathers that clog nozzles, overload digesters, and impede proper pump operation.
3. Rags that clog pumps and valves and interfere with proper operation.
4. Acids and alkalis that may corrode pipes, pumps, and treatment units, interfere with settling, upset the biological purification of sewage, release odors, and intensify color.
5. Flammables that cause fires and may lead to explosions.
6. Fat that clogs nozzles and pumps and overloads digesters.
7. Noxious gases that present a direct danger to workers.
8. Detergents that cause foaming.
9. Phenols and other toxic organic material.

For additional details, refer to: G. Burke, R. Singh, and L. Theodore, *Handbook of Environmental Management and Technology*, 2nd ed., Wiley-Interscience, NYC, 2001.

LCWA.11 BIOCHEMICAL OXYGEN DEMAND (BOD) AND CHEMICAL OXYGEN DEMAND (COD)

Discuss the difference between BOD and COD.

SOLUTION: BOD and COD are two important water quality parameters in wastewater engineering. The accurate measurement of these parameters is essential in the proper design of wastewater treatment systems and in the study of the transport and fate of contaminants in the aquatic environment.

Biochemical oxygen demand, or BOD, is the quantity of dissolved oxygen required to biochemically stabilize substrate materials in water. BOD is a measure of the oxygen demand of wastewater because it provides an approximate amount of oxygen needed in aerobic biological treatment. BOD is a measure of treated wastewater quality because its discharge into a natural receiving water exerts oxygen demand and is therefore a critical factor in the viability of the aquatic system's ecology. The EPA and state environmental regulatory agencies routinely require monitoring of BOD in all municipal and industrial discharges.

BOD is generally measured in a three-step process. First, the sample is quantitatively diluted so that there is an appropriate concentration of oxidizable substrate and dissolved oxygen in the sample volume to be measured. For well-treated or "polished" wastewaters (5–30 mg/L BOD), dilution is small or may not be necessary at all. For raw, untreated wastewaters (100–400 mg/L BOD), a high level of dilution is necessary. Second, the sample is inoculated with microorganisms (seed), sealed, and incubated in a controlled temperature environment for a set period of time. During that time, dissolved oxygen (DO) and substrate are consumed through microbial activity. Third, DO is measured, typically once at the start of incubation and again at the end. The BOD is equal to the difference in DO concentration after it has been adjusted for the dilution factor. There is also typically an adjustment factor necessary to compensate for extra BOD introduced to the sample with the inocculating culture material (seed).

BOD is most commonly expressed as "five-day BOD" (BOD_5) and ultimate BOD (BODU). BOD_5 is the oxygen demand that is exerted after a standard period of 5 days of incubation and is the value required by regulatory agencies. BODU represents the oxygen demand that would be exerted if incubation was allowed to occur long enough for virtually all the biologically oxidizable substrate to be consumed. Often, BODU is approximated by allowing incubation to occur for a standard period of 20 days, or longer.

Typically, BOD_5 is equal to approximately two-thirds of the BODU. Although this ratio is a good approximation for typical domestic wastewaters, it can also vary significantly depending on the nature of the wastewater source. Lower ratios are typical of industrial wastewaters with a lower relative composition of readily oxidizable substrate. For example, such a wastewater typically might contain higher colloidal or suspended solids substrates (such as petroleum refinery or paper and pulp mill effluents) or higher nitrogenous substrates (such as meat processing effluents). Higher ratios are indicative of industrial wastewaters with a higher relative composition of readily oxidizable soluble substrate (such as brewery or carbohydrate food processing wastewaters). Substrates such as these are often referred to by wastewater engineers as "jellybeans" because they are readily consumed during treatment by the process microorganisms.

BOD is often distinguished by the type of oxidizable substrate that is consumed during the incubation period. Carbonaceous BOD (CBOD) represents the oxygen demand exerted by the carbon-based components of the substrate. Nitrogenous BOD (NBOD) represents oxygen demand exerted by the process of nitrification, or the oxidation of ammonia to nitrite and nitrate. To measure CBOD only, an inhibitory chemical is added to the sample to stop the nitrification portion of oxidation from occurring during the incubation period. To measure NBOD, it is necessary to measure both total BOD (TBOD) and CBOD in samples that have incubated 5 or more days. NBOD is calculated as the difference between total BOD and CBOD. Thus,

$$\text{Total BOD} = \text{CBOD} + \text{NBOD} \tag{4.1}$$

$$\text{TBOD (without inhibitor)} - \text{CBOD (with inhibitor)} = \text{NBOD} \tag{4.2}$$

Since CBOD is the primary substrate consumed during the first 5 days of incubation, BOD_5 is approximately equal to $CBOD_5$, except in highly treated waters. The measurement of NBOD with unacclimated organisms typically requires longer incubation periods (such as 20 days) because nitrifying organisms are slower growing and do not start consuming oxygen to a measurable degree until well after carbonaceous oxidation has begun. This "delay" is generally considered to be approximately 8–10 days in raw municipal wastewater.

Chemical oxygen demand (COD) is equal to the equivalent oxygen concentration required to chemically oxidize substrate materials in water. Because the test can be completed in a little over 2 hours, it very quickly provides the concentration of the wastewater chemical oxygen demand. It also is an indicator of nonbiodegradable organics. To chemically oxidize all the substrate in wastewater, a strong chemical oxidant (such as dichromate in sulfuric acid) is used as the standard oxidant and a catalyst. The COD is reported as the dissolved oxygen concentration equivalent to the decrease in the acidic dichromate concentration. The level of oxidation provided by this reagent is generally sufficient to oxidize almost all of the oxidizable substrates in water. Since there are additional (albeit less commonly used) chemical oxidants other than acidic dichromate, which can serve as COD standard oxidants, COD measured using dichromate is often referred to as "dichromate COD."

COD is an important water quality parameter because for similar wastewaters it often correlates well with CBOD. COD often serves as a reliable measure of the CBOD once the wastewater being examined has been characterized. It is a faster, less expensive analysis than CBOD, and is therefore used where frequent, routine process control analyses are required. A wastewater that is well suited to biological treatment generally has a CBODU concentration approximately equal to the COD concentration. In effect, the bulk of the substrate that can be oxidized ultimately by process microoganisms is also oxidized by a strong chemical oxidant. CBODU approximately equals COD for typical, primary-treated domestic sewage. Industrial wastewater, or domestic sewage with a significant industrial component, will often have CBODU/COD ratios less than 1 (unity) if they contain substrates that are chemically reactive, but biologically refractive. This type of wastewater is typical of several industries. Chemical oxygen demand is of direct significance in industrial treatment when there are relatively few biologically oxidizable components in the wastewater. COD then can be used to directly monitor those components.

Usually, regulatory agencies will not accept COD concentrations for permit limits, but COD/BOD relationships can be very useful in predicting effluent BOD_5 values from COD values. COD measurements provide immediate knowledge of a process efficiency, whereas BOD requires five days.

The theoretical chemical (or ultimate) oxygen demand, or ThCOD (BODU), can be calculated for a wastewater with relatively few oxidizable components. The procedure requires one to calculate the equivalent oxygen concentration necessary for the complete stoichiometric oxidation of each of the wastewater components to its corresponding highest oxidation state and sum them for the total wastewater ThCOD. Carbonaceous components of wastewater are generally considered to be fully oxidized to carbon dioxide. Typically, a laboratory measured COD is very close or equal to ThCOD.

Calculation details on all of the above terms are provided in the TCWA Problems.

LCWA.12 CLEAN WATER ACT AND PRIORITY WATER POLLUTANTS (PWPS)

Describe the interrelationship between the *Clean Water Act* (CWA) and *priority water pollutants* (PWPs).

SOLUTION: The Clean Water Act addresses a large number of issues related to water pollution management, including the control of industrial wastewaters. Any municipality or industry that discharges wastewater in the United States must obtain a discharge permit under the regulations set forth by the aforementioned *National Pollutant Discharge Elimination System* (NPDES). Under this system, there are three classes of pollutants (conventional pollutants, priority pollutants, and nonconventional/nonpriority pollutants). Conventional pollutants are substances such as biochemical oxygen demand (BOD), suspended solids (SS), pH, oil and grease, and coliforms. Priority pollutants were so designated on a list of 129 substances originally set forth in a consent decree between the Environmental Protection Agency and several environmental organizations. This list was incorporated into the 1977 amendments to the Clean Air Act and has since been reduced to 126 substances. Most of the substances on this list are organics, but it does include most of the heavy metals. These substances are generally considered to be toxic. However, the toxicity is not absolute; it primarily depends on the concentration. In recent years, pollution prevention programs have been implemented to reduce their use in industrial processes. *Regulatory Chemicals Handbook*, by Spero et al., Marcel Dekker (recently acquired by CRC Press/Taylor & Francis Group, Boca Raton, FL), New York City, 2000 provides extensive details on these priority (water) pollutants. The third class of pollutants could include any pollutant not in the first two categories. Examples of substances that are presently regulated in the third category are nitrogen, phosphorus, and sodium.

LCWA.13 PRIORITY WATER POLLUTANTS (PWPS) LIST

Provide a list of the priority water pollutants.

SOLUTION: The list in Table 4.2 below was adapted from J. Spero et al., "Regulatory Chemicals Handbook," Marcel Dekker (recently acquired by CRC Press/Taylor & Francis Group, Boca Raton, FL), New York City, 2000.

LCWA.14 PWP DEFINITIONS

With regard to priority water pollutants (PWPs), briefly describe the following four terms:

1. Biological Properties
2. Bioaccumulation
3. Probable Fate
4. Treatability/Removability

TABLE 4.2 Priority Water Pollutants (PWPs)

Chemical or Trade Name	CAS No.	Chemical or Trade Name	CAS No.
Acenaphthene	83-32-9	2-Chlorophenol	7005-72-3
Acenaphylene	208-96-8	4-Chlorophenyl-	7440-47-3
Acrolein	107-02-8	phenylether	
Acrylonitrile	107-13-1	Chromium	218-01-9
Aldrin	309-00-2	Chrysene	7440-50-8
Anthracene	120-12-7	Cyanide	57-12-5
Antimony	7440-36-0	4-4′-DDD	75-54-8
Arsenic	7440-38-2	4-4′-DDE	72-55-9
Asbestos	1332-21-4	4-4′-DDT	50-29-3
Benzene	71-43-2	Dibenz(a,h)Anthracene	53-70-3
Benzidine	92-87-5	Dibutylphthalate	84-74-2
Benzo(a)Anthracene	56-55-3	1,2-Dichlorobenzene	95-50-1
Benzo(a)Pyrene	50-32-8	1,3-Dichlorobenzene	541-73-1
Benzo(g,h,I)Perylene	191-24-2	1,4-Dichlorobenzene	106-46-7
3,4-Benzofluoranthene	205-99-2	3,3-Dichlorobenzidene	91-94-1
Benzo(k)Fluoranthene	207-08-9	Dichlorobromomethane	75-27-4
Beryllium	7440-41-7	1,1-Dichloroethane	75-34-3
Alpha-BHC	319-84-6	1,2-Dichloroethane	107-06-2
Beta-BHC	319-85-7	1,1-Dichloroethylene	75-35-4
Delta-BHC	319-86-8	1,2-trans-	156-60-5
Gamma-BHC	58-89-9	Dichloroethylene	
bis(2-chloroethoxy)	111-91-1	2,4-Dichlorophenol	120-83-2
Methane		1,2-Dichloropropane	78-87-5
bis(2-chloroethyl)	111-44-4	1,3-Dichloropropylene	542-75-6
Ether		Di-n-Octyl Phthalate	117-84-0
bis(2-chloroisopropyl)	108-60-1	Dieldrin	60-57-1
ether		Diethylphthalate	84-55-2
bis(2-ethylhexyl)	117-81-7	2,4-Dimethylphenol	105-67-9
Phthalate		4,6-Dinitro-o-cresol	534-52-1
Bromoform	75-25-2	2,4-Dinitrophenol	51-28-5
4-Bromophenyl-	101-55-3	2,4-Dinitrotoluene	121-14-2
phenylether		2,6-Dinitrotoluene	606-20-2
Butylbenzylphthalate	85-68-7	Dimethyl Phthalate	131-11-3
Cadmiun	7440-43-9	1,2-Diphenylhydrazine	122-66-7
Carbon Tetrachloride	56-23-5	Endosulfan I	959-98-8
Chlordane	57-74-9	Endosulfan II	33213-65-9
Chlorobenzene	108-90-7	Endosulfan sulfanate	1031-07-8
Chlorodibromomethane	124-48-1	Endrin	72-20-8
Chloroethane	75-00-3	Endrin aldehyde	7421-93-4
2-Chloroethylvinylether	67-66-3	Ethylbenzene	100-41-4
Chloroform	74-87-3	Fluoranthene	206-44-0
Chloromethane	91-58-7	Fluorene	86-73-7
2-Chloronaphthalene	95-57-8	Heptachlor	76-44-8

(*Continued*)

TABLE 4.2 *Continued*

Chemical or Trade Name	CAS No.	Chemical or Trade Name	CAS No.
Heptachlor epoxide	1024-57-3	PCB-1242	53469-21-9
Hexachlorobenzene	118-74-1	PCB-1248	12672-29-6
Hexachlorobutadiene	8-768-3	PCB-1254	11097-69-1
Hexachloro-	77-47-4	PCB-1260	11096-82-5
cyclopentadiene		Pentachlorophenol	87-86-5
Hexachloroethane	67-72-1	Phenanthrene	85-01-8
Indeno(1,2,3-cd)	193-39-5	Phenol	108-95-2
Pyrene		Pyrene	129-00-0
Isophorone	78-59-1	Selenium	778249-2
Lead	7439-92-1	Silver	7440-22-4
Mercury	7439-97-6	2,3,7,8-	1746-01-6
Methyl bromide	74-83-9	Tetrachlorodibenzo-	
Methylene Chloride	75-09-2	p-dioxin	
N-Nitroso-Di-n-	621-64-7	1,1,2,2-	
propylamine		Tetrachloroethane	79-34-5
N-Nitrosodimethylamine	62-75-9	Tetrachloroethene	127-18-4
N-Nitrosodiphenylamine	86-30-6	Thallium	7440-28-0
Naphthalene	91-20-3	Toluene	108-88-3
Nickel	7440-02-0	Toxaphene	8001-35-2
Nitrobenzene	98-95-3	1,2,4-Trichlorobenzene	120-82-1
2-Nitrophenol	88-75-5	1,1,1-Trichloroethane	71-55-6
4-Nitrophenol	100-02-7	1,1,2-Trichloroethane	79-00-5
para-Chloro-meta-cresol	59-50-7	Trichloroethene	79-01-6
PCB-1016	12674-11-2	2,4,6-Trichlorophenol	88-06-2
PCB-1221	11104-28-2	Vinyl Chloride	75-01-4
PCB-1232	11141-16-5	Zinc	7440-66-6

1. Biological Properties

These are properties listed for the Priority Water Pollutants. This section includes descriptive information on the biodegradability of the chemical, including information on biological systems used, measurement methods, removals achieved, and rate of removal. Biodegradation is the degradation of chemicals via biological pathways. Microorganisms such as bacteria and fungus account for the majority of the biodegradation; however, some plants have also been identified that biodegrade chemicals. This section also includes the method by which a Priority Water Pollutant can be detected in water; these are primarily EPA methods.

2. Bioaccumulation

Some of the regulated priority water pollutants concentrate in biological tissues or in the fatty tissue of organisms. Bioaccumulation refers to the fact that the chemical is not broken down by the metabolism of an organism and it tends to remain in the body tissue for a long time (bioconcentrate). The log of the Bioconcentration Factor (log BCF) is used to describe the ratio between the concentration of a chemical in biota living in the river (e.g., fish) and the concentration in the surrounding water. The Bioconcentration Factor (BCF) can be estimated from Kow or the water solubility of the chemical.

3. Probable Fate

This source group is separated into several fate processes. They are: photolysis, oxidation, hydrolysis, volatilization, sorption, biological processes, and in some cases, other reactions/interactions. Photolysis is the chemical reaction either directly or indirectly mediated by light (photons). An example is the dechlorination of chlorinated organics. Oxidation involves the removal of electrons. Hydrolysis is the addition of water molecules to outer shells of compounds. Hydrolysis can destabilize chemicals and can cause them to form insoluble precipitates. Volatilization is the transfer from (aqueous) solutions to the atmosphere (vapors). Adsorption is the physical and/or chemical process in which a substance is accumulated at an interface between phases (e.g., solid to liquid). Biological processes include biodegradation, bioaccumulation, and biomagnification. Biodegradation and bioaccumulation were discussed in the previous Problems. Biomagnification involves increasing concentrations up the food chain.

4. Treatability/Removability

For each alternative standard treatment process for the Priority Water Pollutants, removal ranges and achievable concentrations for each priority water pollutant are present in actual wastewater samples and in some cases in synthetic wastewater samples, where appropriate data are available.

LCWA.15 IMPACT OF THE CLEAN AIR ACT ON WASTEWATER TREATMENT PLANTS

Discuss the impact of the Clean Air Act on wastewater treatment plants.

SOLUTION: A wide range of residential, commercial, and industrial dischargers contribute VOC and toxic pollutants to publicly owned treatment works (POTWs). An even wider range of pollutants is potentially discharged to industrial wastewater treatment plants, depending on the specific type of industrial activity generating the wastewater. Limited information on air emissions of VOCs and air toxics from industrial wastewater plants and POTWs is currently available. However, more extensive information on treatment plant wastewater influent quality is more readily available in the literature, particularly for POTWs, due to the monitoring requirements for all wastewater treatment plants under the NPDES Program. In addition to those pollutants commonly present in the influent of

POTWs, byproducts of various wastewater treatment processes considered to be VOCs or toxic air pollutants can be potentially emitted from POTWs. For example, chloroform is usually formed as a byproduct of wastewater chlorination.

The provisions of the Clean Air Act dealing with the reduction of VOCs for reducing urban smog and the control of air toxics significantly affect the water pollution control field in the areas of water quality and wastewater treatment. Although the Clean Air Act does not specifically require that industrial or municipal wastewater treatment plants control VOC and air toxics emissions, federal, state, and local air quality laws and regulations, developed as a result of the Clean Air Act, focus on stationary sources of ozone precursors. State air pollution control agencies in ozone nonattainment areas, particularly areas classified as extreme and severe, may require that large wastewater treatment plants install air pollution control devices.

In addition, it is possible that some of the large POTWs may fall into the category of a major source (i.e., emitting greater than 25 tons/year) if all 189 pollutants are included in the calculations. In Los Angeles, the only extreme ozone nonattainment area, the local air quality agency adopted Rule 1401, which regulates emissions of known or suspected carcinogenic air toxics based on a cancer risk assessment. Local POTWs are covered under this rule.

LCWA.16 BIOLOGICAL TREATMENT

Briefly describe biological processes that are discussed in the regulations.

SOLUTION: Biological processes also involve chemical reactions, but are differentiated from the chemical category in that these reactions take place in or around microorganisms. The most common use of biological processes in waste treatment is for the decomposition of organic compounds.

The different biological processes described below are activated sludge, aerated lagoons, anaerobic digestion, composting, enzyme treatment, trickling filter, and waste stabilization ponds. All of these, except enzyme treatment, use microorganisms to decompose the waste. Enzyme treatment generally involves extracting the enzyme from the microorganism and using it to catalyze a particular reaction. With proper control, these processes are reliable and environmentally sound; additive chemicals are usually not needed and operational expenses are relatively low.

Activated Sludge

The *activated sludge* process uses microorganisms to decompose organics in aqueous waste streams. The microorganisms take the organics into the cell, through the cell wall, and into the cytoplasm where the organics are broken down by enzyme oxidation and hydrolysis to produce energy and other cellular material. Besides taking in the organics as food, the biomass acts as a filter to collect colloidal matter and suspended solids. Volatile organics can be driven off somewhat by the aeration process. Some metals are collected in the organisms and in the sludge.

Aerated Lagoon

The *aerated lagoon* is an earthen basin that is artificially aerated; the basin is generally lined to make it impermeable. The microbial reactions are the same as those that take place in the activated sludge process except the biological sludge is not recycled. In this process microorganisms are used to decompose organics in waste streams containing <1% solids; solid contents >1% usually result in the settling of some solid matter because of limited mixing and aeration. These solids often undergo anaerobic microbial decomposition.

Anaerobic Digestion

In *anaerobic digestion*, microorganisms that do not require oxygen for respiration are employed. This process is useful for the degradation of simple organics. The microbiology is not well understood, but essentially the cells use part of the organic compounds for cell growth and part is converted to methane and carbon dioxide. A delicate equilibrium is required, and this makes the process less suitable for industrial waste streams.

Composting

Composting is aerobic digestion by microorganisms in the soil. The organisms decompose organics and multiply. This process, unlike other biological processes, can tolerate some toxicants and metals. This is accomplished essentially by piling waste in the ground and aerating it occasionally by turning and moving the soil. The collection of leachate and runoff is normally required to protect the groundwater.

Enzyme Treatment

Enzyme treatment involves the application of specific proteins (simple or combined), which act as catalysts in degrading the waste. Enzymes work on specific types of compounds, specific molecules, or a specific bond in a particular compound. Enzymes are inhibited by the presence of insoluble inorganics, are sensitive to pH and temperature fluctuations, and do not adapt to variable concentrations.

Trickling Filters

Trickling filters are also used for the decomposition of organic waste streams. The microbiological reactions are similar to those that occur with activated sludge and aerated lagoons. The microorganisms are held on a support media.

LCWA.17   PROCESSES INVOLVING NITROGEN

Describe the nitrification, denitrification, and ammonia stripping processes.

SOLUTION: Nitrification converts ammonia to the nitrate form, thus eliminating toxicity to fish and other aquatic life and reducing the nitrogenous oxygen demand. Ammonia is first oxidized to nitrate and then to nitrite by autotrophic bacteria. The reactions are:

$$NH_3 + 3O_2 + \text{biomass} \rightarrow NO_2^- + H^+ + H_2O + \text{more biomass} \qquad (4.3)$$

$$NO_2^- + \tfrac{1}{2}O_2 \rightarrow NO_3^- + \text{biomass} \qquad (4.4)$$

The term "biomass" will be neglected in later reactions, since all biochemical reactions require and produce biomass. Temperature, pH, dissolved oxygen, and the ratio of BOD to total Kjeldahl nitrogen (TKN) are important factors in nitrification.

Nitrite and nitrate are reduced to gaseous nitrogen by a variety of facultative heterotrophs in an anoxic environment. An organic source, such as acetic acid, sewage, acetone, ethanol, methanol, or sugar is needed to act as a hydrogen donor (oxygen acceptor) and to supply carbon for synthesis. Methanol is used as the donor, as it is frequently the least expensive. The basic reactions take the form:

$$3O_2 + 2CH_3OH \rightarrow 2CO_2 + 4H_2O \qquad (4.5)$$

$$6NO_3^- + 5CH_3OH \rightarrow 3N_2 + 5CO_2 + 7H_2O + 6OH^- \qquad (4.6)$$

$$2NO_2^- + CH_3OH \rightarrow N_2 + CO_2 + H_2O + 2OH^- \qquad (4.7)$$

Biological phosphorus and nitrogen removal have received considerable attention in recent years. Basic benefits reported for biological nutrient removal includes monetary saving through reduced aeration capacity and the obviated expense for chemical treatment. Biological nutrient removal involves anaerobic and anoxic treatment of return sludge prior to discharge into the aeration basin. Based on the anaerobic, anoxic, and aerobic treatment sequence and internal recycling, several processes have been developed. Over 90% phosphorus and high nitrogen removal (by nitrification and denitrification) has been reported using biological means.

Ammonia gas can be removed from an alkaline solution by air stripping. This operation can be expressed in equation form as:

$$NH_4^+ + OH^- \rightarrow NH_3 \uparrow + H_2O \qquad (4.8)$$

The basic equipment for an ammonia-stripping system includes chemical feed, a stripping tower, a pump and liquid spray system, a forced-air draft, and a recarbonation system. This process requires raising the pH of the wastewater to about 11, the formation of droplets in the stripping tower, and providing air-water contact and droplet agitation by countercurrent circulation of large quantities of air through the tower. Ammonia-stripping towers are simple to operate and can be very effective in ammonia removal, but the extent of their efficiency is dependent on the air temperature. As the air temperature decreases, the ammonia removal efficiency drops significantly (as expected). This process, therefore, is not recommended for use in a cold climate. A major operational disadvantage of stripping is the need for neutralization and the prevention of calcium carbonate scaling within the tower. Also, there is some concern over discharge of ammonia into the atmosphere. The process has not found much use in the United States.

LCWA.18   CHEMICAL TREATMENT PROCESSES

Describe the two primary chemical treatment processes.

SOLUTION: *Coagulation–Precipitation*   The nature of an industrial wastewater is often such that conventional physical treatment methods will not provide an adequate level of treatment. In particular, ordinary settling or flotation processes will not remove ultrafine colloidal particles and metal ions. In these instances, natural stabilizing forces (such as electrostatic repulsion and physical separation) predominate over the natural aggregating forces and mechanisms, namely, van der Waals forces and Brownian motion, which tend to cause particle contact. Therefore, to adequately treat such particles in industrial wastewaters, coagulation–precipitation may be warranted.

The first and most important part of this technology is coagulation, which involves two discrete steps. Rapid mixing is employed to ensure that the chemicals are thoroughly dispersed throughout the wastewater flow for uniform action. Next, the wastewater undergoes flocculation, which provides for particle contact at a slow mix so that particles can agglomerate to a size large enough for removal. The second part of this technology involves precipitation, which is the separation of solids from liquid and thus can be performed in a unit similar to a clarifier.

Coagulation–precipitation is capable of removing pollutants such as the aforementioned biochemical oxygen demand (BOD), chemical oxygen demand (COD), and total suspended solids (TSS) from industrial wastewater. In addition, depending upon the specifics of the wastewater being treated, coagulation–precipitation can remove additional pollutants such as phosphorus, organic nitrogen and metals. This technology is attractive to industry because a high degree of clarification and toxic pollutant removal can be combined in one treatment process. Disadvantage of this process are the substantial quantity of sludge generated, which presents a sludge disposal problem. Also, all industrial wastewaters do not respond to this treatment.

Neutralization   In virtually every type of manufacturing industry, chemicals play a major role. Whether they result from the raw materials or from the various processing agents used in the production operation, some residual compounds will ultimately end up in a process wastewater. Thus, it can generally be expected that most industrial waste streams will deviate from the neutral state (i.e., will be acidic or basic in nature).

Highly acidic or basic wastewaters are undesirable for two reasons. First, they can adversely impact the aquatic life in receiving waters; second, they might significantly effect the performance of downstream biological treatment processes at the plant site or at a POTW. Therefore, to rectify these potential problems, one of the most fundamental treatment technologies, neutralization, is employed at industrial facilities. Neutralization involves adding an acid or a base to a wastewater to offset or neutralize the effects of its counterpart in the wastewater flow, i.e., adding acids to alkaline wastewaters and bases to acidic wastewaters.

The most important considerations in neutralization are a thorough understanding of the wastewater constituents so that the proper neutralizing chemicals are used, and proper monitoring to ensure that the required quantities of these chemicals are employed and that the effluent is in fact neutralized. For acid waste streams, lime, soda ash, and caustic soda are the most common base chemicals used in neutralization. In the case of alkaline waste streams, sulfuric, hydrochloric, and nitric acid are generally used. Some industries have operations that separate acid and alkaline waste streams. If properly controlled, these waste streams can be mixed to produce a neutralized wastewater with less or no additional neutralizing chemicals.

Neutralizing treats the pH level of a wastewater flow. Although most people do not think of pH as a pollutant, it is in fact designated by the EPA as such. Since many subsequent treatment processes are pH-dependent, neutralization can be considered as a preparatory step in the treatment of all pollutants.

Eliminating the adverse impacts on water quality and wastewater treatment system performance is not the only benefit of neutralization. Acidic or alkaline wastewaters can be very corrosive. Thus, by neutralizing its wastewaters, a plant can protect its treatment units and associated piping. The major disadvantage of neutralization is that the chemicals used in the treatment process are often themselves corrosive and can be dangerous.

LCWA.19   PHYSICAL TREATMENT

Describe some of the key physical treatment processes.

SOLUTION: Physical treatment may be separated into two categories: *phase separation* and *component separation* processes; in the latter, a particular species is separated from a single-phase, multicomponent system. The various physical treatments may fall into one or both of these categories. Sedimentation and centrifugation are used in phase separation; liquid ion exchange and freeze crystallization are used for component separation; and, distillation and ultrafiltration are used in both.

Component separation processes remove particular ionic or molecular species without the use of chemicals. These include such techniques as: liquid ion exchange, reverse osmosis, ultrafiltration, air stripping, and carbon adsorption. The first three are used to remove ionic and inorganic components; the last two techniques are used to remove volatile components and gases.

Many factors need to be considered when selecting a particular type of physical treatment. These include: the characteristics of the waste stream and the desired characteristics of the output stream, the technical feasibility of the different physical treatments when applied to a particular case, regulatory constraints, and economic, environmental, and energy considerations.

The more established types of physical treatment used today include: carbon adsorption, resin adsorption, centrifugation, distillation, electrodialysis, evaporation, liquid–liquid extraction, filtration, flocculation, flotation, freeze crystallization, high-gradient magnetic separation, ion exchange, liquid solidification, air stripping, steam stripping, and ultrafiltration. Details on these physical separation processes have been provided by J. Santoleri, J. Reynolds and L. Theodore, "Introduction to Hazardous Waste Incineration, 2nd edition," John Wiley & Sons, Hoboken, NJ, 2000.

LCWA.20 ROLE OF PHYSICAL TREATMENT FOR WASTEWATER

Describe physical treatment and physical treatment equipment in relation to the overall wastewater treatment process.

SOLUTION: The overall wastewater treatment process consists of the following:

1. Wastewater treatment can be made up of roughly three (consecutive) steps with a preliminary process called pretreatment.
2. Pretreatment is the removal of stones, sand, grit, fat/grease, rags, plastics, etc. using mechanical processes such as screening, settling, or flotation.
3. Primary settling is the removal of suspended solids by passing wastewater through settling tanks and with accompanying skimming and removal of floatables.
4. In secondary treatment, a biological process is used where wastewater passes through tanks in which bacteria consume pollutants and transform them into carbon dioxide, water, and more biological cells. Following the biological process, settling is required, which is a physical treatment process.
5. Tertiary or advanced treatment includes nitrogen and phosphorus removal, disinfection by means of chlorination, ultraviolet (UV) radiation, or ozone treatment, and filtration.

LCWA.21 NATIONAL STANDARDS FOR SECONDARY TREATMENT

List the national standards for secondary treatment.

SOLUTION: The EPA set minimum standards for secondary treatment in 1973. These standards were amended in 1985 to allow additional flexibility in applying the percent removal requirements of pollutants to treatment facilities serving separate sewer systems. The standards for secondary treatment include three major effluent parameters: 5-day BOD, TSS, and pH. These standards provided the basis for the design and operation of most treatment plants.

LCWA.22 RADIOACTIVE CLASSIFICATION

Provide answers to the following four questions as they may relate to drinking water:

1. What is a radioactive waste?
2. What are the different categories of radioactive wastes?
3. Identify at least three sources of radioactive wastes.
4. Describe methods by which the volume of radioactive waste may be minimized.

SOLUTION:

1. Radioactive wastes are waste materials that consist of unstable isotopes. Over time, the materials decay to a more stable form (or element) emitting potentially harmful energy in the process.
2. Radioactive wastes are present in several forms. These forms are identified as follows:

   - High-level waste (HLW)
   - Low-level waste (LLW)
   - Transuranic waste (TRU)
   - Uranium mine and mill tailings
   - Mixed wastes
   - Naturally occurring radioactive wastes

3. Sources of radioactive wastes include:

   - Nuclear power
   - Government waste (nuclear defense)
   - Medical radiotherapy/hospitals

- Mining waste (particularly phosphate mining)
- Normally occurring radioactive materials
- Industrial waste

4. One method of radioactive waste is to compact the waste into a smaller, more densely packed volume. A second method is by incineration, specifically of exposed organic materials. A third method of waste volume reduction is by either dewatering (filtration) and/or evaporation for water removal and recovery.

LCWA.23 PRIORITY WATER POLLUTANT (PWP) CHEMICAL FORMULAS

With reference to PWPs, provide the chemical formula for the following 10 pollutants:

Anthracene
Chrysene
Diethyl phthalate
Methylene chloride
Naphthalene
2-Nitrophenol
Pentachlorophenol
PCB-1016
Tetrachloroethylene
Vinyl chloride

SOLUTION: The chemical formula for each pollutant (except PCB-1016) is provided below.

Anthracene: $C_{14}H_{10}$
Chrysene: $C_{18}H_{12}$
Diethyl phthalate: $C_{12}H_{14}O_4$
Methylene chloride: CH_2Cl_2
Naphthalene: $C_{10}H_8$
2-Nitrophenol: $C_6H_5NO_3$
Pentachlorophenol: C_6HCl_5O
PCB-1016: This pollutant is a mixture of mono-, di-, and trichloroisomers of the polychlorinated biphenyls. It has no single chemical formula (see next problem)
Tetrachloroethylene: C_2Cl_4
Vinyl chloride: C_2H_3Cl

LCWA.24 POLYCHLORINATED BIPHENYL (PCB)-1016

One of the worst "actors" in the PWPs list is PCB-1016. Provide detailed information for this chemical.

SOLUTION: (Drawn in part from Spero et al., *Regulatory Chemicals Handbook*, Marcel Dekker (recently acquired by CRC Press/Taylor & Francis Group, Boca Raton, FL), New York, 2000 where details on each of the PWPs are provided).

PCB-1016 is a mixture of mono-, di-, and trichloroisomers of the polychlorinated biphenyls (PCBs), 257.9 average molecular weight.

CAS/DOT Identification #: 12674-11-2/UN 2315.

Synonyms: chlorodiphenyl, aroclor 1016; PCPs, polychlorinated biphenyls, aroclors.

Physical Properties: Polychlorinated biphenyl contains $\sim$16% chlorine; the solubility of PCB decreases with increasing chlorination (0.04–0.2 ppm); colorless oil; soluble in oils and organic solvents; odorless; boiling point (BP) 325–356°C); density (DN) (1.33 g/mL at 25°C); specific gravity (SG) (1.4); vapor pressure, (VP) (4×10^{-4} torr at 25°C estimated).

Chemical Properties: Incompatible with strong oxidizers; generally nonflammable; chemically inert and stable to conditions of hydrolysis and oxidation in industrial use; freezing point (FP) (286°F).

Biological Properties: Aerobic degradation in semicontinuous activated sludge process; 30% degradation of $<$1 mg/L concentration after 48 h incubation; partition coefficient between sediment and water: $\sim$105; 15–10% degradation of 1 mg in 48 h, increased chlorine in molecule decreses degradation; catalytic dechlorination to biphenyl was achieved with 5% platinum or palladium on 60/80 mesh glass beads; 48% biodegraded at the end of 28 days at 5 ppm concentration in a static flask screening procedure using BOD dilution water, settled domestic wastewater innoculum; at 10 ppm, 13% biodegraded; can be detected in water by EPA Method 608 (gas chromatography) or EPA Method 625 (gas chromatography plus mass spectrometry).

Bioaccumulation: PCBs in pelagic organisms; a food chain interrelationship study; PCB concentration factor (wet weight); microplankton 170,000, macroplanktonic enphausiid (*Meganyctiphanes norvegica*) 50,000; camicorous decapod shrimp (*Sergestes arcticus*) 47,000; (*Pasiphaea sivado*) 20,000; myctophid fish (*Myctophus glaciale*) 6000; no biomagnification in this food chain, if whole organisms are considered.

Origin/Industry Sources/Uses: Prepared by the chlorination of biphenyl; used in the electrical industry in capacitors and transformers; used in the formulation of lubricating and cutting oils; pesticides; adhesives; plastics; inks; paints; sealants.

Exposure Routes: Inhalation of fume or vapor; percutaneous adsorption of liquid; ingestion; eye and skin contact; landfills containing PCB waste materials and products; incineration of municipal refuse and sewage sludge; waste transformer fluid disposal to open areas.

Regulatory Status: Criterion to protect freshwater aquatic life; 0.014 µg/L 0.24 h avg; criterion to protect saltwater aquatic life: 0.030 µg/L 0.24 h avg.; criterion to protect human health: preferably 0.0; lifetime cancer risk of 1 in 100,000: 0.00079 ng/L; maximum contaminant level in drinking water: 0.5 µg/L (for PCBs as decachlorobiphenyl).

Probable Fate: Photolysis: too slow to be important, vapor-phase reaction with hydroxyl radicals has a half-life of 27.8 days to 3.1 months; oxidation: not important; hydrolysis: not important; volatilization: slow volatilization is the cause of global distribution of PCBs, but is inhibited by adsorption, may be significant over time, rapid volatilization from water in the absence of adsorption, half-life of 2–7 years in typical water bodies; sorption: PCBs are rapidly adsorbed onto solids, especially organic matter, and are often immobilized in sediments, but may reenter solution; biological processes; strong bioaccumulation, monodi-, and trichlorinated biphenyls are gradually biodegraded; biodegradation is probably the ultimate degradation process in water and soil.

In addition to Spero et al, the reader may refer to following reference for additional details on any of the PWPs.

1. R. Lewis, "Sax's Dangerous Properties of Industrial Materials," 9th Edition, Van Nostrand, New York City, 1996.
2. "Suspect Chemicals Sourcebook," Roytech Publications, Bethesda, Maryland, 1996.

LCWA.25 METHYLENE CHLORIDE

Provide detailed information on methylene chloride.

SOLUTION: Refer to the literature citations in Problem LCWA.24.

CAS/DOT Identification #: 75-09-2/UN 1593.

Synonyms: dichloromethane, methane dichloride, methylene bichloride.

Physical Properties: clear liquid; slightly sweet odor; slightly soluble in water; MP (−97°C); BP (39.8–40°C); DN (1.325 g/mL); SG (1.33); ST (26.52 dynes/cm); VS (0.43 cP @ 20°C); VP (349 mmHg @ 20°C); VD (2.93); solubility in water (16,700 mg/L @ 25°C); OT (250 ppm); HV (28.06 kJ/gmol); Log Kow (1.25); H (3.19×10^{-3}/mole); refractive index (1.4242 @ 20°C); MW = 84.93.

Chemical Properties: Nonflammable; HC (−513.9 kJ/mol); very stable; reacts violently with alkali metals, aluminum, and potassium-butoxide; AT (662°C); LEL/UEL (14%, 22%).

Biological Properties: % degraded under aerobic continuous flow conditions: 94.5%; partially leaches into ground water; soil, surface water, and aerobic half-lives: 7 days-4 weeks; ground water half-life; 14 days-8 weeks: anaerobic half-life: 28 days-16 weeks; can be determined in water by EPA Method 601:inert gas purge followed by gas chromatography with halide specific detection, or EPA Method 624; gas chromatography plus mass spectrometry.

Bioaccumulation: not expected to bioconcentrate in aquatic organisms or in the food chain; the concentration found in fish tissues is expected to be about the same as the concentration in the water the fish were taken.

Origin/Industry Sources/Uses: used as a solvent in the manufacture of drugs, paint, plastics, cellulose acetate; laboratory solvent; propellant; refrigerant; parts degreaser; degreasing agent for citrus fruit; blowing agents in foams.

Toxicity: Bacteria (*pseudomonas*): LD_0: 1.0 g/L; Algae (*Scenedesmus*): 125 mg/L; Arthropod (*Daphnia*): 1.25 g/L; Protozoa (*Colpoda*): 1.0 g/L; threshold concentration of cell multiplication inhibition of the protozoan *Uronema parduczi Chatton-L-woff*: >16,000 mg/L; guppy (*Poecilia reticulata*): 14 d LC_{50}: 294 ppm; fathead minnow: 96 hr LC_{50} (F): 193 mg/L, 96 hr LC_{50} (S): 310 mg/L.

Exposure Routes: primarily by inhalation; adsorption through skin and eyes; ingestion; drinking water; ambient air; occupational or consumer exposure from indoor spray painting or other aerosol uses.

Regulatory Status (as of 2002): Criterion to protect freshwater aquatic life: Criterion to protect saltwater aquatic life: Criterion to protect human health: preferably 0; lifetime cancer risk of 1 in 100,000: 1.9 μg/L; USSR MAC: 7.5 mg/L; the following are guidelines in drinking water set by some states: 2 μg/L (New Jersey); 100 μg/L (New Mexico); 4.7 μg/L (Arizona); 25 μg/L (Connecticut); 40 μg/L (California); 48 Mg/L (Minnesota and Vermont); 50 μg/L (Kansas); 150 μg/L (Maine).

Probable Fate: photolysis: photochemical reactions in aqueous media are probably unimportant, slow decomposition in the troposphere in the presence of nitrogen oxides is possible, appreciable photodissociation may occur in stratosphere, photooxidation half-life in air: 19.1–191 days; oxidation: probably unimportant, in troposphere, oxidation by hydroxyl radicals to CO_2, CO, and phosgene is important fate mechanism; *hydrolysis*: not an important fate process, first-order hydrolytic half-life: 704 yrs; *volatilization*: due to high vapor pressure, volatilization to the atmosphere is rapid and is a major transport process; sorption: sorption to inorganic and organic materials is not expected to be an important fate mechanism; *biological processes*: bioaccumulation is not expected, biodegradation may be possible but is very slow compared with evaporation.

LCWA.26 TYPES OF SLUDGE

Identify two types of sludges and three classes of sludge generated by wastewater treatment processes. Also describe the characteristics of wastewater sludge.

SOLUTION: Primary settled or settleable sludge consists of thickened material that is generated in a primary sedimentation tank. Primary settled sludge is readily dewaterable and can be further thickened for wastewater volume reduction. The second type is the biological and chemical sludges produced in secondary and biological treatment processes. These sludges contain a significant level of biological mass that is more difficult to settle and dewater. A discussion of the three classes of sludge follows.

Primary Sludge Primary sludge is generated during primary wastewater treatment which removes those solids that settle out readily. Primary sludge contains 2–8% solids; usually thickening or other dewatering operations can easily reduce its water content.

Secondary Sludge This sludge is often referred to as biological/process sludge because it is generated by secondary biological treatment processes, including activated sludge systems and attached growth systems such as trickling filters and RBCs (rotating biocontactors). Secondary sludge has a low solids content (0.5–2%) and is more difficult to thicken and dewater than primary sludge.

Tertiary Sludge This is produced in advanced wastewater treatment processes such as chemical precipitation and filtration. The characteristics of tertiary sludge depend on the wastewater treatment process that produced it. Chemical sludges result from treatment processes that add chemicals, such as lime, organic polymers, and aluminum and iron salts, to wastewater. Generally, lime or polymers improve the thickening and dewatering characteristics of a sludge, whereas iron or aluminum salts may reduce its dewatering and thickening capacity by producing very hydrous sludges that bind water.

Settled sludge is a by product of wastewater treatment. It usually contains 95–99.5% water as well as solids and dissolved substances that were present in the wastewater or were added during wastewater treatment processing. Usually these wastewater solids are further dewatered to improve their characteristics prior to ultimate use/disposal.

The characteristics of a sludge depend on both the initial wastewater composition and the subsequent wastewater and sludge treatment processes used. Different treatment processes generate radically different types and volumes of sludge. For a given application, the characteristics of the sludge produced can vary annually, seasonally, or even daily because of variations in incoming wastewater composition and variations in the treatment processes. This variation is particularly pronounced in wastewater systems that receive a large proportion of industrial discharges.

The characteristics of a sludge can significantly affect its suitability for the various use/disposal options. Thus, when evaluating sludge use/disposal alternatives, one should first determine the amount and characteristics of the sludge and the degree of variation in these characteristics.

LCWA.27 WASTEWATER SLUDGE MANAGEMENT

List and describe the three methods involved for wastewater sludge management.

SOLUTION: Three methods are widely employed to use or dispose of wastewater sludge: land application, landfilling, and incineration. Their applicability depends on many factors, including the source, quantity and quality of wastewater sludge, geographic location of the community, hydrogeology of the region, land use, economics, public acceptance, and regulatory framework. Often, one must select and implement more than one sludge use/disposal option and must develop contingency and mitigation plans to ensure reliable capacity and operational flexibility.

Determining which of the use/disposal options is most suitable for a particular application is a multistage process. The first step is to define the needs, i.e., to determine the quantity and quality of sludge that must be handled and estimate future sludge loads based on growth projections. Alternative sludge use/disposal options that meet these needs and that comply with applicable environmental regulations must then be broadly defined. Unsuitable or noncompetitive alternatives must be weeded out in a preliminary evaluation based on readily available information. For example, a rural midwestern agricultural community seldom would elect to incinerate its wastewater sludge. Resources are then focused on a more detailed definition of the remaining alternatives and on their evaluation. Final selection of an option may require a detailed feasibility study.

One planning approach for tallying the important factors is a "system of criteria," which allows the proposed alternative to be evaluated from different criteria including annualized costs compatibility with land uses in close proximity to the proposed site (including public acceptance), recovery of resources (including energy), and reliability. In this approach, the different criteria selected for evaluation are scored numerically to tally the comparative strengths and weaknesses of the various options. This overall approach is described in Problem TCWA.20.

LCWA.28 HEXAVALENT CHROMIUM

List the possible sources of highly toxic hexavalent chromium (Cr^{6+}) and methods to remove it from a wastewater stream.

SOLUTION: Hexavalent chromium-bearing wastewater is produced in chromium electroplating, chromium conversion coating, etching with chromic acid, and in metal-finishing operations carried out using chromium as the base material.

Chromium wastes are commonly treated in a two-stage batch process. The primary stage is used to reduce the highly toxic hexavalent chromium to the less toxic trivalent chromium. There are several ways to reduce the hexavalent chrome to trivalent chrome, including the use of sulfur dioxide, bisulfite, or ferrous sulfate. The trivalent chrome is then removed by hydroxide precipitation. Most processes use caustic soda (NaOH) to precipitate chromium hydroxide. Hydrated lime [$Ca(OH)_2$] may also be used. The chemistry of the reactions is described as follows:

Using SO_2 to convert Cr^{6+} to Cr^{3+} yields:

$$SO_2 + H_2O \quad \rightarrow H_2SO_3 \tag{4.9}$$
$$3H_2SO_3 + 2H_2CrO_4 \rightarrow Cr_2(SO_4)_3 + 5H_2O \tag{4.10}$$

Using NaOH for hydroxide precipitation gives

$$6NaOH + Cr_2(SO_4)_3 \rightarrow 2Cr(OH)_3 + 3Na_2SO_4 \tag{4.11}$$

QUANTITATIVE PROBLEMS (TCWA)

PROBLEMS TCWA.1–33

TCWA.1   SUSPENDED PARTICULATE CONCENTRATION UNITS

The suspended particulate concentration in an aqueous industrial stream has been determined to be 27.6 mg/L. Convert this value to units of $\mu g/L$, g/L, lb/ft^3, and lb/gal.

SOLUTION:

$$C = 27.6 \frac{mg}{L}$$

$$= \left(27.6 \frac{mg}{L}\right)\left(10^3 \frac{\mu g}{mg}\right)$$

$$= 2.76 \times 10^4 \frac{\mu g}{L}$$

$$C = 27.6 \frac{mg}{L}$$

$$= \left(27.6 \frac{mg}{L}\right)\left(10^{-3} \frac{g}{mg}\right)$$

$$= 2.76 \times 10^{-2} \frac{g}{L}$$

$$C = 27.6 \frac{mg}{L}$$

$$= \left(27.6 \frac{mg}{L}\right)\left(\frac{1 lb}{454,000 \, mg}\right)\left(\frac{1000 \, L}{35.3 \, ft^3}\right)$$

$$= 1.72 \times 10^{-3} \frac{lb}{ft^3}$$

$$C = 27.6 \frac{mg}{L}$$

$$= \left(27.6 \frac{mg}{L}\right)\left(\frac{1 lb}{454,000 \, mg}\right)\left(\frac{1000 \, L}{264 \, gal}\right)$$

$$= 2.30 \times 10^{-4} \frac{lb}{gal}$$

TCWA.2 TRACE CONCENTRATION

Some wastewater and water standards and regulations are based on a term defined as *parts per million*, ppm, or *parts per billion*, ppb. Define the two major classes of these terms and describe the interrelationship from a calculational point of view. Also convert 5.0 calcium parts per million parts of water on a mass basis to parts per million on a mole basis.

SOLUTION: Water streams seldom consist of a single component. It may also contain two or more phases (a dissolved gas and/or suspended solids), or a mixture of one or more solutes. For mixtures of substances, it is convenient to express compositions in mole fractions or mass fractions. The following definitions are often used to represent the composition of component. A in a mixture of components:

$$w_A = \frac{\text{Mass of A}}{\text{Total mass of water stream}} = \text{Mass fraction of A} \qquad (4.12)$$

$$y_A = \frac{\text{Moles of A}}{\text{Total moles of water stream}} = \text{Mole fraction of A} \qquad (4.13)$$

Trace quantities of substances in water streams are often expressed in parts per million by weight (ppmw) or as parts per billion (ppbw) on a mass basis. These concentrations can also be provided on a mass per volume basis for liquids and on a mass per mass basis for solids. Gas concentrations are usually represented on a mole or volume basis (e.g., ppmm or ppmv, respectively). The following equations apply:

$$\text{ppmw} = 10^6 \, w_A \qquad (4.14)$$

$$= 10^3 \, \text{ppbw} \qquad (4.15)$$

$$\text{ppmv} = 10^6 \, y_A \qquad (4.16)$$

$$= 10^3 \, \text{ppbv} \qquad (4.17)$$

The two terms ppmw and ppmm are related through the molecular weight. To convert 10 ppmw Ca to ppmm, select a basis of 10^6g of solution.

The mass fraction of Ca is first obtained by the following equation:

$$\text{Mass of Ca} = 5.0 \, \text{g}$$

$$\text{Moles Ca} = \frac{5.0 \, \text{g}}{40 \, \text{g/mol}} = 0.125 \, \text{mol}; \text{MW(Ca)} = 40$$

$$\text{Moles } H_2O = \frac{10^6 \text{ g} - 10 \text{ g}}{18 \text{ g/mol}} = 55{,}555 \text{ mol}; \text{ MW}(H_2O) = 18$$

$$\text{Mole fraction Ca} = y_{Ca} = \frac{0.125 \text{ mol}}{0.125 \text{ mol} + 55{,}555 \text{ mol}} = 2.25 \times 10^{-6}$$

$$\text{ppmm of Ca} = 10^6 \, y_{Ca}$$
$$= (10^6)(2.25 \times 10^{-6})$$
$$= 2.25$$

TCWA.3   EXPRESSING CONCENTRATION TERMS

Express the concentration for the solutions given below in terms of percentage by weight, ppmw, and molarity.

36 g of HCl in 64 cm^3 of water.
0.003 g of ethanol in 1 kg of water.
34 g of ammonia in 2000 g of water.

Note that molarity is defined as the moles of solute per volume of solutions.

SOLUTION: This calculation is left as an exercise for the reader. The concentration conversions for HCl, ethanol and ammonia are given in Table 4.3 below.

TABLE 4.3   Concentration Terms

	% Weight	ppmw	Molarity
1. HCl	36	562,500	15.40
2. Ethanol	3×10^{-4}	3	6.5×10^{-5}
3. Ammonia	1.67	17,000	1.0

TCWA.4   ULTIMATE OXYGEN DEMAND

A wastewater contains 250 mg/L benzoic acid, C_6H_5COOH. What is the ultimate oxygen demand (BODU) of this water? The molecular weight of benzoic acid is 122. As noted earlier the ultimate oxygen demand is a calculated oxygen demand that assumes oxidation of all species to their most highly oxidized stable form: CO_2, H_2O, etc.

SOLUTION: The number of moles of benzoic acid in 1 liter is as follows:

$$n = 250/122$$
$$= 2.05 \text{ mgmol}$$

Note that the units are milligram moles since the mass is given in milligrams.

Convert all of the species (in this case, only benzoic acid) to their (its) most highly oxidized stable form:

$$C_6H_5COOH + \left(\frac{15}{2}\right)O_2 \rightarrow 7CO_2 + 3H_2O \qquad (4.18)$$

The ultimate oxygen demand (BODU) for 2.05 mgmol of benzoic acid is calculated employing Equation (4.18)

$$BODU = [(15/2)O_2/1 \; C_6H_5COOH](2.05 \text{ mgmol } C_6H_5COOH)$$
$$= 15.4 \text{ mgmol } O_2$$
$$= (15.4)(32)$$
$$= 493 \text{ mg } O_2$$

The BODU (or ThCOD) for this wastewater is therefore 493 mg/L.

TCWA.5   THEORETICAL CARBON OXYGEN DEMAND (ThCOD)

Calculate the ThCOD of a 100 mg/L solution of glucose.

SOLUTION: First, balance the chemical equation for the reaction of glucose (MW = 180) and oxygen:

$$C_6H_{12}O_6 + 6O_2 \rightarrow 6CO_2 + 6H_2O \qquad (4.19)$$

Second, determine the oxygen/substrate stoichiometric ratio for the oxygen reaction using the above equation:

$$\text{Ratio} = (6)(32)/[(1)(180)]$$
$$= 1.067 \text{ mg } O_2/\text{mg glucose}$$

Third, calculate the ThCOD for a 100 mg/L solution of glucose. This is equal to the product of the mass concentration of glucose and the stoichiometric ratio:

$$ThCOD = (100 \, mg \; glucose/L)(1.067 \, mgO_2/mg \; glucose)$$
$$= 106.7 \;\; mgO_2/L, \, or \, 106.7 \;\; mgCOD/L$$

TCWA.6 TOTAL ORGANIC CARBON (TOC) OF A GLUCOSE SOLUTION

With reference to the previous problem TCWA.5, calculate the TOC.

SOLUTION: As mentioned earlier, TOC is another important water quality parameter. TOC is equal to the total concentration of organic carbon in water and is reported as milligrams carbon/liter (mg C/L). It serves as a general indicator of the overall quantity of carbon-based matter contained in water. TOC frequently correlates to BOD and COD but is not always a reliable measure. If a repeatable empirical relationship is established between TOC and BOD or TOC and COD for a particular wastewater, TOC can be used as an estimate of the corresponding BOD or COD. The theoretical TOC (ThTOC) can be calculated stoichiometricaly in similar manner to the ThCOD.

First, establish the ratio of mg carbon/mg substrate. In the case of glucose, the ratio is as follows:

$$Molar \; ratio = 6 \, mol \, C/1 \, mol \, C_6H_{12}O_6$$
$$Mass \; ratio = (6)(12) \, g \, C/(1)(180) \, of \, C_6H_{12}O_6$$
$$- \, 0.40 \, mg \, C/mg \; glucose$$

Second, calculate the ThTOC concentration for the particular wastewater. For a 100 mg glucose/L solution, the ThTOC is:

$$ThTOC = (100 \, mg \; glucose/L)(0.40 \, mgC/mg \; glucose)$$
$$= 40 \, mg \, C/L$$
$$= 40 \, mg \, TOC/L$$

TCWA.7 BIOLOGICAL OXYGEN DEMAND (BOD) DISCHARGE

A typical city of 55,000 people has a wastewater treatment discharge of 6.7 million gal/day (MGD) and a BOD_5 in the raw wastewater of 225 mg/L. What is the total discharge of BOD in lb/day? What is the BOD discharge in lb/person · day?

SOLUTION: The total BOD_5 produced per day is (applying appropriate unit conversions)

$$\text{Total } BOD_5/\text{day} = (6.7\,MGD)(225\,mg/L)(8.34\,lb\,L/MG \cdot mg)$$
$$= 12{,}570\,lb\,BOD_5/\text{day}$$

The BOD_5 discharge in lb/person-day is

$$\text{BOD discharge} = \frac{12{,}570\ \text{lb}\,BOD_5/\text{day}}{55{,}000\ \text{people}}$$

$$= 0.23\ \frac{\text{lb}\,BOD_5}{\text{people} \cdot \text{day}}$$

TCWA.8   NITROGEN CONCENTRATION IN ACID

A wastewater contains 0.013% by mass of aminobenzoic acid. Calculate the percent by mass of nitrogen in the acid. Also calculate the percent by mass of nitrogen in the wastewater.

SOLUTION: Although any basis may be chosen, it is convenient to select either 137 gmol (the approximate molecular weight) or 1.0 gmol of the acid. Choose 1 gmol of acid.

$$\text{Moles of acid} = 1\,\text{gmol}$$

$$\text{Molecular formula of aminobenzoic acid} = C_7H_7NO_2$$

Moles N $= 1$

Moles H $= 7$

Moles C $= 7$

Moles O $= 2$

The corresponding mass of each component is

$$\text{Mass N} = (1)(14) = 14\,g$$

$$\text{Mass H} = (7)(1) = 7\,g$$

$$\text{Mass C} = (7)(12) = 84\,g$$

$$\text{Mass O} = (2)(16) = 32\,g$$

$$\text{Total} = 137\,g$$

$$\text{MW of acid} = 137\,g/\text{mol; balance satisfied}$$

$$\text{Mass of acid} = (137\,g/\text{mol})(1\,\text{mol}) = 137\,g$$

The percent nitrogen in the acid is therefore

$$\%N_2 \text{ (by mass of acid)} = \text{(mass N/mass acid)}100$$
$$= (14/137)100 = 10.2\%$$

The calculation for the percent nitrogen in the wastewater is slightly more complicated. Select a basis of 100 g of wastewater. For the acid

$$\% \text{ acid} = \text{fraction acid}$$
$$0.013\% = 0.00013$$
Therefore, $$\text{Mass of acid} = (0.00013)(100) = 0.013 \text{ g}$$
$$\text{Mass of water} = 100 \text{ g}$$
$$\text{Mass of nitrogen in acid} = (0.102)(0.013) = 0.001326 \text{ g}$$

Thus, the mass percent of nitrogen in the wastewater is

$$\text{Mass \%} = (0.001326 \text{ g}/100 \text{ g}) \ 100\%$$
$$= 0.00\,1326\%$$

TCWA.9 pH CALCULATIONS

Calculate the hydrogen ion and the hydroxyl ion concentration of an aqueous solution if the pH of the solution is 1, 3, 5, 7, 8, 10, 12, and 14.

SOLUTION: An important chemical property of an aqueous solution is its pH. The pH measures the acidity or basicity of the solution. In a neutral solution, such as pure water, the hydrogen (H^+) and hydroxyl (OH^-) ion concentrations are equal. At ordinary temperatures, this concentration is

$$C_{H^+} = C_{OH^-} = 10^{-7} \text{g} \cdot \text{ion/L} \tag{4.20}$$

where C_{H^+} = hydrogen ion concentration

C_{OH^-} = hydroxyl ion concentration

The unit g·ion stands for gram·ion which represents an Avogadro number of ions. In all aqueous solutions, whether neutral, basic, or acidic, a chemical equilibrium or balance is established between these two concentrations, so that

$$K_{eq} = C_{H_+} \cdot C_{OH^--} = 10^{-14} \tag{4.21}$$

where K_{eq} is the quilibrium constant.

The numerical value for K_{eq} given above holds for room temperature and only when the concentrations are expressed in gram · ion per liter (g · ion/L). In acid solutions, C_{H^+} is $> C_{OH}$; in basic solutions, C_{OH^-} predominates.

The pH is a direct measure of the hydrogen ion concentration and is defined by

$$pH = -\log(C_{H^+}) \qquad (4.22)$$

Thus, an acidic solution is characterized by a pH below 7 (the lower the pH, the higher the acidity); a basic solution has a pH above 7; and, a neutral solution possesses a pH of 7.

Regarding the problem statement, for a pH of 1.0,

$$pH = -\log(C_{H^+})$$

$$C_{H^+} = 10^{-pH} = 10^{-1} = 0.1\,g \cdot ion/L$$

$$C_{H^+} \times C_{OH^-} = 10^{-14}$$

$$C_{OH^-} = \frac{10^{-14}}{C_{H^+}}$$

$$= 10^{-13} g \cdot ion/L$$

The remaining results are calculation in a similar fashion and are presented in the Table 4.4. A more complex calculation is presented in Problem TCWA.32.

TABLE 4.4   Hydrogen Ion Concentrations

pH	C_{H^+} g·ion/L	C_{OH^-} g·ion/L
1	10^{-1}	10^{-13}
3	10^{-3}	10^{-11}
5	10^{-5}	10^{-9}
7	10^{-7}	10^{-7}
8	10^{-8}	10^{-6}
10	10^{-10}	10^{-4}
12	10^{-12}	10^{-2}
14	10^{-14}	1.0

It should be pointed out that the above equation employed is not the exact definition of pH but is a close approximation to it. Strictly speaking, the *activity* of the hydrogen ion, a_{H^+}, and not the ion concentration, C_{H^+}, belongs in the equation.

TCWA.10   pH OF WASTEWATER

An industrial wastewater has an alkalinity of 60 mg/L and a CO_2 content of 7.0 mg/L.

Determine the pH. The first ionization constant of H_2CO_3, is

$$K_{i,1} = \frac{(H^+)(HCO_3^-)}{(H_2CO_3)} \qquad (4.23)$$

$$= 3.98 \times 10^{-7}$$

Assume that the alkalinity is all in the HCO_3^- formed. Note that the parentheses represent the concentration in gmol/L.

SOLUTION: "Alkalinity" is measured as equivalents of $CaCO_3$ which has a molecular weight of 100 and an equivalent weight of 50. For example, a water solution of 17 mg/L of OH^- (equivalent weight $= 17$) contains 10^{-3} equivalent weights of OH^-, which is equivalent to 10^{-3} equivalents of $CaCO_3$. The alkalinity of that solution is therefore $(50)(10^{-3})$ g/L or 50 mg/L.

For an alkalinity of 60 mg/L, the concentration of HCO_3^- is

$$(HCO_3^-) = \frac{60\,mg/L}{50\,mg/meq}$$

$$= 1.2\,meq/L$$

$$= 1.2\,mmol/L$$

The unit "meq" represents 10^{-3} equivalents.

For the CO_2,

$$(CO_2) = \frac{7\,mg/L}{44\,mg/mmol}$$

$$= 0.159 \ mmol/L$$

Note that $(CO_2) = (H_2CO_3)$.

Substitution in Equation (4.23) gives

$$3.98 \times 10^{-7} = \frac{(H^+)(1.2 \times 10^{-3})}{(0.159 \times 10^{-3})}$$

$$(H^+) = 5.27 \times 10^{-8}$$

The pH is therefore given by

$$pH = -\log(H^+)$$

$$= 7.28$$

TCWA.11 CHEMISTRY FUNDAMENTALS

An elementary school science teacher found an old bottle containing 2 L of concentrated sulfuric acid in the science room storage closet. Not having any use for the acid, the teacher considered disposing of it by pouring it down the drain with the faucet running. The principal happened to stop by and informed the teacher that Federal regulations required that any material put down the drain must have a pH greater than 2.0 and less than 12.5 in order to protect the aquatic environment and the bacteria in the local wastewater treatment plant, not to mention the pipe in the sewerage system. The teacher considered three methods of bringing the acid to a pH of 2.1 to comply with the regulations and to dispose of the acid safely and quickly:

1. Dilution
2. Neutralization to Na_2SO_4 and H_2O with NaOH
3. Neutralization to $CaSO_4$ and H_2O with slaked lime, $Ca(OH)_2$

Answer the following four questions.

1. Find the pH of the concentrated H_2SO_4.
2. Find the volume to which the 2 L of the 18 M H_2SO_4 would have to be diluted in order to raise the pH to 2.1.
3. Calculate the quantity of NaOH needed to neutralize the 2 L of 18 M H_2SO_4.
4. Calculate the quantity of $Ca(OH)_2$ needed to neutralize the 2 L of 18 M H_2SO_4.

Note that $pH = -log[H^+]$, and that concentrated H_2SO_4 is 18 M (molarity, gmol/L) and 100% ionized.

SOLUTION: 1. With the strong acid ionizing 100%, the acid dissociates completely as follows:

$$H_2SO_4 \rightarrow 2H^+ + SO_4^{-2}$$

Therefore, 18 gmol/L H_2SO_4 would ionize to yield 36 gmol/L H^+ to yield a pH of:

$$pH = -log[H^+] = -log(36) = -1.6$$

Note: a negative pH is possible at very high acid concentrations.

2. Next, the $[H^+]$ in a sulfuric acid solution of pH = 2.1 is calculated as:

$$[H^+] = 10^{-pH} = 10^{-2.1} = 7.94 \times 10^{-3}$$

The molarity of this solution is $7.94 \times 10^{-3}/2 = 3.97 \times 10^{-3}$. Thus, in order to raise the pH of the 2 L of concentrated sulfuric acid from -1.6 to 2.1, the molarity of the solution must be lowered from 18 M to approximately 0.004 M.

The volume of the diluted acid can be calculated from a mass balance for hydrogen ions, i.e., $M_1V_1 = M_2V_2$:

$$18 \, M \, (2L) = 0.004 \, M \, (V_2)$$
$$V_2 = 9,000L$$

Note: Since 9,000 L equals approximately 2,250 gallons of solution, this is clearly not a practical method of disposal.

3. The number of moles of NaOH required to neutralize the sulfuric acid is based on the number of moles of sulfuric acid in the 18 M solution and the stoichiometric relationship describing this neutralization reaction, or:

2 L (18 gmol/L) = 36 gmol to be neutralized according to the stoichiometry

$$H_2SO_4 + 2NaOH \rightarrow Na_2SO_4 + 2H_2O$$

This requires 2 gmol of NaOH for each gmol of sulfuric acid neutralized for a total of 72 gmol of NaOH for complete neutralization.

Converting gmol to pounds of NaOH required yields:

$$\frac{(72\,gmol\ NaOH)\ (40\,g/gmol)}{454\,g/lb} = 6.3\ lb\,NaOH\ required$$

Note: As a laboratory scale reaction, this would be a very large reaction, generating a significant amount of heat with considerable chance of splattering and injury.

4. The number of moles of Ca(OH)$_2$ required to neutralize the sulfuric acid is again based on the number of moles of sulfuric acid in the 18 M solution and the stoichiometric relationship describing this neutralization reaction, or:

2 L (18 gmol/L) = 36 gmol to be neutralized according to the stoichiometry

$$H_2SO_4 + Ca(OH)_2 \rightarrow CaSO_4 + 2H_2O$$

This requires 1 gmol of Ca(OH)$_2$ for each mol of sulfuric acid neutralized for a total of 36 gmol of Ca(OH)$_2$ for complete neutralization.

Converting gmol to pounds of Ca(OH)$_2$ required yields:

$$\frac{(36\ gmol\ Ca(OH)_2\ (74\ g/gmol)}{454\,g/lb} = 5.9\,lb\ Ca(OH)_2\ required$$

Note: Again, as a laboratory scale reaction, this would still be a very large reaction, generating a significant amount of heat with considerable chance of splattering and injury.

TCWA.12 CHLORINE DISINFECTION

A biologically contaminated hospital liquid waste requires disinfection with chlorine prior to discharge into a nearby lake. Given a contact time of 30 minutes, a required chlorine dose for a 99.99% kill of the pathogenic organisms = 3.0 mg/L, and a 0.5 mg/L chlorine requirement to carry out oxidation-reduction reactions with compounds in the liquid waste, how many pounds of pure chlorine (in the gaseous form) would be required daily to disinfect a 100,000 gal/d waste flow?

SOLUTION: The total chlorine concentration required to meet the oxidation-reduction reaction requirement in the water is 0.5 mg/L while 3.0 mg/L chlorine is required for the disinfection reaction. The total chlorine dose then is 3.5 mg/L.

The required chlorine dosage (D) in terms of lb/d is found by employing appropriate conversion factors:

$$D = (3.5 \text{ mg/L}) (100,000 \text{ gal/d}) (3.785 \text{ L/gal}) (2.2 \text{ lb/kg}) (1 \text{ kg/}10^6 \text{ mg})$$
$$= 2.92 \text{ lb chlorine gas/d}$$

A handy, engineering equation to remember for dose calculations is:

$$\text{lb/d} = (\text{concentration, mg/L})(\text{flow, MGD}) (8.34 \text{ lb/MG/mg/L})$$
$$\text{lb/d} = (3.5 \text{ mg/L}) (0.1 \text{ MGD}) (8.34 \text{ lb/MG/mg/L}) = 2.92 \text{ lb chlorine gas/d}$$

TCWA.13   RETURN ACTIVATED SLUDGE (RAS) FLOW

Determine the RAS flow in MGD when the influent flow is 8 MGD and the sludge settling volume (SV) in 30 minutes is 280 mg/L. Note: 1 mg/L = 1 ml/L

SOLUTION: The following equation may be employed for RAS.

$$\text{The fraction of RAS flow in the influent} = \frac{SV(\text{ml/L})}{1,000(\text{ml/L}) - SV(\text{ml/L})} \quad (4.24)$$

$$= \frac{280 \text{ ml/L}}{1000 \text{ ml/L} - 280 \text{ ml/L}} = \frac{280}{720}$$

$$= 0.389 \text{ of influent flow}$$

Therefore,

$$\text{RAS flow, MGD} = \text{RAS flow fraction} \times \text{Influent flow (MGD)}$$
$$= (0.389) (8 \text{ MGD})$$
$$= 3.11 \text{ MGD}$$

TCWA.14   SLUDGE VOLUME INDEX (SVI)

SOLUTION: A calculation is used to indicate the settling ability of activated sludge (aerated solids) in the secondary clarifier to the SVI. The calculation is a measure of the volume of sludge compared with its weight. Generally, the sludge sample from the aeration tank is allowed to settle for 30 minutes. The calculated SVI is obtained by dividing the volume (mL) of wet settled sludge by the weight (mg) of that sludge after it has been dried. Sludge with an SVI of one hundred or greater will not settle as readily as desirable because it is as light or lighter than water. The RAS suspended solids (mg/L) is calculated from:

$$\text{RAS Suspended Solids, mg ss/L} = \frac{1,000,000}{\text{SVI}} \qquad (4.25)$$

Calculate the RAS suspended solids based on a sludge volume index (SVI) of 120.

SOLUTION: Substituting into the above equation gives

$$\text{RAS (mg ss/L)} = 10^6/\text{SVI}$$
$$= 10^6/120$$
$$= 8,333 \, \text{mg/L}$$

TCWA.15 POLYMER DOSE IN FLOCCULATION

Polymers are flocculation aids for water treatment which are classified on the basis of the type of electrical charge on the polymer chain. Polymers possessing negative charges are called "anionic," positive charged polymers are called "cationic," and polymers that carry no electrical charge are called "nonionic." Polymers cause the suspended material to stick together by chemical bridging or chemical enmeshment when contact is made in the filter influent flow. Generally only the anionic polymers are used.

Polymer is supplied to a plant at a concentration of 0.5 pounds per gallon. The polymer feed pump delivers a flow of 0.1 gpm and the flow to the filter is 3,000 gpm. Calculate the dose (in mg/L) of polymer in the water applied to the filter.

SOLUTION: The describing equation is:

$$\text{Dose (mg/L)} = \frac{(\text{Flow of polymer, gpm})\,(\text{Concentration, lbs polymer/gal})}{(\text{Flow to filter, gpm})\,(8.34\,\text{lbs water/gal})}$$
$$(4.26)$$

Substituting yields

$$\text{Dose} = \frac{(0.1\,\text{gpm})\,(0.5\,\text{lbs polymer/gal})}{(3000\,\text{gpm})\,(8.34\,\text{lbs/gal})}$$
$$= 2 \times 10^{-6} \, \text{lbs polymer/lbs water}$$

The conversion factor 10^6 is employed to convert the dose to mg/L.
$$= 2\,\text{mg/L}$$

TCWA.16   ALUMINUM DOSE

Liquid alum is usually available at a concentration of 5.4 lbs alum/gallon. In one application, the alum feed pump delivers 88 ml/min and the flow to the filter is 3,000 gpm. Calculate the dose of alum in the water applied to the filter.

SOLUTION: Use the equation in the previous problem (TCWA.15) but maintain dimensional consistency.

$$\text{Dose, mg/L} = \frac{\text{(Flow, ml/min)(Concentration, lbs alum/gal)(0.00026 gal/ml)}}{\text{(Flow, gpm)(8.34 lbs water/gal)}}$$

(4.27)

Substituting yields

$$\text{Dose} = \frac{\text{(88 ml/min)(5.4 lbs alum/gal)(0.00026 gal/ml)}}{\text{(3000 gpm)(8.34 lbs water/gal)}}$$

$$= 4.94 \times 10^{-6} \text{ lbs alum/lbs water}$$

$$= 4.94 \text{ mg/L}$$

TCWA.17   FOOD TO MICROORGANISM (F/M) RATIO

The range of organic loadings of activated sludge plants is described by the F/M ratio. Different ranges of organic loadings are necessary for conventional, extended aeration, and high rate type of activated sludge systems. The F/M ratio is calculated from the amount of COD or BOD applied each day and from the solids inventory in the aeration tank, i.e., the rate of BOD or COD flowing into the activated sludge unit divided by the mass of mixed liquor volatile suspended solids, MLVSS (not the mass of mixed liquor suspended solids, MLSS) retained in the unit.

Determine the F/M ratio for an activated sludge plant with a COD of 100 mg/L applied to the aeration tank, an influent flow of 7.5 MGD and 33,075 lbs of solids under aeration. Assume 70% by mass of the mixed liquor suspended solids (MLSS) are volatile matter.

SOLUTION: By dimensional analysis, the describing equation is

$$\text{F/M}, \frac{\text{lb COD/day}}{\text{lb MLVSS Solids}} = \frac{\text{(Flow, MGD)(COD, mg/L)(8.34 lbs/gal)}}{\text{(Solids under aeration, lbs)(VM fractional portion)}}$$

(4.28)

Substituting, gives

$$\text{F/M} = \frac{\text{(7.5 MGD)(100 mg/L)(8.34 lbs/gal)}}{\text{(33,075 lbs)(0.7)}}$$

$$= 0.27 \text{ lbs COD/lb MLVSS} \cdot \text{day}$$

TCWA.18 TRICKLING FILTER LOADING

A filter 100 feet in diameter and 6 feet deep is receiving 0.5 mgd of primary tank effluent with a BOD of 150 ppm. Calculate the BOD loading in lb BOD/100 ft$^3 \cdot$ day.

SOLUTION:

$$\text{BOD applied} = (0.5\,\text{mgd})(10^6\,\text{gal/mgal})(8.34\,\text{lbs/gal})(150 \times 10^{-6})$$

$$= 625\,\text{lbs/day}$$

$$\text{The volume of filter medium} = \frac{\pi \times (100)^2\,\text{ft}^2}{4} \times 6$$

$$= 47{,}124\,\text{ft}^3$$

Therefore,

$$\text{BOD loading} = (625\,\text{lbs/day}/47{,}124\,\text{ft}^3)(1000)$$

$$= 13.3\text{lbs BOD per 1000 cubic feet of filter medium per day}$$

TCWA.19 SLUDGE AGE

Sludge age is a measure of the length of time a particle of suspended solids has been undergoing aeration in the activated sludge process. It may be calculated from the ratio between the mass of solids in the aeration tank and the solids rate in the primary effluent. In most activated sludge plants, sludge age ranges from 3 to 8 days.

Determine the sludge age for an activated sludge plant with an influent flow of 7.5 MGD. The primary effluent suspended solids concentration is 100 mg/L. Two aeration tanks have a volume of 0.6 MG each and a mixed liquor suspended solids (MLSS) concentration of 2,200 mg/L. Calculate the solids loading in the aeration tanks effluent, and the sludge age.

SOLUTION:

1. Calculate the mass of solids in the aeration tanks. Employ the appropriate conversion factor.

$$(1.0\,\text{lb/gal}) = (454\,\text{g/lb})(1000\,\text{mg/g})(264\,\text{gal/m}^3)(\text{m}^3/1000\,\text{L})$$

$$= 120{,}000\,\text{mg/L}$$

$$= 1.2 \times 10^5\,\text{mg/L}$$

$$\text{Solids loading in aeration} = (2)(0.6\,\text{MG})(2200\,\text{mg/L})$$

$$= (2)(600{,}000)(2200)/(1.2 \times 10^5)$$

$$= 22{,}000\,\text{lbs}$$

2. Calculate the solids loading in the primary effluent, lb/day

$$\text{Solids loading in effluent} = (7.5\,\text{MGD})(100\,\text{mg/L})$$

$$= 6255\,\text{lb/day}$$

3. Finally, determine the sludge age.

$$\text{Sludge Age(day)} = (\text{Solids under aeration, lb})/(\text{solids in effluent, lb/day})$$
$$= 22{,}000\,\text{lb}/(6255\,\text{lb/day}) = 3.5\,\text{days}$$

See also 40 CFR 403.8

TCWA.20 ANALYSIS OF SLUDGE MANAGEMENT OPTIONS

The LSLT Corporation has instituted a new program that is concerned with the management of solid and/or hazardous sludge waste. Three process/treatment options are under consideration:

A. land applications
B. landfill
C. incineration

Based on the data provided below, you have been asked—as the environmental engineer assigned to this project—to use a "system of criteria" method to determine which option is most attractive (L. Theodore: personal notes, 1981).

LSLT Corporation has determined that annualized cost is the most important criterion, with a weight factor of 100 (from a maximum of 100). Other significant criteria include regulations (weight of 80), recovery/reuse (weight of 70), and reliability (weight of 50). Options A, B, and C have also been assigned effectiveness factors. Option A is given a rating of 80 for annualized cost, 60 for regulations, 40 for recovery/reuse and 20 for reliability. The corresponding effectiveness factors for B and C are 60, 30, 40, 20 and 30, 80, 50, 80, respectively.

SOLUTION: In solving this example, first generate the overall rating (OR) for option A:

$$OR_A = (100)(80) + (80)(60) + (70)(40) + (50)(20)$$
$$= 16{,}600$$

Generate the overall rating for option B:

$$OR_B = (100)(60) + (80)(30) + (70)(40) + (50)(20)$$
$$= 12{,}200$$

Generate the overall rating for option C:

$$OR_C = (100)(30) + (80)(80) + (70)(50) + (50)(80)$$
$$= 16{,}900$$

From this screening, option C rates the highest with a score of 16,900.

The reader should note that this problem is revisited in Chapter 6, Problem TRCRA.8 as applied to sludge management.

TCWA.21   NATION POLLUTION DISCHARGE ELIMINATION SYSTEM (NPDES) PERMIT LIMITATION

The following 5-day biochemical oxygen demand (BOD_5) data were collected at a local municipal wastewater treatment plant over a 7-day period. The NPDES permit limitations for BOD_5 effluent concentrations from this wastewater treatment plant is 45 mg/L on a 7-day average. Based on this information, is the treatment plant within its NPDES permit limits?

TABLE 4.5   Daily BOD_5 Concentration Data Collected Over a 7-day Period at a Municipal Wastewater Treatment Plant

Day	BOD (mg/L)
1	45
2	79
3	64
4	50
5	30
6	25
7	21

SOLUTION: The BOD_5 7-day average concentration based on the data tabulated in the problem statement is

$$(BOD_5)_7 = (45 + 79 + 64 + 50 + 30 + 25 + 21)/7$$
$$= 44.9 \, mg/L$$

The wastewater treatment plant is still within its NPDES permit limit (but only marginally) of an average 7-day maximum concentration of 45 mg/L for BOD_5.

TCWA.22   TOTAL MAXIMUM DAILY LOAD (TMDL) DEFINITION

Quantitatively discuss TMDL.

SOLUTION: As noted earlier, TMDL or Total Maximum Daily Load is a calculation of the maximum amount of a pollutant that a waterbody can receive and still meet water quality standards, and an allocation of that amount to the pollutant's sources. Water quality standards are set by States, Territories, and Tribes and identify the uses for each waterbody. The Clean Water Act, Section 303, establishes the water quality standards and TMDL programs.

The TMDL is the sum of the allowable loads of a single pollutant from all contributing point (e.g., wastewater treatment plants) and nonpoint (e.g., agricultural land drainage, storm runoff, etc.) sources. The calculation must include a margin of safety to ensure that the waterbody can be used for the purposes the State has designated. The calculation must also account for seasonal variation in water quality. Details (W. Matystik: personal notes, 2007) are provided below.

1. The describing equation is

$$TMDL = \sum WLA + \sum LA + MOS \tag{4.29}$$

where
$$WLA = \text{point sources}$$
$$LA = \text{non point sources}$$

(MOS) Margin of Safety = typically 5–10% (prescribed by state).

2. Point sources are included at permitted flows and concentrations.
3. A suitably long time period is chosen to account for a range of hydrologic events, usually 10–20 years.
4. The impact of future growth must be included in the analysis.

TCWA.23   WASTELOAD ALLOCATION METHODS

With regard to TMDL, provide a list, including equations where applicable, for wasteload allocation methods.

SOLUTION:

1. Percent removal (equal percent treatment)
   Example: 20% nitrogen removal from all municipal point sources
2. Effluent concentrations (see earlier Problems)
   mg/L; μg/L; ppm; ppb
3. Total mass discharge per day
   Waste stream concentration is ___mg/L with a flow rate of ___gpm/cfs. Present mass discharge rate is:

   (concentration)(flowrate); consistent units

4. Mass discharged per capita per day
5. Reduction of raw load (pounds per day)
6. Ambient mean annual quality (mg/L)
   Water quality shall not exceed ___mg/L. Volume of lake is ___, etc.
7. Cost per pound of pollutant removed
8. Treatment cost per unit of production
9. Mass discharge per unit of raw material used
10. Mass discharged per unit of production
11. Percent removal proportional to raw load per day
12. Larger facilities to achieve higher removal rates

13. Percent removal proportional to community income
14. Effluent charges (pounds per week)
15. Effluent charge above some load limit
16. Seasonal limits on cost-effectiveness analysis
17. Minimum total treatment cost
18. Best Available Technology (BAT for industry) plus some level for municipal inputs
19. Assimilative capacity divided to require an "equal effort among dischargers"
20. Municipal treatment level proportional to plant size
21. Industrial treatment: equal percent between best practicable technology (BPT) and BAT
22. Industrial discharges given different treatment levels for different stream flows and seasons

TCWA.24 WASTE LOAD ALLOCATION CALCULATION

Develop a conversion factor that allows the product of concentration (mg/L) and flow rate (gpm) to be converted to lb/day.

SOLUTION: Mass discharge = (stream concentration, mg/L)(flow rate, gpm) To convert to consistent units:

$$\frac{\text{lb}}{\text{day}} = \left(\frac{\text{mg}}{\text{L}}\right)\left(10^{-6}\frac{\text{L}}{\text{mg}}\right)\left(\frac{\text{gal}}{\text{min}}\right)\left(\frac{8.34\,\text{lb}}{\text{gal}}\right)\left(1440\frac{\text{min}}{\text{day}}\right)$$

$$= \left(\frac{\text{mg}}{\text{L}}\right)\left(\frac{\text{gal}}{\text{min}}\right)(8.34 \times 10^{-6})(1440);\frac{\text{lb}}{\text{day}}$$

$$= 1.2 \times 10^{-2}\left(\frac{\text{mg}}{\text{L}}\right)\frac{\text{gal}}{\text{min}}$$

Therefore, if the concentration and flowrate are given with units of mg/L and gal/min, respectively, multiply the product of both terms by 1.2×10^{-2} to obtain the total daily mass discharge in lb/day.

TCWA.25 SIZING OF AN AEROBIC DIGESTER

A municipality generates 1000 lb of solids daily. Size an aerobic digester to treat the solids. The following design parameters and information are provided:

Detention (retention) time, hydraulic, $t_H = 20$ days
Detention time, solids, $t_s = 20$ days
Temperature = 95°F
Organic loading (OL) = 0.2 lbVS/(ft^3 · day)

Volatile solids (VS) = 78% of total solids

Percentage solids (TS) entering digester = 4.4%

VS destruction = 62%

SOLUTION: Check the design based on the organic load and the hydraulic load. The volume based on the organic load V_{OL}, is

$$V_{OL} = (1000)(0.78)/(0.2)$$
$$= 3900 \, \text{ft}^3$$

Based on the hydraulic load the volume, V_H, is

$$V_H = \frac{(1000)(20)}{(0.044)(8.33)(7.48)} \, ; 8.33 \, \mu/\text{gal}, 7.48 \, \text{gal/ft}^3$$
$$= 7300 \, \text{ft}^3$$

Since 7300 > 3900, the hydraulic detention time controls and the required design volume is 7300 ft^3.

TCWA.26   ANALYSIS OF AN ANAEROBIC CONTACT PROCESS

A 100,000 gal per day protein-contaminated wastewater with a COD of 4000 mg/L is generated by a meat processing plant. This waste is to be treated by the anaerobic contact process using a loading of 0.15 lb COD/ft^3 · day and a sludge detention time of 20 days. Assuming 80% efficiency, you have been asked to determine/provide the following as part of a permit to be submitted to the state.

1. A flow diagram of the process
2. Hydraulic detention time, hours
3. Daily solids accumulation, pounds (assume a 6% synthesis of solids from BOD)
4. Nitrogen and phosphorous requirements, pounds (10 and 1.5%, respectively, in solids)
5. Mixed liquor suspended solids concentration

SOLUTION:

1. A very simple flow diagram of the process is provided in Figure 4.1. The daily COD removed is

$$COD = (4000 \, \text{mg/L})(0.1 \times 10^6 \text{gal/day}) \, (10^{-6} \text{L/mg})(8.34 \, \text{lb/gal})$$
$$= 3336 \, \text{lb COD/day}$$

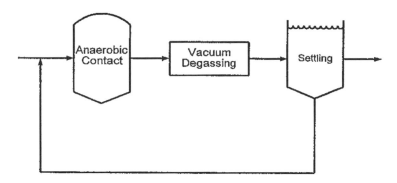

FIGURE 4.1   Flow diagram for anaerobic process.

2. The volume of the anaerobic contact tank is

$$V = (3336\,\text{lb COD/day})(7.48\,\text{gal/ft}^3)(\text{ft}^3 \cdot \text{day}/0.15\,\text{lb})$$
$$= 166,000\,\text{gal}$$

The hydraulic detention time, t_H, is then

$$t_\text{H} = (166,000\,\text{gal})\left(\frac{\text{day}}{0.1 \times 10^6\,\text{gal}}\right)$$

$$= 1.66\,\text{days}$$
$$= 39.8\,\text{h}$$

3. The solids growth (SG) may be calculated using the 6% synthesis figure and 80% efficiency:

$$\text{SG} = (3336)(0.8)(0.06)$$
$$= 160\,\text{lb/day}$$

The SG represents the mass of suspended solids and is the active cell population.

4. Since 10% of the solids are nitrogen,

$$N = (160)(0.1)$$
$$= 16\,\text{lb N/day}$$

Since 1.5% of solids is phosphorous,

$$P = (160)(0.015)$$
$$= 2.4\,\text{lb P/day}$$

Since the wastewater is from a meat processing plant, it is proteinaceous (rich in nitrogen and phosphorous) and N and P will not be required.

5. The concentration in (3) may be converted to mg/L as follows:

$$(1.0 \text{ lb/gal}) = (454 \text{ g/lb})(1000 \text{ mg/g})(264 \text{ gal/m}^3)(\text{m}^3/1000 \text{ L})$$
$$= 120,000 \text{ mg/L}$$

This number is equivalent to the conversion constant

$$\left(\frac{10^6 \text{ mg/L}}{8.34 \text{ lb/gal}}\right) \approx 120,000 \left(\frac{\text{mg/L}}{\text{lb/gal}}\right)$$

The latter conversion constant (120,000) is most often used in industry. Thus,

$$\text{MLVSS} = \frac{(160)(20)}{166,000}(120,000)$$
$$= 2300 \text{ mg/L}$$

TCWA.27 LAND TREATMENT OF INDUSTRIAL WASTEWATER

Land treatment of industrial wastewater is a process in which wastewater is applied directly to the land. This type of treatment is most common for food processing including meat, poultry, dairy, brewery, and winery wastes. The principal rationale of this practice is that the soil is a highly efficient biological treatment reactor, and food processing wastewater is highly degradable. This treatment practice is usually carried out by distributing the wastewater through spray nozzles onto the land or letting the water run through irrigation channels.

Suppose that the rate of the wastewater flowing to a land application site is 178 gal/acre/min and the irrigated land area is 5.63 acres. The entire irrigation process lasts 7.5 hr/d. If the wastewater has a BOD_5 concentration of 50 mg/L, what is the mass of BOD_5 remaining in the soil after the land treatment process is complete (assume a BOD_5 removal efficiency by the land treatment process is 95%)?

SOLUTION: The amount of the wastewater (q) flowing to the land is:

$$q = (178 \text{ gal/acre/min})(60 \text{ min/hr})(7.5 \text{ hr/d})(5.63 \text{ acre})$$
$$= 450,963 \text{ gal/d}$$

The amount of BOD_5 remaining in the soil is:

$$BOD_5 = (450,963 \text{ gal/d}) (1.00 - 0.95) (50 \text{ mg/L}) (3.785 \text{ L/gal}) = 4,267,688 \text{ mg/d}$$
$$= (4,267,688 \text{ mg/d}) (1 \text{ g}/1,000 \text{ mg}) (454 \text{ g/lb})$$
$$= 9.4 \text{ lb/d}$$

TCWA.28 CONTAMINANT TRANSPORT

The Perfect Plating Company is planning to build a plant on the Rugged River, discharging its wastewater into the river. The waste stream is produced 24 hours per day, has a cyanide concentration of 3.0 mg/L, and has a flow rate of 0.3 ft^3/s. The maximum allowable concentration of cyanide in the river below the outfall is 3.5 μg/L. The river flow is seasonal and the monthly average flow rates from the previous 5 year record are as follows:

Month	Oct	Nov	Dec	Jan	Feb	Mar	Apr	May	Jun	Jul	Aug	Sep
Flow (ft^3/s)	146	254	300	395	690	607	333	386	321	177	185	154

1. For an average day in the month of October, show that discharging all of the waste would produce a down stream concentration of cyanide above the maximum allowable value. Assume no background cyanide concentration in the river.
2. Show that over the entire year, the discharge of the waste would not violate the permit.
3. Establish a discharge schedule assuming that, in the months of low river flow, the maximum amount of waste is discharged, with any excess flow diverted to storage, and that waste from storage is discharged with process waste during months of high flow.
4. Calculate the amount of storage required. Assume storage is covered so that there is no change in the concentration or volume of the stored waste.

SOLUTION:

1. Assume perfect mixing. For an average day in the month of October, the waste stream is diluted by the river so that the river concentration is the following mass fraction of the original waste stream concentration:

   Waste flow/(Waste + River flow) = 0.3 ft^3/s$(146 + 0.3$ ft^3/s$) = 0.00205$

   The cyanide concentration in the river is then:

   river concentration $= 0.00205$ (3 mg/L) $= 6.1 \times 10^{-3}$ mg/L $= 6.1$ μg/L

   This is above the permitted level of 3.5 μg/L.

2. The yearly flow of the river and of waste stream are calculated as:

   Waste stream:

   Total flow $= 0.3$ ft^3/s (365 d/yr) (86,400 s/d) $= 9.46 \times 10^6$ ft^3/yr

River stream:

Monthly flow = average flow (ft^3/s) (d/month) (86,400 s/d)

Total flow = $\sum$ monthly flows = 102.72×10^8 ft^3/yr (see Table 4.6 below)

River + Waste flow:

$$102.72 \times 10^8 \text{ ft}^3/\text{yr} + 9.46 \times 10^6 \text{ ft}^3/\text{yr} = 102.81 \times 10^8 \text{ ft}^3/\text{yr}$$

Yearly average dilution is:

$$\text{Wasteflow}/(\text{Total flow}) = (9.46 \times 10^6 \text{ ft}^3/\text{yr})/(102.81 \times 10^8 \text{ ft}^3/\text{yr})$$
$$= 9.20 \times 10^{-4}$$

The effective river cyanide concentration is then:

$$\text{River concentration} = (9.20 \times 10^{-4})(3.0 \text{ mg/L}) = 2.76 \times 10^{-3} \text{ mg/L}$$
$$= 2.76 \text{ } \mu\text{g/L}$$

Therefore, this discharge level is acceptable since it is below 3.5 μg/L.

3. A proposed disposal schedule is based on the monthly average river discharge, and the dilution provided to the waste stream by the river. As shown in Table 4.6, the excess flow is diverted to a covered storage area and held until river flow is high enough to allow its discharge.

TABLE 4.6 Calculations for Storage and Dilution Requirements to Maintain Allowable Cyanide Levels in River

Month	Flow (ft^3/s)	Days per Month	Volume $\times 10^{-8}$ (ft^3)	Allowable Cyanide Discharge (g/s)	Volume to Storage (ft^3)	Discharge Waste Stream	(ft^3/s) From Storage
Oct	146	31	3.91	0.0145	347,000	0.171	–
Nov	254	30	6.58	0.0252	9,000	0.296	–
Dec	300	31	8.04	0.0297	–	0.300	0.049
Jan	395	31	10.58	0.0392	–	0.300	0.161
Feb	690	28	16.69	0.0684	–	0.300	0.505
Mar	607	31	16.26	0.0602	–	0.300	0.408
Apr	333	30	8.31	0.0330	–	0.300	0.088
May	386	31	10.34	0.0383	–	0.300	0.150
Jun	321	30	8.32	0.0318	–	0.300	0.074
Jul	177	31	4.74	0.0175	252,000	0.206	–
Aug	185	31	4.96	0.0183	227,000	0.215	–
Sep	154	30	3.99	0.0153	311,000	0.180	–
			102.72		1,146,000		

The allowable cyanide discharge rate in each month is calculated as follows:

$$\text{Allowable discharge} = \text{River flow (discharge limit)}(1\,\text{L}/0.0353\,\text{ft}^3)$$

For October this becomes:

$$\text{Allowable discharge} = 146\,\text{ft}^3/\text{s}\,(3.5 \times 10^{-6}\,\text{g/L})\,(1\,\text{L}/0.0353\,\text{ft}^3)$$
$$= 0.0145\,\text{g/s}$$

The waste flow discharge rate of cyanide is calculated as:

$$\text{Discharge} = (0.3\,\text{ft}^3/\text{s})\,(3.0 \times 10^{-3}\,\text{g/L})\,(1\,\text{L}/0.0353\,\text{ft}^3)$$
$$= 0.0255\,\text{g/s}$$

The discharge schedule is based on the relationship between monthly allowable cyanide discharge rates, and the waste stream cyanide discharge rate. Discharge is allowed from the waste stream only if the allowable rate is less than the waste stream discharge rate. If this is not the case, as in October, only a fraction of the waste stream may be discharged, the balance being diverted to storage. For October:

$$\text{Allowable discharge rate} = (\text{Waste flow})$$

$$\times \frac{\text{Allowable mass discharge rate}}{\text{Actual waste stream mass discharge rate}}$$

$$(4.30)$$

$$\text{October discharge rate} = 0.3\,\text{ft}^3/\text{s}\,(0.0145/0.0255)$$
$$= 0.171\,\text{ft}^3/\text{s}$$

If the allowable mass rate of discharge of cyanide is greater than the waste flow rate, the discharge of the waste stream is supplemented with discharge from storage, and the storage discharge rate is calculated as follows:

Storage discharge rate

$$= (\text{Waste flow})\,\frac{\text{Allowable mass discharge rate} - \text{Waste mass discharge rate}}{\text{Waste mass discharge rate}}$$

$$\text{December storage discharge rate} = (0.30\,\text{ft}^3/\text{s})\,(0.0297 - 0.0255)/(0.0255)$$

$$= 0.049\,\text{ft}^3/\text{s} \qquad (4.31)$$

If the accumulation period runs from July through November, storage will be empty from March until June.

4. If the allowable rate of cyanide discharge is less than the waste discharge rate, then waste flow must be diverted to storage. The required storage volume is calculated as:

$$\text{Storage} = (\text{Waste flow}) \; \frac{\text{Waste mass flow rate} - \text{Allowable mass flow rate}}{\text{Waste mass flow rate}}$$

$$\times \, (\text{days}) \, (86{,}400 \, \text{s/day}) \tag{4.32}$$

October required storage volume is calculated as:

$$(0.3 \, \text{ft}^3/\text{s}) \, [(0.0255 - 0.0145)/(0.0255)] \, (31 \, \text{d}) \, (86{,}400 \, \text{s/d}) = 347{,}000 \, \text{ft}^3$$

Since the months in which the storage is required are consecutive, the total storage requirement is the sum for these months or 1,146,000 ft³. A volume of 1.25×10^6 ft³ would provide approximately 10% excess storage.

TCWA.29 MATERIAL BALANCES ON AN INDUSTRIAL WASTEWATER SYSTEM

A regulatory official has been requested to answer/solve the following two questions as part of a permit review.

1. Consider the flow diagram in Figure 4.2 for a wastewater treatment system. The following flowrate data are given:

$$\dot{m}_1 = 1000 \, \text{lb/min}$$
$$\dot{m}_2 = 1000 \, \text{lb/min}$$
$$\dot{m}_4 = 200 \, \text{lb/min}$$

Find the amount of water lost by evaporation in the operation.

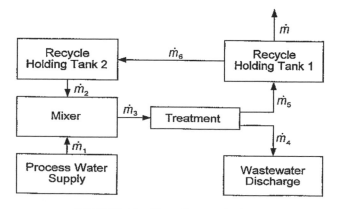

FIGURE 4.2 Flow diagram for Part 1.

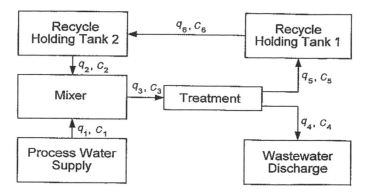

FIGURE 4.3 Flow diagram for Part 2.

2. Consider the same system shown in Figure 4.3.

The following volumetric flowrate and phosphate concentration data have been provided by the plant manager.

$$q_1 = 1000 \text{ gal/day} \quad C_1 = 4 \text{ ppm}$$
$$q_2 = 1000 \text{ gal/day} \quad C_2 = 0 \text{ ppm}$$
$$q_3 = 2000 \text{ gal/day} \quad C_3 = 2 \text{ ppm}$$
$$q_4 = 200 \text{ gal/day} \quad C_4 = 20 \text{ ppm}$$
$$q_5 = 1800 \text{ gal/day} \quad C_5 = 0 \text{ ppm}$$
$$q_6 = 1000 \text{ gal/day} \quad C_6 = 0 \text{ ppm}$$

Are the data correct and/or consistent?

SOLUTION:

1. Apply a material balance around the treatment system to determine the value of $\dot{m}_5$. The value of $\dot{m}_5$ equals the number of gallons of water being turned into steam:

$$\dot{m}_3 = \dot{m}_4 + \dot{m}_5$$
$$\dot{m}_1 + \dot{m}_2 = \dot{m}_4 + \dot{m}_5$$
$$1000 + 1000 = 200 + \dot{m}_5$$
$$\dot{m}_5 = 1800 \text{ lb/min}$$

Similarly (for tank 2),

$$\dot{m}_6 = \dot{m}_2$$
$$\dot{m}_6 = 1000 \text{ lb/min}$$

Thus (for tank 1),

$$\dot{m}_6 = \dot{m}_5 + \dot{m}$$
$$1000 = 1800 - \dot{m}$$
$$\dot{m} = 800 \text{ lb/min}$$

One sees that 800 lb of water per minute are lost in the operation.

2. A componential balance around the mixer gives

$$C_1 q_1 + C_2 q_2 = C_3 q_3$$

$$\left(\frac{4}{120,000}\right)(1000) + \left(\frac{0}{120,000}\right)(1000) = \left(\frac{2}{120,000}\right)(2000)$$

$$4000 = 4000; \quad \text{OK}$$

A balance around the treatment tank gives

$$C_3 q_3 = C_4 q_4 + C_5 q_5$$

$$\left(\frac{2}{120,000}\right)(2000) = \left(\frac{20}{120,000}\right)(200) + \left(\frac{0}{120,000}\right)(1800)$$

$$4000 = 4000; \quad \text{OK}$$

A balance around hold tank 1 gives

$$C_5 q_5 = C_6 q_6$$
$$(0)(1800) = (0)(1000)$$
$$0 = 0; \quad \text{OK}$$

A balance around hold tank 2 gives

$$C_2 q_2 = C_6 q_6$$
$$(0)(1000) = (0)(1000)$$
$$0 = 0; \quad \text{OK}$$

The data are consistent.

TCWA.30   HALF LIFE DETERMINATION

Determine the time, in hours, required for the concentration of toluene in a wastewater treatment process to be reduced to one-half of its initial value. Assume the first-order reaction velocity constant for toluene is 0.07/h.

Calculate the time for the toluene to be reduced to 99% of its initial value. Also calculate the time for 99.99% of the toluene to react.

SOLUTION: The decay (disappearance) of a chemical can often be described as a first-order function:

$$\frac{dC}{dT} = -kC \qquad (4.33)$$

where C = concentration at time t
  t = time
  k = first-order reaction rate constant
  The integrated form of this equation is

$$\ln \frac{C_0}{C} = kt$$

where C_0 is the concentration at time zero.

When half of the initial material has decayed (reacted), C_0/C is equal to 2; the corresponding time is given by the following expression:

$$t_{1/2} = \frac{\ln(2)}{k}$$

$$- \frac{0.69}{0.07/h}$$

$$= 9.86\,h$$

When 99% of the initial material has decayed, C_0/C is equal to 100 and

$$t = \frac{\ln(100)}{k}$$

$$= \frac{4.605}{0.07/h}$$

$$- 65.79\ h$$

When 99.99% of the toluene has, reacted, C_0/C is equal 10,000. For the case,

$$t = \frac{\ln(10,000)}{0.07}$$

$$= 131.6\,h$$

TCWA.31   FEEDSTOCK pH CONTROL BY BLENDING

Raw materials inherently have some variation in composition. Depending on the nature of the industry, the variation can be large or small. Waste treatment plants must be able to handle large variations in "feedstocks" since waste materials may come from many sources or process streams. However, most systems can only tolerate a specific range in composition and still operate efficiently and within required norms. Waste treatment plants can deal with this problem by blending out-of-range supplies in storage tanks until an acceptable feedstock to the plant/system results.

An important chemical property of an aqueous waste that affect system performance is its pH. The pH measures the acidity or basicity of the solution. In a neutral solution, such as pure water, the hydrogen (H^+) and hydroxyl (OH^-) ion concentrations are equal. At ordinary temperatures, this concentration is

$$[H^+] = [OH^-] = 10^{-7} \text{ g-ion/L} \qquad (4.34)$$

where $[H^+]$ = hydrogen ion concentration
   $[OH^-]$ = hydroxyl ion concentration

The unit *g-ion* stands for gram-ion, which represents an Avogadro number of ions. In all aqueous solutions, whether neutral, basic, or acidic, a chemical equilibrium or balance is established between these two concentrations, so that

$$K_{eq} = [H^+][OH^-] = 10^{-14} \qquad (4.35)$$

where K_{eq} = equilibrium constant

The numerical value for K_{eq} given above holds for room temperature and only when the concentrations are expressed in gram-ion per liter (g-ion/L). In acid solutions, $[H^+]$ is $> [OH^-]$; in basic solutions $[OH^-]$ predominates. The pH is a direct measure of the hydrogen ion concentration and is defined by

$$pH = -\log[H^+] \qquad (4.36)$$

Thus, an acidic solution is characterized by a pH below 7 (the lower the pH, the higher the acidity); a basic solution, by a pH above 7; and a neutral solution by a pH of 7. Additional details are available in Problem TCWA.9 and TCWA.10.

In attempting to reduce the generation of pollutant downstream in a process, the environmental engineering department of a company has suggested feedstock blending to control the pH in a 50,000 gallon storage tank. Normally, the tank is kept at a neutral pH of 7. However, operations can tolerate pH variations only from 6 to 8. Feedstock arrives in 5,000 gallon shipments. Assume that the tank is completely mixed, contains 35,000 gallons when the shipment arrives, the incoming acidic waste is fully dissociated, and there is negligible buffering capacity in the tank.

1. What is the pH of the most acidic single feedstock shipment that can be handled without neutralization?
2. What is the pH of the most acidic feedstock shipment that can be handled without neutralization if the storage tank volume is at 45,000 gallons?

SOLUTION:

1. Determine the maximum ion concentration tolerable for a pH = 6.

$$-\log[H^+] = 6$$

$$[H] = 10^{-6}$$

2. Determine the concentration of H^+ ions currently present in the neutrally pH tank for a pH of 7.

$$-\log[H^+] = 7$$
$$[H^+] = 10^{-7}$$

3. Calculate the concentration of ions (X) which can be added and the corresponding pH.

$$[H] = 10^{-6} = [(5000)(X) + (35,000)(10^{-7})]/40,000$$

$$X = [(40,000)(10^{-6}) - (35,000)(10^{-7})]1/5,000$$

$$= 7.3 \times 10^{-6}$$

$$pH = \log[7.3 \times 10^{-6}]$$

$$= 5.14$$

4. Determine the minimum pH of the shipment if the storage tank is 45,000 gal full. With a tank volume of 45,000 gallons, the calculation is performed in an identical manner. Steps 3 and 4 become:

$$X = [(50,000)(10^{-6}) - (45,000)(10^{-7})]/5,000$$

$$= 9.1 \times 10^{-6}$$

$$pH = -\log [9.1 \times 10^{-6}]$$

$$= 5.04$$

Some flexibility exists in blending shipments to keep the feed stream within requirements and thus avoiding the costly step of neutralization. This flexibility increases when several storage tanks can be used or if the sequence of blending of shipments can be varied based on composition.

TCWA.32   EFFECTS OF VOCs FROM PUBLICLY OWNED TREATMENT WORKS (POTWs)

In the mid 1990s, EPA announced plans to regulate POTWs under the new (1990) Clean Air Act. They estimated that POTWs annually release 11,000 tons of air pollutants. The vast majority of these pollutants are 31 VOCs. While the 11,000 tons/yr represent only 0.1% of the nationwide emissions, municipalities continue to brace for expected regulatory action.

In controlling VOC emissions, it is important to address both capture and treatment control methods as well as source reduction. A debate is raging among engineers as to which method is most cost effective within an acceptable risk limitation. Table 4.7 provides some of the proposed control technologies.

TABLE 4.7   Source Reduction/Treatment Alternative

Source Reduction	Treatment Alternatives
Use of iron salt precipitation; advanced treatment; other chemical addition methods	Cover all tanks, capture and treatment emitted gas
Use ozonation or UV disinfection instead of chlorine	Ensure complete combustion of digester gas
Do not use aerated grit chambers (use velocity or vortex instead)	Use air stripping and vapor phase controls
Industrial pretreatment	

It is apparent from the table the source reduction techniques would most likely be less expensive than the construction of additional treatment facilities. The only remaining question is, will the source reduction method meet the acceptable risk level. The answer to this question depends on who is asked. EPA estimates 1.5 cancer cases per year for the most exposed individuals due to the POTW emissions. Given this low number, if VOC reduction is to be implemented at municipal treatment plants, source reductions will be the best control technology.

If advanced primary treatment is implemented (addition of ferric chloride, $FeCl_3$, and polymer), the biochemical oxygen demand (BOD) removed in primary treatment increases from 30% to 45%. This decrease in the loading to the aeration tank causes less oxygen to be delivered and therefore fewer VOCs discharged to the environment.

Calculate the number of cancer cases per year for the most exposed individuals if ferric chloride is added to the wastewater during primary treatment. Assume the reduction in VOCs is reduced linearly with the oxygen supplied, and the cancer cases are reduced at twice the percent VOC reduction (due to decreased concentrations being less harmful). Also calculate the annual chemical cost and the cost/less cancer case. Data is provided below

> Flow, $q = 50$ million gallons per day (MGD)
> Influent BOD $= 175$ mg/L
> $FeCl_3 = 20$ mg/L @ \$0.12/lb
> Polymer $= 0.2$ mg/L; @ \$1.70/lb

SOLUTION:

1. Calculate the inlet BOD load in lb/day. Apply the appropriate conversion factor.

$$\dot{m} = QC = (50 \times 10^6\,\text{gal/day})(175\,\text{mg/L})(8.34 \times 10^{-6}\,\text{lb}\cdot\text{L/gal}\cdot\text{mg})$$
$$= 72{,}975\,\text{lb/day}$$

2. Calculate the BOD load to the aeration basin, with and without chemical addition.

$$W/O \text{ chem: assume } 30\% \text{ removal per load} = (0.7)(72{,}975 \text{ lb/day})$$
$$= 51{,}082 \text{ lb/day}$$
$$W/\text{chem: assume } 45\% \text{ removal per load} = (0.55)(72{,}95 \text{ lb/day})$$
$$= 40{,}136 \text{ lb/day}$$

3. Calculate the percent BOD reduction with chemical addition.

$$\%\text{load reduction} = [(51{,}082 - 40{,}136)/51{,}082](100\%)$$
$$= 21.4\% \text{ reduction}$$

4. Calculate the cancer case reduction and the new cases/year.

$$\text{Cancer reduction} = (2)(\text{VOC reduction})$$
$$= (2)(21.4\%)$$
$$= 42.8\%$$
$$\text{new cases/yr} = (1.5/\text{yr})[1 - (42.8/100)]$$
$$= 0.86/\text{yr}$$

5. Calculate the annual chemical costs.

$$FeCl_3: (50 \times 10^6 \text{ gal/day})(20 \text{ mg/L})(8.34 \times 10^{-6} \text{lb. L/gal} \cdot \text{mg})$$
$$= 8340 \text{ lb/day}$$
$$(8340 \text{ lb/day})(\$0.12/\text{lb})(365 \text{ day/yr}) = \$365{,}300$$

$$\text{Polymer: } (50 \times 10^6 \text{ gal/day})(0.2 \text{ mg/L})(8.34 \times 10^{-6} \text{ lb. L/gal} \cdot \text{mg})$$
$$= 83.4 \text{ lb/day}$$
$$(83.4 \text{ lb/day})(1.70/\text{lb})(365 \text{ day/yr}) = \$51{,}700$$
$$\text{annual chemical cost} = 365{,}300 + 51{,}700$$
$$= \$417{,}000$$

This preliminary analysis does not directly show the cost benefit for preventing the VOC pollution. However, it is easy to see that the recommended VOC treatment methods would be considerably more expensive. The area of VOC management at POTWs will continue to be a major issue as the cost/benefit of source and treatment control technologies are researched.

**TABLE 4.8   Daily TSS Effluent Concentration
Data Collected Over a 7-Day Period at a
Municipal Wastewater Treatment Plant**

Day	TSS (mg/L)
1	20
2	100
3	50
4	42
5	33
6	25
7	15

TCWA.33   NPDES PERMIT LIMIT CALCULATION

The following total suspended solids (TSS) data were collected from a clarifier at a local municipal wastewater treatment plant over a 7-day period (see Table 4.8). The NPDES permit limitations for TSS effluent concentrations from this wastewater treatment plant is 45 mg/L on a 7-day average. Based on this information, is the treatment plant within its NPDES permit limits? This information is being requested since concern has arisen regarding the potential of the waste water discharge affecting local drinking water supplies.

SOLUTION: The 7-day average concentration for TSS is

$$(TSS)_7 = (20 + 100 + 50 + 42 + 33 + 25 + 15)/7$$
$$= 40.7 \, mg/L$$

The wastewater treatment plant is still within its NPDES permit limit (but only marginally) for an average 7-day maximum concertration of 45 mg/L for TSS.

CHAPTER 5

SAFE DRINKING WATER ACT (SDWA)

QUALITATIVE PROBLEMS (LSDWA)

PROBLEMS LSDWA.1–21

LSDWA.1 UNDERSTANDING THE SAFE DRINKING WATER ACT (SDWA)

Provide an overview of the Safe Drinking Water Act (SDWA).

SOLUTION: The SWDA was originally passed by Congress in 1974 to protect public health by regulating the nation's public drinking water supply. The law was amended in 1986 and 1996 and requires many actions to protect drinking water and its sources: rivers, lakes, reservoirs, springs, and ground water wells. (SDWA does not regulate private wells which serve fewer than 25 individuals.) SDWA authorizes the EPA to set national health-based standards for drinking water to protect against both naturally-occurring and man-made contaminants that may be found in drinking water. EPA, states, and water systems then work together to make sure that these standards are met.

Millions of Americans receive high quality drinking water every day from their public water system, which may be publicly or privately owned. Nonetheless, drinking water safety cannot be taken for granted. There are a number of threats to

Environmental Regulatory Calculations Handbook, by Leo Stander and Louis Theodore
Copyright © 2008 John Wiley & Sons, Inc.

drinking water: improperly disposed of chemicals; animal wastes; pesticides; human wastes; wastes injected deep underground; and, naturally-occurring substances can all contaminate drinking water. Likewise, drinking water that is not properly treated or disinfected, or which travels through an improperly maintained distribution system, may also pose a health risk.

Originally, SDWA focused primarily on treatment as the means of providing safe drinking water at the tap. The 1996 amendments significantly enhanced the existing law by recognizing source water protection, operator training, funding for water system improvements, and public information as important components of safe drinking water. This approach ensures the quality of drinking water by protecting it from source to tap.

See also: http://www.epa.gov/safewater/sdwa/30th/factsheets/understand.html

LSDWA.2 SDWA ROLES AND RESPONSIBILITIES

Describe the roles and responsibilities of the SDWA.

SOLUTION: SDWA applies to every public water system in the United States. There are currently more than 160,000 public water systems providing water to almost all Americans at some time in their lives. The responsibility for making sure these public water systems provide safe drinking water is divided among EPA, states, tribes, water systems, and the public. SDWA provides a framework in which these parties work together to protect this valuable resource.

EPA sets national standards for drinking water based on sound science to protect against health risks, considering available technology and costs. The National Primary Drinking Water Regulations set enforceable maximum contaminant levels for particular contaminants in drinking water or require ways to treat water to remove contaminants. Each standard also includes requirements for water systems to test for contaminants in the water to make sure standards are achieved. In addition to setting these standards, EPA provides guidance, assistance, and public information about drinking water, collects drinking water data, and oversees state drinking water programs.

The most direct oversight of water systems is conducted by state drinking water programs. States can apply to EPA for "primacy," the authority to implement SDWA within their jurisdictions, if they can show that they will adopt standards at least as stringent as EPA's and make sure water systems meet these standards. All states and territories, except Wyoming and the District of Columbia, have received primacy. While no Indian tribe has yet applied for and received primacy, four tribes currently receive "treatment as a state" status, and are eligible for primacy. States, or EPA acting as a primacy agent, make sure water systems test for contaminants, review plans for water system improvements, conduct on-site inspections and sanitary surveys, provide training and technical assistance, and take action against water systems not meeting standards.

To ensure that drinking water is safe, SDWA sets up multiple barriers against pollution. These barriers include: source water protection, treatment, distribution system integrity, and public information. Public water systems are responsible for ensuring that contaminants in tap water do not exceed the standards. Water systems treat the water, and must test their water frequently for specified contaminants and report the results to states. If a water system is not meeting these standards, it is the water supplier's responsibility to notify its customers. Many water suppliers now are also required to prepare annual reports for their customers. The public is responsible for helping local water suppliers to set priorities, make decisions on funding and system improvements, and establish programs to protect drinking water sources. Water systems across the nation rely on citizen advisory committees, rate boards, volunteers, and civic leaders to actively protect this resource in every community in America.

LSDWA.3   COMMUNITY VS. NON-COMMUNITY WATER SYSTEM

Describe the difference between a community and a non-community water system.

SOLUTION: All public water systems must have at least 15 service connections or serve at least 25 people per day for 60 days of the year. In addition drinking water standards apply to water systems differently based on their type and size.

Community Water System (there are approximately 54,000): A Public water system that serves the same people year-round. Most residences including homes, apartments, and condominiums in cities, small towns, and mobile home parks are served by Community Water Systems.

Non-Community Water System: A public water system that serves the public but does not serve the same people year-round.

LSDWA.4   TRANSIENT AND NON-TRANSIENT NON-COMMUNITY WATER SYSTEMS

Detail the difference between a transient and non-transient non-community water system.

SOLUTION: There are two types of non-community systems:

Non-transient Non-Community Water System (There are Approximately 20,000)

A non-community water system serves the same people more than six months per year, but not year-round. For example, a school with its own water supply is considered a non-transient system.

Transient Non-Community Water System (There are Approximately 89,000)

A non-community water system serves the public but not the same individuals for more than six months. For example, a rest area of campground may be considered a transient water system.

LSDWA.5 PROTECTION AND PREVENTION ASPECTS OF THE SDWA

Describe the protection and prevention aspects of the SDWA.

SOLUTION: Essential components of safe drinking water include protection and prevention. States and water suppliers must conduct assessments of water sources to see where they may be vulnerable to contamination. Water systems may also voluntarily adopt programs to protect their watershed or wellhead and states can use legal authorities from other laws to prevent pollution. SDWA mandates that states have programs to certify water system operators and make sure that new water systems have the technical, financial, and managerial capacity to provide safe drinking water. SDWA also sets a framework for the Underground Injection Control (UIC) program which sets standards for safe waste injection practices and bans certain types of injection altogether. All of these programs help prevent the contamination of drinking water.

LSDWA.6 SETTING NATIONAL DRINKING WATER STANDARDS

Provide some details on the national drinking water standards.

SOLUTION: EPA sets national standards for tap water which help ensure consistent quality in this nation's water supply. EPA prioritizes contaminants for potential regulation based on risk and how often they occur in water supplies. (To aid in this effort, certain water systems monitor for the presence of contaminants for which no national standards currently exist and collect information on their occurrence.) EPA sets a health goal based on risk (including risks to the most sensitive people, e.g., infants, children, pregnant women, the elderly, and the immuno-compromised). EPA then sets a legal limit for the contaminant in drinking water or a required treatment technique. This limit or treatment technique is set to be as close to the health goal as feasible. EPA also performs a cost-benefit analysis and obtains input from interested parties when setting standards. EPA is currently evaluating the risks from several specific health concerns, including: microbial contaminants (e.g., cryptosporidium); the byproducts of drinking water disinfection; radon; arsenic; and, water systems that do not currently disinfect their water but obtain it from a potentially vulnerable groundwater source.

LSDWA.7 PROCEDURE FOR SETTING DRINKING WATER STANDARDS

Describe the process of setting drinking water standards.

SOLUTION: EPA sets primary drinking water standards through a three-step process. First, EPA identifies contaminants that may adversely affect public health and occur in drinking water with a frequency and at levels that pose a threat to public health. EPA identifies these contaminants for further study, and determines the contaminants to potentially regulate. Second, EPA determines a maximum contaminant level goal for contaminants it decides to regulate. This goal is the level of a contaminant in drinking water below which there is no known or expected risk to health. These goals allow for a margin of safety. Third, EPA specifies a maximum contaminant level, the maximum permissible level of a contaminant in drinking water which is delivered to any user of a public water system. These levels are enforceable standards and are set as close to the goals as "feasible."

SDWA defines "feasible" as the level that may be achieved with the use of the best technology, treatment techniques, and other means which EPA finds (after examination for efficiency under field conditions) are available, taking cost into consideration. When it is not economically or technically "feasible" to set a maximum level, or when there is no reliable or economic method to detect contaminants in the water, EPA instead sets a required treatment technique which specifies a way to treat the water to remove contaminants.

LSDWA.8 SDWA COMPLIANCE AND ENFORCEMENT

Describe compliance enforcement provisions of the SDWA.

SOLUTION: National drinking water standards are legally enforceable, which means that both EPA and states can take enforcement actions against water systems not meeting safety standards. EPA and states may issue administrative orders, take legal actions, or fine utilities. EPA and states also work to increase water systems.

LSDWA.9 PUBLIC INFORMATION

Describe the modification provisions of the SDWA.

SOLUTION: SDWA recognizes that since everyone drinks water, everyone has the right to know what is in it and where it comes from. All water suppliers must notify consumers quickly when there is a serious problem with water quality. Water systems serving the same people year-round must provide annual consumer confidence reports on the source and quality of their tap water. States and EPA must prepare annual summary reports of water system compliance with drinking water safety standards and make these reports available to the public. The public must have a chance to be involved in developing source water assessment programs, state plans to use drinking water state revolving loan funds, state capacity development plans, and state operator certification programs.

LSDWA.10   1996 SDWA AMENDMENT

Provide the highlights of the 1996 SDWA Amendment.

SOLUTION:

1. *Consumer Confidence Reports*
   All community water systems must prepare and distribute annual reports about the water they provide, including information on detected contaminants, possible health effects, and the water's source.

2. *Cost-Benefit Analysis*
   EPA must conduct a thorough cost-benefit analysis for every new standard to determine whether the benefits of a drinking water standard justify the costs.

3. *Drinking Water State Revolving Fund*
   States can use this fund to help water systems make infrastructure or management improvements or to help systems assess and protect their source water.

4. *Microbial Contaminants and Disinfection Byproducts*
   EPA is required to strengthen protection for microbial contaminants, including cryptosporidium, while strengthening control over the byproducts of chemical disinfection. The "Stage 1 Disinfectants and Disinfection Byproducts Rule" and the "Interim, Enhanced Surface Water Treatment Rule" together address these risks.

5. *Operator Certification*
   Water system operators must be certified to ensure that systems are operated safely. EPA issued guidelines in February 1999 specifying minimum standards for the certification and recertification of the operators of community and non-transient, noncommunity water systems. These guidelines apply to state Operator Certification Programs. All States are currently implementing EPA-approved operator certification programs.

6. *Public Information and Consultation*
   SDWA emphasizes that consumers have a right to know what is in their drinking water, where it comes from, how it is treated, and how to help protect it. EPA distributes public information materials (through its Safe Drinking Water Hotline, Safewater web site, and Water Resource Center) and holds public meetings. EPA works with states, tribes, water systems, and environmental and civic groups, to encourage public involvement.

7. *Small Water Systems*
   Small water systems are given special consideration and resources under SDWA to make sure they have the managerial, financial, and technical ability to comply with drinking water standards.

8. *Source Water Assessment Programs*

Every state must conduct an assessment of its sources of drinking water (rivers, lakes, reservoirs, springs, and ground water wells) to identify significant potential sources of contamination and to determine how susceptible the sources are to these threats.

LSDWA.11 SAFE DRINKING WATER ACT REQUIREMENTS

Provide information on the SDWA requirements.

SOLUTION: The Safe Drinking Water Act (SDWA) was originally passed in 1974 to ensure that public water supplies are maintained at high quality. This is accomplished by setting national standards for levels of contaminants in drinking water by regulating underground injection wells and by protecting sole source aquifers.

The SDWA requires the EPA to establish Maximum Contamination Level Goals (MCLGs) and National Primary Drinking Water Regulations (NPDWR) for contaminants that, in the judgement of the EPA Administrator, may cause any adverse effect on the health of persons and that are known or anticipated to occur in public water systems. The NPDWRs are to include Maximum Contamination Levels (MCLs) and "criteria and procedures to assure a supply of drinking water which dependably complies" with such MCLs. If it is not feasible to ascertain the level of a contaminant in drinking water, the NPDWRs may require the use of a treatment technique instead of an MCL. The EPA is mandated to established MCLGs and promulgate NPDWRs for 83 contaminants in public water systems. (The SDWA was amended in 1986 by establishing a list of 83 contaminants for which EPA is to develop MCLGs and NPDWRs). MCLGs and MCLs were to be promulgated simultaneously.

MCLGs do not constitute regulatory requirements that impose any obligation on public water systems. Rather, MCLGs are health goals that are based solely upon consideration of protecting the public from adverse health effects of drinking water contamination. The MCLGs reflect the aspirational health goals of the SDWA that the enforceable requirements of NPDWRs seek to attain. MCLGs are to be set at a level where "no known or anticipated adverse effects on the health of persons occur and which allows an adequate margin of safety."

The House Report on the bill that eventually became the SDWA of 1974 provided congressional guidance on developing MCLGs:

[T]he recommended maximum contamination level [renamed maximum contamination level goal in the 1986 amendments to the SDWA] must be set to prevent the occurrence of any known or anticipated adverse effect. It must include an adequate margin of safety, unless there is no safe threshold for a contaminant. In such a case, the recommended maximum contamination level would be set at the zero level (40 CFR, Parts 141 and 142, June 7, 1991).

NPDWRs include either MCLs or treatment technique requirements as well as compliance monitoring requirements. The MCL for a contaminant must be set as close to the MCLG as feasible. Feasible means "feasible with the use of the treatment techniques and other means which the Administrator of the EPA finds, after examination for efficacy under field conditions and not solely under laboratory conditions, are available (taking cost into consideration)." A treatment technique must "prevent known or anticipated adverse effects on the health of a person to the extent feasible." A treatment technique requirement can be set only if the EPA Administrator makes a finding that "it is not economically or technically feasible to ascertain the level of the contaminant". Also the SDWA requires the EPA to identify the best available technology (BAT) for meeting the MCL for each contaminant.

EPA sets national secondary drinking water regulations (NSDWRs) to control water color, odor, appearance, and other characteristics affecting consumer acceptance of water (40 CFR Parts 141 and 142). The secondary regulations are not federally enforceable but are considered guidelines for the states (40 CFR Part 143).

LSDWA.12   OIL POLLUTION CONTROL ACT

Describe the Oil Pollution Control Act.

SOLUTION: The Oil Pollution Control Act of 1990 (OPA) was enacted to expand prevention and preparedness activities, improve response capabilities, ensure that the spill does not affect drinking water, ensure that shippers and oil companies pay the costs of spills that do occur, and establish and expand research and development programs. This was all in response to the Exxon Valdez oil spill in Prince William Sound in 1989.

The OPA establishes a new Oil Spill Liability Trust Fund, administered by the United States Coast Guard. This fund replaces the fund established under the CWA and other oil pollution funds. The Act mandates prompt and adequate compensation for those harmed by oil spills, and an effective and consistent system of assigning liability. The Act also strengthens requirements for the proper handling, storage, and transportation of oil and for the full and prompt response in the event discharges occur. The Act does so in part by amending section 311 of the CWA.

There are eight titles codified under the Act, details of which are available in the literature. Regulations to implement the oil pollution control act are found at 40 CFR 112.

LSDWA.13   SDWA PURPOSE

What is the purpose of the *Safe Drinking Water Act* (SDWA)?

SOLUTION: The purpose of the SDWA is to establish the aforementioned *maximum containment level goals* (MCLGs) and *national primary drinking water regulations* (NPDWRs) that are designed to protect the public from the contamination of drinking water.

LSDWA.14 NANOTECHNOLOGY REGULATIONS: WATER

Comment on how and to what degree new legislation and rulemaking will be necessary for environmental control/concerns with nanotechnology from a "water" perspective.

SOLUTION: Completely new legislation and regulatory rulemaking will almost certainly be necessary for environmental concerns with nanotechnology. However, in the meantime, one may speculate on how the existing regulatory framework might be applied to the nanotechnology area as this emerging field develops over the next several years. Regarding the "water" area, no activities seem to be in the works at this time. Most of the interest appears to be concerned with air (see Problem LCAA.3) and worker health and safety (see Problem LOSHA.6).

The reader is referred to the following two texts for additional information:

1. L. Theodore and R. Kunz, "Nanotechnology: Environmental Implications and Solutions," John Wiley & Sons, Hoboken, NJ, 2005.
2. L. Theodore, "Nanotechnology: Basic Calculations for Engineers and Scientists," John Wiley & Sons, Hoboken, NJ, 2006.

LSDWA.15 SOURCES OF WATER POLLUTION

Qualitatively discuss some of the sources of water pollution.

SOLUTION: Protecting the nation's drinking water, coastal zone waters, and surface waters is made complex by the variety of sources that affect them. Groundwater is being contaminated by leaking underground storage tanks, fertilizers and pesticides, uncontrolled hazardous wastes sites, septic tanks, drainage wells, and other sources, threatening 50% of the nation's drinking water supplies for half of the nation's population. Many coastal towns along the Atlantic Coast and Gulf of Mexico have to close beaches one or more times during summer months because of shoreline pollution. In Puget Sound, fecal coliform bacteria contaminate oysters, and the harbor seals have higher concentrations of PCBs than do almost any other seal population in the world.

Pollution of groundwater by pesticides and nitrates due to the application of agricultural chemicals is a major environmental concern in many parts of the country, particularly in the Midwest. In Iowa, where agricultural chemicals are used in 60% of the state, some private and public drinking water wells have exceeded public health standards for nitrates. Pesticides also have been found in groundwater. About 30 towns in Nebraska have excessive amounts of nitrate in their drinking water. Bottled water must be provided to infants, and monthly well testing is required.

The United States is losing one of its most valuable, and perhaps irreplaceable, resources—the nation's wetlands. Once regarded as wastelands, wetlands are now recognized as an important resource to people and the environment. Wetlands are among the most productive of all ecosystems. Wetland plants convert sunlight into plant material or biomass, which in turn serves as food for many types of aquatic and terrestrial animals. The major food value of wetland plants occurs as they break down into small particles to form the base of an aquatic food chain.

Wetlands are habitats for many forms of fish and wildlife. Approximately two-thirds of this of nation's major commercial fisheries use estuaries and coastal marshes as nurseries or spawning grounds. Migratory waterfowl and other birds also depend on wetlands, some spending their entire lives in wetlands and others using them primarily as nesting, feeding, or resting grounds.

The roles of wetlands in improving and maintaining water quality in adjacent water bodies is increasingly being recognized by the scientific community. Wetlands remove nutrients such as nitrogen and phosphorus, and thus help prevent overenrichment of water (eutrophication). Also they filter harmful chemicals, such as pesticides and heavy metals, and trap suspended sediments which otherwise would produce turbidity (cloudiness) in water. This function is particularly important as a natural buffer for nonpoint pollution sources.

Wetlands also have socioeconomic values. They play an important role in flood control by absorbing peak flows and releasing water slowly. Along the coast, they buffer land against storm surges resulting from hurricanes and tropical storms. Wetland vegetation can reduce shoreline erosion by absorbing and dissipating wave energy and encouraging the deposition of suspended sediments. Also, wetlands contribute $20–40 billion annually to the nation's economy–for example, through recreational and commercial fishing, hunting of waterfowl, and the production of cash crops such as wild rice and cranberries.

Unfortunately, the natural heritage of swamps, marshes, bogs, and other types of wetlands is rapidly disappearing. Once there were over 200 million acres of wetlands in the lower 48 states; by the mid-1970s, only 99 million acres remained. Between 1955 and 1975, more than 11 million acres of wetlands—an area three times the size of the state of New Jersey—were lost entirely. The average rate of wetland loss during this period was 458,000 acres per year: 440,000 acres of inland wetlands and 18,000 acres of coastal wetlands. Agricultural development involving drainage of wetlands was responsible for 87% of the losses during those two decodes. Urban and other development caused 8% and 5% of losses, respectively. In addition to the physical destruction of habitat, wetlands are also threatened by chemical contamination and other types of pollution.

Ocean dumping of dredged material, sewerage sludge, and industrial wastes are major sources of ocean pollution. Sediments dredged from industrialized urban harbors are often highly contaminated with heavy metals and toxic synthetic organic chemicals such as PCBs and petroleum hydrocarbons. When these sediments are dumped in the ocean, the contaminants can be taken up by marine organisms.

Persistent disposal of plastics from land and ships at sea has also become a serious problem. The most severe impact of this biodegradable debris floating in the ocean is causing the injury and death of fish, marine mammals, and birds. Debris on beaches from sewer and storm drain overflows or from mismanagement of trash poses public safety and aesthetic concerns and can result in major economic losses for coastal communities during the tourist season.

LSDWA.16 INORGANIC CHEMICALS OF CONCERN

Provide a list of inorganic chemicals and their accompanying drinking water maximum contamination level.

SOLUTION: Maximum contamination levels are provided below in Table 5.1.

TABLE 5.1 Drinking Water Regulations: Inorganic Chemicals

Name of Chemical	MCL (mg/L unless otherwise noted)
Antimony	0.006
Arsenic	0.010
Asbestos (fiber length >10 μm)	7 MFL
Barium	2
Beryllium	0.004
Cadmium	0.005
Chromium (total)	0.1
Copper	TT (AL $=$ 1.3)
Cyanide	0.2
Fluoride	4.0
Lead	TT (AL $=$ 0.015)
Mercury (inorganic)	0.002
Nitrate (as N)	10
Nitrite (as N)	1
Radionuclides	
Combined Radium-226 and Radium-228	5 pCi/L
Gross Alpha (excluding radon and uranium)	15 pCi/L
Beta Particles and Photon Emitters	4 mrem/year
Uranium	0.030
Selenium	0.05
Thallium	0.002

LSDWA.17 ORGANIC CHEMICALS OF CONCERN

Provide a list of organic chemicals and their accompanying drinking water maximum contamination level.

SOLUTION: Maximum contamination level levels are provided below in Table 5.2.

TABLE 5.2   Drinking Water Regulations: Organic Chemicals

Synthetic Organic Chemicals	MCL (mg/L unless otherwise noted)
2,3,7,8-TCDD (Dioxin)	0.00000003
2,4,5-TP (Silvex)	0.05
2,4-D	0.07
Acrylamide	TT
Alachlor	0.002
Arrazine	0.003
Carbofuran	0.04
Chlordane	0.002
Dalapon	0.2
Di(2-ethylhexyl) adipate	0.4
1,2-Dibromo-3-choloropropane(DBCP)	0.0002
Di(2-ethylhexyl) phthalate (DEHP)	0.006
Dinoseb	0.007
Diquat	0.02
Endothall	0.1
Endrin	0.002
Epichlorohydrin	TT
Ethylene Dibromide (EDB)	0.00005
Glyphosate	0.7
Heptachlor	0.0004
Heptachlor Epoxide	0.0002
Hexachlorobenzene	0.001
Hexachlorocyclopentadiene (HEX)	0.05
Lindane	0.0002
Methoxychlor	0.04
Oxamyl (Vydate)	0.2
Benzo(a)pyrene (PAHs)	0.0002
Polychlorinated Biphenyls (PCBs)	0.0005
Pentachlorophenol	0.001
Picloram	0.5
Simazine	0.004
Toxaphene	0.003
Volatile Organic Chemicals	
1,1,1-Trichloroethane	0.2
1,1,2-Trichloroethane	0.005
1,1-Dichloroethylene	0.007
1,2,4-Trichlorobenzene	0.07
1,2-Dichloroethane	0.005
1,2-Dichloropropane	0.005
Benzene	0.005
Carbon Tetrachloride	0.005
Chlorobenzene	0.1
Cis-1,2-Dichloroethylene	0.07
Dichloromethane	0.005

(Continued)

TABLE 5.2 *Continued*

Synthetic Organic Chemicals	MCL (mg/L unless otherwise noted)
Ethylbenzene	0.7
Ortho-Dichlorobenzene	0.6
Para-Dichlorobenzene	0.075
Styrene	0.1
Tetrachloroethylene(PCE)	0.005
Toluene	1
Trans-1,2-Dichloroethylene	0.1
Trichloroethylene (TCE)	0.005
Vinyl Chloride	0.002
Xylenes (total)	10

Note: TT (treatment technique)

LSDWA.18 MICROBIOLOGICAL CONTAMINANTS OF CONCERN

Provide a list of microbiological contaminants and their accompanying drinking water maximum contamination levels.

SOLUTION: Maximum contamination levels are provided below in Table 5.3.

TABLE 5.3 Drinking Water Regulations: Microbiological Contaminants

	Comment
Total Coliform Rule (TCR)	
Total Coliform	Variable
Fecal Coliform	
E. coli	
Surface Water Treatment Rule (SWTR)	
Turbidity	TT
Giardia lamblia	TT
Enteric Viruses	TT
Legionella	TT
Heterotrophic Plate Count (HPC)	TT
Interim Enhanced Surface Water Treatment Rule (IESWTR)	
Turbidity	TT
Cryptosporidium	TT
Filter Backwash Rule	
Cryptosporidium	TT
Long Term 1 Enhanced Surface Water Treatment Rule (LTIESWTR)	
Turbidity	TT
Cryptosporidium	TT

Note: TT (treatment technique)

LSDWA.19 TURBIDITY (TU) AND TOTAL SUSPENDED SOLIDS (TSS)

Answer the following three questions.

1. How is turbidity related to total suspended solids (TSS)?
2. Why is it important to maintain low turbidity levels in community water systems?
3. How does the EPA regulate water turbidity?

SOLUTION:

1. *Turbidity* is a measure of the degree to which light is scattered by suspended particulate material and soluble colored compounds in water. It provides an estimate of the cloudiness or muddiness of the water due to silt, clay, finely divided organic or inorganic matter, soluble colored organic compounds, plankton, and microscopic organisms. Turbidity is typically provided in units of TU's (Turbidity Unit).

   Total suspended solids (TSS) refer to the total mass of solids per volume of water. This figure is typically provided as a concentration (ppm or mg/L).

   The correlation between TSS and turbidity will vary according to the type of contamination. For example, when water contains a low density contaminate such as algae, the turbidity could be very high yet the TSS will be very low.

2. Turbidity does not typically result in direct health effects. However, turbidity can interfere with disinfection processes and provide a medium for microbial growth. Turbidity may also indicate the presence of harmful organisms. These organisms include bacteria, viruses, and parasites.

3. The EPA regulates the maximum level of turbidity in community water systems in the Code of Federal Regulations, Subpart B "Maximum Contaminate Levels," Section 141.13. This section of the CFR specifies that the level of turbidity must not exceed one Turbidity Unit (TU).

See also 40 CFR § 141.22 and 141.13.

LSDWA.20 FILTRATION OPTIONS FOR A PUBLIC WATER SUPPLY

Public water systems that use a surface water source or a ground water source under the direct influence of surface water, and do not meet all the criteria in 40 CFR 141.71 (a) and (b) for avoiding filtration, must provide a filtration treatment that complies with 40 CFR 141.73. Briefly describe the EPA approved filtration methods.

SOLUTION:

Conventional filtration treatment: This treatment option includes several types of filtration including direct filtration, packaged filtration, and membrane filtration. Typically, the filtration process will include one or all of the following steps: Chemical treatment, sedimentation, flocculation, and filtration. This method is often used to treat water used by small communities or recreational areas.

Slow sand filtration: Sand filters represent the oldest EPA approved filtration system. The filter consists of a bed of fines and approximately 3 to 4 feet deep supported by a 1-foot layer of gravel and an underdrain system. These systems are typically simple, cheap, and reliable. By itself, these filtration systems are not suitable for high turbidity feeds. Conventional chemical pretreatment methods can be used in applications where the turbidity of the feed is high or variable.

Diatomaceous earth filtration: Diatomaceous Earth (DE) filtration is a process that uses diatoms or diatomaceous earth, which are the skeletal remains of small single-celled organisms, as the filter media. DE filtration relies upon a layer of diatomaceous earth placed on a filter element or septum and is frequently referred to as pre-coat filtration. DE filters are simple to operate and are effective in removing cysts, algae, and asbestos from water. For the last seventy plus years, DE filtration has been used to produce high-quality, low cost drinking water.

Other filtration methods: A public water system may use a filtration technology not listed above if it can demonstrate to the State that the alternative technology consistently meets the requirements of 40 CFR 141.72. This demonstration can be provided by pilot process data or other means.

LSDWA.21   ACID RAIN

How does acid rain form and what are the causes and effects of acid rain?

SOLUTION: Acid rain with a pH below 5.6 is formed when certain anthropogenic air pollutants travel into the atmosphere and react with moisture and sunlight to produce acidic compounds. Sulfur and nitrogen compounds released into the atmosphere from different sources are believed to play the biggest role in formation of acid rain. The natural processes which contribute to acid rain include lightning, ocean spray, decaying plant and bacterial activity in the soil, and volcanic eruptions. Anthropogenic sources include those utilities, industries, businesses, and homes that burn fossils fuels, plus motor vehicle emission. Sulfuric acid is the type of acid most commonly formed in areas that burn coal for electricity, while nitric acid is more common in areas that have a high density of automobiles and other internal combustion engines.

There are several ways that acid rain affects the environment:

1. Contact with plants can harm plants by damaging outer leaf surfaces and by changing the root environment.
2. Contact with soil and water resources. Due to the acid in the rain, fishkills in ponds, lakes and oceans, as well as effects on aquatic organisms, are common.

Acid rain can cause minerals in the soil to dissolve and be leached away. Many of these minerals are nutrients for both plants and animals.
3. Acid rain mobilizes trace metals, such as lead and mercury. When significant levels of these metals dissolve from surface soils they may accumulate elsewhere, leading to poisoning.
4. Acid rain may damage building structures and automobiles due to accelerated corrosion rates.

The general chemical formulae for the formation of acid rain are as follows:

$$SO_x + O_2 \longrightarrow SO_2 + H_2O \longrightarrow H_2SO_4 \tag{5.1}$$

$$NO_x + O_2 \longrightarrow NO_2 + HNO_3 \tag{5.2}$$

$$CO_2 + H_2O \longrightarrow H_2CO_3 \tag{5.3}$$

QUANTITATIVE PROBLEMS (TSDWA)

PROBLEMS TSDWA.1–28

TSDWA.1 CHEMICAL CONCENTRATION CONVERSION

A regulatory agency stipulates that the maximum concentration of benzo(a)pyrene in drinking water should not exceed 200 ng/L. Express this concentration in lb/10,000 U.S. gal.

SOLUTION: This unit conversion may be carried out as follows:

$$(200 \, ng/L) \, (1 \, g/10^9 \, ng) \, (1 lb/453.6 \, g) \, (3.785 \, L/1 \, gal)$$

$$= 1.67 \times 10^{-9} \, lb/gal \, (10{,}000 \, gal/10{,}000 \, gal)$$

$$= 1.67 \times 10^{-5} \, lb/10{,}000 \, gal$$

Alternatively, knowing 1 mg/L = 1 ppm = 8.34×10^{-6} lb/gal, the solution is:

$$200 \, ng/L \, (1 \, mg/10^6 \, ng) \, ((8.34 \times 10^{-6} \, lb/gal)/(mg \, L))(10{,}000 \, gal/10{,}000 \, gal)$$
$$= 1.67 \times 10^{-5} \, lb/10{,}000 \, gal$$

TSDWA.2 CONVERTING CONCENTRATION UNITS

Express the concentration 72 g of HCl in 128 cm^3 of water into terms of fraction and percent by weight (mass), parts per million by weight, and molarity.

SOLUTION: The fraction by weight can be calculated as follows:

$$72 \, g/(72 + 128) \, g = 0.36; \, 128 \, cm^3 \, water = 128 \, g \, water$$

The percent by weight can be calculated from the fraction by weight.

$$(0.36)(100\%) = 36\%$$

The ppmw (parts per million by weight) can be calculated as follows:

$$(72\,\text{g}/128\,\text{g})(10^6) = 562{,}500 \text{ ppmw}$$

The molarity (M) is defined as follows:

$$M = \text{moles of solute}/\text{volume of solution (L)}$$

Using atomic weights,

$$\text{MW of HCl} = 1.0079 + 35.453 = 36.4609$$

$$M = \left[(72\,\text{g HCl}) \left(\frac{1\,\text{mol HCl}}{36.4609\,\text{g HCl}} \right) \right] \Big/ \left(\frac{128\,\text{cm}^3}{1000\,\text{cm}^3/\text{L}} \right) = 15.43 \,\text{mol/L}$$

TSDWA.3 SUSPENDED PARTICULATE CONCENTRATION

The suspended particulate concentration in an aqueous stream has been determined to be 27.6 mg/L. Convert this value to units of µg/L, g/L, lb/ft^3, and lb/gal.

SOLUTION: Apply the appropriate conversion factors.

$$C = 27.6 \frac{\text{mg}}{\text{L}}$$

$$= \left(27.6 \frac{\text{mg}}{\text{L}} \right) \left(10^3 \frac{\mu\text{g}}{\text{mg}} \right)$$

$$= 2.76 \times 10^4 \frac{\mu\text{g}}{\text{L}}$$

$$C = 27.6 \frac{\text{mg}}{\text{L}}$$

$$= \left(27.6 \frac{\text{mg}}{\text{L}} \right) \left(10^{-3} \frac{\text{g}}{\text{mg}} \right)$$

$$= 2.76 \times 10^{-2} \frac{\text{g}}{\text{L}}$$

$$C = 27.6 \frac{\text{mg}}{\text{L}}$$

$$= \left(27.6 \frac{\text{mg}}{\text{L}} \right) \left(\frac{1\,\text{lb}}{454{,}000\,\text{mg}} \right) \left(\frac{1000\,\text{L}}{35.3\,\text{ft}^3} \right)$$

$$= 1.72 \times 10^{-3} \frac{\text{lb}}{\text{ft}^3}$$

$$C = 27.6 \frac{\text{mg}}{\text{L}}$$

$$= \left(27.6 \frac{\text{mg}}{\text{L}}\right)\left(\frac{1 \text{ lb}}{454{,}000 \text{ mg}}\right)\left(\frac{1000 \text{ L}}{264 \text{ gal}}\right)$$

$$= 2.30 \times 10^{-4} \frac{\text{lb}}{\text{gal}}$$

TSDWA.4   TRACE CONCENTRATION

Refer to Problem TCWA.2. Convert a calcium concentration of 10 parts per million parts of water on a mass basis to mole fraction.

To convert 10 ppmw Ca to ppmm, select a basis of 10^6 g of solution. The mass fraction of Ca is first obtained by the following equation:

$$\text{Mass of Ca} = 10 \text{ g}$$

$$\text{Moles Ca} = \frac{10 \text{ g}}{40 \text{ g/mol}} = 0.25 \text{ mol}$$

$$\text{Moles H}_2\text{O} = \frac{10^6 \text{ g} - 10 \text{ g}}{18 \text{ g/mol}} = 55{,}555 \text{ mol}$$

$$\text{Mole fraction Ca} = y_{\text{Ca}} = \frac{0.25 \text{ mol}}{0.25 \text{ mol} + 55{,}555 \text{ mol}} = 4.5 \times 10^{-6}$$

TSDWA.5   PARTS PER MILLION MOLE BASIS

Convert a calcium concentration of 10 parts per million parts of water on a mass basis to parts per million on a mole basis.

SOLUTION: Refer to the solution of Problem TSDWA.4. Since

$$y_{\text{Ca}} = 4.5 \times 10^{-6}$$

One may write

$$\text{ppmm of Ca} = 10^6 y_{\text{Ca}}$$

$$= (10^6)(4.5 \times 10^{-6})$$

$$= 4.5$$

TSDWA.6 SUSPENDED SOLIDS IN AN ESTUARY

The following suspended solids and daily volumetric average flows were obtained from an estuary that has served as a source of drinking water.

Using the data by Table 5.4, determine the average suspended solids in the estuary during a 30-day period when the daily flowrates were 80, 200, 280 (10 days), 300 (10 days), 360, 400 (2 days), 500, 600 (2 days), 620, and 800 cfs. Assume the suspended solids (SS) are related to the volume rate volume rate of flow of water (q) see Table 5.4 through the relationship:

$$SS = aq^b$$

SOLUTION: A linear regression of the Table 5.4 data leads to

$$SS = 0.0205q^{0.644} \qquad (5.4)$$

where $SS = mg/L$
    $q = ft^3/s$

The average SS mass load for each period is obtained by multiplying q by SS and converting to lb/day. For example, for the 10-day period with a flow of $300 \ ft^3/s$,

$$m_{ss} = \left[(300)(0.0205)(300^{0.644})(10)\right]\left[(28.3)(3600)(24)/(1000)(454)\right]$$
$$= \left[(300)(0.0205)(300^{0.644})(10)\right][5.4]$$
$$= 13,080 \ lb$$

Note that (5.4) is the conversion factor employed to convert $(ft^3/s)(mg/L)$ to lb/day.

TABLE 5.4 Suspended Solids Data as a Function of Flowrate

Suspended Solids (mg/L)	Flowrate (cfs)
0.11	20
0.30	40
0.30	70
0.42	100
0.70	300
1.40	440
1.50	1000

Similarly, one can generate the following results:

q (ft^3/s)	m_{ss} (lb)
80	150
200	670
280	11,600
300	13,080
360	1,770
400	4,200
500	3,020
600	9,160
620	4,320
800	6,570
$\sum m_{ss} =$	53,540

The average daily flow of SS is therefore

$$m(\text{average}) = \frac{\sum m_{ss}}{30}$$
$$= \frac{53,540}{30}$$
$$= 1785 \, \text{lb/day}$$

TSDWA.7   ESTIMATING GROUND WATER NITRATE CONCENTRATION (STATE WATER QUALITY—VIRGINIA)

Nitrates are ground water pollutants that often arise from sewage disposal. The concentration of nitrate in ground water from a residential drainage field can be calculated using the following procedure:

Step 1: Determine the number of inches per year of rain that infiltrates into the ground.

Step 2: Use the following equation to calculate the average daily dilution from rainwater

Daily Dilution (gallons/acre · day)

$$= (\text{Rain Infiltration}) (\text{Acres Available for Infiltration}) (74) \quad (5.5)$$

Step 3: Use the following equation to calculate the nitrate concentration leaving the site.

Aquifer Nitrate Concentration (mg/L) = [Waste/(Waste + Dilution)]

(Wastewater Concentration)   (5.6)

Estimate the given ground water nitrate concentration given the following data:

Rainfall = 40 inches per year
Rainwater infiltration = 50%
Dilution area = 5 acres
Source: 12 bedroom residential complex into common drainfield
Hydraulic loading rate = (150 gal/bedroom)(12 bedroom) = 1800 gal
Equivalent nitrate loading rate = (130)(12) = 1560 gal
Typical nitrate concentration for waste = 30 mg/L (residential)

SOLUTION:

Employ the equations provided above.
Rain water infiltration = (40 inch/yr)(0.5) = 20 inch/yr
Daily dilution = (20 inch/yr)(5 acres)(74) = 7400 gallons/acre · day
Aquifer nitrate concentration = [1560 gal/(1560 gal + 7400 gal)] (30 mg/L) = 5.2 mg/L.
If the EPA ground water standard nitrate concentration is 10 mg/L, this design will not exceed the standard.

TSDWA.8 CHLORIDE CONCENTRATION CONTROL

An upstream flow of 25 cfs (previously serving as potential source of drinking water) with background level of chlorides (nonreactive), of 30 mg/L is supplemented with

1. An industrial discharge (I) of 6.5 MGD with 1500 mg chlorides/L
2. A municipal discharge (M) of 2 MGD with 500 mg chlorides/L
3. A runoff (R) of 0.5 MGD with 100 mg chlorides/L
4. A downstream tributary (T) of 5 cfs with the same background chlorides concentration of 30 mg/L

To maintain a maximum chloride concentration of 250 mg/L at the water intake location, determine

1. The required industrial (I) reduction in flowrate, or
2. The required increase in the secondary tributary (T) flowrate, or
3. A combination of steps 1 and 2.

SOLUTION: The system plus data are pictured in Figure 5.1. In Figure 5.1, subscripts I, M, R, T refer to industrial, municipal, runoff, and tributary, respectively.

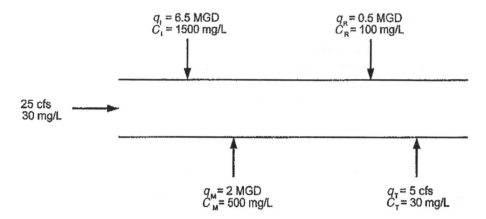

FIGURE 5.1   System and data (flowrate and chloride concentration) for Problem TSDWA.8.

There are five input chlorine loads to the system. These are calculated from the equation

$$\dot{m} = (q)(C)$$

$$= (25)(30)(5.4); \ 5.4 = \text{conversion factor}$$

$$= 4050 \ \text{lb/day}$$

$$\dot{m}_I = (6.5)(1500)(8.34); \ 8.34 = \text{conversion factor}$$

$$= 81,315 \ \text{lb/day}$$

$$\dot{m}_M = (2)(500)(8.34)$$

$$= 8340 \ \text{lb/day}$$

$$\dot{m}_R = (0.5)(100)(8.34)$$

$$= 4170 \ \text{lb/day}$$

$$\dot{m}_T = (5)(30)(5.4)$$

$$= 810 \ \text{lb/day}$$

The allowable downstream chloride concentration is specified as 250 mg/L. Converting volume flows to ft^3/s leads to

$$q_1 = (6.5)(1.547)$$

$$= 10 \ \text{ft}^3/\text{s}$$

$$q_M = (2)(1.547)$$

$$= 3.1 \ \text{ft}^3/\text{s}$$

$$q_R = (0.5)(1.547)$$

$$= 0.77 \ \text{ft}^3/\text{s}$$

Assuming complete mixing at the downstream location, the allowable concentration in terms of the industrial loading, is given by

$$C = \frac{4050 + \dot{m}_I + 8340 + 4170 + 810}{(25 + 10 + 3.1 + 0.77 + 5)(5.4)} = 250$$

Solving for $\dot{m}_I$,

$$\dot{m}_I = 42{,}000 \text{ lb/day}$$

Since the present flow is 81,300 lb/day, a reduction of approximately 40,000 lb/day in the industrial loading must occur to achieve the required chloride concentration.

The increase in the tributary flow may be similarly calculated:

$$C = \frac{4050 + 81{,}315 + 8340 + 417 + (q_T)(30)(5.4)}{(25 + 10 + 3.1 + 0.77 + q_T)(5.4)} = 250$$

$$q_T = 35.0 \text{ ft}^3/\text{s}$$

As expected, the tributary flow must be significantly increased from 5 to 35 ft^3/s.

Obviously, an infinite number of solutions are possible. One may simply select values of the industrial loading between 41,000 and 81,000 lb/day and proceed to calculate the corresponding (increased) value of q_T.

TSDWA.9 RAINFALL RUNOFF

A watershed has an area of 8 mi^2. Rainfall occurs at a rate of 0.06 in/hr and part of the rainfall serves as a supply of drinking water. Approximately 50% of the rain runoff reaches the sewers and contains an average total nitrogen concentration of 9.0 mg/L. In addition, the city wastewater treatment plant discharges 10 MGD with a total nitrogen concentration of 35 mg/L. Compare the total nitrogen discharge with runoff from the watershed with that of the city's sewage treatment plant.

SOLUTION: First calculate the total nitrogen discharge, $\dot{m}_W$, from the treatment plant:

$$\dot{m}_W = (q)(C)$$
$$= (10)(35)(8.34)$$
$$= 2919 \text{ lb/day}$$

The volumetric flow of the runoff, q, is

$$q = (0.5)(0.06)(8)(5280)^2/(3600)(12)$$
$$= 155 \text{ ft}^3/\text{s}$$

The total nitrogen discharge, $\dot{m}_R$, from runoff is then

$$\dot{m}_R = \left(155\frac{\text{ft}^3}{\text{s}}\right)\left(9\frac{\text{mg}}{\text{L}}\right)\left(10^{-6}\frac{\text{L}}{\text{mg}}\right)\left(3600 \times 24\frac{\text{s}}{\text{day}}\right)\left(62.4\frac{\text{lb}}{\text{ft}^3}\right)$$

$$= 7521 \text{ lb/day}$$

During rain, the runoff is over 2.5 times that for the treatment plant.

TSDWA.10   RADIOACTIVE DECAY

As described earlier, exponential first order decay can be described using either a reaction rate coefficient (k) or a half-life (τ). The equation that relates these two parameters is as follows:

$$\tau = \frac{0.693}{k} \tag{5.7}$$

Exponential decay can be expressed by the following equation:

$$N = N_o e^{-kt} \tag{5.8}$$

where k = reaction rate coefficient $(\text{time})^{-1}$
$\quad N_o$ = initial amount
$\quad N$ = amount at time t

Determine how much of a 100-g sample of Po-210 is left after 5.52 days using:

1. The reaction rate coefficient
2. The half-life
3. Calculate the percent error between the two methods.

The half-life for Po-210 is 1.38 days.

SOLUTION:

1. Determine the reaction rate coefficient:

$$k = \frac{0.693}{\tau} = \frac{0.693}{1.38 \text{ day}}$$

$$= 0.502/\text{day}$$

The amount of substance left after 5.52 days is as follows:

$$N = N_o e^{-kt} = (100)e^{-(0.502)(5.52)}$$
$$= 6.26\,g$$

2. The first step is to determine how many half-lives the 100 g sample has undergone in the given time period:

$$\text{Number of half-lives} = (5.52\ \text{days})/(1.38\ \text{days}) = 4.0$$

Therefore, in a 5.52-day period, the 100 g sample has undergone four half-lives.

The amount of the substance left after one half-life is calculated as follows:

$$(0.5)^1(100\,g) = 50\,g$$

Therefore, the amount of substance left after four half-lives is

$$(0.5)^4(100\,g) = 6.25\,g$$

3. Since the two equations are, in principle, identical to each other, there is no difference between the two values. The small difference between the two results arises because of roundoff errors.

TSDWA.11 PROBABILITY OF EXCESSIVE Pb CONCENTRATION IN TRANSPORT DRUMS

Concern has arisen regarding the potential of radioactive lead leakage into a nearby water table. Tests indicate that radioactive sludge waste arrive in 55-gal drums to a storage facility that has a mean lead content of 15 ppm with a standard deviation of

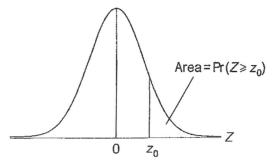

FIGURE 5.2 Standard normal, cumulative probability in right-hand tail (for negative values of z, areas are found by symmetry).

TABLE 5.5 Standard Normal Distribution Values

					Next Decimal Place of z_0					
z_0	0	1	2	3	4	5	6	7	8	9
0.0	0.500	0.496	0.492	0.488	0.484	0.480	0.476	0.472	0.468	0.464
0.1	0.460	0.456	0.452	0.448	0.444	0.440	0.436	0.433	0.429	0.425
0.2	0.421	0.417	0.413	0.409	0.405	0.401	0.397	0.394	0.390	0.386
0.3	0.382	0.378	0.374	0.371	0.367	0.363	0.359	0.356	0.352	0.348
0.4	0.345	0.341	0.337	0.334	0.330	0.326	0.323	0.319	0.316	0.312
0.5	0.309	0.305	0.302	0.298	0.295	0.291	0.288	0.284	0.281	0.278
0.6	0.274	0.271	0.268	0.264	0.261	0.258	0.255	0.251	0.248	0.245
0.7	0.242	0.239	0.236	0.233	0.230	0.227	0.224	0.221	0.218	0.215
0.8	0.212	0.209	0.206	0.203	0.200	0.198	0.195	0.192	0.189	0.187
0.9	0.184	0.181	0.179	0.176	0.174	0.171	0.189	0.166	0.164	0.161
1.0	0.159	0.156	0.154	0.152	0.149	0.147	0.145	0.142	0.140	0.138
1.1	0.136	0.133	0.131	0.129	0.127	0.125	0.123	0.121	0.119	0.117
1.2	0.115	0.113	0.111	0.109	0.107	0.106	0.140	0.102	0.100	0.099
1.3	0.097	0.095	0.093	0.092	0.090	0.089	0.087	0.085	0.084	0.082
1.4	0.081	0.079	0.078	0.076	0.075	0.074	0.072	0.071	0.069	0.068
1.5	0.067	0.066	0.064	0.063	0.062	0.061	0.059	0.058	0.057	0.056
1.6	0.055	0.054	0.053	0.052	0.051	0.049	0.048	0.047	0.046	0.046
1.7	0.045	0.044	0.043	0.042	0.041	0.040	0.039	0.038	0.038	0.037

z_0	0	1	2	3	4	5	6	7	8	9
1.8	0.036	0.035	0.034	0.034	0.033	0.032	0.031	0.031	0.030	0.029
1.9	0.029	0.028	0.027	0.027	0.026	0.026	0.025	0.024	0.024	0.023
2.0	0.023	0.022	0.022	0.021	0.021	0.020	0.020	0.019	0.019	0.018
2.1	0.018	0.017	0.017	0.017	0.016	0.016	0.015	0.015	0.015	0.014
2.2	0.014	0.014	0.013	0.013	0.013	0.012	0.012	0.012	0.011	0.011
2.3	0.011	0.010	0.010	0.010	0.010	0.009	0.009	0.009	0.009	0.008
2.4	0.008	0.008	0.008	0.008	0.007	0.007	0.007	0.007	0.007	0.006
2.5	0.006	0.006	0.006	0.006	0.006	0.005	0.005	0.005	0.005	0.005
2.6	0.005	0.005	0.004	0.004	0.004	0.004	0.004	0.004	0.004	0.004
2.7	0.003	0.003	0.003	0.003	0.003	0.003	0.003	0.003	0.003	0.003
2.8	0.003	0.002	0.002	0.002	0.002	0.002	0.002	0.002	0.002	0.002
2.9	0.002	0.002	0.002	0.002	0.002	0.002	0.002	0.001	0.001	0.001

Detail of Tail ($0_{-2}135$, for example, means 0.00135)

z_0	0	1	2	3	4	5	6	7	8	9
2.0	$0_{-1}228$	$0_{-1}179$	$0_{-1}139$	$0_{-1}107$	$0_{-2}820$	$0_{-2}621$	$0_{-2}466$	$0_{-2}347$	$0_{-2}256$	$0_{-2}187$
3.0	$0_{-2}135$	$0_{-3}968$	$0_{-3}687$	$0_{-3}483$	$0_{-3}337$	$0_{-3}233$	$0_{-3}159$	$0_{-3}108$	$0_{-4}723$	$0_{-4}481$
4.0	$0_{-4}317$	$0_{-4}207$	$0_{-4}133$	$0_{-5}854$	$0_{-5}541$	$0_{-5}340$	$0_{-5}211$	$0_{-5}130$	$0_{-6}793$	$0_{-6}479$
5.0	$0_{-6}287$	$0_{-6}170$	$0_{-7}996$	$0_{-7}579$	$0_{-7}333$	$0_{-7}190$	$0_{-7}107$	$0_{-8}599$	$0_{-8}332$	$0_{-8}182$

[a]From R. J. Woonacott and T. H. Woonacott, *Introductory Statistics*, 4th edn., Wiley, New York, 1985.

12 ppm. The drums are unloaded into a 300-gal receiving tank. The storage facility is required to keep the lead concentration at or below 20 ppm in order to meet the required drinking water standard. Assume that the lead content from one drum to the next are not correlated, and that the tank is nearly full. What is the probability that the lead content in any drum exceeds 45 ppm?

SOLUTION: For a standard deviation of 12 ppm and a mean of 15 ppm, a lead concentration of 45 ppm represents $(45-15)/12$ or 2.5 standard deviations above (displaced from) the mean.

In Figure 5.2, the area in the "right-hand tail" (above z_0) of the normal distribution curve represents the probability that the variable (an event) is z_0 standard deviations above the mean. Applied to this problem, the area in the right-hand tail (above a z_0 of 2.5) is the probability that the lead content in any drum exceeds 45 ppm. From Table 5.5, the area in the right-hand tail corresponding to a z_0 of 2.5, is 0.006 or 0.6% of the total area under the curve. The probability that lead content in a drum exceeds 45 ppm is therefore 0.6%.

The reader is referred to S. Shaefer and L. Theodore, "Probability and Statistics: Applications for Environmental Science," CRC Press/Taylor & Francis, Boca Raton, FL, 2007, for additional details.

TSDWA.12 LEAD CONTENT IN TANK

With reference to the previous problem TSDWA.11, what is the probability that the lead content in the receiving tank exceeds 20 ppm? State the limitations of the analysis, and how these limitations could be accommodated.

SOLUTION: To determine the standard deviation of the mean lead concentration in the receiving tank, a basic theorem in probability called the *central limit theorem* must be employed. If σ represents the standard deviation of the lead concentrations in the drums, then the standard deviation of the mean lead concentration in the receiving tank is given by $\sigma/\sqrt{n}$ where n is the number of drums:

$$n = 300 \, \text{gal}/(55 \, \text{gal/drum}) = 5.45 \, \text{drums}$$

The standard deviation of the mean is then

$$\frac{\sigma}{\sqrt{n}} = \frac{12 \, \text{ppm}}{\sqrt{5.45}} = 5.1 \, \text{ppm} \tag{5.9}$$

For a standard deviation of 5.1 ppm and a mean of 15 ppm, a lead concentration of 20 ppm is $(20-15)/5.1$ or 0.98 standard deviations above the mean. Again, from the normal distribution table (see Table 5.5), for $z_0 = 0.98$, the probability that the mean lead content in the receiving tank exceeds 20 ppm is 0.164 or 16.4%.

The limitations in the analysis and possible accommodations are:

a. The critical assumption made is that the lead content of the drums and waste shipments are not correlated with time. In other words, it has been assumed that it is unlikely that the facility will receive consecutive shipments of waste with a lead content higher than the mean. With multiple drums from a given source, it is highly likely, however, that the drum contents will be correlated.

b. The distribution of the lead content in the problem was assumed to be normal. Concentrations often have skewed distributions, e.g., log-normal.

c. With additional information, e.g., autocorrelation and skewness of the concentration distribution, much more complex and realistic problems can be solved using queuing or inventory techniques. Oversizing the tank would provide a margin of safety.

The reader is referred to S. Shaefer and L. Theodore, "Probability and Statistics: Applications for Environmental Science," CRC Press/Taylor & Francis Group, Boca Raton, FL, 2007, for additional details.

TSDWA.13 LEAKING UNDERGROUND STORAGE TANK

A total of 400 L of pure benzene leaks from an underground storage tank before the leak is discovered. The water table lies a few feet below the tank. Discuss the following issues related to this release:

1. What is the possibility of recovering some of the pure product benzene, and how might this product recovery be accomplished?

2. What is the maximum benzene concentration expected in the groundwater?

SOLUTION:

1. The possibility exists for recovering some of the benzene, but the success of this recovery effort depends on how far the benzene has spread and how much of the benzene has volatilized. Benzene's density is less than that of water so it will float on top of the groundwater as it dissolves. This fact allows contact of the free product with the air in the soil pores above the water table, and since benzene has a high vapor pressure (≈ 0.1 atm at $20°C$), over time, a significant portion of the residual benzene will volatilize into the unsaturated zone of the soil. In this case, some of the benzene may be recovered using soil vapor extraction. Once the free product phase is gone, volatilization will be reduced due to benzene's high water solubility and resulting low Henry's (air to water distribution) constant. Free product may be recovered from the groundwater table by pumping if product recovery is initiated before the

product has dissipated by volatilization or dissolution; otherwise, only the dissolved plume remains and significant mass recovery through pumping and treating the contaminated groundwater would not be expected to be highly effective.

2. The maximum concentration expected in the groundwater in the presence of significant amounts of free product benzene would be approximately equal the aqueous solubility of benzene, or approximately 1,780 mg/L. After the source is removed, the concentration would decline due to the dilution by groundwater flow and by natural attenuation due to biodegradation.

This problem will be revisited in Problem TSUP.19, Chapter 8.

TSDWA.14 ION EXCHANGE PRINCIPLES

Describe the mechanism of ion exchange.

SOLUTION: Ion exchange involves the displacement of ions of given species from insoluble exchange materials by ions of different species when solutions of the latter are brought into contact with the exchange materials. The exchange can be expressed by the general equilibria,

$$B_1^+ + R^- B_2^+ \rightleftharpoons B_2^+ + R^- B_1^+ \qquad (5.10)$$
$$A_1^- + R^+ A_2^- \rightleftharpoons A_2^- + R^+ A_1^- \qquad (5.11)$$

where B_1^+, B_2^+ = cations of two different species
A_1^-, A_2^- = anions of two different species
R^-, R^+ = cationic and anionic exchange materials, respectively

The exchange behaves as a chemically reversible interaction between a fixed, ionized exchange site on the exchange material and ions in solution. The equilibrium expression for the first equation is:

$$K = \frac{[B_2^+][R^- B_1^+]}{[B_1^+][R^- B_2^+]} \qquad (5.12)$$

where K = an equilibrium constant sometimes referred to as the equilibrium selectivity coefficient. A similar expression can be written for the second equation.

Although ion exchange has been used to concentrate valuable ionized materials from dilute solutions and for the selective fractionation of certain ionized solutes in solution, its widest application is found in the removal of objectionable ions from water used for domestic (primarily) and industrial purposes.

When the objective in water treatment is the removal of the hardness ions Ca^{++} and Mg^{++} only, exchange materials are used which operate on a sodium ion cycle, that is, which exchange Na^+ for Ca^{++} and Mg^{++}. The equilibrium equation for the removal of calcium with such exchange materials is (see first equation)

$$Ca^{++} + R^=(Na^+)_2 \rightleftharpoons 2Na^+ + R^-Ca^{++} \tag{5.13}$$

When the exchange material is "spent," that is, when equilibrium is reached, the exchange material in the form of R^-Ca^{++} is regenerated to $R^-(Na^+)_2$ so that it can be reused. Regeneration is accomplished by contacting the exchange material with a solution of sodium ions. Since the equilibrium selectivity coefficients for equilibriums such as the above equation have large values, at equilibrium almost all of the exchange material is in the R^-Ca^{++} form and solutions with high concentrations of Na^+ are required for regeneration. Normally, regeneration is accomplished using solutions of 5 to 10 percent NaCl.

TSDWA.15 STRIPPER PERFORMANCE

An atmospheric packed tower air stripper is used to clean contaminated groundwater with a concentration of 100 ppm trichloroethylcne (TCE). The stripper was designed such that packing height is 13 ft, the diameter (D) is 5 ft, and the height of a transfer unit (HTU) is 3.25 ft. Assume Henry's law applies with a constant (H) of 324 atm at 68°F. Also, at these conditions the molar density of water is 3.47 lbmol/ft^3 and the air–water mole flowrate ratio (G/L) is related to the air–water volume ratio (G''/L'') through $G''/L'' = 130 \ G/L$, where the units of G'' and L'' are ft^3/(s · ft^2).

1. If the stripping factor (R) used in the design is 5.0, what is the removal efficiency?
2. If the air blower produces a maximum air flow (q) of 106 acfm, what is the maximum water flow (in gpm) that can be treated by the stripper?

For purposes of this problem, the height of a packed tower can be calculated by

$$Z = (N_{OG})(H_{OG}) = (NTU)(HTU) \tag{5.14}$$

In addition, the following equation has been employed for the calculation of the number of transfer units (NTU) for an air–water stripping system and is based on the stripping factor, R, and the inlet/outlet concentrations.

$$NTU = \frac{R}{R-1} \ln\left(\frac{C_{in}/C_{out}(R-1)+1}{R}\right) \tag{5.15}$$

where C_{in} = inlet contaminant concentration, ppm
C_{out} = outlet contaminant concentration, ppm
R = stripping factor

Note also that

$$R = \frac{H}{P}\frac{G}{L} \qquad (5.16)$$

where H = Henry's law constant, atm
$\quad P$ = system pressure, atm
$\quad G$ = gas (air) loading rate (or flux) lbmol/ft$^2 \cdot$ s
$\quad L$ = liquid loading rate (or flux) lbmol/ft$^2 \cdot$ s

SOLUTION: The number of transfer units (NTU) is first calculated:

$$Z = (N_{OG})(H_{OG}) = (NTU)(HTU)$$
$$NTU = Z/HTU$$
$$= 13/3.25$$
$$= 4$$

Rearranging Equation 5.15 above,

$$C_{out} = \frac{C_{in}(R-1)}{R \, \exp[(NTU)(R-1)/R] - 1}$$
$$= \frac{(100)(5.0-1)}{(5.0) \exp[(4.0)(5.0-1)5.0] - 1}$$
$$= 3.3$$

The removal efficiency (RE) is then

$$RE = [(C_{in} - C_{out})/C_{out}]100\%$$
$$= [(100 - 3.3)/100]100\%$$
$$= 96.7\%$$

The air–water mole ratio, G/L is

$$G/L = (P)(R)/H$$
$$= (1 \text{ atm})(5)(324 \text{ atm})$$
$$= 0.0154$$

In addition,

$$G''/L'' = 130 \, G/L$$
$$= 130(0.0154)$$
$$= 2.0$$

The tower cross-sectional area, S, in ft^2, is

$$Area = S = \pi D^2/4$$
$$= \pi (5\,ft)^2/4$$
$$= 19.63\,ft^2$$

the air volumetric loading rate, G'', in $ft^3/(ft^2 \cdot min)$ is then

$$G'' = (106\;ft^3/\,min\,)/(19.63\,ft^2)$$
$$= 5.4\;ft^3/(\,ft^2 \cdot min\,)$$

Also the water volumetric loading rate, L'', in $ft^3/(ft^2 \cdot min)$ is

$$L'' = 2G''$$
$$= 2(5.4\;ft^3/\,ft^2 \cdot min\,)$$
$$= 10.8\;ft^3/(\,ft^2 \cdot min\,)$$

This can be converted to gpm:

$$L = \frac{[10.8\;ft^3/(\,ft^2 \cdot min\,)](3.47\;lbmol/ft^3)(18\;lb/lbmol)(19.63\;ft^2)}{8.33\;lb/gal}$$
$$= 1590\;gpm$$

Once the volatile organic compounds (VOCs) have been recovered from a ground-water stream, the off-gas (air/VOC mixture) usually needs to be treated. This entails the installation of other equipment to handle the disposal of the VOCs. Some typical methods include flaring, carbon adsorption, and incineration. Flaring is potentially hazardous due to the oxygen that is allowed to enter the flare header. Carbon adsorption can be an efficient means of recovering the VOCs from the off-gas, but it can generate large quantities of solid hazardous waste and is expensive. A utility boiler (incineration) is probably the best alternative since VOC destruction is typically more than 99%, and it is safer and less expensive. In addition, catalytic incinerators can achieve high VOC destruction efficiencics.

TSDWA.16 COOLING WATER FROM A RECREATIONAL RIVER

Determine the percentage of a recreational river stream's flow available to a water treatment facility for cooling such that the river temperature does not increase more than 10°F. Fifty percent of the industrial withdrawal is lost by evaporation and the industrial water returned to the river is 60°F warmer than the river.

SOLUTION: Draw a flow diagram representing the process as shown in Figure 5.3. Express the volumetric flow lost by evaporation from the process in terms of that entering the process:

$$q_{lost} = 0.5q_{in}$$

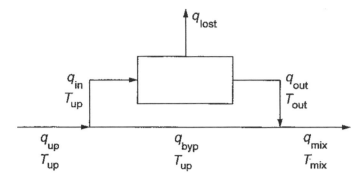

FIGURE 5.3   Flow diagram for Problem TSDWA.17.

Express the process outlet temperature and the maximum river temperature in terms of the upstream temperature:

$$T_{out} = T_{up} + 60°F$$
$$T_{mix} = T_{up} + 10°F$$

Using the conservation law for mass, express the process outlet flow in terms of the process inlet flow. Also express the flow bypassing the process in terms of the upstream and process inlet flows:

$$q_{out} = 0.5q_{in}$$
$$q_{byp} = q_{up} - q_{in}$$
$$q_{mix} = q_{up} - 0.5q_{in}$$

The flow diagram with the expressions developed above are shown in Figure 5.4.

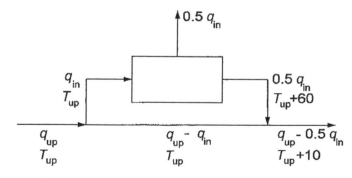

FIGURE 5.4   Flow diagram after applying mass balances.

Noting that the enthalpy of any stream can be represented by $qc_p\rho(T-T_{ref})$, an energy balance around the downstream mixing point leads to

$$(q_{up} - q_{in})c_p\rho(T_{up} - T_{ref}) + 0.5q_{in}c_p\rho(T_{up} + 60 - T_{ref})$$
$$= (q_{up} - 0.5q_{in})c_p\rho(T_{up} + 10 - T_{ref})$$

Note that T_{ref} is arbitrary and indirectly defines a basis for the enthalpy. Setting $T_{ref} = 0$ and assuming that density (ρ) and heat capacity (c_p) are constant yields

$$(q_{up} - q_{in})T_{up} + 0.5q_{in}(T_{up} + 60) = (q_{up} - 0.5q_{in})(T_{up} + 10)$$

The equation may now be solved for the inlet volumetric flow to the process in terms of the upstream flow:

$$q_{up}T_{up} - q_{in}T_{up} + 0.5q_{in}T_{up} + 30q_{in} = q_{up}T_{up} + 10q_{up} - 0.5q_{in}T_{up} - 5q_{in}$$

Canceling terms produces

$$35q_{in} = 10q_{up}$$
$$q_{in} = 0.286q_{up}$$

Therefore, 28.6% of the original flow, q_{up}, is available for cooling.

Note that the problem can also be solved by setting $T_{ref} = T_{up}$. Since (for this condition) $T_{ref} - T_{up} = 0$, the solution to the problem is greatly simplified.

TSDWA.17 REACTING CHEMICAL CONCENTRATION

Consider the water system pictured in Figure 5.5. The upstream flowrate at point 1 and downstream flowrate at point 2 are 20 cfs and 28 cfs, respectively. If the decay (reaction) rate of the chemical in the drinking water can be described by a first-order reaction with a reaction velocity constant of 0.02 day^{-1}, determine the concentration profile of the chemical if both the upstream and infiltrating flows do not contain the chemical.

Assume the cross-sectional area available for flow is constant at 400 ft^2 and that the flow variation with distance is a linear relationship. Perform the calculation for the following cases:

1. Use the inlet velocity and assume it to be constant. Neglect concentration variations arising due to the infiltration.

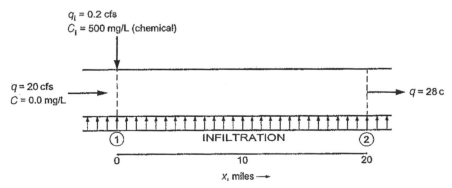

FIGURE 5.5 Water system for Problem TSDWA.17.

2. Use the average velocity (inlet, outlet) and assume it to be constant. Neglect concentration variations arising due to the infiltration.
3. Account for both velocity and concentration variation due to infiltration.

SOLUTION:

1. The inlet water velocity is

$$v = q/\text{ft}$$

$$= \frac{20.2}{400}$$

$$= 0.0505 \ \text{ft/s}$$

$$= 0.828 \ \text{mpd (miles per day)}$$

The velocity at point 2 is

$$v = \frac{28}{400}$$

$$= 0.07 \ \text{ft/s}$$

$$= 1.15 \ \text{mi/d (mpd)}$$

Since the cross-sectional area for flow is constant, one may employ an average velocity flow model based on a constant inlet velocity (0,828 mpd). For this case,

$$\frac{dC}{dx} = -\frac{kC}{v} \tag{5.17}$$

with,

$$C_0 = \frac{(0.02)(500)}{20.2}$$

$$= 0.495 \ \text{mg/L}$$

Integrating the above,

$$C = C_0 e^{-k(x/v)}$$

$$= 0.495 e^{-0.02(x/0.828)}$$

$$= 0.495 e^{-0.0241x}$$

2. Using the average velocity of 0.989 mpd results in the following concentration profile equation:

$$C = 0.495 e^{-0.0202x}$$

3. Theodore (personal notes, 1970) has provided an approach to account for these two variations. Assume the increase in velocity (and volumetric

flowrate) occurs linearly down the system. For this condition

$$v = 0.828(1 + 0.0161x)$$

Since the volumetric flowrate correspondingly increases, a concentration reduction will occur due to its "dilution" effect. This accounting is given by dividing the concentration by the factor $(1 + 0.0161x)$. The describing equation now becomes

$$\frac{dC}{dx} = -k\left(\frac{C}{1 + 0.0161x}\right)\left(\frac{1}{1 + 0.0161x}\right)$$

Separating the variables leads to

$$\frac{dC}{C} = -\frac{k\,dx}{(1 + 0.0161x)^2}$$

Integrating gives

$$\ln\frac{C}{C_0} = k\left[-\frac{1}{(0.0161)(1 + 0.0161x)}\right]_0^x$$

$$C = C_0\exp\left\{-k\left[\frac{1}{(0.0161)(1 + 0.0161x)} - \frac{1}{0.0161}\right]\right\}$$

$$= (0.495)\exp\left(1.242 - \frac{1.242}{1 + 0.0161x}\right)$$

TSDWA.18   PCB LAKE CONTAMINATION

Lake Pristine is a small, shallow lake in the Adirondack Mountains which is a source of drinking water and home to thousands of fish, having escaped the ravages of acid rain. Unfortunately, a waste dump is sited only 0.5 miles upstream from the lake. On January 1, 1982, PCBs (polychlorinated biphenyls) began to enter the lake and began to accumulate there. The lake has a surface area of $9 \times 10^5\,m^2$ and an average depth of 3.0 m. The average annual flow rate into the lake from the surrounding drainage area is $80\,m^3/h$, while the annual loss due to evaporation averages 20% of its depth.

1. Calculate the average annual outflow of the lake and the annual PCB influx (in kg), given that the PCB concentration in the contaminated water is 0.10 mg/L, but that only 10% of the incoming water is contaminated.

2. If the lake volume is assumed constant, calculate the concentration of PCBs (mg/L) in the lake water after one year if there is no outflow.

3. Derive an equation which describes the mass and concentration of PCBs in the lake with respect to time. Assume perfect mixing.

4. In what year will the concentration of PCBs become 10 times that of Year One?
5. In what year will the lake become saturated with PCBs? (Solubility of PCBs in water ≈ 0.20 mg/L).
6. How many kg of PCBs will have been lost from the dump (into the lake) when the lake becomes saturated?

SOLUTION:

1. Calculate the volume of the lake:

$$\text{Volume} = (\text{surface area})(\text{depth}) = (9 \times 10^5 \text{ m}^2)(3.0 \text{ m})$$
$$= 2.7 \times 10^6 \text{ m}^3$$

Calculate the volume of water lost annually due to evaporation:

$$\text{Evaporation Loss} = (0.2)(3.0 \text{ m})(\text{surface area})$$
$$= (0.60 \text{ m})(9 \times 10^5 \text{ m}^2)$$
$$= 5.4 \times 10^5 \text{ m}^3/\text{yr}$$

Calculate the volume of water flowing into the lake annually:

$$\text{Volume} = (\text{water rate})(\text{time}) = (80 \text{ m}^3/\text{h})(24 \text{ h/d})(365 \text{ d/yr})$$
$$= 7.0 \times 10^5 \text{ m}^3/\text{yr}$$

Calculate the volume of water flowing out of the lake annually:

$$\text{Volume In} = \text{Volume Out}$$
$$7.0 \times 10^5 \text{ m}^3/\text{yr} = \text{Evaporation} + \text{Outflow} = 5.4 \times 10^5 \text{ m}^3/\text{yr} + \text{Outflow}$$
$$\text{Outflow} = 1.6 \times 10^5 \text{ m}^3/\text{yr}$$

Calculate the volume of contaminated water entering the lake annually and then, using the inlet concentration of PCBs, the annual kg of PCBs in the inflow.

$$\text{Volume of Contaminated Water} = 10\% \text{ of total Inflow} = 7.0 \times 10^4 \text{ m}^3/\text{yr}$$

$$\text{Concentration of PCBs in Inflow} = 0.10 \text{ mg/L} = 1.0 \times 10^{-4} \text{ kg/m}^3$$

$$\text{Inflow kg PCBs annually} = (\text{Volume})(\text{Concentration})$$
$$= (7.0 \times 10^4 \text{ m}^3/\text{yr})(1.0 \times 10^{-4} \text{kg/m}^3)$$
$$= 7.0 \text{ kg/yr}$$

2. Year One PCB Inflow to the lake $= 7.0 \, \text{kg}$
3. The equation is derived by applying the conservation law for the mass of PCB on a time rate basis.

(mass rate PCB in) $-$ (mass rate PCB out) $=$ (rate of accumulation of PCB)

Substituting (in kg/yr) gives

$$7.0 - (1.6 \times 10^5)(c) = (\text{V lake})\frac{dc}{dt} = 2.7 \times 10^6 \frac{dc}{dt};$$

where $c =$ concentration of PCB in the lake at time t, kg/m^3
   This equation reduces to

$$0.26 \times 10^{-5} - 0.059c = \frac{dc}{dt}; \quad c = 0 \text{ at } t = 0$$

This is an equation of the form

$$\frac{dx}{dt} = a + bx$$

which may be integrated to give

$$c = 1.0 \times 10^{-5}(1 - e^{-0.26t})$$

The mass M of PCB in the lake at any time is then

$$\text{M} = (c)(\text{V lake}) = 2.7 \times 10^6 \, c$$

4. After year one ($t = 1$),

$$c = 10^{-5}(1 - e^{-0.26})$$
$$= 0.23 \times 10^{-5} \, \text{kg/m}^3 = 0.0023 \, \text{mg/L}$$

For 3 times this concentration,

$$c = 0.69 \times 10^{-5} \, \text{kg/m}^3 = 0.0069 \, \text{mg/L}$$

and
$$t = 4.5 \text{yr}$$

5. For $c = 0.2 \, \text{mg/L} = 0.2 \times 10^{-3} \, \text{kg/m}^3$, one notes that this concentration will never be achieved. The maximum concentration that can be achieved ($t = \infty$) in

$$c = 1.0 \times 10^{-5} \, \text{kg/m}^3 = 0.01 \, \text{mg/L}$$

6. The total mass of PCBs in the lake when the lake is saturated is:

(Concentration of PCBs)(Volume of lake) $= (0.20 \text{ mg/L})(2.7 \times 10^6 \text{ m}^3)$

$$= 2.0 \times 10^{-7} \text{ kg/L}(2.7 \times 10^9 \text{ L})$$
$$= 540 \text{ kg}$$

TSDWA.19   DRINKING WATER EQUIVALENT LEVEL

A term employed for non-carcinogens is the drinking water equivalent level (DWEL). This term can be calculated from

$$\text{DWEL} = \frac{(\text{NOAEL})(w)}{(\text{Qw})(\text{UF})} \tag{5.18}$$

where NOAEL = no observed adverse effect level, $\text{mg/kg} \cdot \text{d}$
       w = human weight; 70 kg
       Qw = water consumption of human; 2 L/d
       UF = uncertainity of correction factor, 5–100
   Calculate the DWEL for fluorides if

$$\text{NOAEL} = 11.4 \text{ mg/kg} \cdot \text{d}$$
$$\text{UF} = 100 \text{ (a conservative value)}$$

SOLUTION: Apply Equation (5.18).

$$\text{DWEL} = \frac{(\text{NOAEL})(w)}{(\text{Qw})(\text{UF})}$$

substitution gives

$$\text{DWEL} = \frac{(11.4)(70)}{(2)(100)}$$
$$= 3.99 \text{ mg/L}$$

TSDWA.20   MCLG-RfD RELATIONSHIP

In lieu of information from sources other than water, the MCLG may be assumed equal to the RfD. If data is available,

$$\text{MCLG} = \text{RFD} - \text{(other sources); consistent units} \tag{5.19}$$

If the RfD for fluorine is 5.0 mg/L and the contribution from other source is 20% of the RfD, estimate the MCLG for fluorides.

SOLUTION: Employ Equation (5.19),

$$MCLG = RfD - \text{other sources}$$

Since,

$$\text{other source} = 0.2\,RfD$$
$$MCLG = RfD - 0.2\,RfD$$
$$= (0.8)(5.0)$$
$$= 4.0\,mg/L$$

TSDWA.21   NOAEL CALCULATION

Estimate the NOAEL for a cyanide. Employ an uncertainty factor of 35.

SOLUTION: Employ Equation (5.18).

$$DWEL = MCLG = \frac{(NOAEL)(w)}{(Qw)(UF)};\ \ UF = 35 \qquad (5.20)$$

The MCLG for cyanide is 0.2. Therefore

$$0.2 = \frac{(NOAEL)(70)}{(2)(35)}$$
$$NOAEL = 0.2\,mg/k \cdot d$$

TSDWA.22   SAFETY FACTOR

Refer to the previous Problem. Is the proposed UF value within guidelines?

SOLUTION: Equation (5.18) indicates that the normal range for UF is $5-100$. A value of 35 is within the guidelines and is often arbitrarily selected.

TSDWA.23   FLUORIDE FEED RATE

The feed rate of certain treating chemicals to a water plant can be calculated employing the following equation

$$F = \frac{(D)(Q)}{(CC)};\ \text{consistent units} \qquad (5.21)$$

where F = feed rate of chemicals
  D = dosage of chemical
  Q = water flow (plant) rate
  CC = correction factor, units of D

The term CC may obtained firm

$$CC = (PC)(ACI) \qquad (5.22)$$

where PC = percent chemical (e.g., F, fluorine) available for intended purpose
  ACI = available chemical ion

Calculate the chemical feed rate if the purity of the treating chemical is 98.5% and the ACI is 0.60. A concentration of 1.25 mg/L of fluoride is required for a 50,000 gpd water rate.

SOLUTION: Employ Equation (5.21) to calculate CC.

$$CC = (98.5)(0.60) = 59.1$$

The required flowrate can now be calculated from Equation (5.20)

$$FR = \frac{(1.25)(50,000)}{59.1}$$

$$= 1,057.5 \text{ gal/d (gpd)}$$

TSDWA.24   FLUORIDE FEED RATE CONVERSION

Refer to the previous problem. Calculate the required fluoride feed rate in the following units

1. gpm
2. lb/d
3. lb/min

SOLUTION: Employ appropriate conversion factors.

For (1),

$$F = (1057.5)/1440; \ 1440 \text{ min/day}$$

$$= 0.734 \text{ gpm}$$

For (2),

$$F = (1,057.5) \ (8.34); \ 8.34 \text{ lb/gal}$$

$$= 8,820 \text{ lb/d}$$

For (3),

$$F = 8,820/1440$$

$$= 6.125 \text{ lb/min}$$

TSDWA.25 FLUORIDE DOSAGE

Refer to the two previous problems. Calculate the dosage of the fluoride in the plant water if the flowrate is 2,000 gpm and the feed rate is 50,000 gal/d.

SOLUTION: Once again, employ Equation (5.21).

$$F = \frac{(D)(Q)}{(CC)}$$

Rearranging,

$$D = \frac{(F)(CC)}{(Q)}$$

substituting

$$D = \frac{(50,000)(59.1)}{(2,000)(1440)}$$

$$= 1,026 \ \mu g/L$$

TSDWA.26 SUGGESTED WATER FLUORIDE CONCENTRATION

The suggested fluoride concentration, F, for fluoridated water supply system can be estimated from

$$F = 0.34/E \tag{5.23}$$

where F = fluoride concentration, mg/L and

$$E = 0.038 + 0.0062 \ T \tag{5.24}$$

where T = yearly average atmosphere temperature, °F

Based on the above equations, estimate the fluoride water concentration in Cary, NC (the home of one of the authors). The average annual air temperature in Cary is 63°F.

SOLUTION: Employ Equation (5.24) to calculate E.

$$E = 0.038 + 0.0062 \ T$$

$$= 0.038 + 0.0062(63)$$

$$= 0.4286$$

The fluoride concentration is now obtained from Equation (5.23).

$$F = 0.34/E$$
$$= 0.34/0.4286$$
$$= 0.793 \text{ mg/L}$$

TSDWA.27 TOXIC WASTEWATER

The part-per-billion concentration of a particular toxic in a groundwater stream occasionally used for drinking purposes is known to be normally distributed with mean, $m = 100$ and a standard deviation $\sigma = 2.0$. Calculate the probability that the toxic concentration, C, is between 98 and 104.

SOLUTION: If C is normally distributed with mean m and standard deviation σ; then the random variable $(C - m/\sigma)$ is normally distributed with mean 0 and standard deviation 1. The term $(C - m/\sigma)$ is called a "standard normal variable," and the graph of its probability distribution function (PDF) is called a "standard normal curve." (The probability distribution of a random variable is specified by the probability distribution function, or PDF.) Referring to Figure 5.6, areas under a standard normal curve to the right of z_0 for non-negative values z_0 can be found in any standard statistics book. Such an area is referred to as a "right hand tail" (see Figure 5.6). Probabilities about a standard normal variable Z can be determined from Table 5.5. For example,

$$P(Z > 1.54) = 0.062$$

Since C is normally distributed with mean $m = 100$ and standard deviation $\sigma = 2$, then $(C - 100)/2$ is a standard normal variable and

$$P(98 < C < 104) = P\{-1 < [(C - 100)/2] < 2\} = P(-1 < Z < 2)$$

From Table 5.5,

$$P(98 < C < 104) = 0.341 + 0.477 = 0.818 = 81.8\%$$

TSDWA.28 TOXIC ASH

Concerns regarding a toxic ash in a water sludge waste that could potentially impact a drinking water supply calls for a level of 1.0 ppm or less of the toxic. Earlier observations of the concentration of the ash indicates a normal distribution with a mean of 0.60 ppm and a standard deviation of 0.20 ppm. Estimate the probability that ash will exceed the regulatory limit.

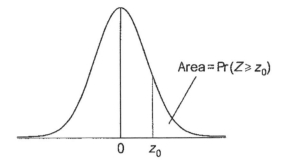

FIGURE 5.6 Standard normal, cumulative probability curve.

See also earlier problem. The reader is referred to S. Shaefer and L. Theodore, "Probability and Statistics: Application for Environmental Science," CRC Press/ Taylor & Francis Group, Boca Raton, FL, 2007, for additional details.

SOLUTION: Refer to Problem TSDWA.27. This problem requires the calculation of $P(C > 1.0)$. Normalizing the variable C,

$$P\{[(C - 0.6)/0.2] > [(1.0 - 0.6)/0.20]\}$$
$$P(Z > 2.0)$$

From the standard normal table (see Table 5.5),

$$P(Z > 2.0) = 0.0228$$
$$= 2.28\%$$

For this situation, the area to the right of the 2.0 is 2.28% of the total area. This represents the probability that ash will exceed the regulatory limit of 1.0 ppm.

CHAPTER 6

RESOURCE CONSERVATION AND RECOVERY ACT (RCRA)

QUALITATIVE PROBLEMS (LRCRA)

PROBLEMS LRCRA.1–31

LRCRA.1 EARLY LEGISLATION

Describe early legislation as it applies to solid waste.

SOLUTION: Prior to the passage of RCRA, little solid waste legislation existed. Practices that permitted technologies to remove pollutants efficiently from air and water streams, only to dispose of them indiscriminately on land, were commonplace. Legislative authority to control solid waste disposal dates as far back as the 1899 *River and Harbors Act*, the main concern of which was keeping the waterways *navigable*, rather than keeping them *clean*. The following is a quotation from the *Refuse Act*, which comprise Section 13 of the River and Harbors Act:[1]

> ... it shall not be lawful to throw, discharge or deposit, or cause, suffer or procure to be thrown, discharged or deposited either from or out of any ship, barge or other floating craft of any kind, or from the shore, wharf, manufacturing establishment, or mill of any kind, any refuse matter of any kind or description whatever other than that flowing from streets and sewers and passing therefrom in a liquid state, into any tributary of any navigable water from which the same shall float or be washed into such navigable

Environmental Regulatory Calculations Handbook, by Leo Stander and Louis Theodore
Copyright © 2008 John Wiley & Sons, Inc.

water; and it shall not be lawful to deposit, or cause, suffer, or proceed to deposit material of any kind in any place on the bank of any navigable water, or on the bank of any tributary of any navigable water, where the same shall be liable to be washed into such navigable water, either by ordinary or high tides, or by storms or floods, or otherwise, whereby navigation shall or may be impeded or obstructed: provided that nothing herein contained shall extend to, apply to, or prohibit the operations in connection with the improvement of navigable waters or construction of public works, considered necessary and proper by the United States officers supervising such improvement or public work: and provided further, that the Secretary of War, whenever in the judgment of the Chief of Engineers, anchorage and navigation shall not be injured thereby, may permit the deposit of any material above mentioned in navigable waters within limits to be defined and under conditions to be prescribed by him, provided application is made to him prior to depositing such material; and whenever any permit is so granted the conditions thereof shall be strictly complied with, and any violation thereof shall be unlawful.

The first specific solid waste legislation aimed at protecting the environment did not pass through Congress until the mid-1960s. The *Solid Waste Disposal Act* of 1965 fostered a research program concerned with utility-generated solid waste and the recovery of valuable fractions of municipal solid waste. This research and development (R&D) program concentrated on the improvement of collection, transport, recycling, reuse, processing, and disposal methods of solid waste. The program was jointly administered by the Department of Interior (Bureau of Mines) and the Department of Health, Education, and Welfare (Bureau of Solid Waste Management).

The *Resource Recovery Act* of 1970 provided a new directive to the 1965 Solid Waste Disposal Act. While the 1965 Act was limited mainly to R&D, the 1970 Act expanded the federal role to the planning of solid waste management programs and training; and while the 1965 Act made no mention of *hazardous wastes*, the 1970 Act registered concern for the adverse health effects of such wastes and directed the Department of Health, Education, and Welfare to study the problem of storage and disposal of hazardous wastes.

The 1970 Act initiated a change in the nation's attitude toward waste disposal. For one thing, the emphasis shifted from simply waste disposal to include waste management and resource recovery; for another, it became evident that the existing regulations were inadequate to assure the proper management of hazardous wastes. The result of this latter concern was the passage by Congress in 1976 of the RCRA, an amendment to the Resource Recovery Act of 1970.

In 1976, the Toxic Substances and Control Act (TSCA) prohibited the further manufacture of polychlorinated biphenyls (PCBs) after July 2, 1979, established limits on PCB use in commerce, and established regulations for proper disposal of PCBs. In 1980, the Comprehensive Environmental Response, Compensation and Liability Act (CERCLA) established a National Fund (Superfund) to assist in remedial actions of cleanup of the uncontrolled waste sites created by the poor disposal practices of the past. In 1984, the Hazardous and Solid Waste Act (HSWA) amended the RCRA and established a strict timeline for restricting untreated hazardous waste from land disposal. The Superfund Amendments and Reauthorization Act (SARA) of 1986 not only reauthorized the Superfund program, but greatly expanded the provision and funding of the initial act. In 1988, the Medical Waste Tracking Act established a 2-year demonstration program to track regulated medical waste from the point of generation to the point of disposal in five states. The Clean Air Act Amendments (CAAA) of 1990, as well as numerous state and county regulations, regulated emissions from various incineration sources.

Additional details and recent changes receives treatment in the next problem.

LRCRA.2   RESOURCE CONSERVATION AND RECOVERY ACT

Describe the Resource and Recovery Act as it applies to the treatment of hazardous waste.

SOLUTION: The RCRA completely replaced the previous language of the Solid Waste Disposal Act of 1965 to address the enormous growth in the production of waste. It provided "cradle to grave" provisions for controlling the storage, transport, treatment, and disposal of hazardous waste. The objectives of this act were to promote the protection of health and the environment and to conserve valuable materials and energy resources by (see also Chapter 2):

1. Providing technical and financial assistance to state and local governments and interstate agencies for the development of solid waste management plans (including resource recovery and resource conservation systems) that promote improved solid waste management techniques (including more effective organizational arrangements), new and improved methods of collection, separation and recovery of solid waste, and the environmentally safe disposal of nonrecoverable residues.

2. Providing training grants in occupations involving the design, operation, and maintenance of solid waste disposal systems.

3. Prohibiting future open dumping on the land and requiring the conversion of existing open dumps to facilities that do not pose danger to the environment or to health.

4. Regulating the treatment, storage, transportation, and disposal of hazardous wastes that have adverse effects on health and environment.

5. Providing for the promulgation of guidelines for solid waste collection, transport, separation, recovery, and disposal practices and systems.

6. A national research and development program for improved solid waste management and resource conservation techniques, more effective organizational arrangements, new and improved methods of collection, separation, recovery, and recycling of solid wastes and environmentally safe disposal of nonrecoverable residues.

7. Promoting the demonstration, construction, and application of solid waste management, resource recovery, and resource conservation systems that preserve and enhance the quality of air, water and land resources.

8. Establishing a cooperative effort among federal, state, and local governments and private enterprises in order to recover valuable materials and energy from solid waste.

The Act provides, in broad terms, general guidelines for the waste management program envisioned by Congress (e.g., EPA is directed to develop and promulgate criteria for identifying hazardous waste). The Act also provides the EPA Administrator (or his or her representative) with the necessary authority to develop these broad standards into specific requirements for the regulated community.

What is commonly known as RCRA, or the Act, is actually a combination of the aforementioned first federal solid waste statutes and all subsequent amendments. The Act was amended in 1976 by RCRA, which substantially remodeled the nation's solid waste management system and laid out the basic framework of the current hazardous waste management program.

The Act, which has been amended several times since 1976, continues to evolve as Congress alters it to reflect changing waste management needs. The Act was amended significantly on November 8, 1984, by the Hazardous and Solid Waste Amendments (HSWA), which expanded the scope and requirements of RCRA. HSWA was created largely in response to citizen concerns that existing methods of hazardous waste disposal, particularly land disposal, were not safe. Congress also revised RCRA in 1992 by passing the Federal Facility Compliance Act, which strengthened the authority to enforce RCRA at federal facilities. In addition, the Land Disposal Program Flexibility Act of 1996 amended RCRA to provide regulatory flexibility for the land disposal of certain wastes.

Today, the Act consists of 10 subtitles (see Table 6.1). Subtitles A, B, E, F, G, H, and J outline general provisions; authorities of the Administrator; duties of the

TABLE 6.1   Outline of the Act

Subtitle	Provisions
A	General Provisions
B	Office of Solid Waste; Authorities of the Administrator and Interagency Coordinating Committee
C	Hazardous Waste Management
D	State or Regional Solid Waste Plans
E	Duties of the Secretary of Commerce in Resource and Recovery
F	Federal Responsibilities
G	Miscellaneous Provisions
H	Research Development, Demonstration and Information
I	Regulation of Underground Storage Tanks
J	Standards for the Tracking and Management of Medical Waste

Secretary of Commerce; federal responsibilities; miscellaneous provisions; research, development, demonstration, and information requirement; and, medical waste tracking. Other subtitles out the framework for the three major programs that comprise RCRA include Subtitle C (the hazardous waste management program), Subtitle D (the solid waste program), and Subtitle I (the UST program).

The text of the Act can be found at www.epa.gov/epahome/laws.htm.

LRCRA.3   RCRA SUBTITLES A AND C

Provide the key portions and a general description of subtitles A and C of RCRA.

SOLUTION: Subtitle A: General Provisions

In Section 1004, legal definitions for many key terms used in waste management are presented. Some of the more pertinent of these are:

1. *Hazardous Waste*: A solid waste or combination of solid wastes, which because of its quantity, concentration, or physical, chemical or infectious characteristics may (a) cause, or significantly contribute to an increase in mortality or an increase in serious irreversible, or incapacitated reversible illness, or (b) pose a substantial present or potential hazard to human health or the environment when improperly treated, stored, transported or disposed of, or otherwise managed.

2. *Hazardous Waste Generation*: The act or process of producing hazardous waste.

3. *Hazardous Waste Management*: The systematic control of collection, source separation, storage, transportation, processing, treatment and recovery, and disposal of hazardous waste.

4. *Solid Waste*: Any garbage, refuse, sludge from a waste treatment plant, water supply treatment plant, or air pollution control facility and other discarded material, including solid, liquid, semisolid or contained gaseous material resulting from industrial, commercial, mining and agricultural operations and from community activities, but does not include dissolved material in domestic sewage or solid or dissolved material in irrigation return flows or industrial discharges which are point sources subject to permits under the federal *Clean Water Act* or source, special nuclear or byproduct material as defined by the *Atomic Energy Act*.

5. *Storage*: When used in connection with hazardous waste, means the containment of hazardous waste either on a temporary basis, or for a period of years, in such a manner as not to constitute disposal of such hazardous waste.

6. *Disposal*: The storage, deposit, injection, dumping, spilling, leaking or placing of any solid waste or hazardous waste into or in any land or water so that such solid waste or hazardous waste or any constituent thereof may not enter the

environment or be emitted into the air or discharged into any waters including ground waters.

7. *Treatment*: When used in connection with hazardous waste, means any method, technique, or process, including neutralization designed to change the physical, chemical, or biological character or composition of any hazardous waste so as to neutralize such waste or as to render such waste non-hazardous, safer for transport, amenable for recovery, amenable for storage or reduced in volume. Such terms include any activity or processing designed to change the physical form or chemical composition of hazardous waste so as to render it non-hazardous.

Subtitle C: Hazardous Waste Management

Subtitles C and D generate the framework for regulatory control programs for the management of hazardous and solid nonhazardous wastes, respectively. The hazardous waste program outline under Subtitle C is the one most people associate with the term *RCRA*. It is divided into 11 subsections, each of which is summarized here, using much of the language it contained in the original document.

1. *Section 3001. Identification and Listing of Hazardous Waste*   Section 3001 required EPA to promulgate the criteria to be used in determining how hazardous a particular waste is. Factors such as flammability (ignitability), corrosivity, toxicity, persistence, and degradability in the environment as well as potential for accumulation in tissue are to be considered in defining the criteria. In addition, other hazardous characteristics may be added to the list in the future if the need arises.

2. *Section 3002. Standards Applicable to Generators of Hazardous Waste*   The administrator (EPA) is required to promulgate regulations, after consultation with appropriate federal and state agencies, establishing standards for generators of hazardous waste. These standards are to establish record-keeping practices that accurately identify the waste stream's constituents and quantities; the disposition of the waste; labeling the waste for storage, transport, or disposal identification; use of appropriate containers; furnishing information to treatment facilities, storage, or disposal facilities on the chemical composition of the waste; use of a manifest system to reconcile the disposition of the waste and to assure waste only proceeds to a permitted facility; submitting reports to EPA regarding the quantities of waste generated during this reporting period; and, the disposition of all hazardous wastes during that same period.

3. *Section 3003. Standards Applicable to Transporters of Hazardous Waste*   Under Section 3003, standards for the transporters of hazardous wastes are established. The standards involve recordkeeping of the waste transported, including source and delivery points; proper labeling of the wastes; compliance with the prescribed manifest system; and, limiting transportation of all

such hazardous waste only to permitted facilities designated on the manifest form. This section also stipulates that EPA must coordinate its efforts with the Department of Transportation in those cases where the hazardous waste fall under the Hazardous Materials Transportation Act.

4. *Section 3004. Standards Applicable to Owners and Operators of Hazardous Waste Treatment, Storage, and Disposal Facilities*   This section provides for the establishment of standards and regulations for facilities that treat, store, and dispose of hazardous wastes. These standards and regulations pertain to maintaining records of all hazardous waste handled with respect to treatment, storage, and disposal; reporting, monitoring, and inspection of manifests for wastes received to assure compliance; maintaining and operating facilities in accordance with procedures satisfactory to EPA; designating the location, design, and construction of the facility; maintaining contingency plans to minimize unanticipated damage; maintaining the operation of facilities to provide training, continuity of ownership, and to provide standards for financial responsibility.

5. *Section 3005. Permits for Treatment, Storage, and Disposal of Hazardous Waste*   This section covers permits for the treatment, storage, or disposal of hazardous waste. A permit application must include estimates of the quantities, compositions, and concentrations of the waste anticipated, as well as the time frequency, or rate of receiving and handling the waste, and description of the site. (The permit process is discussed later in Problem LRCA.5.)

6. *Section 3006. Authorized State Hazardous Waste Program*   State hazardous waste programs under Section 3006 will receive approval from EPA unless the state program (a) is not equivalent to the federal programs, (b) is not consistent with federal or state programs of other states, or (c) does not provide adequate enforcement of compliance. Any action taken under a hazardous waste program authorized under this section has the same force and effect as actions taken by EPA under this subtitle.

7. *Section 3007. Inspections*   This section provides one of the means of monitoring compliance. Through it, EPA is given the right to gain entry to facilities and records where hazardous wastes are generated, treated, stored, or disposed of for the purpose of inspection and sampling. Records, reports, and information resulting from these inspections are available to the public unless the operator can demonstrate that release of these data would divulge information privileged to protection.

8. *Section 3008. Federal Enforcement*   When EPA determines that a facility is in violation of the regulations, a failure-to-comply notification is issued. If the violation extends beyond 30 days after notification, EPA may (a) issue a second notification requiring compliance within a specified time limit, (b) assess civil penalties, or (c) revoke the RCRA permit. Criminal penalties may be assessed to transporters delivering to nonpermitted facilities, persons disposing of waste without a permit, or persons found to have falsified information on permit applications.

9. *Section 3009. Retention of State Authority*   In order to make the federal programs the minimum standard for hazardous waste regulation, states are prohibited through Section 3009 from enacting regulations less stringent than those promulgated under RCRA.

10. *Section 3010. Effective Date*   Section 3010 requires that 90 days after the promulgation of Section 3001, all generators, transporters, and disposers must have completed the hazardous waste notification procedure. This section also sets a period of 6 months between effective date and promulgation date of Subtitle C regulations.

11. *Section 3011. Authorization of Assistance to States*   This section authorizes funding to assist states in the development and implementation of hazardous waste programs at the state level.

The RCRA regulations are continuously being developed and published according to the following procedure (see Chapter 2 for more details). When a regulation is first proposed, it is published in the *Federal Register* as a proposed regulation along with a discussion of EPAs rationale for the regulation. For a period of time, normally 60 days, public reaction is invited. After this period, EPA revises the proposed regulation and finalizes it by again publishing it in the *Federal Register.* Annually, all such regulations are complied and placed in the *Codes of Federal Regulations* (CFR); this process is called *codification.* Most of RCRA has been codified in this manner and can be found in the CFR under Title 40 Chapter 1 EPA Subchapter I (Solid Waste) Parts 239–299, or more simply [40 CFR Parts 239–299].

LRCRA.4   RCRA SUBTITLES D AND I

Outline key features of subtitles D and I.

SOLUTION: Subtitle D: Managing Solid Waste

1. RCRA's solid waste management program, Subtitle D, encourages environmentally sound solid waste management practices that maximize the reuse of recoverable material and foster resource recovery.

2. The terms solid waste is very broad, including not only the traditional nonhazardous solid wastes, such as municipal garbage, but also some hazardous wastes. RCRA Subtitle D addresses solid wastes, including those hazardous wastes that are excluded from the Subtitle C regulations (e.g., household hazardous waste), and hazardous waste generated by conditionally exempt small quantity generators (CESQGs).

3. The solid waste management program also addresses municipal solid waste, which is generated by business and households and is typically collected and disposed of in municipal solid waste landfills (MSWLFs).

4. EPA recommends an integrated, hierarchical approach to managing municipal solid waste that includes: source reduction, recycling, combustion, and land-filling. Source reduction and recycling are preferred elements of the system.

5. The Subtitle D program includes technical criteria for MSWLFs to ensure that such landfills will be fully protective of human health and the environment.

6. EPA has a number of programs to encourage sound waste management— Wastewise, the Jobs Through Recycling program, unit pricing, and full cost accounting for municipal solid waste.

Subtitle I: Managing Underground Storage Tanks (UST)

1. The RCRA Subtitle I UST regulatory program regulates underground tanks storing petroleum or hazardous substances.

2. In order to protect human health and the environment from threats posed by releases from such tanks, the program governs tank design, construction, installation, operation, release detection, release response, corrective action, closure, and financial responsibility.

3. Many UST owners and operators must secure loans from financial and other institutions to comply with environmental regulations, such as UST upgrading and maintenance requirements. The Subtitle I program contains specific pro-visions to protect lending institutions from liability that they might incur from extending these loans.

4. Similar to RCRA Subtitle C, Subtitle I contains provisions that allow EPA to approve state government implementation and enforcement of UST regulatory programs.

5. The expense and threats of contamination from leaking USTs necessitate efficient, effective, and thorough cleanups. To guarantee that such cleanups will be conducted in an efficient and protective manner, Subtitle I also established a Leaking Underground Storage Tank (LUST) Trust Fund. The Fund facilitates cleanup oversight and guarantees cleanups when the responsible owner and operator cannot take action, or when the situation requires emergency response.

LRCRA.5   THE PERMIT PROCESS

Describe the permit process.

SOLUTION: The permitting of any hazardous or toxic waste management site or process is a highly complex, lengthy and costly procedure. The RCRA imposes legally enforceable requirements on owners or operators of *TSD* facilities (TSD is an acronym for the treatment, storage, or disposal of solid wastes considered hazardous under RCRA). A permit is granted only after compliance with federal standards has been demonstrated. Under this permit, the facility has legal authorization to treat hazardous waste within the limits specified in the permit.

A RCRA permit application is composed of two parts: A and B. To satisfy application requirements for Part A, Consolidated Permit Application Forms 1 (EPA Form 3510-1) and 3 (EPA Form 3510-3) must be completed and submitted to the appropriate federal or state agency. To satisfy application requirements for Part B, a wide variety of procedures to safeguard human health and the environment, e.g., groundwater monitoring, contingency plans, emergency response plans, and risk assessment must be described in detail. The EPA has provided listings and general descriptions of what must be included in this part. However, because there is no specific form to complete, the success of the Part B application depends on how well the applicant can follow and interpret EPA directives to the satisfaction of the agency (federal, state, and/or local) that has RCRA permitting authority over the facility. It should also be noted that some states have their own RCRA permit forms and the authority to issue permits.

LRCRA.6   THE TRIAL BURN

Describe the trial burn and include quantitative details.

SOLUTION: As part of the RCRA Part B permitting process, an incineration facility is required to perform a *trial burn*. A trial burn is performed to demonstrate conformity with the regulations. In the case of a hazardous waste incinerator, the specific regulations are as follows:

1. 40 CFR 264.343(a)(1), *Standards to Control Organic Emissions*, require a destruction and removal efficiency (DRE) of selected principal organic hazardous constituents (POHCs) of 99.99 percent (four nines DRE). The DRE for an incinerator/air pollution control system is defined by the following formula:

$$\text{DRE}(\%) = [(\dot{m}_{in} - \dot{m}_{out})/\dot{m}_{in}] \times 100 \tag{6.1}$$

where DRE = destruction and removal efficiency, percent

$\dot{m}_{in}$ = mass feed rate of POHC fed to the incinerator (lb/h)

$\dot{m}_{out}$ = mass emission rate of POHC to the atmosphere (lb/h)

2. 40 CFR 264.343(b), *Standards to Control Hydrogen Chloride*, requires the control of hydrogen chloride emissions such that the rate of emission is no greater than the larger of either 1.8 kg/h (4 lb/h) or 1% of the hydrogen chloride in the stack gas prior to entering any pollution control equipment.

3. 40 CFR 264.343(c), *Standards to Control Particulate Matter*, mandates that the incinerator may not emit particulate matter in excess of 180 mg per dry standard cubic meter (mg/dscm) or 0.08 grains per dry standard cubic foot

(gr/dscf) after correction to 7% O_2 dry volume in the stack gas. The correction factor (CF) is defined as follows:

$$\text{Correction factor (CF)} = (21 - 7)/(21 - Y) = 14/(21 - Y) \qquad (6.2)$$

where $Y =$ measured O_2 concentration in the stack gas on a dry basis (expressed as percentage).

When systems operate with excess air that results in stack O_2 above 7% d.v. (dry volume), the CF is less than one. If the system operates with stack O_2 less than 7% d.v., the CF is greater than one. This impacts systems that dilute the combustion gases with excess air. This CF is used when air (21% O_2 by volume) is used as the combustion air. If enriched air ($>21\%$ O_2 up to 100% O_2) is used as combustion air, this CF does not apply.

A *trial burn plan* must be included as part of the engineering description section of the RCRA Part B permit application. The trial burn plan describes the goals and procedures of the trial burn. The plan includes discussions on the waste characteristics and the system equipment, including controls and monitoring devices, descriptions of the conditions that will be demonstrated during the test, and descriptions of the sampling and analysis methods that will be used to collect the data.

To verify compliance with the DRE performance standard during the trial burn, it is not required that the incinerator DRE for every principal organic hazardous constituent (POHC) identified in the waste be measured. Typically, two or three compounds are chosen as representative POHCs for the waste stream. Interestingly, the designated trial burn POHCs do not have to be actually present in the normal waste; they do, however, have to be considered more difficult to incinerate than any POHCs in the normal waste. The substitute POHCs in this strategy are refered to as *surrogate* POHCs. The surrogate POHCs would be added to the waste stream during the trial burn testing.

Two primary ranking hierarchies have been used as criteria in the selection of POHCs to ensure that the POHCs chosen represent the widest range of compounds expected to be burned. The first ranking is the USEPA's incinerability list, included in USEPA's *Guidance Manual for Hazardous Waste Incinerator Permits*, July 1983 (EPA/SW-966), which ranks compounds by heat of combustion. This approach to POHC selection is based on thermodynamic and kinetic equilibrium theories that claim that the primary concern in evaluating the difficulty of destruction for a compound is the amount of energy necessary to complete the combustion process. The second POHC ranking is the thermal stability index, which is included in the USEPA's handbook *Guidance on Setting Permit Conditions and Reporting Trial Burn Results*, January 1989 (EPA/625/6-89/019). This ranking of 40 CFR Part 261, Appendix VIII, compounds is based on their thermal stability, with the most stable being considered the most difficult to burn. These two ranking systems are used with other factors, such as waste characteristics and physical and chemical characteristics of compounds, to choose an appropriate POHC for the trial turn demonstration. The EPA permit review process works with the owners/operators incinerator facility in determining which POHCs in a given waste should be designated for sampling and analysis during the trial burn.

The trial burn plan also describes the operating conditions(s) to be demonstrated during the testing. The test conditions should be developed to demonstrate the worst-case operating conditions of the system. Factors that should be considered include the following:

1. Maximum capacity (heat release) of the system
2. Design capacity for each waste stream
3. Worst-case composition of each waste stream, e.g., maximum chlorine and ash concentrations
4. Minimum combustion temperature and minimum combustion gas oxygen concentration to be demonstrated

The complexity of a trial burn will depend on the number of operating limits that must be demonstrated during the test. More than one test condition may be required to demonstrate all limits.

The trial burn plan is submitted with the Part B permit application or, occasionally, will be submitted prior to the application. This is done to give the agency sufficient time to review the plan and provide comments to the facility. After the agency has approved the trial burn plan, the facility must prepare for and execute the trial burn. In some cases, it is advisable to plan and conduct a pretrial burn to ensure the owner/operator of proper performance before final trial burn.

In conducting the trial burn, only approved EPA sampling and analysis methods or their equivalents may be employed. The current EPA sampling methods, Volatile Organic Sampling Train (VOST) (SW-846 Method 0030) and Modified Method 5 (MM-5) (SW-846 Method 0010) are commonly used for sampling of POHCs. USEPA Method 5 sampling protocol is used to sample for particulate, and SW-846 Method 0050 is used to sample for hydrogen chloride. Analytical methods for hazardous organic compounds include gas chromatography (GC), flame ionization detection (FID), photoionization detection (PID), Hall electrolytic conductivity detection (HECD), electron capture detection (ECD), gas chromatography/mass spectrometry (GCMS), high-performance liquid chromatography (HPLC), and atmospheric pressure chemical ionization mass spectrometry (APCI-MS). Information on EPA sampling and analysis methods are available in EPA publication SW-846 "Test Methods in Evaluating Solid Waste, Physical/Chemical Methods," http://www.epa.gov/epaoswer/hazwaste/test/sw846.htm.

The POHC compounds normally have to be concentrated from large amounts of the flue gas sample on special polymer resins: a 3-hour timeframe is typical. The extraction and concentration techniques required to prepare the sample for analysis are complicated and time consuming. As a result, trial burns are costly, and it usually requires 30–60 days before the final laboratory analysis results are known.

Because of the complexity of the sampling and analysis, trial burns can be very costly. The costs associated with the trial burn normally include expenditures for surveying, agency meetings, equipment, sampling, analysis, cleanup, and report preparation.

LRCRA.7 BOILER AND INDUSTRIAL FURNACE (BIF) REGULATIONS

Provide details on the boiler and industrial furnace regulations.

SOLUTION: Boilers and industrial furnaces burning hazardous waste were initially exempt from the RCRA incinerator requirements. Under RCRA, these units are considered to be "resource recovery" units, which do not fall under the definition of incinerators. As burning hazardous waste in boilers and industrial furnaces became common practice, the EPA studied these units and determined that there was a need for regulations. These units were burning the same wastes as permitted hazardous waste incineration facilities but were not required to meet any of the same performance standards.

In 1991, rules governing the operation of boilers and industrial furnaces (BIF) burning hazardous waste were promulgated. The BIF regulations are found in the [40 CFR Part 266 Subpart H]. These rules cover the burning of hazardous wastes in boilers, industrial furnaces, and kilns for cement, lime, and aggregate. The rule requires that these facilities obtain RCRA Part B permits to burn hazardous waste in resource recovery units. Important changes to the regulations which impact BIFs were promulgated in 1996 and 1998. Details on these rules are provided below. Additional details can be found in the "RCRA Orientation Manual 2006; Resource Conservation and Recovery Act" at http://www.epa.gov/epaoswer/general/orientat/.

Performance Standards

BIFs are required to meet performance standards similar to those of RCRA incinerators. In addition, the BIF regulations set limits on other pollutants, such as heavy metals and chlorine, which are not regulated by RCRA.

Destruction and Removal Efficiency A boiler or industrial furnace burning hazardous waste must achieve DRE of 99.99% for all organic hazardous constituents in the waste feed. To demonstrate conforming with this requirement, 99.99% DRE must be demonstrated during a trial burn for each designated waste feed stream.

BIF units burning hazardous waste containing (or derived from) EPA Hazardous Waste Nos. F020, F021, F022, F023, F026, or F027 must achieve a DRE of 99.9999% for each designated POHC. (These substances are identified in 40 CFR 261 Appendix VII.) This performance must be demonstrated on POHCs that are more difficult to burn than tetra-, penta-, and hexachlorodibenzo-p-dioxins and dibenzofurans. DRE is determined using the same calculation method as that for incinerators.

Carbon Monoxide/Hydrocarbons The stack gas concentration of carbon monoxide (CO) from a BIF unit must not exceed 100 parts per million by volume (ppm$_v$) on an hourly rolling average basis (i.e., over any 60-min period), continuously corrected to 7% O_2, on a dry gas basis. The CO and O_2 must be continuously monitored.

As an alternate to the CO standard, the CO concentration from a BIF unit burning hazardous waste may exceed the 100 ppmv limit provided that stack gas concentrations of hydrocarbons (HC) do not exceed 20 ppmv. The HC limits must be on an hourly rolling average basis (i.e., over any 60-min period), reported as propane and continuously corrected to 7% O_2 dry gas basis.

If the alternative standard is used, HC emissions must be continuously monitored. CO and O_2 must also be continuously monitored. The alternative CO standard is established based on CO data during the trial burn or the compliance test. The alternative CO standard is the average over all valid runs of the highest hourly average CO level for each run. The CO limit is implemented on an hourly rolling average basis and continuously corrected to 7% O_2 dry gas basis.

Dioxins and Furans   BIF units that are equipped with a dry particulate matter control device that operates within the temperature range of 450–750°F, and industrial furnaces operating under an alternative hydrocarbon limit must conduct a site-specific risk assessment to demonstrate that emissions of polychlorinated dibenzo-p-dioxins (PCDD) and polychlorinated dibenzofurans (PCDF) do not result in an increased lifetime cancer risk to the hypothetical maximum exposed individual (MEI) exceeding 1 in 100,000.

Particulate Matter   A BIF unit may not emit particulate matter in excess of 180 mg/dscm (0.08 gr/dscf) after correction to a stack gas concentration of 7% O_2. The method of calculating the corrected concentration is identical to that for incinerators.

Hydrogen Chloride and Chlorine   Hydrogen chloride (HCl) and chlorine gas (Cl_2) emissions are regulated using a tiered set of standards. The standards have been developed to ensure that emissions of HCl and Cl_2 do not exceed ambient health-based levels. The standards involve the use of dispersion modeling or screening limits based on worst-case facilities. Typically, a facility has the option to choose any of the four tiers to demonstrate compliance. For some facilities, Tier III or Adjusted Tier I must be used because of site-specific conditions such as neighboring building heights, ground level rise within 5 km of stack, bodies of water, and so forth.

Tier I   Tier I is the simplest approach to determine HCl and Cl_2 limits. This approach also produces the most conservative feed rate limits. Feed rate limits are determined using a set of tables developed by the EPA (see 40 CFR 266 Appendix II and Appendix III). The feed rate limits are a function of the stack height, surrounding land use, and the flue gas flow rate. By limiting the HCl and Cl_2 feed rates, the facility can eliminate the need for stack testing. However, with this approach, no credit can be taken for any HCl or Cl_2 removed from the gas stream by downstream air pollution control equipment.

Tier II   Tier II also involves the use of USEPA-developed tables. The Tier II tables set limits on stack emission rates that are also functions of stack height, surrounding land use, and the flue gas flowrate. In order to demonstrate compliance with Tier II, a facility is required to perform a stack test. By measuring the HCl and Cl_2 at the stack, the facility can take credit for any removal efficiency of the air pollution control system. For this reason, Tier II can be less restrictive than Tier I if a facility utilizes a HCl/Cl_2 control device.

Tier III Tier III involves the use of site-specific risk assessment. With this tier, a facility can choose to perform its own dispersion modeling to set emission limits. In a Tier III risk assessment, the calculated ambient concentration of HCl and Cl_2 is compared to the allowable reference air concentration (RAC). The RACs are concentration levels that are considered to pose an acceptable level of risk to the surrounding population. Typically, a facility will use a Tier III analysis to back-calculate for the RAC to determine the maximum allowable emission rate for the unit. This tier is usually the least restrictive for a facility. Stack sampling is required to demonstrate compliance.

Adjusted Tier I Adjusted Tier I is a derivative of the Tier III procedure. This involves all the same procedures as the Tier III risk assessment and includes a final step to back-calculate from stack emission rate limits to feed rate limits. With this tier, no stack testing would be required to demonstrate compliance. However, as with Tier I, there is no allowance in this tier for any removal efficiency associated with a control device.

Metals Emissions Standards Metals emissions for BIF units are limited using the same tiered approach as is used for HCl and Cl_2. There are 12 metals regulated under the BIF regulations. Four are considered to be carcinogens: arsenic, beryllium, cadmium, and chromium. The other 8 are noncarcinogenic: antimony, barium, lead, mercury, nickel, selenium, silver, and thallium.

Tables (see 40 CFR 266 Appendix I) have been developed for Tiers I and II, just as for HCl and Cl_2. Tier III and Adjusted Tier I involve site specific dispersion modeling. For the noncarnogenic metals, a facility is required to limit the emissions (or feed rate) below the values in the tables for Tiers I and II. For Tier III and Adjusted Tier I, the facility is required to maintain the emission rates of the metals at a level that would produce ambient concentrations below the individual RACs of the metals.

For the carcinogenic metals, the process is slightly different. For Tiers I and II, the facility is required to maintain the emission rates (or feed rates) at some fraction of the values given in the tables. The sum of the individual fractions must be less than or equal to one. For example, the limit for each of the four metals could be set to 25% of the values given in the table. In Tier III, each carcinogen has been assigned a unit risk factor (URF). The BIF regulations state that the aggregate risk from emissions of carcinogenic metals must be less than 1 in 100,000. To determine risk, the ambient concentration is multiplied by the URF. The four individual risks are then summed to determine the total aggregate risk. This value must be less then 1 in 100,000.

LRCRA.8 TOXIC WASTE CLASSIFICATION

What is a toxic waste?

SOLUTION: A waste is toxic if, after undergoing one of two specified leaching procedures, the leachate contains any one of a set of listed chemicals at a concentration above a given threshold value. In a leaching procedure, a specified quantity of waste is extracted in a known volume of acidic solution for a specified period of time. The resulting "soup" of water and substances that leached from the waste is then assayed for the concentrations of a set of harmful chemicals that are known to be very toxic to animal and plant life.

There are two leaching procedures in use: the EP (Extraction Procedure) test and the TCLP (Toxic Compound Leaching Procedure) test. As an example, after undergoing a TCLP procedure, a waste's leachate contains 3.0 mg/L of cadmium. Since this value exceeds the specified 1.0 mg/L threshold value for cadmium in TCLP leachate, the waste is classified as hazardous.

LRCRA.9 HAZARDOUS WASTE COMBUSTOR MAXIMUM ACHIEVABLE CONTROL TECHNOLOGY (MACT) STANDARDS

Provide details on the MACT standards for hazardous waste incinerators, cement kilns, and lightweight aggregate kilns.

SOLUTION: Under joint authority of RCRA and the Clean Air Act, EPA promulgated emission standards for hazardous waste incinerators (HWIs), cement kilns (CKs) and lightweight aggregate kilns (LWAKs) in 40 CFR 63.1200 (or 40 CFR 63 Subpart EEE).

The standards limit emissions from both new and existing facilities in each equipment category. Pollutants regulated under the rule include dioxins and furans (D/F), mercury (Hg), total chlorine (HCl/Cl$_2$), semi-volatile metals (SVM) including lead and cadmium, low-volatility metals (LVM) including arsenic, beryllium, and chromium, particulate matter (PM), carbon monoxide (CO), and hydrocarbons (HC). The standards also require continuous emissions monitoring systems (CEMS) and performance testing. See EPA web page at http://www.epa.gov/hwcmact for links to regulatory and supporting documents.

The MACT standards include emission limits for both new and existing hazardous waste incinerators (Table 6.2). The standards require a minimum *destruction and removal efficiency* (DRE) of 99.99% for each POHC and a minimum DRE of 99.9999% for dioxin-listed wastes. These requirements are identical to the existing hazardous waste incinerator performance standards found in 40 CFR 264 subpart O, except that the POHCs are to be chosen from the hazardous air pollutants established by the CAAA instead of the hazardous constituents provided in 40 CFR 261 Appendix VIII.

Cement kilns have the same DRE requirements as incinerators under the MACT standards. Other emission standards are similar to those for incinerators with some exceptions due to differences in equipment and operations.

TABLE 6.2   Emissions Limits from Incinerators[a]

Pollutant	Units	Existing	New
DF	ng TEQ/dscm	0.2 or 0.4[b]	0.2
Mercury	μg/dscm	130	45
SVM	μg/dscm	240	24
LVM	μg/dscm	97	97
HCl/Cl$_2$	ppm$_v$ (dry)	77	21
PM	mg/dscm (gr/dscf)	34 (0.015)	34 (0.015)
CO	ppm$_v$ (dry)	100	100
HC	ppm$_v$ (dry)	10	10

[a]All emission rates corrected to 7% oxygen dry volume.
[b]If the gas temperature at the inlet of the initial particulate control device is maintained below 400°F.

In addition to the requirements above, new facilities, which began kiln construction after April 19, 1996, at a plant site where a cement kiln had not previously existed, must continuously monitor HC and meet an HC limit of 50 ppm by volume, over a 30-day block average, dry basis, corrected to 7% oxygen.

LWAKs have the same DRE requirements as incinerators under the MACT standards. Other emission standards are similar to those for incinerators with some exceptions due to differences in equipment and operations.

LRCRA.10   NANOTECHNOLOGY REGULATION(S)

Discuss the role RCRA might play with respect to nanotechnology regulation(s). Comment on how and to what degree new and existing legislation/rulemaking will be necessary for environmental control/concerns of nanotechnology.

SOLUTION: A waste from a commercial-scale nanotechnology facility would be "captured" under RCRA, provided that it meets the criteria for a RCRA waste. RCRA requirements could be triggered by a listed manufacturing process or the RCRA's specified hazardous waste characteristics. The type and extent of regulation would depend on how much hazardous waste is generated and whether the wastes generated are treated, stored, or disposed of onsite.

The reader is referred to the following two texts for additional information:

1. L. Theodore and R. Kunz, "Nanotechnology: Environmental Implications, and Solutions," John Wiley & Sons, Hoboken, NJ, 2005.
2. L. Theodore, "Nanotechnology: Basic Calculations for Engineers and Scientists," John Wiley & Sons, Hoboken, NJ, 2006.

LRCRA.11 SOLID WASTE DEFINITION

Define a solid waste.

SOLUTION: In the regulations for implementing RCRA, (40 CFR 261.4(a)) a solid waste is defined as any material that is discarded, abandoned, recycled, or inherently waste-like. To gain further insight into the meaning of solid waste, the RCRA Section 1004(27) definition is:

> The term "solid waste" means any garbage, refuse, sludge from a waste treatment plant, water supply treatment plant, or air pollution control facility and other discarded material, including solid, liquid, semisolid, or contained gaseous material resulting from industrial, commercial, mining, and agricultural operations, and from community activities, but does not include solid or dissolved material in domestic sewage, or solid or dissolved materials in irrigation return flows or industrial discharges which are point sources subject to permits under Section 1342 of Title 33, or source, special nuclear, or byproduct material as defined by the Atomic Energy Act of 1954, as amended (68 Stat. 923).

A material can be considered a solid waste if it is recycled in a manner constituting disposal. This type of recycling includes materials used to produce products that are applied to the land, burned for energy recovery, or accumulated speculatively before recycling. A material that fits the definition of a solid waste may be regulated as a hazardous waste if it poses a threat to human health or the environment. The definitions of solid waste and hazardous waste interlock, which results in EPA regulating a plethora of materials that may not be commonly thought of as wastes for certain industries.

LRCRA.12   HAZARDOUS WASTE DEFINITIONS

List the general criteria that are used to define the types of wastes that are hazardous.

SOLUTION: As defined in Section 1004-Definitions, of the Solid Waste Disposal Act a waste or combination of wastes is "hazardous" if its quantity, concentration, or physical, chemical or infectious characteristics may either: (a) cause, or significantly contribute to an increase in mortality or an increase in serious irreversible or incapacitating reversible illness, or (b) pose a substantial present or potential hazard to human health or the environment when improperly treated, stored, transported disposed of, or otherwise managed. The RCRA definition can be found in the regulations at [40 CFR 240.101 (m)].

See also G. Burke et al., *Handbook of Environmental Management and Technology*, 2nd Edition, Wiley-Interscience, New York, 2000. p. 365.

LRCRA.13   SOLID/HAZARDOUS WASTE COMPARISON

When is a solid waste not a hazardous waste?

SOLUTION: Once a material is found to be a solid waste, the next question is whether it is a hazardous waste. The EPA automatically exempts certain solid waste from being hazardous waste. A few examples are; household waste; including household waste that has been collected, transported, stored, treated, disposed, recovered (e.g., refuse-derived fuel), or reused; solid waste generated in the growing and harvesting of agricultural crops that is returned to the soil as fertilizer; fly ash waste; bottom ash waste, slag waste; flue gas emissions control waste generated primarily from the combustion of fossil fuels; and, cement kiln dust.

The EPA may also grant a variance from classification as a solid waste on a case-by-case basis. Eligible materials include those that are reclaimed and then reused within the original primary production process in which they were generated, reclaimed partially, but require further processing by being completely recovered, and accumulated speculatively with less than 75 percent of the volume having the potential for recycling.

The rule entitled "Identification and Listing of Hazardous Waste" can be found at [40 CFR Parts 261.2 and .3]. The discussion of variance for classification as a solid waste is located at [40 CFR 260.30 and .31].

LRCRA.14 SOLID WASTE EXCLUSIONS

List materials that are excluded from the definition of solid waste.

SOLUTION: The regulation 40 CFR 261.4 contains an extensive list of these materials. A sampling is provided below. See also previous Problem.

1. Domestic sewage
2. Industrial wastewater discharges that are point source discharges subject to regulation under section 402 of the Clean Water Act. (This exclusion applies only to point source discharges. It does not exclude industrial wastewaters while they are being collected, stored, or treated before discharge, nor does it exclude sludges that are generated by industrial wastewater treatment.)
3. Irrigation return flows
4. Source, special nuclear, or byproduct material as defined by the Atomic Energy Act
5. Materials subjected to in-situ mining techniques that are not removed from the ground as part of the extraction process
6. Pulping liquors (i.e., black liquor) that are reclaimed in a pulping liquor recovery furnace and then reused in the pulping process, unless it is accumulated speculatively
7. Spent sulfuric acid used to produce virgin sulfuric acid, unless it is accumulated speculatively
8. Secondary materials that are reclaimed and returned to the original process or processes in which they were generated where they are reused in the production process provided. (This exemption has limitations stipulating closed processes, reclamation not involving controlled flame combustion, accumulation, use as fuel, and not used in a manner constituting disposal.)

Obviously, solid waste is far less regulated, which, when treated by hazardous waste generators, means a savings on disposal costs. Certain materials are not regulated as a solid waste, when recycled. These include materials shown to be recycled by being:

1. Used or reused as ingredients in an industrial process to make a product, provided the materials are not being reclaimed
2. Used or reused as effective substitutes for commercial products
3. Returned to the original process from which they are being generated, without first being reclaimed. The material must be returned as a substance for raw material feed stock, and the process must use raw materials as principal feed stock.

Certain materials may still be considered solid wastes, i.e., RCRA-regulated, even if the recycling involves use, reuse, or return to the original process. This includes:

1. Materials used in a manner constituting disposal, or used to produce products that are applied to the land; or materials burned for energy recovery, used to produce a fuel, or contained in fuels
2. Materials accumulated speculatively
3. Hazardous waste numbers F020, F021 (unless used as an ingredient to make a product at the site of generation), F022, F023, F026, and F028. (These substances are identified in 40 CFR 261 Appendix VII.)

LRCRA.15   LISTED WASTES

Describe the three listed wastes.

SOLUTION: The USEPA has established three lists, Type F, K, and P and U wastes are codified and listed in [40 CFR 261.31], [40 CFR 261.32], and [40 CFR 261.33], respectively. The discussion of the various types of wastes can also be found on page III-18 of the "RCRA Orientation Manual." Abbreviated details are provided below.

1. "K" Listed Wastes—These hazardous wastes are from specific sources, i.e., those wastes generated in a specific process that is specific to an industry group. Examples include wastewater treatment sludge from the production of chrome yellow and orange pigments, and still bottoms from the distillation of benzyl chloride.
2. "F" Listed Wastes—These hazardous wastes are from nonspecific sources, i.e., those wastes that are generated by a nonspecific industry that are generated from a standard operation that is part of a particular manufacturing process. Examples include spent solvent mixtures and blends used in degreasing containing, before use, and a total of 10 percent or more (by volume) of various solvents.

3. "P" and "U" Listed Wastes—The third list has been broken into two distinct subsets. The "U" list contains chemicals that are deemed toxic and the "P" list contains chemicals that are deemed acutely hazardous. These hazardous wastes are discarded commercial chemical products, off-specification products, container residues, and spill residues. Examples include beryllium, fluorine, and methyl isocyanate.

Those wastes that are not listed in either [40 CFR Parts 261.31–261.33] the F, K, P, U lists may still be a hazardous waste if they exhibit one or more of the four following characteristics: reactivity, ignitability, corrosivity, or toxicity. These are defined below.

1. Reactivity—The waste will react violently with or release toxic gases or fumes when mixed with water, or is susceptible to explosions or detonations
2. Ignitability—A solid, liquid, or gas that is easily ignitable
3. Corrosivity—Alkaline or acidic material normally having a pH in the range of less than 2 and greater than 12.5
4. Toxicity—Wastes that have the ability to bioaccumulate in various aquatic species

Additional details follow in the next Problem.

LRCRA.16 HAZARDOUS WASTE CLASSIFICATION

Federal regulations provide a classification of hazardous waste. "D" wastes, or characteristic wastes, are defined by [40 CFR Part 261.22 Subpart C.] These are classified on the basis of the following characteristics; ignitability, toxicity, reactivity, and corrosivity. If the waste is classified by its ignitability it carries the "D001" code. Some common D codes are:

D code	CFR	Classification Basis
D001	[40 CFR 261.21]	Ignitability
D002	[40 CFR 261.22]	Corrosivity
D003	[40 CFR 261.23]	Reactivity
D004	[40 CFR 261.24]	Toxicity characteristic leaching procedure (TCLP)

Here are some waste streams. Identify the waste classification for each:

1. A concentrated sodium hydroxide solution.
2. A waste stream that is packed in light drums. If exposed to air, it will undergo an exothermic reaction and catch fire.
3. A waste containing high concentrations of leachable polychlorinated biphenyls (PCBs).

Note: Additional waste characteristics can also be found in the "RCRA Orientation Manual" referred to in the previous Problem.

SOLUTION:

1. D002. Sodium hydroxide solution is corrosive.
2. D001 and D003. This waste is both ignitable and extremely reactive with the potential to explode if exposed to air.
3. D004. PCBs are classified *toxic* by law.

LRCRA.17 LIABILITY DEFINITION

The term liability is closely tied to the system of values and ethics developed in this country. In concise language, define and give an example of the terms *liability* and *strict liability*. Explain how the interpretation of liability affects waste incineration permitting, application, and design.

SOLUTION: Liability implies responsibility for an action. An individual maybe held liable for a result if, in the mind of the normal, prudent person, the individual failed to exercise due caution. Examples include driving too fast, losing control of a car, and causing damage to property or persons.

Strict liability implies responsibility without regard to prudence or care, i.e., without regard to negligence. Such standards are imposed for a variety of activities, such as handling dynamite, statutory rape, or hazardous waste management. These standards require that proper caution be exercised at all times. Defenses available, if harm results, are limited. These standards are the basis for training requirements imposed upon the permittee who is an owner/operator of hazardous waste treatment, storage, or disposal facilities.

See also Problem LSUP.15.

LRCRA.18 "DERIVED FROM" AND "MIXTURE" RULE APPLICATION

You are Director of Environmental Affairs for Bogus Chemical Co. One of your engineers rushes into your office to tell you that he has the solution to all of Bogus' hazardous waste disposal problems. This engineer has just returned from lunch with an equipment vendor who has a proprietary process proven to render non-hazardous virtually any EPA-listed hazardous waste. Prior to firing the employee (or at least before administering the breathalizer test), please explain why this is impossible.

SOLUTION: EPA's "Derived-From" and "Mixture" Rules under RCRA [page III-24 "RCRA Orientation Manual"] essentially state that any mixture of a listed hazardous waste and a nonhazardous solid waste is itself a RCRA hazardous waste and that any waste derived from the treatment, storage, or disposal of a listed waste is deemed hazardous. Both the "derived-from" and "mixture" rules apply regardless of how small the concentration of the listed waste ("the solution to pollution is not dilution") is in the final mixture. Likewise, mitigation of the characteristics which prompted the waste to be listed in the first place does not magically purify the material. In the case of listed hazardous wastes, the only method of removing the taint is by the EPA's delisting procedure, a process that has only rarely been successfully used.

LRCRA.19 COMPOSTING

Composting is the biological decomposition of organic waste material by micro-organisms. It is one of the major treatment alternatives for municipal solid waste, especially in developing countries where solid wastes contain mostly organic carbon substances and are rich in nutrients. During the process, certain physical, chemical and biological changes take place which alter the character of the waste material. Describe these physical, chemical, and biological changes as municipal solid waste is converted to humus in the composting process.

SOLUTION: Changes taking place in the physical, chemical, and biological quality of municipal solid waste during the composting process are summarized in Table 6.3.

TABLE 6.3 Composting Process Information

Changes	Solid Waste	Compost
Physical changes:		
Color	Variable	Dark black
Particle size	Irregular, large	Humus-like, small
Water content	Seasonal, variable	Generally drier
Chemical changes:		
C/N ratio	Variable from 20 to 78/1	$\approx 20/1$
Organic substrate	Unstable	Stable
Amount of NO_3 and NH_4	Relatively low	Increased due to degradation of proteins
Biological changes:		
Microbial activity	High	Stabilized, decrease in
Number of pathogens	High	both microbial activity and pathogens due to aerobic pile conditions and heat generated during decomposition

LRCRA.20 TREATMENT OPTIONS

Discuss other treatment options to incineration as well as pretreatment and process modification techniques associated with incineration and pollution prevention.

SOLUTION: These options are:

1. Chemical treatment
2. Biological treatment
3. Physical treatment
4. Landfilling and other disposal methods
5. Process modification(s)/pollution prevention

Typically, when options other than incineration are chosen, more than one unit process (or step) is required. For example, a typical process sequence that could serve as an alternative to incineration might be sedimentation with neutralization, biodegradation followed by sludge separation, and finally landfilling the waste. These alternative options are often less costly and are possibly easier to permit and start up than incineration.

Additional details are available in the Pollution Prevention Chapter (10). The reader is also referred to the work of J. Santoleri, J. Reynolds and L. Theodore, "Introduction to Hazardous Waste Incineration," 2nd Edition, John Wiley & Sons, Hoboken, NJ, 2000.

LRCRA.21   MUNICIPAL SOLID WASTE CHARACTERISTICS

Besides economic considerations and site specifications, what are the major characteristics of municipal solid waste that are employed for the selection of the treatment and disposal processes listed below?

a. Composting.
b. Incineration.
c. Sanitary landfill.

SOLUTION: The characteristics relevant to the treatment and disposal options listed in the problem statement are summarized below:

1. Composting—Biodegradable/non-biodegradable fraction, moisture content, carbon to nitrogen ratio.
2. Incineration—Heating value, moisture content, inorganic fraction, metal content, alkali earth metal content.
3. Sanitary landfill—Biodegradable to non-biodegradable fraction, liquid content, hazardous material content.

LRCRA.22   SMALL QUANTITY GENERATORS

Define small quantity generators.

SOLUTION: *Small Quantity Generators (SQGs)* are defined as facilities generating between 100 and 1,000 kg of hazardous waste per calendar month, while a *Conditionally Exempt Small Quantity Generator* generates by definition less than 100 kg of hazardous waste per calendar month. These sources represent a wide variety of industrial groups, and present unique challenges in terms of regulation and technical assistance. These generators generally have fewer than 5 to 10 employees and are managed and staffed by individuals with limited training in identification and management of hazardous wastes or pollution prevention.

While these approximately 600,000 to 700,000 SQGs in the U.S. do not individually contribute significant quantities of materials to the waste stream, the nature of their waste materials and their aggregate are a significant part of the total hazardous waste stream, generating as much as 1 million metric tons of hazardous wastes annually.

LRCRA.23 IMPACT OF THE 1990 CLEAN AIR ACT AMENDMENTS (CAAA)

Discuss the impact of the 1990 CAAA on Hazardous Waste Incinerators.

SOLUTION: The MACT rules for Hazardous Waste Incinerators have been promulgated and the ozone requirements are in place. A RCRA training module has been developed and can be viewed at:

http://www.epa.gov/epaoswer/hotline/training/incin.pdf. Other information can also be viewed at http://www.epa.gov/epaoswer/hazwaste/hazcmbst.htm#emissions

LRCRA.24 UNDERGROUND STORAGE TANKS (USTs)

Define the term underground storage tank (UST). List the types of tanks which are not regulated by the Resources Conservation and Recovery Act (RCRA) UST program. Also list the major causes of tank failures.

SOLUTION: An underground storage tank (UST) is defined as any storage tank with at least 10% of its volume buried below ground, including pipes attached to the tank. Above ground tanks with extensive piping may be regulated under the Resource Conservation and Recovery Act (RCRA) Sub-Title I, UST regulations. Types of tanks to which the UST program does not apply include:

1. Farm and residential tanks which store less than 1100 gallons of motor fuel
2. On-site heating oil storage tanks
3. Septic tanks and sewers
4. Pipelines for gas or liquid
5. Surface impoundments
6. Flowthrough process tanks

Major UST tank failures are due to:

1. Corrosion
2. Faulty installation
3. Pipe failure
4. Overfills

LRCRA.25 CLASSIFICATION OF MEDICAL WASTE

Answer the following medical waste questions:

1. Describe the difference between a medical waste and an infectious waste.
2. Identify the seven classes of regulated medical wastes.
3. By what means are hospital wastes disposed?
4. What steps can be taken by hospital operators to minimize occupational hazards, contamination, and infection?
5. Provide information in medical waste regulations and background documents.

SOLUTION:

1. A medical waste is any solid waste that is generated in the diagnosis, treatment, or immunization of human beings or animals. A medical waste can be infectious or noninfectious. Infectious waste is a medical waste that contains pathogenic microorganisms that can cause disease.
2. The seven classes of regulated medical wastes are:

   a. Cultures and stocks
   b. Pathological waste
   c. Blood and blood products
   d. Sharps
   e. Animal waste
   f. Isolation wastes
   g. Unused sharps

3. The predominant means of disposing of hospital waste is incineration (almost 35% of all hospital waste is incinerated). Other means of treatment or disposal include autoclaving, sanitary landfilling, chemical, disinfection, thermal inactivation, ionizing radiation, gas vapor sterilization, segregation, and bagging.
4. The steps that may be taken by hospital operators to minimize occupational hazards, contamination, and infection include:

   a. Enclose medical waste at point of generation.
   b. Construct carts and equipment that have sanitary construction.

c. Construct and operate chutes to minimize microbiological contamination.

d. Reduce dangers of medical waste handling by using personal protective equipment.

e. Require higher training and qualifications for personnel handling medical waste.

f. Improve incinerator operation through training.

g. Implement safe management of hazardous wastes.

5. For information about the regulations and the background and implementation documents the reader should go to the EPA website at: http://www.epa.gov/ttn/atw/129/hmiwi/rihmiwi.html.

LRCRA.26 METALS

Provide introductory information on the four major metals of environmental concerns.

SOLUTION: Since heavy metals are strongly attracted to biological tissues and to the environment, and remain in them for long periods of time, metal pollution is a serious issue. Overall, metals are quite abundant and persistent in the environment. The four major metals of concern are lead, mercury, cadmium, and arsenic.

LRCRA.27 4,4-DDT

Provide physical chemical and health related data on 4,4-DDT.

SOLUTION: Adapted from A. Spero, B. Devito, and L. Theodore, "Regulatory Chemicals Handbook," Marcel Dekker (recently acquired by CRC Press/Taylor & Francis Group, Boca Raton, FL), New York City, 2000.

4,4-DDT ($C_{14}H_9Cl_5$, 354.5)

CAS/DOT Identification #: 50-29-3/UN 2761.

Synonyms: p-p'-DDT, DDT, 2,2-bis(p-chlorophenyl)-1,1,1-trichloroethane, dichloro-diphenyltrichlororethane, dicophane, chlorophenothane, Gesarol®, Neocid®.

Physical Properties: waxy solid; weak, chemical odor; tasteless; sinks in water; MP ($107-109°C$); BP ($185°C$) SG (1.56 @ $15°C$); VP (1.9×10^{-7} torr @ $20°C$); solubility in water (0.006 mg/L @ $25°C$).

Chemical Properties: incompatible with strong oxidizers; may react with iron, aluminum, aluminum and iron salts, and alkalies; incompatible with ferric chloride and aluminum chloride; FP ($72-75°C$).

Biological Properties: long term persistence in soil and water; sticks strongly to soil particles and does not leach rapidly into underground water; soil half-life: 2–15 yrs; aerobic half-life: 2–15.6 yrs; anaerobic half-life: 16–100 days; surface water half-life: 7–350 days; ground water half-life: 16–31.3 days; can be detected in water by EPA Method 608: gas chromatography, or EPA Method 625: gas chromatography plus mass spectrometry.

Bioaccumulation: concentrated accumulation in the fat of wildlife and humans as a result of low water solubility and high lipophilicity; builds up in plants and in the fatty tissues of fish, birds and animals.

Origin/Industry Sources/Uses: manufactured chemical; used to control insects on agricultural crops and insects that carry diseases like malaria and typhus; ectoparasiticide; use was banned in USA in 1972.

Toxicity: no data available.

Exposure Routes: inhalation; skin adsorption; ingestion eye and skin contact; eating contaminated foods such as root and leafy vegetables, meat, fish, and poultry; inhalation of contaminated air or drinking contaminated water near waste sites and landfills; swallowing soil particles near waste sites or landfills.

Regulatory Status: Criterion to Protect Freshwater Aquatic Life: 0.0010 μg/L/ 24 hr avg., concentration not to exceed 1.1 μg/L any time; *Criterion to protect saltwater aquatic life*: 0.0010 μg/L/24 hr avg., concentration not to exceed 0.13 μg/L any time; *Criterion to protect human health*: preferably 0; concentration calculated to keep the lifetime cancer risk level below 10^{-5} is 0.24 ng/ L; USSR MAC; 0.1 mg/L in water used for domestic purposes, 0 in surface water for fishing; Mexico MPC: 0.042 mg/L in drinking water, 0.006 mg/L in estuaries, 0.6 μg/L in coastal waters; the following are guidelines in drinking water set by some states: 0.83 μg/L (Maine), 0.42 μg/L (Kansas), 1.0 μg/L (Minnesota), 50 μg/L (Illinois).

Probable Fate: Photolysis: photooxidation to DDE occurs slowly, indirect photolysis may be important; *oxidation*: photoxidation occurs, photooxidation half-life in water: 7–350 days, photoxidation half-life in air: 7.4 days; *hydrolysis*: may be an important process under certain conditions, first-order hydrolytic half-life: 22 yrs; *volatilization*: is an important process, some will evaporate from soil and surface water into the air; *sorption*: is an important process, will adsorb very strongly to soil if released to the soil, will adsorb very strongly to sediments if released to water; *biological processes*: biotransformation and bioaccumulation are important processes, may be subject to biodegradation in flooded soils or under anaerobic conditions, may be significant sediments.

Treatability/Removability (Process, Removable (%), Avg. Achievable Conc. (μg/L)): *Sedimentation with chemical addition (alum, lime)*, >52, <1; *Activated sludge* (based on synthetic wastewater), 0, not available; *Powdered activated carbon adsorption* (based on synthetic wastewater), ∼100, 0.008;

continuous activated sludge biological treatment simulator: removal was 100%; *Wet and dry deposition* will be major removal mechanisms from the air.

In addition to Spero et al., the reader may refer to following references for additional details on these chemicals

1. R. Lewis, "Sax's Dangerous Properties of Industrial Materials," 9th Edition, Van Nostrand Reinhold, New York City, 1996.
2. "Suspect Chemicals Sourcebook," Roytech Publications, Bethesda, Maryland, 1996.

LRCRA.28 SELECTING HOSPITAL INCINERATION SYSTEM

A city has decided to incinerate all hospital and infectious wastes at a central location due to new stringent air emission regulations. Two hospitals, A and B, each containing 500 beds, along with a 50-bed nursing home and an animal research center, contribute their waste to the central facility. Hospital A contributes 20 lb/bed/day of waste while hospital B contributes 15 lb/bed/day. The nursing home contributes 10 lb/bed/day of waste, and the animal research center provides 250 lb/day of waste to the incineration facility. The compositions by weight of the wastes from the contributing facilities are as follows (see Table 6.4):

TABLE 6.4 Hospital Waste Composition Data

Waste Type	Hospital A	Hospital B	Nursing Home	Animal Research
Trash	55	55	65	30
Plastic	30	15	20	15
Garbage	10	10	10	5
Pathological	5	20	5	50

Hospital A and the nursing home have a high percentage of plastics, and hospital B and the animal research facility have high percentages of pathological (infectious) waste.

1. What are some considerations that are necessary before an incineration system is selected?
2. Compare controlled-air incineration and rotary kiln incineration for the central facility.
3. Would heat recovery be viable for this central facility?

SOLUTION:

1. The total quantity of waste received must first be determined. Continuous or batch operation of the waste must be chosen. Biomedical waste incinerators are normally batch-operated, but since the quantity of waste received at this facility is large, continuous operation may make more sense.

    Since the wastes from hospital A and the nursing home contain a high percentage of plastics, excessive acid gas emissions are possible. Therefore, acid gas emissions control should be considered.

    The high percentage of pathological waste from hospital B and the animal research facility would result in lower waste heating values. Operators must be trained to account for variations in the heating value of the feed to the incinerator. Batch operation should still be considered an option due to the expected variation of the incinerator feed heating value.

    The POHCs of the waste must be destroyed to the required destruction and removal efficiency. HCl and particulate emissions must comply with state and federal regulations for hospital waste incinerators.

    Since a high percentage of plastics occurs in the overall waste stream fed to the incinerator, a suitable method of treating the acid gases produced must be chosen. This may entail reacting the acid gases with an alkaline material. Therefore, a spray dryer should be considered to neutralize the acid gases.

2. Controlled-air incinerators have a lower capital cost since they do not normally require air pollution control equipment (unless acid gas emissions are excessive). However, since this facility must comply with new stringent regulations and acid gas emissions may be excessive, a rotary kiln system may be more advantageous.

    It is expected that the heating value of the waste fed to the incinerator may vary substantially. Therefore, a rotary kiln incinerator would be more appropriate since it can handle a large range of waste heating values. The residence time of a rotary kiln incinerator may be changed by adjusting the rotational speed of the kiln.

    Since this facility is large compared to traditional hospital waste incinerators, the incinerator may be operated continuously. Controlled-air incinerators are normally operated in batch mode.

    All factors suggest that a rotary kiln incineration unit would be more appropriate for this facility than a controlled-air incineration system.

3. The facility generates over 9 tons of waste per day. While this is not large compared to municipal solid waste incineration facilities, if the facility expands in the future, heat recovery may be a viable option. But heat recovery is only economically advantageous if customers exist to purchase the recovered heat (normally steam) or if there is an onsite need for the recovered heat.

Biomedical wastes are not only generated by hospitals. Animal research facilities, research centers, universities, rest homes, and veterinary clinics also generate pathological (infectious) waste. Pathological waste includes animal carcasses, contaminated laboratory wastes, hypodermic needles, contaminated food and equipment, blood products, and even dialysis unit wastes. Normally, biomedical wastes are incinerated along with other wastes generated by the facilities such as paper and plastic.

In addition to the above, an increase in plastics in hospital waste streams has occurred during the last decade. Plastics may account for as much as 30% of a hospital waste stream. Unfortunately, incinerating plastics normally increases the chlorine content of the exiting flue gas. This creates a need for air pollution control devices to remove chlorine compounds.

In the past, controlled-air incinerators have been the most popular incinerators for biomedical waste destruction. A controlled-air incinerator is a two-chamber, hearth-burning, pyrolytic unit. The primary chamber receives the waste and burns it with less than stoichiometric air. Volatiles released in the primary chamber are burned in the secondary combustion chamber. These units result in low fly ash generation and low particulate emissions. In addition, they have a low capital cost and may be batch operated. As noted above, they normally do not require air pollution control equipment unless acid gas emissions are excessive.

LRCRA.29   DISPOSAL

Discuss the disposal of wastes.

SOLUTION: *Disposal* is described by many as the final process in the treatment and management of hazardous wastes. Before disposal, most wastes undergo the various treatments previously described (biological, chemical, and/or physical) in order to concentrate, detoxify, and reduce the volume of the wastes. This has not always been the case. Before the early 1970s, most wastes were haphazardly land-filled or ocean dumped with little concern to the environmental effects of the practices. The passage of certain environmental protection laws has helped to eliminate the irresponsible handling of hazardous wastes by requiring that proper measures be taken before they are disposed.

Landfarming, deep-well injection, landfilling, and ocean dumping are four methods used for disposal of hazardous wastes. Theodore (personal notes, 1993) has recently added atmospheric dispersion to this list of four.

LRCRA.30   LANDFILLING

Describe landfilling.

SOLUTION: *Landfilling* is the third disposal method and is generally used on wastes in the form of sludges. There are two types of landfilling: area fill and trenching. *Area fill* is essentially accomplished above ground while *trenching* involves burying the waste. Trenching is the more established and popular form of the two. Since trenching requires excavation, area fill has the advantage that it requires less manpower and machinery. Area fill is also less likely to contaminate ground-water since the filling is above ground. Trenching, however, may be used for both stabilized and unstabilized sludges and makes more efficient use of the land. Both techniques require the use of lime and other chemicals to control odors and with both methods, cold and wet weather can cause problems. Both methods also produce gas, which can cause explosions or harm vegetation, and leachate, which can contaminate ground and surface water.

Most hazardous wastes must be subjected to one or more pretreatments such as solidification, degradation, volume reduction, and detoxification before being land-filled. This practice stabilizes the waste and helps decrease the amount of gas and leachate produced from the landfill. Landfilling is similar to landfarming in that both disposal methods combine wastes and soil. Landfarming, however, involves the biochemical reaction between soil nutrients and wastes to degrade and stabilize the waste; as a result, only specific types of wastes can be landfarmed. A larger variety of wastes may be handled by landfilling.

Area fill may be done in one of three ways: by mixing the waste with soil and forming a mound with the mixture and covering with soil; by spreading alternate layers of soil and soil-waste mixture over the area; or, by filling a containment area surrounded by dikes with the waste and then covering with a soil layer. These forms of area fill are known as area fill mound, area fill layer, and dike containment, respectively.

In *area fill mounds*, the sludge–soil mixture may be stacked in mounds as high as 6 ft. A soil covering, usually 3–5 ft, is placed on top of the mound. The amount of covering and soil-to-sludge ratio are dependent on the concentration of solids in the waste and on the degree of mound stability required. An earthern containment may be constructured for better mound stability.

In an *area fill layer*, the sludge-soil mixture is spread out over an area in consecutive layers each having a thickness from 0.5 to 3 ft. Layers of soil 0.5–1 ft thick are often placed between the mixture layers. Final heights for these landfills will vary depending upon the amount of waste being disposed and the size of the site. Sites for these landfills are usually flat or slightly sloped.

Dike containment landfills employ dikes to surround the four sides of the containment area. The site is often on a hill, which can serve as one of the containment sides. In this methods, access roads are needed in order to bring the waste to the top of the landfill. Layers of soil, 1–3 ft thick, are usually placed at certain levels in the landfill, as well as a range from 10 to 30 ft with widths of 50–100 ft and lengths of 100–200 ft.

Trenching involves placing the waste in an evacuated trench and covering it with one or more layers of soil. There are two types of trenching: *narrow* trenching operations, with trench widths of 10 ft or less, and *wide* trenching operations with trench widths greater than 10 ft. The type of operation used depends upon the solids content of the waste. In either case, it is necessary that the width between trenches is large enough to achieve sidewall stability and space for soil stockpiles and equipment. There must be a soil thickness of 2–5 ft between the trench bottom and nearest groundwater level in order to provide adequate leachate control.

Wastes may be placed in *narrow* trenches in one application, with one soil layer placed on top of the waste. The soil removed from the trench may be used for the covering. Narrow trenches of 2–3 ft are best for sludges with low solids

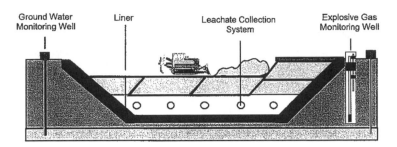

FIGURE 6.1   Cross-Section of a Municipal Solid Waste Landfill.

content (15–20%) because the support from the side walls prevents the soil layer from sinking to the bottom. Sludges with higher solids content should be placed in wider trenches.

In *wide trench* operations, excavating and filling equipment are placed within and at the top of the trench. The excavated soil is used to cover the sludge and to provide an intermediate soil layer in order to support the equipment. The sludge must have a fairly high solids content in order to support the equipment and soil covering at the top of the trench. If the trench is very wide, dikes may be used to confine the sludge to certain areas.

Interestingly, municipal solid waste landfill (MSWLF) is defined as a discrete area of land or excavation that receives household waste. A MSWLF may also receive other types of RCRA Subtitle D wastes, such as commercial solid waste, nonhazardous sludge, and industrial nonhazardous solid waste. There are approximately 2,000 MSWLFs in the United States. A frontal diagram of a MSWLF is provided in Figure 6.1.

When a landfill site is to be selected, a number of factors must be evaluated; these factors fall into three categories: technical, economic, and public acceptance considerations. The following factors are technical considerations: haul distance, site size and life, topography, surface water, soils and geology, groundwater, soil quantity and suitability, vegetation, environmentally sensitive areas, archaeological or historical significance, site access, and land use. Economic considerations involve site capital cost, site operating cost, and hauling cost. Public acceptance considerations include local laws and public opinion; a public hearing is usually held in order to obtain local government and public input.

In comparing the costs of the different types of landfills, wide trench landfilling has the lowest capital and operating costs; area fill layer has the highest capital cost; and area fill mound has the highest operating costs. The other types of landfills fall somewhere in between these extremes. Capital costs in this case involve land, site preparation, equipment purchase, and engineering. Overall costs for landfills are dependent upon the efficiency of the operation with respect to land use, equipment, and manpower. Hauling costs are independent of the landfill method and hence do not play a role in an economic comparison of the different techniques.

Once a landfill has been completed and closed, it must still be maintained in order to prevent soil erosion and other damaging effects of weather. It must also be monitored for gas and leachate as well as for any other potential environmental hazards. For example, closed landfills are vented and can be sources of landfill gas which is combustible.

LRCRA.31 DEEP-WELL INJECTION IN SALT BEDS

As described in Problem LRCRA.29, deep-well injection is a disposal method that transfers liquid wastes far underground and away from freshwater sources. Like landfarming, this disposal process has been used for many years by the petroleum industry. It is also used to dispose of saltwater in oil fields. When the method first came into use, the injected brine would often eventually contaminate groundwater and freshwater sands because the site was poorly chosen. The process has since been improved, and laws such as the Safe Drinking Water Act ensured that sites for potential wells are better surveyed.

Many factors are considered in the selection of a deep-well injection site. For example, the rock formation surrounding the disposal zone must be strong but permeable enough to absorb liquid wastes, and the site must be far enough from drinking water sources to prevent contamination. Once a site is selected, it must be tested by drilling a pilot well. The performance data from the pilot well, besides testing permeability and water quality, also aid in the design of the final well and in determining the proper injection rate.

The type of waste injected into the well is a determinant in how deep the injection will be made. The more toxic the waste, the farther down the disposal zone must usually be. Disposal zones have been classified into five different types:

1. *Zone of Rapid Circulation.* This designation describes the area that runs from the soil surface to only a few hundred feet below. Waste is not injected into this zone.

2. *Zone of Delayed Circulation.* This zone contains circulating fresh water and may be used for certain wastewaters if properly monitored. The water circulation is slow enough so that residence times of a few decades to a few centuries can be achieved for the waste.

3. *Subzone of Lethargic Flow.* The liquid flowing in this zone is very slow moving and saline. More concentrated wastes are injected here.

4. *Stagnant Subzone.* The liquid contained in this region is hydrodynamically trapped, and the zone is generally several thousand feet below the soil surface. Highly toxic wastes may be injected here if the zone can properly accept and keep the waste.

5. *Dry Subzone.* Salt beds fall in this classification. This zone does not contain water and is nearly impermeable. Because there is a possibility that liquid movement could occur through hydrofractures, this zone must be monitored.

More recently, salt beds have been considered for storing nuclear and solid wastes. One possibility is to inject water to remove some of the salt and thereby form an underground cavern. The cavern could then be filled with the waste, plus possibly a solidifying agent, and then concrete-sealed.

QUANTITATIVE PROBLEMS (TRCRA)

PROBLEMS TRCRA.1–22

TRCRA.1 AVERAGE WASTE DENSITY

An analysis of a solid waste generator has revealed that the waste is composed (by volume) of 20% supermarket waste, 15% plastic-coated paper waste, 15% polystyrene, 20% wood, 10% vegetable food waste, 10% rubber, and 10% hospital waste. What is the average density of this solid waste in lb/yd^3 and lb/ft^3? Use the following as discarded waste densities (in lb/yd^3) for each of the components of the generator's waste: supermarket waste, 100; plastic-coated paper waste, 135; polystyrene, 175; wood, 300; vegetable food waste, 375; rubber-synthetics, 1,200; and hospital waste, 100.

SOLUTION: Based on the volume percent composition and waste densities given in the problem statement, the following average discarded waste density is estimated to be

$$\text{Average density} = (0.2)(100) + (0.15)(135) + (0.15)(175)$$
$$+ (0.2)(300) + (0.1)(375) + (0.1)(1200) + (0.1)(100)$$
$$= 294\,lb/yd^3$$
$$= 294/27$$
$$= 10.9\,lb/ft^3$$

TRCRA.2 COMPACTED WASTE

The Jefferson County Solid Waste Management Corporation analysis of its solid waste includes the following major components: weight, volume, and compaction factors. Determine the density of well-compacted waste as delivered to a landfill. (*Note*: waste components and compaction factors vary from place to place and may not be indentical to those shown in Table 6.5 for Jefferson County.)

SOLUTION: The compacted volume of each component in the landfill can be determined using the following equation:

$$\text{Compacted volume} = (\text{Discarded volume})(\text{Compaction factor}) \qquad (6.3)$$

TABLE 6.5   Composition, Weight, Volume, and Compaction Factor of Solid Waste from Jefferson County, Mississippi

Component	Weight (kg)	Volume as Discarded (m³)	Compaction Factor
Food waste	250	1.00	0.33
Paper	300	3.75	0.15
Garden waste	250	1.18	0.20
Plastic	50	0.80	0.10
Textiles	60	9.61	0.15
Wood	50	0.34	0.30
Glass	20	0.10	0.40
Metals	20	0.10	0.30
Total	1000		

Note: Compaction factor represents the ratio of the resultant volume to the original volume.

TABLE 6.6   Composition, Weight, Volume, Compaction Factor, and Compacted Volume of Solid Waste from Jefferson County, Mississippi

Component	Weight (kg)	Volume as Discarded (m³)	Compaction Factor	Compacted Volume (m³)
Food waste	250	1.00	0.33	0.33
Paper	300	3.75	0.15	0.56
Garden waste	250	1.18	0.20	0.24
Plastic	50	0.80	0.10	0.08
Textiles	60	9.61	0.15	1.44
Wood	50	0.34	0.30	0.10
Glass	20	0.10	0.40	0.004
Metals	20	0.10	0.30	0.018
Total	1000	—	—	2.77

Based on the weight, discarded volume, and compaction factor data provided in the problem statement, the calculations presented in Table 6.6 result for each waste component. Finally, the average density of the well-compacted mixed waste delivered to the landfill can be determined as:

$$\text{Average density} = (\text{Discarded weight})/(\text{Total compacted volume}) \qquad (6.4)$$

The two key results are provided in Table 6.6.

$$\text{Average waste density} = 1000\,\text{kg}/2.77\,\text{m}^3$$
$$= 360.75\,\text{kg/m}^3$$

TRCRA.3   MOISTURE CONTENT

The design of garbage collection vehicles differs from county to county and country to country based on location, local culture, etc. Many factors influence the design of these vehicles, with moisture content being one of the most important considerations in the hauling of municipal waste. The general formula for calculating the moisture content of solid waste is as follows:

$$\text{Moisture content} = \left(\frac{A - B}{A}\right)(100\%) \qquad (6.5)$$

where A = weight of sample as delivered, kg

B = weight of sample after drying, kg

Use the data provided in Tables 6.7 and 6.8 for component moisture content and the weight composition of Theodoraki County, Greece, waste to determine the average moisture content of the municipal solid waste. Base the calculation on a 100-kg sample.

SOLUTION: The dry weight of each component can be determined using the following equation:

$$\text{Dry weight} = (\text{Discarded weight})(100 - \%\text{Moisture})/100 \qquad (6.6)$$

The discarded weight for each component can be determined using the following equation:

$$\text{Discarded weight} = (\text{Total waste weight})(\text{wt\% of component}) \qquad (6.7)$$

TABLE 6.7   Typical Moisture Content of Municipal Solid Waste

Component	Moisture Content (wt%)
Food waste	70
Paper	6
Garden waste	60
Plastic	2
Textiles	10
Wood	20
Glass	2
Metals	3

TABLE 6.8   Composition of Theodoraki County, Greece, Solid Waste on a Weight Percent Basis

Component	Composition (wt%)
Food waste	20
Paper	22
Garden waste	18
Plastic	3
Textiles	7
Wood	25
Glass	2
Metals	3

Finally, the average moisture content of the waste is determined as:

$$\text{Avg.\%moisture} = [(\text{Discarded weight} - \text{Dry weight})/100 \text{ kg}](100\%) \qquad (6.8)$$

Based on the %moisture content data and waste composition data provided in the problem statement, and assuming a total waste weight of 100 kg, the calculations presented below in Table 6.9 result for each waste component given in the problem statement.

TABLE 6.9 Composition, Weight Percent, Component Discarded Weight and Component Dry Weight for the Municipal Solid Waste from Theodoraki County, Greece

Component	Wt%	Component Discarded Weight (kg)	Component Dry Weight (kg)
Food waste	20	20	$(1.0 - 0.7)(20) = 6.00$
Paper	22	22	$(1.0 - 0.06)(22) = 20.68$
Garden waste	18	18	$(1.0 - 0.6)(18) = 7.20$
Plastic	3	3	$(1.0 - 0.02)(3) = 2.91$
Textiles	7	7	$(1.0 - 0.1)(7) = 6.30$
Wood	25	25	$(1.0 - 0.2)(25) = 20.00$
Glass	2	2	$(1.0 - 0.02)(2) = 1.96$
Metals	3	3	$(1.0 - 0.03)(3) = 2.91$
Total	100		$67.96 = 68$
	Average %moisture $= [(100 - 68)/100](100\%) = 32\%$		

TRCRA.4 WASTE GENERATED BY A MUNICIPALITY

A municipality in the Midwest has a population of 50,000 and generates 100,000 yd³ of municipal waste annually. The waste is made up of 30% compacted waste and 70% uncompacted waste. Assume that the waste has a density of 1,000 lb/yd³ compacted, and 400 lb/yd³ uncompacted. How many pounds of waste are generated by this city each year? By each person each year?

SOLUTION: Based on the waste densities given in the problem statement, the following generation rates are determined:

$$\text{Waste generated/yr} = (0.3)(100,000 \text{ yd}^3)\left(\frac{1,000 \text{ lb}}{\text{yd}^3}\right)$$

$$+ (0.7)(100,000 \text{ yd}^3)\left(\frac{400 \text{ lb}}{\text{yd}^3}\right)$$

$$= 30,000,000 \text{ lb} + 28,000,000 \text{ lb}$$

$$= 58,000,000 \text{ lb/yr}$$

$$\text{Per capita generation rate} = \frac{58{,}000{,}000\,\text{lb/yr}}{50{,}000\,\text{people}} = 1160\,\text{lb/person-yr}$$

$$= 3.2\,\text{lb/person-day}$$

TRCRA.5 LANDFILL AREA

Estimate the required daily landfill area for a community with a population of 260,000. Assume that the following conditions apply:

1. Solid waste generation $= 7.6$ lb/capita $\cdot$ day
2. Compacted specific weight of solid wastes in landfill $= 830$ lb/yd^3
3. Average landfill depth of compacted solid wastes $= 20$ ft

SOLUTION: Determine the daily solid wastes generation rate in tons per day:

$$\text{Generation rate} = \frac{(260{,}000\,\text{people})(7.6\,\text{lb/capita} \cdot \text{day})}{2000\,\text{lb/ton}}$$

$$= 988\,\text{ton/day}$$

The required area is determined as follows:

$$\text{Volume required/day} = \frac{(988\,\text{ton/day})(2000\,\text{lb/ton})}{830\,\text{lb/yd}^3}$$

$$= 2381\,\text{yd}^3/\text{day}$$

$$\text{Area required/yr} = \frac{(2381\,\text{yd}^3/\text{day})(365\,\text{day/yr})(27\,\text{ft}^3/\text{yd}^3)}{(20\,\text{ft})(43{,}650\,\text{ft}^2/\text{acre})}$$

$$= 26.88\,\text{acre/yr}$$

$$\text{Area required/day} = \frac{(2381\,\text{yd}^3/\text{day})(27\,\text{ft}^3/\text{yd}^3)}{(20\,\text{ft})(43{,}650\,\text{ft}^2\,\text{acre})}$$

$$= 0.074\,\text{acre/day}$$

The actual site requirements will be greater than the value computed because additional land is required for a buffer zone, office and service buildings, access roads, utility access, and so on. Typically, this allowance varies from 20 to 40%. Thus, if an allowance of 40% is employed, the daily area requirement becomes

$$\text{Area required/day} = (0.074)(1.4)$$

$$= 0.104\,\text{acre/day}$$

A more rigorous approach to the determination of the required landfill area involves consideration of the contours of the completed landfill and the effects of any gas production and overburden compaction.

TRCRA.6   FILTER PRESS APPLICATION

A plate and frame filter press is to be employed to filter a slurry containing 10% by mass of solids. If 1 ft^2 of filter cloth area is required to treat 7.5 lb/h of solids, what cloth area, in ft^2, is required for a slurry flowrate of 6000 lb/min?

SOLUTION:

Convert the slurry flowrate, $\dot{m}$, to lb/h:

$$\dot{m}(\text{slurry}) = (6000\,\text{lb/min})(60\,\text{min/h}) = 360{,}000\,\text{lb/h}$$

Calculate the solids flowrate in the slurry:

$$\dot{m}(\text{solids}) = (0.1)(360{,}000\,\text{lb/h}) = 36{,}000\,\text{lb/h}$$

Calculate the filter cloth area, A, requirement:

$$A = (36{,}000\,\text{lb/h})\left(\frac{1}{7.5}\frac{\text{h}\cdot\text{ft}^2}{\text{lb}}\right) = 4800\ \text{ft}^2$$

TRCRA.7   TREATED SLUDGE

Sludge generated in a water or wastewater treatment plant usually contains substantial amounts of water and therefore needs to be processed to reduce the water content (the process is called sludge dewatering) for disposal or landfilling. A municipal water treatment plant in Kansas produces 1.05 tons of sludge every day. The wet sludge (before dewatering) has a density of 1.05 g/cm^3, which increases to 1.65 g/cm^3 after being treated. How much additional space would be needed in a landfill site annually if the wet sludge was dumped directly into the landfill site without dewatering? Note: this practice is no longer permitted by law because of leachate and gas production concerns in sanitary landfills.

SOLUTION: The change of volume of sludge before and after dewatering can be expressed in terms of the inverse ratio of densities of the materials as follows:

$$V_{\text{wet}}/V_{\text{treated}} = \text{treated density}/\text{wet density} = 1.65/1.05 = 1.57$$

The increased volume would be:

$$(0.571) \ (365 \, \text{d/yr}) \ (1.05 \, \text{T/d}) \ (2{,}000 \, \text{lb/T}) \ (454 \, \text{g/lb})/(1.65 \, \text{g/cm}^3)$$

$$= 120{,}425{,}976 \, \text{cm}^3/\text{yr} = 120{,}426 \, \text{L/year}$$

$$= (120{,}426 \, \text{L/year}) \ (1 \, \text{gal}/3.785 \, \text{L}) \ (1 \, \text{ft}^3/7.48 \, \text{gal}) = 4{,}254 \, \text{ft}^3/\text{yr}$$

The problem may alternatively be solved as follows:
The volume of wet sludge before dewatering is:

$$(1.05 \, \text{T/d}) \ (2{,}000 \, \text{lb/T}) \ (454 \, \text{g/lb})/(1.05 \, \text{g/cm}^3)$$

$$= 908{,}000 \, \text{cm}^3/\text{d} = 908 \, \text{L/d}$$

$$= (908 \, \text{L/d}) \ (1 \, \text{gal}/3.785 \, \text{L})(1 \, \text{ft}^3/7.48 \, \text{gal}) = 32.1 \, \text{ft}^3/\text{d}$$

The volume of treated sludge after dewatering is therefore:

$$(1.05 \, \text{T/d})(2{,}000 \, \text{lb/T})(454 \, \text{g/lb})/(1.65 \, \text{g/cm}^3)$$

$$= 577{,}818 \, \text{cm}^3/\text{d} = 578 \, \text{L/d}$$

$$= (578 \, \text{L/d})(1 \, \text{gal}/3.785 \, \text{L})(1 \, \text{ft}^3/7.48 \, \text{gal}) = 20.4 \, \text{ft}^3/\text{d}$$

The additional space requirement is therefore $32.1 - 20.4$ or $11.7 \, \text{ft}^3/\text{d}$.

TRCRA.8 OPTION SELECTION PROCESS

Refer to Problem TCWA.20. Comment on the results.

SOLUTION: From this screening calculation provided in Problem TCWA.20, option C rated the highest with a score of 16,900. Option A's score was 16,600 and option B's score was 12,200. In this case, both option C and option A should be selected for further evaluation because their scores are high and close to each other. Consideration should also be given to combining the "best" features of options A and C.

The system of criteria is a valuable tool for demonstrating the objectivity of the planning to groups that may oppose project alternatives.

TRCRA.9 REFUSE BURNING EMISSIONS

Refuse (garbage) burning is a common method of trash disposal/reduction in many states. While this practice can reduce the quantity of solid waste requiring disposal, it produces air particulate and gaseous contaminants. Although the quantity of air contaminants for one household is relatively small, the combined contaminate generation from an entire state's population can be very large. The quantity of air contaminants can be estimated using emission factors. (The emission factor is an experimentally determined quantity that relates the quantity of air contaminant generated per mass of source; in this case, the source is burning refuse.)

Given the following data, determine the yearly emission of benzene and carbon monoxide from refuse burning:

Population which can burn refuse = 443,704

Annual refuse generation = 0.67 tons/person/yr

% combustible refuse = 80%

CO Emission Factor (EF_{CO}) = 85 lb carbon monoxide/ton burned

Benzene Emission Factor (EF_B) = 2.48 lb benzene/ton burned

SOLUTION: The population which can burn refuse originated from census data corrected for the number of people that live within areas where burning refuse is permitted.

The annual refuse generation for this population is:

$$\text{Annual refuse} = (\text{Generation per person})(\text{Population})$$
$$= (443,704)(0.67 \text{ ton/ person/yr})$$
$$= 297,282 \text{ ton/yr}$$

The amount of refuse burned is:

$$\text{Refuse combusted} = (\text{Annual Refuse})(\text{fraction combustible})$$
$$= (297,282)(0.8) = 237,825 \text{ ton/yr}$$

Carbon monoxide emission:

$$\text{CO emission} = (\text{Refuse Combusted})(EF_{CO})$$
$$= (237,825 \text{ ton/yr})(85 \text{ lbCO/ton burned})$$
$$= 20,215,125 \text{ lb/yr}$$

Benzene emission:

$$\text{Benzene emission} = (\text{Refuse Combusted})(EF_B)$$
$$= (237,825 \text{ ton/yr})(2.48 \text{ lb benzene/ton burned})$$
$$= 589,806 \text{ lb/yr}$$

TRCRA.10   DESTRUCTION EFFICIENCY CALCULATION

The waste flow rate into an incinerator is 1000 lb/h. Calculate the waste flow rate leaving the unit to achieve a destruction efficiency of

1. 95%
2. 99%

3. 99.9%

4. 99.99%

5. 99.9999%

SOLUTION: One may rearrange Equation (6.1) and solve for W_{out},

$$W_{out} = W_{in}\left[1 - \left(\frac{DRE}{100}\right)\right]$$ (6.9)

1. For a DRE of 95%:

$$W_{out} = 1000\left[1 - \left(\frac{95}{100}\right)\right]$$

$$= 50 \text{ lb/h}$$

2. For a DRE of 99%:

$$W_{out} = 1000\left[1 - \left(\frac{99}{100}\right)\right]$$

$$= 10 \text{ lb/h}$$

3. For a DRE of 99.9%:

$$W_{out} = 1.0 \text{ lb/h}$$

4. For a DRE of 99.99%:

$$W_{out} = 0.1 \text{ lb/h}$$

5. For a DRE of 99.9999%:

$$W_{out} = 0.001 \text{ lb/h}$$

TRCRA.11 PACKED COLUMN PERFORMANCE

A packed column is operating at an efficiency of 85.5% for the removal of HCl from a waste flue gas. Since regulations for existing units require that no more than 4.0 lb/h be emitted into the atmosphere, what is the maximum inlet HCl rate in the flue gas to operate within the regulations? In order to meet the MACT regulations of 21 ppm_v(dry) for new units, determine the compliance status if the stack volume is 8000 dscfm and if the HCl rate is 40 lb/h.

SOLUTION: By definition,

$$E = \frac{W_{in} - W_{out}}{W_{in}} \tag{6.10}$$

where E = efficiency (expressed as a fraction)

W_{in} = inlet mass flow rate

W_{out} = outlet mass flow rate

This equation may be rearranged and solved for w_{in}:

$$W_{in} = \frac{W_{out}}{1 - E}$$

For this problem, W_{out} is the maximum allowable emission rate (4 lb/h) and the fractional efficiency (E) is 0.855. Thus,

$$W_{in} = \frac{4.0}{1.0 - 0.855}$$

$$= 27.58 \text{ lb/h}$$

This represents the maximum allowable mass flow rate of HCl in the inlet flue gas.

To determine compliance status, first calculate the discharge HCl volume rate

$$\text{Volume rate of HCl} = (4 \text{ lb/h})\left(\frac{1 \text{ hr}}{60 \text{ min}}\right)\left(\frac{1}{36.5 \text{ lb/lb mol}}\right)(379 \text{ scf/lb mol})$$

$$= 0.6922 \text{ scfm}$$

If the stack volume is 8000 dscfm, then

$$\text{Concentration} = \frac{0.6922}{8000} = 86.53 \text{ ppm}$$

This indicates noncompliance for the existing incinerator. In order to attain compliance as a new incinerator (21 ppm_v), the emission level must be reduced to

$$W_{out} = \frac{21}{86.5} \times \left(4\frac{\text{lb}}{\text{h}}\right)$$

$$= 0.97 \text{ lb/h}$$

The fractional efficiency, E, must then be increased to

$$E = \frac{W_{in} - W_{out}}{W_{in}} = \frac{40 - 0.97}{40} = 0.976$$

$$= 97.6\%$$

TRCRA.12 OVERALL HCl REMOVAL EFFICIENCY

A quench tower operates at a HCl removal efficiency of 65%. This is then followed by a packed tower absorber. What is the minimum collection efficiency of the absorber if an overall HCl removal efficiency of 99.0% is required?

SOLUTION: Select as a basis, 100 lb/h of HCl entering the unit. The mass of HCl leaving quench tower is calculated from the following equation:

$$E = \frac{W_{in} - W_{out}}{W_{in}} \tag{6.11}$$

$$W_{out} = W_{in}(1 - E) = 100(1.0 - 0.65)$$

$$= 35 \text{ lb/h HCl leaving the quench tower}$$

Use the overall efficiency to calculate the mass flow rate of HCl leaving the packed tower absorber:

$$W_{out} = W_{in}(1 - E) = 100(1.0 - 0.990)$$

$$= 1.0 \text{ lb/h leaving the packed tower}$$

The efficiency of the packed tower can now be calculated:

$$E = \frac{W_{in} - W_{out}}{W_{in}} = \frac{35.0 - 1.0}{35.0}$$

$$= 0.971 = 97.1\%$$

TRCRA.13 PARTICLE COLLECTION EFFICIENCY

Given the following inlet loading and outlet loading of an air pollution particulate control unit, determine the collection efficiency of the unit.

$$\text{Inlet loading} = 2 \text{ gr/ft}^3$$

$$\text{Outlet loading} = 0.01 \text{ gr/ft}^3$$

SOLUTION: Collection efficiency is a measure of the degree of performance of a control device; it specifically refers to the degree of removal of a pollutant and may be calculated through the application of the conservation law for mass. *Loading* refers to the concentration of pollutant, usually in grains (gr) of pollutant per cubic feet of contaminated gas stream.

The equation describing collection efficiency (fractional), E, in terms of inlet and outlet loading is a modification of Equation (6.11).

$$E = \frac{\text{Inlet loading} - \text{Outlet loading}}{\text{Inlet loading}} \quad (6.12)$$

Calculate the collection efficiency of the control unit in percent for the rates provided.

$$E = \left(\frac{2 - 0.01}{2}\right) 100 = 99.5\%$$

$$= 0.995 \text{ (fractional basis)}$$

Note that the term η is also used as a symbol for efficiency E.

The reader should also note that the collected amount of pollutant by the control unit is the product of E and the inlet loading. The amount discharged to the atmosphere is given by the inlet loading minus the amount collected.

TRCRA.14   WASTE MIXTURE INCINERATION

A waste mixture consisting of toluene and chlorobenzene is presently being incinerated at 2050°F with a 2.2 s residence time. The componential waste inlet feed rate and the stack discharge rate (including particulate catch) are given below. The stack gas flow rate has been determined to be 13,250 dscfm (60°F, 1 atm) with an O_2 concentration of 4% O_2 dry volume. Is the unit in compliance with existing regulations? Components (1) and (2) are the designated POHCs.

Compound	Inlet (lb/h)	Outlet (lb/h)
1. C_7H_8 (toluene)	953	0.082
2. C_6H_5Cl (chlorobenzene)	337.5	0.022
3. HCl	—	2.2
4. Particulates	—	4.0

SOLUTION: From Equation 6.1,

$$DRE_i = \frac{(W_{in})_i - (W_{out})_i}{(W_{in})_i} 100$$

$$DRE_{toluene} = \frac{(953) - (0.082)}{(953)} 100 = 0.99.991\%$$

$$DRE_{chloro} = \frac{337.5 - 0.022}{337.5} 100 = 99.993\%$$

Regulations require that the DRE for each designated POHC be at least 99.99%. In this case, the designated POHCs are toluene and chlorobenzene. Both POHCs exhibit a DRE of >99.99%. Therefore, the unit is in compliance with DRE regulations.

The reaction of chlorobenzene in the incinerator may be shown in the following equation:

$$C_6H_5Cl + air \rightarrow 6CO_2 + HCl + 2H_2O$$

Assume all the chlorine in the feed is converted to HCl. The molar feed rate of chlorobenzene $\dot{n}_2$ is

$$\dot{n}_2 = \frac{W_2}{(MW)_2} = \frac{337.5}{112.5} = 3.0 \text{ lbmol/h}$$

where (in general) n_i = molar flow rate of component i

$(MW)_i$ = molecular weight of component i

Each molecule of chlorobenzene contains one atom of chlorine. Therefore,

$$\dot{n}_{HCl} = \dot{n}_2$$
$$= 3.0 \text{ lbmol/h}$$
$$= w_{HCl} = 109.5 \text{ lb/h}$$

The removal efficiency (RE) for HCl is

$$RE = \frac{W_{in} - W_{out}}{W_{in}} 100 = \frac{109.5 - 2.2}{109.5} 100 = 97.99\%$$

Previous RCRA regulations require that HCl emissions must either achieve a 99% HCl scrubbing efficiency or emit <4 lb/h of hydrogen chloride. In this case, emissions are <4 lb/h and therefore in compliance with past regulations.

To determine if the emissions meet the MACT rules, the outlet emission must not exceed 21 ppm corrected to 7% O_2 dry volume.

The stack flow is given at 13,250 dscfm. The HCl emissions are 2.2 lb/h. The HCl (3) volume is calculated as follows:

$$\text{Volume, } \text{cfm} = (W_3)\left(\frac{1}{365}\right)\left(\frac{1\,\text{h}}{60\,\text{min}}\right)\left(\frac{379\,\text{cu ft}}{\text{mol}}\right); \ W_3 = 2.2$$

$$= 0.3807 \text{ cfm}$$

$$O_2 \text{ concentration} = \frac{\text{volume HCl}}{\text{volume flue gases}}$$

$$= \frac{0.3807}{13,250}$$

$$= 281\,\text{ppm at } 4\% \ O_2 \text{ dry volume}$$

$$\text{Correction factor, } CF = \frac{14}{21-4} = 0.8235$$

$$\text{Corrected concentration} = (28)(CF)$$

$$= (28)(0.8235)$$

$$= 23.06 \text{ ppm HCl at } 7\% \ O_2 \text{ dry volume}$$

The removal efficiency must increase (or penetration decrease) to

$$P_{\text{new}} = \frac{21}{23.06}(1 - 0.9799) = 0.0183 = 1.83\%$$

and

$$E_{\text{new}} = 0.9817 = 98.17\%$$

The outlet loading W_{out} (PM) of the particulates is

$$\text{PM} = \frac{\text{PM}_{\text{out}}(7000)}{(\text{dscfm})(60)}$$

$$= \frac{(4.0)(7000)}{(13,250)(60)}; \ 7000 \text{ gr/lb}$$

$$= 0.0352 \text{ gr/dscf} \tag{6.13}$$

$$\text{Corrected PM}_{\text{out}} = (\text{PM})(CF)$$

$$= 0.0352 \times 0.8235$$

$$= 0.029 \text{ gr/dscf}$$

MACT regulations require that particulate emissions must be <0.015 gr/dscf corrected to 7% O_2 dry volume. It is therefore in compliance with regulations at <0.08 gr/dscf but will not meet the new MACT standards for particulate emissions.

The unit is therefore in compliance with the MACT regulations for all but particulates.

TRCRA.15   TRIAL BURN

During a trial burn, an incinerator was operated at a waste feed rate of 5000 lb/h and 7% O_2 dry volume in the stack gases. The gas flow rate measured in the stack was 19,200 dscfm. Under these conditions, the measured concentrations of the principal organic hazardous components were:

Trichloroethylene (1) = 4.9 μg/dscf
1,1,1-Trichloroethane (2) = 1.0 μg/dscf
Methylene chloride (3) = 49 μg/dscf
Perchloroethylene (4) = 490 μg/dscf

Each hazardous component listed here constitutes approximately 5% of the total waste feed rate. Calculate the destruction and removal efficiency of the POHCs. Is the unit in compliance with regulations?

SOLUTION: Each hazardous component constitutes 5% of the total waste feed. Therefore,

$$(W_{in})_1 = (0.05)(\text{total feed rate}) = (0.05)(5000 \text{ lb/h})$$
$$= 250 \text{ lb/h}$$

The mass flow rate of each hazardous component in the stack is given by:

$$(W_{out})_i = \frac{q_s C_i}{7.57 \times 10^6}$$

where q_s = volumetric gas flow rate (dscfm)

C_i = concentration of component i (μg/dscf)

7.57×10^6 = conversion factor that converts μg/min to lb/h

Note that

$$\left(454 \frac{g}{lb}\right)\left(10^6 \frac{\mu g}{g}\right)\left(\frac{hr}{60 \text{ min}}\right) = 7.57 \times 10^6$$

Therefore,

$$(W_{out})_1 = \frac{(19,200)(4.9)}{7.57 \times 10^6}$$
$$= 0.01243 \text{ lb/h}$$

$$(W_{out})_2 = \frac{(19,200)(1.0)}{7.57 \times 10^6}$$
$$= 0.00254 \text{ lb/h}$$

$$(W_{out})_3 = \frac{(19,200)(49)}{7.57 \times 10^6}$$
$$= 0.1243 \text{ lb/h}$$

$$(W_{out})_4 = \frac{(19,200)(490)}{7.57 \times 10^6}$$
$$= 1.243 \text{ lb/h}$$

The destruction and removal efficiency for component i, DRE_i, is given by:

$$DRE_i = \left(\frac{(W_{in})_i - (W_{out})_i}{W_{in}} \right) 100$$

$$DRE_1 = \left(\frac{250 - 0.01243}{250} \right) 100 = 99.995\%$$

$$DRE_2 = \left(\frac{250 - 0.00254}{250} \right) 100 = 99.999\%$$

$$DRE_3 = \left(\frac{250 - 0.1243}{250} \right) 100 = 99.95\%$$

$$DRE_4 = \left(\frac{250 - 1.1243}{250} \right) 100 = 99.55\%$$

Regulations require that 99.99% destruction and removal efficiency be achieved. In this case, only two of the principal organic hazardous components (1 and 2) achieved this status. Assuming that either component 3 or 4 (or both) was a designated POHC, the unit would be out of compliance.

TRCRA.16   CARBON INCINERATION

A coal process waste, containing carbon and free water, is being incinerated by ABC Waste Disposal, Inc. The flue gas leaving the incinerator is at a temperature of 450°F following a waste heat recovery unit, and contains a particulate loading of 0.03 gr/acf. The flue gas analysis shows the following composition:

% O_2 in Stack Gas = 12.5%

% CO_2 in Stack Gas = 12.5%

% N_2 in Stack Gas = 50.0%

% H_2O in Stack Gas = 25.0%

Determine the outlet particulate loading based on a dry, standard basis, corrected to 50% EA. Is the process in compliance with the current particulate state regulations of 0.03 gr/dscf corrected to 50% EA? Use standard conditions of 1 atm pressure and a temperature of 68°F.

SOLUTION: As described earlier, the gas volume must be corrected for water content, temperature and oxygen content since the flue gas does not correspond to the reference conditions. Assume as a basis 100 acf of flue gas.

Correction for water content:

$$V_2 = V_1 \left(\frac{n_2}{n_1}\right) = 1.0 \left(\frac{n_1 - n_{H_2O}}{n_1}\right); \ m = \text{moles dry flue gas}$$

$$= \frac{100 - 25}{100} = 0.75 \text{ dacf}$$

The correction for oxygen content is based on 50% excess air provided above the stoichiometric level required for complete combustion. The stoichiometric oxygen required is based on carbon dioxide formation using the following balanced equation:

$$C + 1.5O_2 \rightarrow CO_2 + 0.5O_2$$

Thus, the flue gas should contain 0.5 volumes of oxygen per volume of carbon dioxide at 50% EA. The actual amount is 1 volume oxygen per volume of carbon dioxide (see flue gas analysis above), and thus the actual volume must be reduced using the ideal gas law by removing half the oxygen. Take a basis of 100 dacf flue gas, n_2.

$$V_3 = V_2 \left(\frac{n_3}{n_2}\right) = 0.75 \left(\frac{n_2 - n_{O_2}}{100}\right)$$

$$= 0.75 \left(\frac{100 - 6.25}{100}\right)$$

$$= 0.70 \text{ dacf corrected to 50\% EA}$$

Temperature corrections are made based on Charles' Law to convert dacf to dscf, both at 50% EA:

$$V_4 = V_3\left(\frac{T_4}{T_3}\right) = 0.70\left(\frac{460 + 68}{460 + 450}\right) = 0.41 \text{ dscf corrected to 50\% EA}$$

The loading (L) under dscf corrected to 50% EA conditions is determined as follows:

$$L_{dscf} = L_{acf}\left(\frac{V_{acf}}{V_{dscf}}\right) = 0.03 \text{ gr/acf}\left(\frac{1.0 \text{ acf}}{0.41 \text{ dscf}}\right)$$

$$= 0.075 \text{ gr/dscf corrected to 50\% EA}$$

This value exceeds the regulatory maximum limit of 0.03 gr/dscf corrected to 50% excess air, and the unit is therefore out of compliance.

TRCRA.17   THERMAL AFTERBURNER DESIGN

Provide a design procedure for thermal afterburners.

SOLUTION: There are two key calculations associated with combustion devices. These include determining:

1. The fuel requirements
2. The physical dimensions of the unit

Both these calculations are interrelated. The general procedure to follow, with pertinent equations, is given below. It is assumed that the process gas stream flowrate, inlet temperature, and the combustion temperature are known. The required residence time is also specified. Primary (outside the process) air is employed for combustion.

1. Calculate the heat load required to raise the process gas stream from its inlet temperature to the operating temperature of the combustion device:

$$\dot{Q} = \Delta\dot{H}; \qquad \dot{H} = \text{enthalpy} \tag{6.14}$$

2. Correct the heat load term for any radiant losses (RL):

$$\dot{Q} = (1 + \text{RL})(\Delta\dot{H}) \qquad \text{RL} = \text{fractional basis} \tag{6.15}$$

3. Assuming natural gas of known heating value, HV_G, is the fuel, calculate the available heat (HA_T) at the operating temperature. For engineering purposes,

one may use a short-cut method that bypasses a detailed calculation.

$$HA_T = (HV_G)(HA_T/HV_G)_{ref} \qquad (6.16)$$

The subscript "ref" refers to a reference fuel. For natural gas with a reference HV_G of 1059. Btu/scf, the available heat (assuming stoichiometric) is given by (L. Theodore; personal notes, 1980):

$$(HA_T)_{ref} = -0.237T + 981 \qquad T = {}^{\circ}F \qquad (6.17)$$

Note that the *available heat* is defined as the quantity of heat released within a combustion chamber minus (1) the sensible heat carried away by the dry flue gases, and (2) the latent heat and sensible heat carried away in water vapor contained in the flue gases. Thus, the available heat represents the net quantity of heat remaining for useful heating purposes.

4. Calculate the flowrate of natural gas required, q_{NG}:

$$q_{NG} = \dot{Q}/HA_T \qquad \text{consistent units} \qquad (6.18)$$

5. Determine the volumetric flowrates of both the process gas stream, q_p, and the flue products of combustion of the natural gas, q_c, at the operating temperature:

$$q_T = q_p + q_c \qquad (6.19)$$

A good estimate for q_c is

$$q_c = (11.5)q_{NG} \qquad (6.20)$$

6. The cross-sectional area of the combustion device is given by

$$S = q_T/v \qquad (6.21)$$

where v is the throughput velocity.

7. The residence time of gases in the combustion chamber may be calculated from

$$t = V/q_T \qquad (6.22)$$

where t is the residence time (s), V is the chamber volume (ft^3), and q_T is the gas volumetric flowrate at combustion conditions in the chamber (acfs).

Adjustments to both the gas volumetric flowrate and fuel rate must be performed if secondary (from process gas stream) air rather than primary (outside) air is added for combustion.

Additional details are available in the following references: L. Theodore, "Ask the Experts: Designing Thermal Afterburners," CEP, New York City, April, 2005 and J. Santoleri et al., "Introduction and Hazardous Waste Incineration," 2nd Edition, John Wiley & Sons, Hoboken, NJ, 2002.

TRCRA.18 MERCURY WASTE

An incinerator burns mercury contaminated medical waste. The waste material has an ash content of 1%. The solid waste feed rate is 1000 lb/h and the gas flow rate is 20,000 dscfm. The average mercury content of the particulate matter is 2.42 μg/g, while the vapor concentration is 0.3 mg/dscm. For the case when the incinerator meets the particulate standard of 0.08 gr/dscf (0.1832 g/dscm) with a 99.5% efficient electrostatic precipitator (ESP), calculate the amount of mercury (lb/d) bound to the fly ash which is captured in the ESP.

SOLUTION: The amount of mercury bound to the fly ash which is captured in the ESP is calculated based on the mass of ash leaving the stack. This mass is calculated as follows:

$$\text{Mass of ash out stack} = (0.08\,\text{gr/dscf})\,(1\,\text{lb/7000 gr})\,(20,000\,\text{dscfm})$$
$$\times\,(60\,\text{min/h})\,(24\,\text{h/d}) = 329\ \text{lb ash/d}$$

The mass exiting the stack represents 0.5% of the ash collected in the ESP, i.e.,

$$\text{Mass in the ESP} = (329\,\text{lb/d})/0.005$$
$$= 65,830\,\text{lb ash/d}$$

The amount of mercury bound to the fly ash in the ESP is calculated based on the mercury content of the fly ash:

$$\text{Mercury in the ESP} = (65,830\,\text{lb ash/d})\,(2.42 \times 10^{-6}\,\text{g mercury/g ash})$$
$$\times\,(454\,\text{g/lb})$$
$$= 72.3\ \text{lb/d}$$

TRCRA.19 MERCURY REMOVAL

Medical sludge containing mercury is burned in an incinerator. The mercury feed rate is 9.2 lb/h. The resulting 500°F product (40,000 lb/h of gas; MW = 32) is quenched with water to a temperature of 150°F. The resulting stream is filtered to remove all particulates. What happens to the mercury? Assume the process pressure is 14.7 psi and that the vapor pressure of Hg at 150°F is 0.005 psi.

SOLUTION: For the mercury to be removed by the filter, it must condense and form particles. Therefore, the first question to be answered relates to the partial pressure of mercury during removal compared to its vapor pressure at 150°F:

$$\text{Molar flowrate of Hg} = (9.2\,\text{lb/h})/200.6\,\text{lb Hg/lbmol})$$
$$= 0.046\,\text{lbmol/h}$$
$$\text{Molar flowrate of gas} = (40{,}000\,\text{lb gas})/(32\,\text{lb gas/lbmol})$$
$$= 1250\,\text{lbmol/h}$$

The mole fraction of mercury may now be calculated

$$y = \frac{\text{lbmol Hg}}{\text{lbmol Hg} + \text{lbmol gas}} = \frac{0.046\,\text{lbmol/h}}{0.046\,\text{lbmol/h} + 1250\,\text{lbmol/h}} = 3.68 \times 10^{-5}$$

The partial pressure is given by:

$$\text{Partial pressure } p_i = y_i p \qquad\qquad (6.23)$$
$$= y_i(14.7\ \text{psia})$$
$$= 3.68 \times 10^{-5}(14.7\ \text{psia})$$
$$= 5.4 \times 10^{-4}\ \text{psia}$$

Since the partial pressure is much less than the vapor pressure, mercury will NOT condense and thus will NOT be removed by the filter.

TRCRA.20 SALT DOME APPLICATION

A large, deep cavern (formed from a salt dome) located north of Houston, TX has been proposed as disposal site for both solid hazardous and municipal wastes. Preliminary geological studies indicate that there is a chance that the wastes and any corresponding leachates will penetrate the cavern walls and contaminate the adjacent soil and aquifers. A risk assessment analysis was also conduced during the preliminary study and the results indicate that there was a greater than 99% probability that no hazardous and/ or toxic material would "meander" beyond the cavern walls during the next 25 years.

The following data and information has been provided by the company preparing the permit application for the Texas Water Pollution Board.

Approximate total volume of cavern $= 0.78(\text{mi})^3$

Approximate volume of cavern available for solid waste depository

$= 75\%$ of total volume

Proposed maximum waste feed rate to cavern $= 20{,}000$ lb/day

Feed rate schedule $= 6$ days/week

Average bulk density of waste $= 30\,\text{lb/ft}^3$

Based on the above data, estimate the minimum amount of time it will take to fill the volume of the cavern available for the waste deposition.

SOLUTION:

1. Calculate the volume of the cavern in $(mi)^3$ available for the solid waste.

$$V = (0.75)(0.78)$$
$$= 0.585 \, mi^3$$

2. Convert the volume in step (1) to ft^3.

$$V = (0.585 \, mi^3)(5280 \, ft/mi)^3$$
$$= 8.61 \times 10^{10} \, ft^3$$

3. Calculate the daily volume rate of solids deposited within the cavern in ft^3/day.

$$q = (20,000 \, lb/day)/(30 \, lb/ft^3)$$
$$= 667 \, ft^3/day$$

4. Convert the solids volume rate in step (3) to $ft^3/year$.

$$q = (667 \, ft^3/day) \, (6 \, days/wk)(52 \, wks/yr)$$
$$= 208,000 \, ft^3/yr$$

5. Determine how long it will take to fill the cavern.

$$t = V/q$$
$$= (8.61 \times 10^{10})/(208,000)$$
$$= 414,000 \, yrs$$

The proposed operation will extend well beyond the 25 years upon which the risk assessment analysis was based. The decision whether to grant the permit is somewhat subjective since there is a finite, though extremely low, probability that the cavern walls will be penetrated. Another more detailed and exhaustive risk analysis study should be considered.

TRCRA.21   HAZARDOUS WASTE INCINERATOR APPLICATION

A hazardous waste incinerator operates at a chemical plant to treat a liquid slurry production waste stream. Your manager requests that in preparation for a trial burn, you institute a procedure for obtaining compliance with emission regulations. The waste stream feed rate is 800 lb/h and the stack gas flowrate is 22,760 scfm (60°F, 1 atm). The incinerator is to operate at a destruction and removal efficiency (DRE) of 99.995% for hexachlorobenzene, the waste feed stream.

The analytical laboratory informs you that the detection limit of hexachlorobenzene is 10 µg/L. If the sample is concentrated in 25 mL of solvent, what is the minimum volume of flue gas that must be collected to detect 99.995% DRE? The feed stream initially contains 1% (by mass hexachlorobenzene). The sample can be collected at 1 L/min (standard conditions).

SOLUTION: Determine the mass flowrate of hexachlorobenzene (1% by mass):

$$\dot{m} = (0.01)(8000 \text{ lb/h}) = 80 \text{ lb/h}$$

Determine the principal organic hazardous constituents (POHC) flowrate in the stack for a 99.995% DRE.

$$\text{DRE} = \left(\frac{\dot{m}_{\text{in}} - \dot{m}_{\text{out}}}{\dot{m}_{\text{in}}}\right)(100) \tag{6.24}$$

$$= 0.99995 = 1 - \frac{\dot{m}_{\text{out}}}{80}$$

Solving for $\dot{m}_{\text{out}}$

$$\dot{m}_{\text{out}} = 0.004 \text{ lb/h}$$

$$= \left(\frac{0.004 \text{ lb}}{h}\right)\left(\frac{454 \times 10^6 \text{ µg}}{\text{lb}}\right)\left(\frac{1h}{60 \text{ min}}\right)$$

$$= 30,300 \text{ µg/min}$$

Determine the POHC concentration in the stack gas with units of µg/scf:

$$\text{POHC} = \frac{30,300 \text{ µg/min}}{22,760 \text{ scf/min}}$$

$$= 1.33 \text{ µg/scf}$$

Determine the sample gas volume required in scfm and the time required for sample collection in minutes:

$$\text{Volume} = \left(\frac{10 \text{ µg}}{L}\right)\left(\frac{1L}{1000 \text{ mL}}\right)(25 \text{ mL})\left(\frac{1 \text{ scf}}{1.33 \text{ µg}}\right)$$

$$= 0.188 \text{ scf}$$

$$\text{Time} = (0.188 \text{ scf})\left(\frac{28.3 \text{ L}}{\text{ft}^3}\right)\left(\frac{1 \text{ min}}{L}\right)$$

$$= 5.3 \text{ min}$$

TRCRA.22   COMPLIANCE DETERMINATION FOR A HOSPITAL INCINERATOR

A hospital in the state of Pennsylvania is currently incinerating its waste in a modular incinerator at a temperature of 1800°F and a residence time of 2 s. Regulations for particulate emissions for hospital waste incinerators in Pennsylvania are given in Table 6.10.

Hydrogen chloride (HCl) emissions must not exceed 4 lb/h for incinerators operating at capacities of 500 lb/h or less. For larger incinerators, HCl emissions must not exceed 30 ppmdv (corrected to 7% O_2).

The hospital generates 20 lb/bed · day of waste of which 10% is infectious (the remaining 90% is paper, cardboard, etc.). It also produces 0.056 lb/bed·day of Resource Conservation and Recovery Act (RCRA) hazardous wastes. Two hundred beds are present in the hospital. The RCRA hazardous wastes are listed in Table 6.11.

A trial burn is conducted with the waste composition in the preceding table. The designated principal organic hazardous constituents (POHC$_S$) for the trial burn are methyl alcohol, polyvinyl chloride, and xylene. The results of the trial burn are given in Table 6.12.

The total stack gas flowrate is 10,000 dscfm. Determine:

1. The hazardous waste generation rate in pounds/month.
2. The regulations the incinerator must comply with (state hospital regulations or combined state and hospital and RCRA regulations).
3. The total incinerator capacity in pounds/hour.
4. Is the incinerator in compliance?

TABLE 6.10   Particulate Emissions Standards

Capacity (lb/h)	Particulate Emissions Standard (gr/dscf)
≤500	0.08 (corrected to 7% O_2)
500–2000	0.03 (corrected to 7% O_2)
≥2000	0.015 (corrected to 7% O_2)

TABLE 6.11   Hospital Waste Components

Hazardous Waste Component	Wt%
Methyl alcohol	12.5
Polyvinyl chloride (PVC)	75
Xylene	12.5

TABLE 6.12 Trial Burn Result

Hazardous Waste Component	Outlet Mass Flowrate (lb/h)
Methyl alcohol	0.0001
Polyvinyl chloride (PVC)	0.0002
Xylene	0.00001
HCl	3.2
Particulates	5.64

SOLUTION:

1. Calculate the hazardous waste generation rate in pounds/month:

$$\text{Hazardous waste rate} = (0.056\,\text{lb/bed·day})(200\text{ beds})$$
$$= 11.2\,\text{lb/day}$$
$$= 11.2\ \text{lb/day (30 days/mo)}$$
$$= 336\ \text{lb/mo}$$

2. Determine if the incinerator must comply with RCRA regulations. Since the hazardous waste generation rate of 336 lb/mo is greater than the 220 lb/mo RCRA regulation, the incinerator must comply with RCRA regulations in addition to state regulations for hospital incinerators.

3. Calculate the amount of nonhazardous waste produced in pounds/day:

$$\text{Nonhazardous waste rate} = (20\ \text{lb/bed·day})\ (200\text{ beds})$$
$$= 4000\,\text{lb/day}$$

Determine the total incinerator capacity in pounds/hour:

$$\text{Total capacity} = \text{Hazardous waste rate} + \text{Nonhazardous waste rate}$$
$$= (11.2\,\text{lb/day} + 4000\,\text{lb/day})(\text{day}/24\,\text{h})$$
$$= 167.1\,\text{lb/h}$$

Determine the inlet mass rate of methyl alcohol to the incinerator in pounds/hour:

$$\dot{m}_{in} = (0.125)(11.2\ \text{lb/day})(\text{day}/24\ \text{h})$$
$$= 0.0583\ \text{lb/h}$$

Calculate the destruction efficiency for the hazardous component methyl alcohol (MeOH):

$$\text{DRE}_{MeOH} = [(0.0583\ \text{lb/h} - 0.0001\ \text{lb/h})/(0.0583\ \text{lb/h})](100)$$
$$= 99.827\%$$

Calculate the destruction efficiency for the hazardous component xylene:

$$\dot{m}_{in} = (0.125)(11.2 \text{ lb/day})(\text{day}/24 \text{ h})$$
$$= 0.0583 \text{ lb/h}$$
$$\text{DRE}_{xylene} = [(0.0583 \text{ lb/h} - 0.00001 \text{ lb/h})/(0.0583 \text{ lb/h})](100)$$
$$= 99.983\%$$

Calculate the destruction efficiency for the hazardous component polyvinyl chloride (PVC):

$$\dot{m}_{in} = (0.75)(11.2 \text{ lb/day})(\text{day}/24 \text{ h})$$
$$= 0.350 \text{ lb/h}$$
$$\text{DRE}_{PVC} = [(0.350 \text{ lb/h} - 0.0002 \text{ lb/h})/(0.350 \text{ lb/h})](100)$$
$$= 99.943\%$$

Since the total incineration capacity is less than 500 lb/h, the 4 lb/h regulation applies. The trial burn resulted in an HCl emission rate of 3.2 lb/h.

Calculate the outlet loading (OL) of the particulates in gr/dscf.

$$\text{OL} = [(5.64 \text{ lb/h})(7000)]/[(10,000 \text{ dscfm})(60)]$$
$$= 0.0658 \text{ gr/dscf}$$

4. Determine if the incinerator is in compliance.

   POHCs: Since all of the POHCs had DREs $< 99.99\%$, the incinerator is out of compliance.
   HCl: The incineration complies with the 4 lb/h limit on HCl emissions.
   Particulates: The outlet loading of 0.0658 gr/dscf is less than the 0.08 gr/dscf limit for incinerators operating at a capacity less than 500 lb/h.

The incinerator is not in compliance.
Note that the pathological waste in the waste stream in this problem is the most difficult waste to destroy since its heating value is low. Hospital waste incinerators must be designed to destroy pathological and infectious waste, not paper waste alone. The contents of a hospital waste stream are normally more complex than shown in this problem. Other hazardous components may include pentane, diethyl ether, acetone, methyl cellosolve, and other laboratory wastes.

Each state has its own regulations concerning hospitals. Other regulations must also be complied with in addition to those stated in this problem. The appropriate state agencies should be contacted for a list of the detailed regulations for hospital waste incinerators.

CHAPTER 7

TOXIC SUBSTANCES CONTROL ACT (TSCA)

QUALITATIVE PROBLEMS (LTSCA)

PROBLEMS LTSCA.1–15

LTSCA.1 THE TOXIC SUBSTANCES CONTROL ACT (TSCA TITLE I)

Provide an overview and history of TSCA.

SOLUTION: In 1970, the President's Council on Environmental Quality developed a legislative proposal to address the increasing problems of toxic substances. After six years of public hearings and debate, Congress enacted the Toxic Substances Control Act (TSCA) in the fall of 1976. EPA/OPPT (Office of Pollution Prevention and Toxics) is charged with implementing TSCA. TSCA (Title I) does not provide opportunities for EPA to authorize state programs to operate in lieu of the federal program, although the office actively collaborates with regions, states and tribal governments. Through the provisions of TSCA, EPA can collect or require the development of information about the toxicity of particular chemicals and the extent to which people and the environment are exposed to them. Such information allows EPA to assess whether the chemicals pose unreasonable risks to humans and the environment, and TSCA provides tools instituting appropriate control actions. TSCA provides the basis for EPA's programs on New and Existing Chemicals,

Environmental Regulatory Calculations Handbook, by Leo Stander and Louis Theodore
Copyright © 2008 John Wiley & Sons, Inc.

the basis for the national programs for major chemicals of concern such as lead (TSCA Title IV) and asbestos (TSCA Title II) and possibly nanomaterials in the future, and the foundation for other OPPT programs such as the voluntary data development activities under the High Production Volume (HPV) Challenge Program.

TSCA §2(b) (1) establishes the underlying national policy that:

> Adequate data should be developed with respect to the effect of chemical substances and mixtures on health and the environment and that the development of such data should be the responsibility of those who manufacture and those who process such chemical substances and mixtures.

EPA has authority under TSCA §6 to regulate the manufacture (including import), processing, use, distribution in commerce, and disposal of chemical substances and mixtures that present or will present an unreasonable risk to human health and the environment. EPA may ban the manufacture or distribution in commerce, limit use, require labeling, or place other restrictions on chemicals that pose unreasonable risks after making certain statutory findings. In order to regulate under §6, EPA must find that there is a reasonable basis to conclude that a chemical substance "presents or will present an unreasonable risk of injury to health or the environment," where "unreasonable risk" is a risk-benefit standard. EPA must consider risks, costs and benefits of a substance to be regulated, including the availability of substitutes. TSCA requires the Administrator to impose the "least burdensome" regulatory measure that provides adequate protections.

TSCA §4 gives EPA broad authority to require manufacturers (includes importers) and processors to test chemicals for health and environmental effects. EPA uses the §4 rulemaking authority only when it can make certain statutory findings about the substance involved, including that there are insufficient data available to determine the effects of the substance on health and/or the environment; testing is necessary to provide such data; and, the chemical may present an unreasonable risk of injury to health or the environment, and/or may be produced at substantial quantities and is reasonably expected to enter the environment in substantial quantities, or may result in significant or substantial human exposure. TSCA §4 has generated data on approximately 200 chemicals since the 1970s.

TSCA §8 has a variety of data-gathering authorities. Under TSCA §8(e) EPA must be notified immediately of new unpublished information on chemicals that reasonably supports a conclusion of substantial risk. TSCA §8(e) has been an important information-gathering tool that serves as an "early warning" mechanism.

TSCA §5 requires manufacturers to give EPA a 90-day advance notice (via a premanufacture notice or PMN) of their intent to manufacture and/or import a new chemical (including microorganisms). The PMN includes information such as specific chemical identity, use, anticipated production volume, exposure and release information, and existing available test data. The information is reviewed through OPPT's new chemicals program to determine whether action is needed to prohibit or limit manufacturing, processing, or use of a chemical. Many PMNs include little or no toxicity or fate data; consequently, OPPT uses several general approaches to address data gaps to rapidly evaluate potential risks and make risk management decisions for new chemicals within the 90-day timeframe prescribed by TSCA. Under TSCA §5(a), EPA is authorized to designate a new use of a new or existing chemical as a Significant New Use Rule (SNUR), based on consideration of several factors, including the anticipated extent and type of exposure to humans and the environment.

TSCA §9 addresses EPA's authority to regulate chemical substances and associated activities that fall under both TSCA and other Federal laws, including laws administered by other Federal agencies and the EPA. It includes procedures under which EPA can refer the regulation of chemicals to other agencies and requirements to coordinate actions taken under activities with other Federal agencies "for the purpose of achieving the maximum enforcement of this act [TSCA] while imposing the least burdens of duplicative requirements on those subject to the Act and for other purposes."

Industry or other submitting companies may also claim certain information as Confidential Business Information (CBI) under TSCA §14(a). The provision prohibits EPA from disclosing trade secrets, or commercial or financial information that is privileged or confidential, to the public (including States, Tribes, local governments), except in certain limited circumstances.

Under TSCA §21, any citizen may petition EPA to take action under TSCA §4 (rules requiring chemical testing), §6 (rules imposing substantive controls on chemicals), or §8 (information gathering rules). TSCA §21 also authorizes a petitioner to request the issuance, amendment, or repeal of orders, including certain orders under §§5 and 6.

OPPT has also the responsibility for implementing other Titles of TSCA, for example, The Residential Lead-Based Paint Hazard Reduction Act of 1992, also known as Title X of the Housing and Community Development Act (TSCA Title IV) and The Asbestos Hazard Emergency Response Act (AHERA) (TSCA Title II).

In summary, TSCA provides EPA with the authority to control the risks of thousands of chemical substances, both new and old, that are not regulated as drugs, food additives, cosmetics, or pesticides. TSCA mandates testing of chemical substances to regulate their uses in industrial, commercial, and consumer products. TSCA fills in the gaps and supplements other laws regulating toxic substances, such as the Clean Air Act, the Occupational Safety and Health Act, and the Federal Water Pollution Control Act. TSCA allows EPA to tailor its regulation to specific sources of risk. TSCA essentially contains two sections: requirements for information on the substance to identify risks to health and the environment from chemical substances (a premanufacturing notification (PMN)); and, regulations on the production and distribution of new chemicals and regulations on the manufacturing, processing, distribution, and use of existing chemicals (recordkeeping and reporting requirements).

LTSCA.2   THE TSCA CHEMICAL SUBSTANCE INVENTORY

Describe the TSCA Chemical Substance Inventory.

SOLUTION: The initial TSCA Chemical Substance Inventory ("Inventory") of existing chemical substances (approximately 61,000 chemicals) was based on information reported to EPA by chemical manufacturers (including importers) and processors from 1975–1978. The Inventory lists all existing chemicals in commerce by chemical name and Chemical Abstracts Service (CAS) Registry numbers or accession numbers (accession numbers are used for chemicals whose identities have been claimed confidential business information (CBI)). The Inventory provides an overall picture of the organic, inorganic, polymers, and UVCB (chemical substances of Unknown, or Variable Composition, Complex Reaction Products, and Biological Materials) chemicals produced, processed or imported for commercial purposes in the United States; it is not a list of chemicals based on toxic or hazardous characteristics.

In 1986, EPA promulgated the Inventory Update Rule (IUR) for the partial updating of the production volume data reported to the Inventory. The rule required manufacturers of nonpolymeric organic chemical substances included on the Inventory to report current data on the production volume, plant site, and site-limited status of these substances if produced or imported at levels of 10,000 pounds or more per year per site. (Inorganic chemicals are defined as any chemical substances that do not contain carbon or contain carbon in specific forms (40 CFR 710.26(a).) Polymers are defined as any chemical substance described with the word "poly," "alkyd," "oxylated" (40 CFR 710.26 (b)). After the initial reporting during 1986, recurring reporting was required every 4 years (1990, 1994, 1998, 2002, 2006). EPA amended the TSCA IUR in a *Federal Register* notice published on January 7, 2003. The IUR Amendments (IURA) update the TSCA Inventory by modifying the reporting threshold from the original 10,000 pounds per year per site to 25,000 pounds per year. In addition, the IURA required reporting of processing and use information for substances above the reporting threshold of 300,000 pounds per year. IURA also includes requirements for the reporting of inorganic chemicals and additional exposure-related information to assist EPA and others in screening potential exposures and risks, modified the IUR reporting and record keeping requirements, removed one reporting exemption and created others, and modified its procedures for making Confidential Business Information claims.

There are currently approximately 82,000 chemical substances on the TSCA Inventory that are or have been produced, processed or imported into the United States. These fall broadly into three types of substances:

1. Discrete chemicals having definite structures (Class 1)
2. Chemical substances having indefinite structures or substances that are of unknown or variable composition, complex reaction products, and biological materials (Class 2)
3. Polymers

LTSCA.3 OFFICE OF POLLUTION PREVENTION AND TOXICS (OPPT's) NEW CHEMICALS PROGRAM

Describe OPPT's new chemicals program.

SOLUTION: Chemicals not on the TSCA Inventory are considered "new" chemicals and are reviewed by EPA before they are produced or imported in the United States. Certain genetically modified microorganisms are also considered "new chemicals." The TSCA New Chemicals Program was established to help manage the potential risk from chemicals new to the marketplace, e.g., nanoparticles. The New Chemicals Program functions as a "gatekeeper" that can identify concerns and impose conditions, up to and including a ban on manufacture, on the commercialization of a new chemical before entry into commerce, or on a "significant new use" of an existing or new chemical. The New Chemicals Program also serves as an advocate for environmental stewardship in encouraging the development and introduction of safer or "green" new chemicals.

To implement TSCA requirements for new chemicals, OPPT developed the Premanufacture Notification (PMN) Review Process. Manufacturers and importers of new chemicals must give EPA a 90-day advance (premanufacture) notification of their intent to manufacture and/or import a new chemical. The PMN, which includes information such as specific chemical identity, use, anticipated production volume, exposure and release information, and existing available test data, is reviewed by OPPT to determine whether action is needed to prohibit or limit manufacturing, processing, or use of a chemical.

The PMN review process is designed to accommodate the large number of PMNs received (approximately 1500 annually), while adequately assessing the risks posed by each substance within the 90-day timeframe prescribed by TSCA. The information included in PMNs is limited: 67% of PMNs include no test data and 85% include no health data. Consequently, OPPT uses several general approaches to address data gaps to rapidly evaluate potential risks and make risk management decisions for new chemicals. For example, OPPT has developed and relies on Structure-Activity Relationship (SAR) analyses to estimate or predict physical-chemical properties, environmental fate, and human and environmental effects. A SAR is the relationship between the chemical structure of a molecule and its properties, including any possible interaction with the environment or organisms; the type of analyses will probably not apply to nanoparticles. EPA's New Chemicals Program has established 55 chemical categories to facilitate the PMN review process.

Every PMN that is submitted to OPPT goes through a streamlined initial review process. The first of four review phases is essentially a chemistry review. On about day 8–12 after receipt of the PMN, OPPT chemists gather at a chemical review/ search strategy meeting (CRSS). During this meeting, EPA staff establish a chemical profile for each PMW. These profiles include: chemical identity, structure and nomenclature, structural analogues and inventory status, notice completeness, synthesis, use/TSCA jurisdiction, and physical-chemical properties. The submitter of the PMN may be contacted at this point in the review if questions about the PMN arise. On approximately days 9–13, the Structure Activity Team (SAT) meeting occurs. At this meeting, additional OPPT experts evaluate the PMNs utilizing SAR data, PMN data, and the information in the chemistry reports compiled at the CRSS meeting. All the PMNs are given hazard potential ratings for health effects, environmental effects and environmental fate. An exposure release profile is developed on days 10–19. Again, OPPT experts look at each PMN, and by using the information in the PMN on process, exposure, and production volume, develop a profile of exposures and releases from manufacture, processing and use, including: occupational exposure/releases, environmental releases, consumer exposure, and ambient or general population exposure. Twice a week a "Focus Meeting" is held and a PMN is reviewed. At this multidisciplinary risk management meeting, decisions are made ranging from "dropping" a chemical from further review to banning a chemical pending further information. Decisions are based on information compiled by the CRSS, SAT, and exposure reviews, as well as consideration of related cases and other relevant factors. If more information is needed to make a decision on a PMN, the submitter may be contacted for questions/clarifications, and/or the PMN may be placed into Standard Review. A standard review goes through days 21–85 of the review period and is a detailed risk assessment of the PMN chemical.

Following the 90-day review period, if EPA takes no action, the submitter may begin manufacturing or importing the chemical. A "Notice of Commencement" (NOC) must be submitted to EPA within 30 days of first manufacture (including importation). Following receipt of the NOC, the chemical substance is added to the inventory. Once a substance is listed on the TSCA Inventory, it is considered an existing chemical.

Other possible outcomes of the PMN review process may include one or more of the following:

1. Voluntary withdrawal of the notice, often (but not always) in the face of possible EPA action.

2. Issuance of TSCA §5(e) Orders. EPA may negotiate a TSCA §5(e) Consent Order to prohibit or limit activities associated with the new chemical if EPA determines that insufficient information exists to evaluate the human health and environmental effects of the substance, and that: (a) it may present an unreasonable risk ("risk-based finding") or (b) be produced in substantial quantities, and substantial or significant exposure/release ("exposure-based finding"). TSCA §5(e) orders typically include: exposure or release mitigation, testing, labeling and hazard communication, and record keeping. Evaluating substitutes for ozone depleting substances (ODSs) is one example where the 5(e) process was applied.

3. TSCA §5(a)(2) Significant New Use Rules (SNURs). §5(e) Consent Orders are only binding on the original PMN submitter that manufactured or imported the substance. Consequently, after signing a §5(e) Consent Order, EPA may promulgate a Significant New Use Rule (SNUR) under TSCA §5(a)(2) that mimics the Consent Order to bind all other manufacturers and processors of

former new chemicals to the terms and conditions contained in the Consent Order. Also, EPA has the authority to issue SNURs without a §5(e) Consent Order. Under TSCA §5(a)(2), EPA can determine that a use of a chemical is a significant new use after considering several factors, including but not limited to the projected production and processing volume of the chemical substance, and the anticipated extent to which the use increases the type, form, magnitude and duration of exposure to humans or the environment associated with the new use. The SNUR requires that manufacturers, importers, and processors of such substances notify EPA at least 90 days before beginning any activity that EPA has designated as a "significant new use" (40 CFR 721). The notification required by SNURs allows EPA to prevent or limit potentially adverse exposure to, or effects from, the new use of the substance. Such a SNUR would require the submission of a Significant New Use Notification (SNUN) 90 days prior to commercial manufacture not conforming to the conditions of the SNUR.

4. TSCA §5(f) actions. If EPA determines that the manufacturing, processing, distribution in commerce or disposal of a substance that is the subject of PMN or SNUR notification requirements presents or will present an unreasonable risk before a TSCA §6 rule can be promulgated under TSCA §5(f), EPA may (a) limit the amount or impose other restrictions on the substance via an immediately effective proposed rule, or (b) prohibit the manufacturing, processing by applying to a U.S. District Court for an injunction.

5. Voluntary Testing Actions (TSCA §5(e) Regulation Pending Development of Information). In a limited number of cases, PMN submitters voluntarily agree to suspend the notice review period and conduct hazard or environmental fate testing in response to a request from EPA. During the PMN review process, OPPT might find risks that cannot be mitigated by controls. The "voluntary" testing is performed during the 90-day review period with a suspension(s) until the testing is completed. The submitter must decide if it is economically feasible to do the testing before going into the marketplace. Submitters may also take the option of withdrawing instead of performing the testing.

EPA has received approximately 36,600 PMNs from 1979 to the present.

The PMN submissions contain information of future commercial activities of new substances; therefore, it is common to find CBI claims in them. For example, in 1990 approximately 90% of the PMNs submitted claimed the chemical identification as CBI. However, for those substances that complete the PMN process and enter in commerce (i.e., those for which NOCs have been received), the chemical identification CBI claim rate drops to approximately 65% (based on NOC statistics from 1995–1999).

There are several exemptions from filing a PMN for inventory listing. Two are required by the statue:

1. The Test Market Exemption (TME) is established at TSCA §5(h)(1), and its implementing regulations are at 40 CFR 720.38;

2. The Research and Development Exemption (R&D) is established at TSCA §5(h)(3), and its implementing regulations are at 40 CFR 720.36 and .78 for commercial R&D, and 40 CFR 720.30(i) for non-commercial R&D.

TSCA §5(h)(4) gives the Administrator the authority to exempt manufacturers from some or all of the requirements of TSCA §5 upon a determination that the intended activities with the substances will not present an unreasonable risk to health or the environment. The Agency has established eligibility criteria for three exemptions based on §5(h)(4): the Low Volume Exemption (LVE), implementing regulations at 40 CFR 723.50(c)(1), the Low Release and Exposure Exemption (LOREX), implementing regulations at 40 CFR 723.50(c)(2), and the Polymer Exemption (PE), implementing regulations at 40 CFR 723.250.

Written submissions and Agency review/approval are required for the TME, the LVE, and the LOREX. If the R&D and Polymer exemptions are based on the user's determination that they meet the requirements of the exemption, no review/approval need be sought from the Agency, though a user is required to report to the Agency that the Polymer Exemption has been used (an earlier version of the Polymer Exemption did require a request for permission, and polymers reported under that program were listed in the inventory with "Y" status).

LTSCA.4 INFORMATION GATHERING AUTHORITY

Discuss TSCA's Section 8(a) Information Gathering Authority.

SOLUTION: TSCA Section 8(a) gives EPA the broad authority to require (by rulemaking) manufacturers (including importers) and processors of chemical substances to maintain records and/or report such data as EPA may reasonably require to carry out the TSCA mandates.

Examples of information that can be required to be reported include:

1. Chemical or mixture identity
2. Categories of use
3. Quantity manufactured or processed
4. By-product description
5. Health and environmental effects information
6. Number of individuals exposed
7. Method(s) of disposal

Section 8(a) regulations can be tailored to meet unique information needs (e.g., via chemical-specific rules) or information can be obtained via the use of "model" or standardized reporting rules. One example of a model TSCA Section 8(a) reporting rule is the "Preliminary Assessment Information Rule" (or PAIR).

Under PAIR, producers and importers of a listed chemical are required to report the following site-specific information on a two page form:

1. Quantity of chemical produced and/or imported
2. Amount of chemical lost to the environment during production or importation
3. Quantity of enclosed, controlled and open releases of the chemical
4. Per release, the number of workers exposed and the number of hours exposed

Exemptions for such reporting are as follows:

1. Production or importation for the sole purpose of research and development (R&D).
2. Production or importation of less than 500 kilograms during the reporting period at single plant site.
3. Companies whose total annual sales from all sites owned by the domestic or foreign parent company are below $30 million for the reporting period and who produced or imported less than 45,400 kilograms of the chemical.
4. Production or importation of the listed chemical solely as an impurity, a non-isolated intermediate, and under certain circumstances as a by-product.

Additional details can be found at:
http://www.epa.gov/oppt/itc/pubs/sect8a.htm.
http://www.epa.gov/oppt/chemtest/pubs/pairform.pdf.

LTSCA.5 ALLEGATIONS OF SIGNIFICANT ADVERSE REACTIONS

Discuss the Allegations of Significant Adverse Reactions under TSCA Section 8(c)

SOLUTION: Under TSCA Section 8(c), companies can be required to record, retain and in some cases report "allegations of significant adverse reactions" to any substance/mixture that they produce, import, process, or distribute. EPA's TSCA Section 8(c) rule requires producers, importers, and certain processors of chemical substances and mixtures to keep records concerning significant adverse reaction allegations and report those records to EPA upon notice in the *Federal Register* or upon notice by letter. The TSCA Section 8(c) rule also provides a mechanism to identify previously unknown chemical hazards in that it may reveal patterns of adverse effects which may not be noticed or detected. Further information is available under 40 CFR Part 717.

An "Allegation" is defined as "a statement, made without formal proof or regard for evidence, that a chemical substance or mixture has caused a significant adverse reaction to health or the environment."

"Significant adverse reactions" are defined as "reactions that a indicate a substantial impairment of normal activities, or long lasting or irreversible damage to health or the environment."

Any person can make a written or verbal allegation. Verbal allegations must be transcribed either by the company or the individual making the allegation (if transcribed by the individual, they must be signed). To be recordable, allegations must implicate a substance that caused the reaction by naming either the specific substance, a mixture or article containing the substance, or a company process in which substances are involved, or by identifying a discharge from a site of manufacture, processing, or distribution of the substance.

Examples of significant adverse reactions include:

1. Long-lasting or irreversible damage to human health
2. Partial or complete impairment of bodily functions
3. Impairment of normal activity by all/most persons exposed at one time/each time an individual is exposed
4. Gradual or sudden changes to animal or plant life in a given geographic area
5. Abnormal numbers of deaths/changes in behavior or distribution of organisms
6. Long lasting or irreversible contamination of the physical environment

Allegations that are "exempt" from the requirement of the TSCA Section 8(c) rule include:

1. Those alleging "known human effects"
2. Allegations involving adverse reactions to the environment if the alleged cause can be directly attributable to an incident of environmental contamination that has already been reported to the U.S. Government under any applicable authority
3. Anonymous allegations

TSCA Section 8(c) records must be kept at a company's headquarters or at a site central to their chemical operations. The record must contain the following information:

1. The original allegation as received
2. An abstract of the allegation
3. The results of any self-initiated investigation regarding the allegation
4. Copies of any further required information regarding the allegations (e.g., copies of any reports required to be made to the U.S. Occupational Safety and Health Administration)

TSCA Section 8(c) records must be retrievable by the alleged cause of the reaction (i.e., specific chemical identity, mixture, article company process or operation, or site operation, or site emission, effluent, or discharge).

An allegation made by an employee must be kept by the company for 30 years while all other allegations (e.g., those made by plant site neighbors or customers) must be kept by the company for 5 years.

Additional details are available at:

http://www.epa.gov/oppt/chemtest/pubssect8c.htm

LTSCA.6 UNPUBLISHED HEALTH AND SAFETY STUDIES

Discuss TSCA Section 8(d) Revisions concerned with Unpublished Health and Safety Studies

SOLUTION: Key elements of these revisions are:

1. Persons who must report (unless otherwise specified) include chemical producers and importers under the NAICS Codes Subsection 325 (chemical manufacturing and allied products) and Industry Group 32411 (petroleum refiners).
2. Reporting period for studies for a listed chemical substance or listed mixture will terminate 60 days after the effective date of the listing.
3. Studies to be reported will be specified by EPA to include the specific type(s) of health and safety data needed; the chemical grade/purity of the test material (studies involving mixtures are not required unless otherwise specified).
4. Initiated studies are reportable only for study initiation that occurs during the 60 day reporting period.
5. Adequate file searches encompass reportable information data on or after January 1, 1977 (unless otherwise specified).

Under TSCA Section 8(d), EPA has the authority to promulgate rules to require producers, importers, and processors to submit lists and/or copies of ongoing and completed, unpublished health and safety studies. EPA's TSCA Section 8(d) "Health & Safety Data Reporting Rule" was developed to gather health and safety information on chemical substances and mixtures needed by EPA to carry out its TSCA mandates (e.g., to support OPPT's Existing Chemicals Program and Chemical Testing Program and to set priorities for TSCA risk assessment/management activities). EPA has also used its TSCA Section 8(d) authority to gather information needed by other EPA Program Offices and other Federal Agencies. Chemicals that are designated or recommended for testing by the TSCA Interagency Testing Committee (ITC) may be added to the rule via immediate final rulemaking (up to 50 substances/year). Non-ITC chemicals can be added to the Section 8(d) rule via notice and comment rulemaking. Further information is available under 40 CFR Part 716.

Persons who must report under the TSCA Section 8(d) rule include:

1. Current as well as prospective producers, importers, and (if specified) processors of the subject chemical(s); and
2. Person(s) who, in the 10 years preceding the effective data that a substance or mixture is added to the rule, either had proposed to produce, import, or (if specified) process, or had produced, imported, or processed (if specified) the substance or listed mixture.

Once a chemical substance or mixture is added to the rule, reporting obligations terminate (i.e., sunset) no later than 2 years after the effective date of the listing of the substance or mixture, or on the removal of the substance or mixture from the rule.

Unpublished studies on listed substances or mixtures are potentially reportable (i.e., studies may be subject to either copy submission requirements or listing requirements). Generally, copies of studies possessed at the time a person becomes subject to the rule must be submitted, and the following categories of studies must be listed:

1. Studies ongoing as of the date a person becomes subject to the rule (copies must be submitted when completed);
2. Studies initiated after the date a person becomes subject to the rule (copies must be submitted when completed);
3. Studies which are known as of the date a person becomes subject to the TSCA Section 8(d) rule, but not possessed; and,
4. Studies previously sent to U.S. Government Agencies without confidentiality claims.

The term "health and safety study" is intended to be interpreted broadly and means "any study of any effect of a chemical substance or mixture on health or the environment or on both," including but not limited to:

1. Epidemiological or clinical studies
2. Studies of occupation exposure
3. In vivo and in vitro toxicological studies, and
4. Ecotoxicological studies

Additional information is available at:
http://www.epa.gov/oppt/chemtest/pubssect8d.htm

LTSCA.7   SUBSTANTIAL RISK INFORMATION

Discuss Substantial Risk Information under TSCA Section 8(e).

SOLUTION: TSCA Section 8(e) is a self-implementing statutory provision that states:

"Any person who manufactures, [(includes imports)] processes or distributes in [U.S] commerce a chemical substance or mixture, and who obtains information which reasonably supports the conclusion that such substance or mixture presents a substantial risk of injury to human health or the environment, shall immediately inform the [EPA] Administrator of such information unless such person has actual knowledge that the Administrator has been adequately informed of such information."

The term "substantial risk" information refers to that information which offers reasonable support for a conclusion that the subject chemical or mixture poses a substantial risk of injury to health or the environment and need not, and typically does not, establish conclusively that a substantial risk exists.

Additional information is available at:

TSCA 8(e) Website http://www.epa.gov/oppt/chemtest/pubssect8e.htm.

LTSCA.8 SPECULATE ON HOW NEW LEGISLATION AND RULEMAKING MAY BE NECESSARY FOR ENVIRONMENTAL CONTROL/CONCERN WITH NANOTECHNOLOGY

Speculate on the impact TSCA may have on nanotechnology.

SOLUTION: Completely new legislation and regulatory rulemaking will almost certainly be necessary for environmental control of nanotechnology. However, in the meantime, one may speculate on how the existing regulatory framework might be applied to the nanotechnology area as this emerging field develops over the next several years.

As indicated earlier in Problems LTSCA.3–7, commercial applications of nanotechnology are likely to be regulated under TSCA, which authorizes EPA to review and establish limits on the manufacture, processing, distribution, use, and/ or disposal of new materials that EPA determines to pose "an unreasonable risk injury to human health or the environment." The term *chemicals* is defined broadly by TSCA. Unless qualifying for an exemption under the law [R&D (a statutory exemption requiring no further approval by EPA), low-volume production, low environmental releases along with low volume, or plans for limited test marketing], a prospective manufacturer is subject to the full-blown PMN procedure. This requires submittal of said notice, along with toxicity and other data to EPA at least 90 days before commencing production of the chemical substance.

Approval then involves record keeping, reporting, and other requirements under the statute. Requirements will differ, depending on whether EPA determines that a particular application constitutes a "significant new use" or a "new chemical substance." EPA can impose limits on production, including an outright ban when it is deemed necessary for adequate protection against "an unreasonable risk of injury to health or the environment." EPA may revisit a chemical's status under TSCA and change the degree or type of regulation when new health/ environmental data warrant. EPA is expected to be issuing several new TSCA test rules in the future. However, given EPA's past history, one of the authors (LT) believes it will probably not occur in this decade.

Key sites and publications that relate to EPA discussions on nanotechnology follow:

1. EPA Nanotechnology and Environment: Applications and Implications STAR Progress Review Workshop, August 28–29, 2002, Arlington, Virginia available at: http://es.epa.gov/ncer/publications/workshop/nano_proceed.pdf.
2. U.S. Environmental Protection Agency: Nanotechnology White Paper (External Review Draft). December 2, 2005. Nanotechnology has the potential to change and improve many sectors of American industry, from consumer products to health care to transportation, energy and agriculture. [http://es.epa.gov/ncer/nano/publications/whitepaper12022005.pdf] (PDF, 134pp., 738KB) [Note appendix C Additional Detailed Risk Assessment Information].
3. Nanotechnology: An EPA Perspective Factsheet available at: http:// es.epa.gov/ncer/nano/factsheet/nano_factsheet.pdf (PDF, 2 pages).
4. National Nanotechnology Initiative at http://www.nano.gov.

The reader is also referred to the following two texts for additional information:

1. L. Theodore and R. Kunz, "Nanotechnology Environmental Implications and Solutions," John Wiley & Sons, Hoboken, NJ, 2005.
2. L. Theodore, "Nanotechnology Basic Calculations for Engineers and Scientists," John Wiley & Sons, Hoboken, NJ, 2006.

LTSCA.9 FORMULAE OF TOXIC AIR COMPOUNDS

What are the structural and molecular formulae for the following toxic air compounds?

1. Tetrachloroethene
2. Formaldehyde
3. Carbon tetrachloride
4. Benzene
5. 2,3,7,8-Tetrachlorodibenzodioxin (TCDD)

SOLUTION: Molecular Formulae:

1. Tetrachloroethene: C_2Cl_4
2. Formaldehyde: CH_2O
3. Carbon tetrachloride: CCl_4

4. Benzene: C_6H_6

5. 2,3,7,8-tetrachlorodibenzodioxin: $C_{12}H_4O_2Cl_4$

Structural Formulae:

1. Tetrachloroethene:

```
Cl   Cl
 \   |
  C=C
 /   \
Cl    Cl
```

2. Formaldehyde:

```
    O
    |
H—C—H
```

3. Carbon tetrachloride:

```
     Cl
     |
Cl—C—Cl
     |
     Cl
```

4. Benzene:

```
     H   H
     |   |
     C—C
    //    \\
H—C       C—H
    \     /
     C=C
     |   |
     H   H
```

5. 2,3,7,8-tetrachlorodibenzodioxin:

```
     Cl    H
     |     |
     C  O  C
    /\\/ \\/ \
Cl—C  C  C  C—Cl
    ||  |  |  ||
H—C  C  C  C—Cl
    \\// \ /\\//
     C  O  C
     |     |
     H     H
```

LTSCA.10 DIOXIN/FURAN CHEMICALS

Describe the dioxin/furan family of chemicals.

SOLUTION: Dioxin is a term used to describe a large group of chemical compounds having a similar basic structure. The most common subgroup of dioxin compounds (of which there are 75 different varieties) comprises those that include chlorine atoms. A few are toxic; the most toxic is 2,3,7,8-tetrachlorodibenzo(para) dioxin (2,3,7,8-TCDD). There are also 135 polychlorinated-dibenzofuran compounds. These are known as "furans."

These compounds are chlorinated tricyclic aromatic compounds. Each of these compounds has a triple-ring structure consisting of two benzene rings interconnected to each other by, respectively, one or two oxygen atoms. The number of chlorine atoms can vary between 1 and 8. The PCDD and PCDF compounds are referred to as "congeners"—a specific member of a group of structurally related compounds.

The most widely discussed chemical, 2,3,7,8-TCDD, is a colorless crystalline solid, is slightly soluble in water, and binds strongly to solids and particulate matter.

The known natural sources of PCDDs and PCDFs are related to fires and combustion processes. Thus, these toxic compounds may be produced from forest fires, by lightning or volcanic action. The formation of PCDD/PCDFs is very dependent on the presence of oxygen, carbon, chlorine, and heat. Since fire produces PCDD/PCDFs, combustion devices may have emissions of these compounds. These include municipal waste incinerators, industrial waste incinerators, medical waste combustors, sewage sludge incineration, wood and coal combustion, oil combustion including motor vehicle engines and accidental fires.

Additional details on 2,3,7,8-TCDD are provided in the next problem.

LTSCA.11   2,3,7,8-TETRACHLORODIBENZO-P-DIOXIN INFORMATION

Provide key information on the above toxic chemical.

SOLUTION: The information below was adapted from J. Spero, B. Devito, and L. Theodore, "Regulatory Chemicals Handbook", Marcel Dekker (recently acquired by CRC Press/Taylor & Francis Group, Boca Raton, FL), New York City, 2000.

2,3,7,8-Tetrachlorodibenzo-p-Dioxin ($C_{12}H_4Cl_4O_2$, 321.96)

CAS/DOT #: 1746-01-6

Synonyms: Dioxin, dioxine, TCBDB, TCDD, 2,3,7,8-TCDD, tetradioxin.

Physical Properties: White, crystalline solid or colorless needles; slightly soluble in water; BP (412.2°C, 774°F); MP (305°C, 581°F); DN (1.827 g/mL at 20°C); VP (1.52E-09 mm Hg at 25°C).

Chemical Properties: Caustic.

Exposure Routes: Inhalation (fly ash, soot particles, flue gases, ambient air, incineration fumes, herbicides and wood dust), ingestion (urban vegetation, fish and cow's milk), occupational exposure in pulp and paper, wood industries.

Human Health Risks: Probable human carcinogen; Acute Risks: irritation of skin and eyes; tightness in chest; dizziness; headache; nausea; allergic dermatitis; hepatic necrosis; thymic atrophy; hemorrhage; chloracne; Chronic Risks: skin lesions; chloracne; severe weight loss; pancreatic, bronchogenic carcinoma; gastric ulcers; delayed death.

Hazard Risk: Most toxic member of the dioxin family; caustic and corrosives.

Measurement Methods: Not available.

Major Uses: Byproduct of herbicides, defoliants and Agent Orange; research chemicals; wood preservative (not commercially).

Storage: Not available.

Fire Fighting: Not available.

Personal Protection: Wear gastight suit and viton7 rubber gloves; wear approved chemical safety goggles; material should be handled or transferred in an approved fume hood or with adequate ventilation; electrically ground all equipment when handling this product; a NIOSH approved air supplied respirator is recommended in absence of proper environmental controls; maintain eyewash baths and safety showers in work area.

Spill Clean-up: Consider evacuation; contain release and eliminate its source, if this can be done without risk; remove any sources of ignition until the area is determined to be free from explosion or fire hazards.

Health Symptons: Inhalation (headache, dizziness, hallucinations, changes in motor activity, nausea, respiratory irritation); skin (prickling, allergic dermatitis); eyes (severe irritation); ingestion (dizziness, headache, nausea, drowsiness, tightness of chest).

General Comments: Oral rat LD_{50} 20 g/kg; First aid: immediately wash eyes with large amounts of water; if skin contact occurs remove clothing and flush skin with large amounts of water and soap; if inhaled, remove to fresh air and provide respiratory assistance as needed.

In addition to Spero et al., the reader may refer to following references for specific details on each of the chemicals.

1. R. Lewis, "Sax's Dangerous Properties of Industrial Materials," 9th Edition, Van Nostrand Reinhold, New York City, 1996.
2. "Suspect Chemicals Sourcebook," Roytech Publications, Bethesda, Maryland, 1996.

LTSCA.12 DIOXIN-FURAN STATEMENTS

Which one of the following statements is true of the organic pollutants 2,3,7,8-tetrachloro-p-dioxin and 2,3,7,8-tetrachlorodibenzofuran?

1. They are formed in post-furnace reactions in cyclones and electrostatic precipitators.
2. They are found in municipal solid waste incinerators at higher levels in summer than in winter.
3. The average ambient air outside the municipal solid waste incinerator facility contains the same amount of these pollutants as the human work area inside the facility.
4. The mechanism of their formation is known.

SOLUTION:

1. *True*. Combustion units are a primary source of dioxins and furans in the workplace. They are probably emitted from the back pressure or leakage from boilers.

2. *False*. There is more ventilation in municipal solid waste incinerators during the summer because of open windows and air conditioners; thus, there is a lower level (lower concentration) of pollution indoors in the summer than in the winter.

3. *False*. The concentration of dioxins and furans inside solid waste incinerator facilities is significantly more than is found in the ambient air sampled in several locations throughout Ohio (where a study was conducted).

4. *False*. The exact mechanism of formation was not known at the time of the preparation of this manuscript.

Therefore, the correct answer is 1.

LTSCA.13 ARSENIC INFORMATION

A priority water pollutant (PWP) of major corcern to the government, industry and public is arsenic. Provide key information including regulatory states, on this toxic chemical.

SOLUTION: Adapted from J. Spero, B. Devito, and L. Theodore, "Regulatory Chemicals Handbook," Marcel Dekker (recently acquired by CRC Press/Taylor & Francis Group, Boca Raton, FL), New York City, 2000.

Arsenic (As, 74.92)

CAS/DOT Indentification #: 7440-38-2UN 1558

Synonyms: Arsen, Fowler's solution, grey arsenic, colloidal arsenic.

Physical Properties: Gray, crystalline material; soluble inorganic arsenate (arsenic trioxide) predominates under normal conditions; MP (817°C @ 28 atm); BP (613°C); SG (5.727); VP (1 mmHg @ 372°F).

Chemical Properties: Forms a complete series of trihalides; can react strongly with strong oxidizers, forms highly toxic fumes on contact with acids or active metals.

Biological Properties: Surface water samples have concentrations ranging from 5 to 336 µg/L; seawater concentration: 3.0 µg/L; highly persistent in water, with a half-life of more than 200 days; can be detected in water by digestion followed by silver diethyldithiocarbamate, atomic adsorption, or inductively coupled plasma optical emission spectrometry.

Bioaccumulation: Bioconcentrates in both fresh and saltwater organisms; does not accumulate in plants to toxic levels; concentration found in fish tissues is expected to be somewhat higher than the average concentration found in the water the fish were taken.

Origin/Industry Sources/Uses: Naturally occurring element; coal fuel power plants; manufacturing of glass, cloth, and electrical semiconductors;

fungicides; wood preservatives; growth stimulates for plants and animals; veterinary medicine.

Toxicity: high acute toxicity to aquatic life.

Exposure Routes: found in air and all living organisms; people living within 12 miles of copper, zinc, and lead smelters may be exposed to 10 times the USA avg. atmospheric levels; 40,000 people living near some copper smelters may be exposed 100 times the national atmospheric avg.; inhalation and ingestion of dust and fumes.

Regulatory Status: Criterion to protect freshwater aquatic life: 57 µg/L/24 hr avg., concentration not to exceed 130 µg/L any time, total recoverable trivalent inorganic arsenic never to exceed 440 µg/L; Criterion to protect saltwater aquatic life: 29 µg/L/24 hr avg., concentration not to exceed 67 µg/L any time, 508 µg/L based on acute toxicity; Criterion to protect human health: maximum allowable level: 50 µg/L, preferably; concentrations calculated to keep the cancer risk level below 10^{-5}, 10^{-6}, and 10^{-7} are 0.02. 0.002, and 0.0002 µg/L respectively; Maine has set a guideline in drinking water of 0.05 mg/L.

Probable Fate: *photolysis*: not an important process; *oxidation*: under reducing conditions, it is a stable solid, dissolved arsenic is present in oxygenated water; *hydrolysis*: all arsenic halides hydrolyze in presence of water; hydrolyzed to arsenious and arsenic acid forms (soluble); *volatilization*: not important under natural redox conditions; *sorption*: removed by clays, iron oxides, manganese oxides, and aluminum; *biological processes*: bioaccumulated, but not biomagnified, biotransformed to organic arsenicals.

Treatability/Removability (Process, Removable Range (%), Avg. Achievable Conc. (µg/L)); Gravity oil separation, not available, 46; Gas flotation with chemical addition (calcium chloride, polymer), > 28–80, <8.5; Gas flotation with chemical addition (alum, polymer), 56, 3.5; Filtration, 31– > 99, 28; Sedimentation, 68– > 99, 72; Sedimentation with chemical addition (lime, polymer), 37–75, 10; Sedimentation with chemical addition (Fe^{2+}, lime), >69–99, <2; Sedimentation with chemical addition (sulfide), >99, 5; Sedimentation with chemical addition ($BaCl_2$), 17– > 33, <8.5; Sedimentation with chemical addition (alum, polymer), 29, 12; Sedimentation with chemical addition (alum), 19– > 37, 32; Sedimentation with chemical addition (lime), 60– > 99, <16; Ozonation, 24–48, 23; Activated sludge, >43– > 96, 35; Granular activated carbon adsorption, 21– > 99, 11; Reverse osmosis, 79– > 99, 7.7.

In addition to Spero et al., the reader may refer to following reference for specific details on additional chemicals.

1. R. Lewis, "Sax's Dangerous Properties of Industrial Materials," 9th Edition, Van Nostrand Reinhold, New York City, 1996.
2. "Suspect Chemicals Sourcebook," Roytech Publications, Bethesda, Maryland, 1996.

LTSCA.14 TOXIC SUBSTANCE DEFINITION

Which of the following statements is *not* a definition of a "toxic substance"?

1. A substance which has an immediate or long-term adverse effect on the environment.
2. A substance that enters a living organism and metabolizes into its component derivatives.
3. A substance constituting or that may constitute a danger to the environment on which human life depends.
4. A substance constituting or that may constitute a danger in the United States to human life or to health.

SOLUTION: The answer is 2. All of the others (1, 3, and 4) are correct definitions of a "toxic substance."

LTSCA.15 TOXIC PROCESS FACTORS

When deciding what process, method, and equipment to use in controlling a specific toxic problem, a number of environmental, engineering, and economic factors should be considered. Give at least two examples for each of these three factors.

SOLUTION: A number of typical responses are provided below:
Environmental Factors to be Considered:

1. The amount of pollutant that the regulations allow to be emitted.
2. Whether or not the control equipment will produce another toxic waste that will require further treatment or disposal.

Engineering Factors to be Considered:

1. The chemical and physical properties of the toxic substance, such as reactivity, flammability, corrosiveness, toxicity and state of matter.
2. Design and operational requirements such as power and water needs, temperature range limits, maintenance needs, reliability and dependability, size and how readily the equipment could be adapted to meet a lower emission limit as might be required by a change in the regulations.

Economic Factors to be Considered:

1. The initial cost of buying and installing the equipment.
2. The cost of operating the equipment.

QUANTITATIVE PROBLEMS (TTSCA)

PROBLEMS TTSCA.1–6

TTSCA.1 TOXIC CONCENTRATIONS

Atmospheric concentrations of toxic pollutants are usually reported using two types of units: 1) mass of pollutant per volume of air, i.e., mg/m^3, $\mu g/m^3$, ng/m^3, etc.; or 2) parts of pollutant per parts of air (by volume), i.e., ppmv, ppbv, pptv, etc. In order to compare data collected at different conditions, actual concentrations are often converted to Standard Temperature and Pressure (STP). According to EPA, STP conditions for atmospheric or ambient sampling are often either 0°C or 25°C and 1 atm.

Assume the concentration of chlorine vapor is measured to be 15 mg/m^3 at a pressure of 600 mmHg and at a temperature of 10°C.

1. Convert the concentration units to ppmv.
2. Calculate the concentration in units of mg/m^3 at STP (0°C, 1 atm).

SOLUTION:

1. To convert the concentration units to ppmv, the molecular weight of chlorine must first be obtained. Chlorine gas is Cl_2 with 2 gmol of Cl atom per Cl_2 molecule so chlorine gas has a molecular weight of:

$$Cl_2(MW) = 2(35.45\,g/gmol) = 70.9\,g/gmol$$

The absolute temperature of the gas is:

$$T = 10°C + 273.2°C = 283.2\ K$$

The pressure of the gas in atmospheres is:

$$P = 600\,mmHg/760\,mmHg = 0.789\,atm$$

The concentration of chlorine gas in ppmv can be determined by taking the mass per unit volume value, converting the chlorine gas mass to volume, and then expressing this concentration as volume per million volumes as follows. Take as a basis 1.0 m^3 of volume of gas and note that 15 mg = 0.015 g. Thus,

$$Cl_2(Volume) = \frac{(Cl_2 Mass)(R)(T)}{(MW)(P)}; \quad R = \text{ideal gas law constant}$$

$$= \frac{(0.015\ g)(0.082\ atm \cdot L/gmol \cdot K)(283.2\ K)}{(70.9\ g/gmol)(0.789\ atm)}$$

$$= 0.00623\ L = 6.23\,mL$$

Since there are 10^6 mL in a m^3, the concentration in ppmv is expressed as mL/m^3 or $mL/10^6$ mL, or:

$$\text{Concentration} = (6.23\,\text{mL})/1\,\text{m}^3 = 6.23\,\text{ppmv}$$

2. Knowing that the concentration of Cl_2 on a volume basis does not change with temperature, at STP there are 6.23 mL $= 0.00623$ L of Cl_2 in 1 m^3 of air. From the ideal gas law, at STP 0°C and (1 atm) these 6.23 mL represent the following mass of Cl_2:

$$Cl_2(\text{Mass}) = \frac{(Cl_2\,\text{Volume})(P)(MW)}{(R)(T)}$$

$$= \frac{(0.00623\,\text{L})\,(1\,\text{atm})\,(70.9\,\text{g/gmol})}{(0.082\,\text{atm}\cdot\text{L/gmol}\cdot\text{K})\,(273.2\,\text{K})}$$

$$= 0.0197\,\text{g} = 19.7\,\text{mg}$$

The concentration of Cl_2 in mass per unit volume units is then:

$$\text{Concentration} = 19.7\,\text{mg}/1\,\text{m}^3 = 19.7\,\text{mg/m}^3$$

The reader is left the exercise of explaining why this result is different from the 15 mg/m^3 value given in the Problem statement.

TTSCA.2 TOXIC SAMPLING

To sample for toxics in the atmosphere, air may be pumped through an adsorption tube during a sampling period. A solvent is then flushed through the tube to desorb the compounds from the adsorbent, and gas chromatography/mass spectrometry (GC/MS) analysis is used to identify and quantity the VOCs dissolved in the solvent.

Assume that a smoker's breath was drawn through an adsorbent tube at a rate of 498 mL/min for a period of 159 s at a temperature of 22°C and at a pressure of 624 mmHg. If the concentration of benzene in the smoker's breath was found to be 10.1 ppbv, what mass of benzene, in ng, was recovered in the solvent?

SOLUTION: First, the volume of the smoker's breath that was sampled is calculated from the product of the sampling rate and the duration of sampling as:

$$\text{Sample volume} = (\text{Sampling rate})(\text{Sampling duration})$$
$$= (498\,\text{mL/min})(159\,\text{s})(1\,\text{min/60 s})$$
$$= 1,320\,\text{mL} = 0.00132\,\text{m}^3$$

The volume of benzene collected is based on the volume concentration of benzene of 10.1 ppbv = 0.0101 ppmv or:

$$0.0101 \ \text{ppmv} = (\text{Benzene volume, mL})/(0.00132 \, \text{m}^3)$$

$$\text{Benzene volume} = (0.0101)(0.00132) = 1.33 \times 10^{-5} \text{mL}$$

$$= 1.33 \times 10^{-8} \text{L benzene vapor}$$

This volume of benzene represents the following weight based on the ideal gas law:

$$\text{Benzene mass} = \frac{\left(\dfrac{624 \, \text{mmHg}}{760 \, \text{mmHg/atm}}\right)(1.33 \times 10^{-8} L)(78 \, \text{g/gmol})}{(0.082 \ \text{atm} \cdot \text{L/gmol} \cdot \text{K})(295.2 \ \text{K})}$$

$$= 3.52 \times 10^{-8} \text{g}$$

$$= 35.2 \, \text{ng}$$

The reader is left the exceruse of explering why the results in different fine the 15 m/m^3 value give in the problem statement.

TTSCA.3 RENOVATION OF AN OLD WAREHOUSE TO ARTISTS' STUDIOS

An old warehouse constructed of concrete walls and floors is being renovated and will be subdivided into artists' studios. The new owners also plan to install a child-care facility for the children of the artists. The concrete floor which is contaminated with PCBs must be cleaned up in compliance with the appropriate cleanup standard prior to use. What clean-up level is required?

SOLUTION: The converted warehouse will be used as a high occupancy area, i.e., the artists and/or children will be occupying the building for 6.7 hours per week or more. The flooring is a porous surface; therefore, the standard applicable for bulk PCB remediation waste applies. The concrete floor must be removed, at least in part, and replaced if it cannot be decontaminated to required levels (i.e., cleaned up to <1 ppm PCBs). The material contaminated with PCBs must be disposed of as PCB remediation waste. Disposal options for non-liquid cleanup wastes at any concentration (e.g., cleaning materials, personal protective equipment, non-porous surfaces, etc.) and bulk PCB remediation wastes including porous surfaces at <50 ppm include: an approved PCB disposal facility, a permitted municipal solid waste or non-municipal non-hazardous waste facility pursuant to EPA requirements for PCB remediation wastes (40 CFR 761.61(a) or (c)), or a RCRA permitted hazardous waste landfill. Disposal of >50 ppm PCB remediation waste is limited to an approved PCB disposal facility or a RCRA permitted hazardous waste landfill. Refer to "Polychlorinated Biphenyl (PCB) Site Revitalization Guidance under the Toxic Substances Control Act (TSCA)," November 2005 for additional details.

TTSCA.4   CALCULATION INVOLVING MCL AND RfD

The drinking water maximum contaminant level (MCL) set by the EPA for atrazine is 0.003 mg/L and its reference dose (RfD) is 3.5 mg/(kg · day). How many liters of water containing atrazine at its MCL would a worker have to drink each day to exceed the RfD for this triazine herbicide?

SOLUTION: As with most of these calculations, it is assumed that those exposed can be represented by a 70-kg individual. The volume rate, q, of drinking water at the MCL to reach the RfD for atrazine is (using the standard RfD equation)

$$
\begin{aligned}
q &= (\text{RfD})(\text{W})/\text{MCL} \\
&= [3.5\,\text{mg}/(\text{kg}\cdot\text{day})](70\,\text{kg})/(0.003\,\text{mg/L}) \\
&= 81{,}700\,\text{L/day}
\end{aligned}
\tag{7.1}
$$

This is a surprisingly high volume of water on a daily basis. This large volume indicates that there is considerable uncertainty (i.e., the product of the uncertainty factors is large) in estimating a reference dose for atrazine. However, based on the MCL and RfD, there appears to be no problem.

TTSCA.5   INHALABLE PCBs AND PAHs

Several industrial facilities emit inhalable pollutants ethylene oxide (EO), polychlorobiphenyls (PCBs) and polycyclic aromatic hydrocarbons (PAHs). The worker's annual average exposure concentration of EO, PCBs, and PAHs are $10\,\mu\text{g/m}^3$, $2\,\mu\text{g/m}^3$, and $5\,\mu\text{g/m}^3$, respectively.

Calculate the cancer risk caused by each pollutant and the total cancer risk. Express results in additional cancer cases per million people. Use the following data in Table 7.1 to solve this problem, along with Equation (7.2).

Cancer risk of pollutant = (annual average concentration) (Unit Risk)   (7.2)

TABLE 7.1   Unit Risk Data

Pollutant	Unit Risk ($\text{m}^3/\mu\text{g}$)
EO	8.8×10^{-5}
PCBs	1.4×10^{-3}
PAHs	1.7×10^{-3}

SOLUTION: The cancer risk caused by ethylene oxide is calculated using Equation 7.1 and data presented in the problem statement:

$$\text{Cancer risk of ethylene oxide} = 10\,\mu g/m^3 (8.8 \times 10^{-5}\,m^3/\mu g) = 8.8 \times 10^{-5}$$
$$= 880 \times 10^{-6} = 880 \text{ excess cancer cases per million people.}$$

The cancer risk caused by PCBs is estimated to be:

$$\text{Cancer risk of PCBs} = 2\,\mu g/m^3 (1.4 \times 10^{-3}\,m^3/\mu g) = 2.8 \times 10^{-3}$$
$$= 2,800 \times 10^{-6} = 2,800 \text{ excess cancer cases per million people.}$$

The cancer risk caused by PAHs is estimated to be:

$$\text{Cancer risk of PAHs} = 5\,\mu g/m^3 (1.7 \times 10^{-3}\,m^3/mg) = 8.5 \times 10^{-3}$$
$$= 8500 \times 10^{-6} = 8,500 \text{ excess cancer cases per million people.}$$

The total cancer risk by inhalation is calculated from the arithmetic sum of individual cancer risks calculated above, assuming no interaction of pollutants in terms of carcinogenic effects.

$$\text{Total Cancer Risk} = 880 + 2,800 + 8,500$$
$$= 12,180 \text{ excess cancer cases per million people}$$

This rate is extremely high, 12,180 times higher than the 1 in a million cancer risk normally used as a basis for management of air toxics. This situation should be rectified as soon as possible by a reduction in air toxics. Also note that PAH contributes the majority of risk to the workers at this facility.

TTSCA.6 LEAKING UNDERGROUND STORAGE TANK

A ground water plume has developed from a leaking underground storage tank (UST) and fumes have migrated to the basement of a nearby factory. A sample of indoor air indicates toluene and benzene concentrations of 50 and 70 $\mu g/m^3$, respectively.

Determine the resulting health risk for a 70 kg adult working in the plant who is exposed to these vapors for 15 years (2000 hr/yr) assuming a breathing rate of 15 m^3/d and 75% absorption of both toluene and benzene. The carcinogenic slope factors-risk specific doses (RSDs) via the inhalation route are 0.021 $(mg/kg \cdot d)^{-1}$ and 0.028 $(mg/kg{\cdot}d)^{-1}$ for toluene and benzene, respectively. Assume the environmental health risks for toluene and benzene are additive.

Note: The RSD is an estimate of the average daily dose of a carcinogen that corresponds to a specified excess cancer risk for a 70-yr lifetime exposure.

SOLUTION: Apply the formulas in the Problem statement.

$$\text{Total Risk} = \text{RISK}_{toluene} + \text{RISK}_{benzene}$$

Using the formula presented in the Problem statement, the environmental risks of toluene and benzene are determined as follows:

$$\text{RISK}_{toluene} = \frac{(50\,\mu g/m^3)(0.021(mg/kg\text{-}d)^{-1})(15\,m^3/d)(15yr)}{(70\,kg)(70\,yr)} \times \left(\frac{1\,mg}{1000\,\mu g}\right)$$

$$\text{RISK}_{toluene} = 1.10 \times 10^{-5}$$

$$\text{RISK}_{benzene} = \frac{(70\,\mu g/m^3)(0.028(mg/kg\text{-}d)^{-1})(15\,m^3/d)(15yr)}{(70\,kg)(70\,yr)} \times \left(\frac{1\,mg}{1000\,\mu g}\right)$$

$$\text{RISK}_{benzene} = 2.05 \times 10^{-5}$$

$$\text{Total Risk} = \text{RISK}_{toluene} + \text{RISK}_{benzene}$$
$$= (1.10 \times 10^{-5}) + (2.05 \times 10^{-5})$$
$$= 3.15 \times 10^{-5}$$

Under the given conditions, one could expect 0.315 additional cases of cancer per 10,000, or 30 additional cases of cancer per 1,000,000 workers exposed to toluene and benzene from this source.

CHAPTER 8

COMPREHENSIVE ENVIRONMENTAL RESPONSE, COMPENSATION AND LIABILITY ACT (CERCLA-SUPERFUND)

QUALITATIVE PROBLEMS (LSUP)

PROBLEMS LSUP.1–35

LSUP.1   CERCLA OVERVIEW

Provide an overview of CERCLA.

SOLUTION: The Comprehensive Environmental Response, Compensation, and Liability Act (CERCLA), commonly known as Superfund, was enacted by Congress on December 11, 1980. This law created a tax on the chemical and petroleum industries and provided broad Federal authority to respond directly to releases or threatened releases of hazardous substances that may endanger public health or the environment. Over its first five years, $1.6 billion was collected and the tax went to a trust fund for cleaning up abandoned or uncontrolled hazardous waste sites. Specifically, CERCLA:

1. Established prohibitions and requirements concerning closed and abandoned hazardous waste sites;
2. Provided for liability of persons responsible for releases of hazardous waste at these sites; and,
3. Established a trust fund to provide for cleanup when no responsible party could be identified.

Environmental Regulatory Calculations Handbook, by Leo Stander and Louis Theodore
Copyright © 2008 John Wiley & Sons, Inc.

The law also authorizes two kinds of response actions:

1. Short-term removals, where actions may be taken to address releases or threatened releases requiring prompt response.
2. Long-term remedial response actions that permanently and significantly reduce the dangers associated with releases or threats of releases of hazardous substances that are serious, but not immediately life threatening. These actions can be conducted only at sites listed on EPA's National Priorities List (NPL). See Problem LSUP, 2-3 for additional information.

CERCLA also enabled the revision to the National Contingency Plan (NCP). See Problem LSUP.4 for additional information. The NCP provided the guidelines and procedures needed to respond to releases and threatened releases of hazardous substances, pollutants, or contaminants. The NCP also established the NPL.

Additional details can be found at: http://www.epa.gov/superfund/action/law/cercla.htm.

LSUP.2   THE NATIONAL PRIORITIES LIST (NPL)

Describe how sites are placed on the NPL.

SOLUTION: Sites are first proposed to the National Priorities List (NPL) in the *Federal Register*. EPA then accepts public comments on the sites, responds to the comments, and places on the NPL those sites that continue to meet the requirements for listing.

Regulations implementing NCP 40 CFR 300.425(c), provide three mechanisms for placing sites on the NPL:

1. The first mechanism is EPA's Hazard Ranking System (HRS). (See Problem LSUP.5.)
2. The second mechanism for placing sites on the NPL allows States or Territories to designate one top-priority site regardless of score.
3. The third mechanism allows listing a site if it meets all three of these requirements:

   a. the Agency Toxic Substances and Disease Registry (ATSDR) of the U.S. public Health Service has issued a health advisory that recommends removing people from the site;
   b. EPA determines the site poses a significant threat to public health; and,
   c. EPA anticipates it will be more cost-effective to use its remedial authority (available only at NPL sites) than to use its emergency removal authority to respond to the site.

LSUP.3 NPL SITE LISTING PROCESS

Describe the NPL site listing process.

SOLUTION: Sites are listed on the National Priorities List (NPL) upon completion of Hazard Ranking System (HRS) screening, public solicitation of comments about the proposed site, and after all comments have been addressed.

The NPL primarily serves as an information and management tool. It is a part of the Superfund cleanup process. The NPL is updated periodically. The "Federal Register Notices for NPL Updates" page provides a list of Federal Register Notices for proposed and final NPL Updates. The list is ordered by year and provides the rule type, rule date, FR citation, and a short content description for each FR.

Section 105(a)(8)(B) of CERCLA (CERCLA Overview) as amended, requires that the statutory criteria provided by the HRS be used to prepare a list of national priorities among the known releases or threatened releases of hazardous substances, pollutants, or contaminants throughout the United States. This list, which is Appendix B of the National Contingency Plan, is the NPL.

The identification of a site for the NPL is intended primarily to guide EPA in:

1. Determining which sites warrant further investigation to assess the nature and extent of the human health and environmental risks associated with a site;
2. Identifying what CERCLA-financed remedial actions may be appropriate;
3. Notifying the public of sites EPA believes warrant further investigation; and
4. Serving notice to potentially responsible parties that EPA may initiate CERCLA-financed remedial action.

Inclusion of a site on the NPL does not in itself reflect a judgment of the activities of its owner or operator; it does not require those persons to undertake any action, nor does it assign liability to any person. The NPL serves primarily informational purposes, identifying for the States and the public those sites or other releases that appear to warrant remedial actions.

Additional details can be found at: http://www.epa.gov/superfund/sites/npl/

LSUP.4 NATIONAL CONTINGENCY PLAN OVERVIEW

Provide an overview of the NCP.

SOLUTION: The National Oil and Hazardous Substances Pollution Contingency Plan, more commonly called the National Contingency Plan or NCP, is the federal government's blueprint for responding to both oil spills and hazardous substance releases. The National Contingency Plan is the result of this country's efforts to develop a national response capability and promote overall coordination among the hierarchy of responders and contingency plans.

The first National Contingency Plan was developed and published in 1968 in response to a massive oil spill from the oil tanker *Torrey Canyon* off the coast of England in 1967. More than 37 million gallons of crude oil spilled into the water, causing massive environmental damage. To avoid the problems faced by response officials involved in this incident, U.S. officials developed a coordinated approach to cope with potential spills in U.S. waters. The 1968 plan provided the first comprehensive system of accident reporting, spill containment, and cleanup, and established a response headquarters, a national reaction team, and regional reaction teams (precursors to the current National Response Team and Regional Response Teams).

Congress has broadened the scope of the National Contingency Plan over the years. As required by the Clean Water Act of 1972, the NCP was revised in 1973 to include a framework for responding to hazardous substance spills as well as oil discharges. Following the passage of Superfund legislation in 1980, the NCP was broadened to cover releases at hazardous waste sites requiring emergency removal actions. Over the years, additional revisions have been made to the NCP to keep pace with the enactment of legislation. The latest revisions to the NCP were finalized in 1994 to reflect the oil spill provisions of the Oil Pollution Act of 1990.

Additional details are available at: http://www.epa.gov/oilspill/ncpover.htm

LSUP.5   INTRODUCTION TO THE HAZARD RANKING SYSTEM (HRS)

Provide details on the HRS.

SOLUTION: The HRS is the principal mechanism EPA uses to place uncontrolled waste sites on the National Priorities List (NPL). It is a numerically based screening system that uses information from initial, limited investigations—the preliminary assessment and the site inspection—to assess the relative potential of sites to pose a threat to human health or the environment. Any person or organization can petition EPA to conduct a preliminary assessment using the Preliminary Assessment Petition.

HRS scores do not determine the priority in funding EPA remedial response actions because the information collected to develop HRS scores is not sufficient to determine either the extent of contamination or the appropriate response for a particular site. The sites with the highest scores do not necessarily come to the EPA's attention first—this would require stopping work at sites where response actions were already underway. EPA relies on more detailed studies in the remedial investigation/feasibility study which typically follows listing.

The HRS uses a structured analysis approach to scoring sites. This approach assigns numerical values to factors that relate to risk based on conditions at the site. The factors are grouped into three categories:

1. Likelihood that a site has released or has the potential to release hazardous substances into the environment;

2. Characteristics of the waste (e.g., toxicity and waste quantity); and,

3. People or sensitive environments (targets) affected by the release.

Four pathways can be scored under the HRS:

1. Ground water migration (drinking water);

2. Surface water migration (drinking water, human food chain, sensitive environments);

3. Soil exposure (resident population, nearby population, sensitive environments); and

4. Air migration (population, sensitive environments).

After scores are calculated for one or more pathways, they are combined using a root-mean-square equation to determine the overall site score.

For more information, consult the publications, *The Hazard Ranking System Guidance Manual; Interim* Final, November 1992, (NTIS PB92-963377, EPA 9345.1-07), and 40 CFR Part 300 Appendix A.

Additional details can be found at:

1. http://www.epa.gov/superfund/programs/npl_hrs/hrsint.htm

2. http://www.epa.gov/superfund/resources/hrstrain/hrstrain.htm

LSUP.6   SARA OVERVIEW

Describe the details of SARA.

SOLUTION: The Superfund Amendments and Reauthorization Act (SARA) amended the Comprehensive Environmental Response, Compensation, and Liability Act (CERCLA) on October 17, 1986. SARA reflected EPA's experience in administering the complex Superfund program during its first six years and made several important changes and additions to the program. Specifically, SARA:

1. Stressed the importance of permanent remedies and innovative treatment technologies in cleaning up hazardous waste sites;

2. Required Superfund actions to consider the standards and requirements found in other State and Federal environmental laws and regulations;

3. Provided new enforcement authorities and settlement tools;

4. Increased State involvement in every phase of the Superfund program;

5. Increased the focus on human health problems posed by hazardous waste sites;

6. Encouraged greater citizen participation in making decisions on how sites should be cleaned up; and,

7. Increased the size of the trust fund to $8.5 billion.

SARA also required EPA to revise the Hazard Ranking System (HRS) to ensure that it accurately assessed the relative degree of risk to human health and the environment posed by uncontrolled hazardous waste sites that may be placed on the National Priorities List (NPL).

Additional details can be found at: http://www.epa.gov/superfund/action/law/sara.htm

LSUP.7 ALL APPROPRIATE INQUIRIES RULE

Describe EPA's Final Rule on all appropriate inquiries.

SOLUTION: The Environmental Protection Agency published a final rule setting federal standards for the conduct of all appropriate inquiries. The rule was published in the Federal Register on November 1, 2005. The final rule and preamble is available in the Federal Register 70 FR 66069.

The final rule establishes specific regulatory requirements for conducting all appropriate inquiries into the previous ownership, uses, and environmental conditions of a property for the purposes of qualifying for certain landowner liability protections under CERCLA. The final rule became effective on November 1, 2006.

Parties must comply with the requirements of All Appropriate Inquiries Rule (40 CFR Part 312), or follow the standards set forth in the ASTM E1527-05 Phase I Environmental Site Assessment Process, to satisfy the statutory requirements for conducting all appropriate inquiries. All appropriate inquiries had to be conducted in compliance with either of these standards to obtain protection from potential liability under CERCLA as an innocent landowner, a contiguous property owner, or a bona fide prospective purchaser.

Additional information is available at the following sites:

1. Fact Sheet on All Appropriate Inquiries Final Rule
   Publication Number: EPA 560-F-05-240
   October 2005 (http://www.epa.gov/swerosps/bf/aai/aai_final_factsheet.htm)
2. Fact Sheet on Definition of Environmental Professional included in the Final Rule
   Publication Number: EPA 560-F-05-241
   October 2005 (http://www.epa.gov/swerosps/bf/aai/ep_deffactsheet.htm)

LSUP.8 EMERGENCY PLANNING AND COMMUNITY RIGHT-TO-KNOW ACT (EPCRA) OVERVIEW

Provide an overview of EPCRA.

SOLUTION: EPCRA was passed in response to concerns regarding the environmental and safety hazards posed by the storage and handling of toxic chemicals. These concerns were triggered by the disaster in Bhopal, India, in which more than 2,000 people suffered death or serious injury from the accidental release of methyl isocyanate. To reduce the likelihood of such a disaster in the United States, Congress imposed requirements on both states and regulated facilities.

EPCRA established requirements for Federal, State, and local governments, Native American Tribes, and industry regarding emergency planning and "Community Right-to-Know" reporting on hazardous and toxic chemicals. The Community Right-to-Know provisions helped increase the public's knowledge and provide access to information on chemicals at individual facilities, their uses, and releases into the environment. States and communities, working with facilities, can use the information to improve chemical safety and protect public health and the environment.

EPCRA has four major provisions each of which is detailed below:

1. Emergency planning (Section 301–303)
2. Emergency release notification (Section 304)
3. Hazardous chemical storage reporting requirements (Sections 311–312), and
4. Toxic chemical release inventory (Section 313).

(Regulations implementing EPCRA are codified in 40 CFR Part 370.)

1. Emergency Planning (EPCRA Sections 301–303, 40 CFR Part 355)

The emergency planning section of the law is designed to help communities prepare for and respond to emergencies involving hazardous substances. Every community in the United States must be part of a comprehensive plan.

What are the SERCs and LEPCs?

The Governor of each state has designated a State Emergency Response Commission (SERC). Each SERC is responsible for implementing EPCRA provisions within their state. The SERCs in turn have designated about 3,500 local emergency planning districts and appointed a Local Emergency Planning Committee (LEPC) for each district. The SERC supervises and coordinates the activities of the LEPC, establishes procedures for receiving and processing public requests for information collected under EPCRA, and reviews local emergency response plans.

The LEPC membership must include, at a minimum, local officials including police, fire, civil defense, public health, transportation, and environmental professionals, as well as representatives of facilities subject to the emergency planning requirements, community groups, and the media. The LEPCs must develop an emergency response plan, review it at least annually, and provide information about chemicals in the community to citizens.

What are the required elements of a community emergency response plan?

1. Identify facilities and transportation routes of extremely hazardous substances;
2. Describe emergency response procedures, on and off site;
3. Designate a community coordinator and facility coordinator(s) to implement the plan;
4. Outline emergency notification procedures;
5. Describe how to determine the probable affected area and population by releases;
6. Describe local emergency equipment and facilities and the persons responsible for them;
7. Outline evacuation plans;
8. Provide a training program for emergency responders (including schedules); and,
9. Provide methods and schedules for exercising emergency response plans.

What do facilities need to report?

This section involves chemicals listed on the extremely hazardous substances list (see 40 CFR Part 355 Appendix A).

Any facility that has any of the listed chemicals at or above its threshold planning quantity must notify the SERC and LEPC within 60 days after it first receives a shipment or produce the substance on site. The facility also must provide the LEPC with the identity of a facility representative who will participate in the emergency planning process. Upon request from the LEPC, the facility shall promptly provide all information to the LEPC that is necessary for developing and implementing the emergency plan. For further information, see 40 CFR Part 355.

2. Emergency Release Notification (ECPRA Section 304, 40 CFR 355)

Who is covered?

Facilities must provide an emergency notification and a written follow-up notice to the LEPCs and the SERCs (for any area likely to be affected by the release) if there is a release into the environment of a hazardous substance that is equal to or exceeds the minimum reportable quantity set in the regulations.

There are two types of chemicals that require reporting under this section:

1. Extremely Hazardous Substances (EHSs); and
2. Comprehensive Environmental Response, Compensation and Liability Act (CERCLA) hazardous substances.

Both the EHSs and the CERCLA hazardous substances are found in "List of Lists"—"Consolidated list of chemicals subject to the Emergency Planning and Community Right-to-know Act (EPCRA) and Section 112(r) of the Clean Air Act," EPA 550-B-01-003 October 2006.

Initial notification can be made by telephone, radio, or in person. In addition, CERCLA spills must also be reported to the National Response Center at (800) 424-8802. Emergency notification requirements involving transportation incidents can be met by dialing 911, or in the absence of a 911 emergency number, calling the local operator.

What must be included in the emergency notification?

1. The chemical name;
2. An indication of whether the substance is extremely hazardous;
3. An estimate of the quantity released into the environment;
4. The time and duration of the release;
5. Whether the release occurred into air, water, and/or land;
6. Any known or anticipated acute or chronic health risks associated with the emergency, and where necessary, advice regarding medical attention for exposed individuals;
7. Proper precautions, such as evacuation or sheltering inplace; and,
8. Name and telephone number of contact person.

What is the written follow-up notice?

A written follow-up notice must be submitted to the SERC and LEPC as soon as practicable after the release. The follow-up notice must update information included in the initial notice and provide information on actual response actions taken and advice regarding medical attention necessary for citizens exposed.

For further information, see 40 CFR Part 355.

3. Hazardous Chemical Storage Reporting Requirements (EPCRA Section 311/312, 40 CFR Part 370)

What are the Material Safety Data Sheet (EPCRA Section 311) submission requirements?

Under Occupational Safety and Health Administration (OSHA) regulations, employers must maintain a material safety data sheet (MSDS) for any hazardous chemicals stored or used in the work place. Approximately 500,000 products have MSDSs.

Section 311 requires facilities that have MSDSs for chemicals held above certain quantities to submit either copies of their MSDSs or a list of MSDS chemicals to SERC, LEPC, and local fire department.

What are the Emergency and Hazardous Chemical Inventory (EPCRA Section 312) reporting requirements?

Facilities that need to report under EPCRA Section 311 must also submit an annual inventory report for the same chemicals. This inventory report must be submitted the SERC, LEPC and local fire department by March 1 of each year. For more information on EPCRA Section 311/312, see 40 CFR 370.

4. Toxic Chemical Release Inventory

Certain facilities are required to complete a Toxic Chemical Release Inventory Form annually for specified chemicals.

Details about reporting requirements can be found at 40 CFR Part 372 or the Toxic Release Inventory website [http://www.epa.gov/tri/]. See the next problem for more information.

LSUP.9   THE TOXICS RELEASE INVENTORY (TRI) PROGRAM

Describe the TRI program.

SOLUTION: EPCRA's primary purpose is to inform communities and citizens of chemical hazards in their areas. Sections 311 and 312 of EPCRA require businesses to report the locations and quantities of chemicals stored on-site to state and local governments in order to help communities prepare to respond to chemical spills and similar emergencies. EPCRA Section 313 requires EPA and the States to annually collect data on releases and transfers of certain toxic chemicals from industrial facilities, and make the data available to the public in the Toxics Release Inventory (TRI). In 1990 Congress passed the Pollution Prevention Act (see Chapter 10) which required that additional data on waste management and source reduction activities be reported under TRI. The goal of TRI is to empower citizens, through information, to hold companies and local governments accountable in terms of how toxic chemicals are managed.

EPA compiles the TRI data each year and makes it available through several data access tools, including the TRI Explorer and Envirofacts. There are other organizations which also make the data available to the public through their own data access tools, including Unison Institute which puts out a tool called "RTKNet" and Environmental Defense which has developed a tool called "Scorecard."

The TRI program has expanded significantly since its inception in 1987. The Agency has issued rules to roughly double the number of chemicals included in the TRI to approximately 650. Seven new industry sectors have been added to expand coverage significantly beyond the original covered industries, i.e., manufacturing industries. Most recently, the Agency has reduced the reporting thresholds for certain persistent, bioaccumulative, and toxic (PBT) chemicals in order to be able to provide additional information to the public on these chemicals.

Armed with TRI data, communities have more power to hold companies accountable and make informed decisions about how toxic chemicals are to be managed. The data often spurs companies to focus on their chemical management practices since they are being measured and made public. In addition, the data serves as a rough indicator of environmental progress over time.

Additional details can be found at: http://www.epa.gov/tri/whatis.htm

LSUP.10 HOW THE EMERGENCY RESPONSE PROGRAM WORKS

Describe how the Emergency Response Program (ERP) Works.

SOLUTION: When a release or spill occurs, the company responsible for the release, its response contractors, the local fire and police departments, and the local emergency response personnel provide the first line of defense. If needed, a variety of state agencies stand ready to support, assist, or take over response operations if an incident is beyond local capabilities. In cases where a local government or Native American tribe conducts temporary emergency measures in response to a hazardous substance release, but does not have emergency response funds budgeted, EPA operates the Local Governments Reimbursement program that will reimburse local governments or Native American tribes up to $25,000 per incident.

If the amount of a hazardous substance release or oil spill exceeds the established reporting trigger, the organization responsible for the release or spill is required by law to notify the federal government's National Response Center (NRC). Once a report is made, the NRC immediately notifies a pre-designated EPA or U.S. Coast Guard On-Scene Coordinator (OSC) based on the location of the spill. The procedure for determining the lead agency is clearly defined so there is no confusion about who is in charge during a response. The OSC determines the status of the local response and monitors the situation to determine whether, or how much, federal involvement is necessary. It is the OSC's job to ensure that the cleanup, whether accomplished by industry, local, state, or federal officials, is appropriate, timely, and minimizes human and environmental damage.

The OSC may determine that the local action is sufficient and that no additional federal action is required. If the incident is large or complex, the federal OSC may remain on the scene to monitor the response and advise on the deployment of personnel and equipment. However, the federal OSC will take command of the response in the following situations:

1. If the party responsible for the chemical release or oil spill is unknown or is not cooperative;
2. If the OSC determines that the spill or release is beyond the capacity of the company, local, or state responders to manage; or,

3. For oil spills, if the incident is determined to present a substantial threat to public health or welfare due to the size or character of the spill.

The OSC may request additional support to respond to a release or spill, such as additional contractors, technical support form EPA's Environmental Response Team, or Scientific Support Coordinators from EPA or the National Oceanic and Atmospheric Administration. The OSC also may seek support from the Regional Response Team to access special expertise or to provide additional logistical support. In addition, the National Response Team stands ready to provide backup policy and logistical support to the OSC and the RRT during an incident.

The federal government will remain involved at the oil spill site following response actions to undertake a number of activities, including assessing damages, supporting restoration efforts, recovering response costs from the parties responsible for the spill, and, if necessary, enforcing the liability and penalty provisions of the Clean Water Act, as amended by the Oil Pollution Act of 1990.

Additional details are available at: http://www.epa.gov/superfund/programs/er/nrsworks.htm

LSUP.11   CERCLA HAZARDOUS SUBSTANCES

Provide an overview of CERCLA hazardous substances.

SOLUTION: CERCLA hazardous substances are defined in terms of those substances either specifically designated as hazardous under the Comprehensive Environmental Response, Compensation, and Liability Act (CERCLA), otherwise known as the Superfund law, or those substances identified under other laws. In all, the Superfund law includes references to four other laws (Legal Authorities Defining Hazardous Substances) to designate more than 800 substances as hazardous, and identify many more as potentially hazardous due to their characteristics and the circumstances of their release.

The Superfund law gives EPA's Superfund Emergency Response program the authority to respond to emergencies involving these substances and to pollutants or contaminants that pose "imminent and substantial danger to public health and welfare or the environment." The term "pollutant or contaminant" includes, but is not limited to, any element, substance, compound or mixture, including disease-causing agents, which after release in the environment and upon exposure, ingestion, inhalation, or assimilation into any organism, will likely cause death, disease, behavioral abnormalities, cancer, genetic mutations, physiological malfunctions (including reproductive), or physical deformations in such organisms or their offspring.

The terms "hazardous substance" and "pollutant or contaminant" do not include petroleum or natural gas. EPA conducts emergency responses to incidents involving petroleum and non-petroleum oils through its Oil Program, which functions under the Clean Water Act authorities. Throughout the Emergency Response Program, the term "hazardous substance" includes pollutants and contaminants.

In addition to the hazardous substances identified under the Superfund law, the Title III amendments to Superfund, identify several hundred hazardous substances for their extremely toxic properties. EPA designated them as "extremely hazardous substances" to help focus initial chemical emergency response planning efforts.

Superfund's definition of a hazardous substance includes the following:

1. Any element, compound, mixture, solution, or substance designated as hazardous under section 102 of CERCLA.

2. Any hazardous substance designated under section 311(b)(2)(a) of the Clean Water Act, or any toxic pollutant listed under section 307(a) of the Clean Water Act. There are over 400 substances designated as either hazardous or toxic under the CWA.

3. Any hazardous waste having the characteristics identified or listed under section 3001 of the Resource Conservation and Recovery Act.

As discussed earlier, hazardous waste is defined under RCRA as a solid waste (or combination of solid wastes) which, because of its quantity, concentration, or physical, chemical, or infectious characteristics, may: (1) cause or contribute to an increase in mortality or an increase in serious irreversible, or incapacitating illness; or (2) pose a substantial present or potential hazard to human health or the environment when improperly treated, stored, transported, disposed of, or otherwise managed. In addition, under RCRA, EPA establishes four characteristics that will determine whether a substance is considered hazardous, including ignitability, corrosiveness, reactivity, and toxicity. Any solid waste that exhibits one or more of these characteristics is classified as a hazardous waste under RCRA and, in turn, as a hazardous substance under Superfund. Also included are:

1. Any hazardous air pollutant under section 112 of the Clean Air Act, as amended. There are over 200 substances listed as hazardous air pollutants under the CAA.

2. Any imminently hazardous chemical substance or mixture which the EPA Administrator has "taken action under" section 7 of the Toxic Substances Control Act.

In all, more than 800 substances have been specifically identified as CERCLA hazardous substances under these laws.

Additional details are available at: http://www.epa.gov/superfund/programs/er/hazsubs/lauths.htm

LSUP.12   CERCLA EXTREMELY HAZARDOUS SUBSTANCES

The emergency planning activities of the local emergency planning committee was to be initially based on, but not limited to the *Extremely Hazardous Substance* found in Appendix A to 40 CFR 355. This table provides a list of extremely hazardous substances and their threshold planning quantities. Provide this information.

SOLUTION: The list and the accompanying reportable quantities and threshold planning quantities are provided in Table 8.1.

LSUP.13   OIL POLLUTION ACT

Provide an overview of the Oil Pollution Act.

SOLUTION: The Oil Pollution Act (OPA) was signed into law in August 1990, largely in response to rising public concern following the *Exxon Valdez* incident. The OPA improved the nation's ability to prevent and respond to oil spills by establishing provisions that expand the federal government's ability, and provide the money and resources necessary, to respond to oil spills. The OPA also created the national Oil Spill Liability Trust Fund, which is available to provide up to one billion dollar per spill incident.

In addition, the OPA provided new requirements for contingency planning both by government and industry. The National Oil and Hazardous Substances Pollution Contingency Plan (NCP) has been expanded in a three-tiered approach: the Federal government is required to direct all public and private response efforts for certain types of spill events; Area Committees—composed of federal, state, and local government officials—must develop detailed, location-specific Area Contingency Plans; and, owners or operators of vessels and certain facilities that pose a serious threat to the environment must prepare their own Facility Response Plans.

Finally, the OPA increased penalties for regulatory noncompliance, broadened the response and enforcement authorities of the Federal government, and preserved State authority to establish law governing oil spill prevention and response.

Additional details are available at: http://www.epa.gov/oilspill/opaover.htm

LSUP.14   SUPERFUND CONCERNS

Detail concerns regarding the effectiveness of Superfund.

SOLUTION: Thank the lawyers. Litigation is the number one reason why so few sites have been cleaned up. These legal actions have generated 50 to 90% of all expenditures in the Superfund program, resulting in only 10 to 50% of the expenditures on actual clean up costs. In addition, states and communities are fighting over which sites shall be cleaned first. Finally, as is the case in most bureaucratic programs, each Superfund site must be carried through a specific, exhaustive, ten-step procedure regardless of whether each of the ten steps is necessary or not. Only after these ten steps have been completed may corrective action officially begin at a Superfund site.

Note: The solution presented above is more opinion (of one of the authors) than fact.

TABLE 8.1 List of Extremely Hazardous Substances and Their Threshold
Planning Quantities

CAS No.	Chemical Name	Notes	Reportable Quantity* (pounds)	Threshold Planning Quantity (pounds)
75-86-5	Acetone cyanohydrin		10	1,000
1752-30-3	Acetone thiosemicarbazide		1,000	1,000/10,000
107-02-8	Acrolein		1	500
79-06-1	Acrylamide		5,000	1,000/10,000
107-13-1	Acrylonitrile		100	1,000
814-68-6	Acrylyl chloride	h	100	100
111-69-3	Adiponitrile		1,000	1,000
116-06-3	Aldicarb	c	1	100/10,000
309-00-2	Aldrin		1	500/10,000
107-18-6	Allyl alcohol		100	1,000
107-11-9	Allylamine		500	500
20859-73-8	Aluminum phosphide	b	100	500
54-62-6	Aminopterin		500	500/10,000
373-497-2	Amiton oxalate		100	100/10,000
78-53-5	Amiton		500	500
7664-41-7	Ammonia		100	500
300-62-9	Amphetamine		1,000	1,000
62-53-3	Aniline		5,000	1,000
88-05-1	Aniline, 2,4,6-trimethyl-		500	500
7783-70-2	Antimony pentafluoride		500	500
1397-94-0	Antimycin A	c	1,000	1,000/10,000
86-88-4	Antu		100	500/10,000
1303-28-2	Arsenic pentoxide		1	100/10,000
1327-53-3	Arsenous oxide	h	1	100/10,000
7784-34-1	Arsenous trichloride		1	500
7784-42-1	Arsine		100	100
2642-71-9	Azinphos-ethyl		100	100/10,000
86-50-0	Azinphos-methyl		1	10/10,000
98-87-3	Benzal chloride		5,000	500
98-16-8	Benzenamine, 3-(trifluoromethyl)-		500	500
100-14-1	Benzene, 1-(chloromethyl)-4-nitro-		500	500/10,000
98-05-5	Benzenearsonic acid		10	10/10,000
3615-21-2	Benzimidazole, 4,5-dichloro-2-(trifluoromethyl)-	g	500	500/10,000
98-07-7	Benzotrichloride		10	100
100-44-7	Benzyl chloride		100	500
140-29-4	Benzyl cyanide	h	500	500
57-57-8	Beta-propiolactone		10	500
15271-41-7	Bicyclo [2.2.1]heptane-2-carbonitrile, 5-chloro-6-(((((methylamino) carbonyl)oxy)imino)-,(1-alpha, 2-beta, 4-alpha, 5-alpha, 6E))-		500	500/10,000

(Continued)

TABLE 8.1 *Continued*

CAS No.	Chemical Name	Notes	Reportable Quantity* (pounds)	Threshold Planning Quantity (pounds)
534-07-6	Bis(chloromethyl) ketone		10	10/10,000
4044-65-9	Bitoscanate		500	500/10,000
353-42-4	Boron trifluoride compound with methyl ether (1:1)		1,000	1,000
10294-34-5	Boron trichloride		500	500
7637-07-2	Boron trifluoride		500	500
28772-56-7	Bromadiolone		100	100/10,000
7726-95-6	Bromine		500	500
2223-93-0	Cadmium stearate	c	1,000	1,000/10,000
1306-19-0	Cadmium oxide		100	100/10,000
7778-44-1	Calcium arsenate		1	500/10,000
8001-35-2	Camphechlor		1	500/10,000
56-25-7	Cantharidin		100	100/10,000
51-83-2	Carbachol chloride		500	500/10,000
26419-73-8	Carbamic acid, methyl-, O-(((2,4-dimethyl-1,3-dithiolan-2-yl) methylene)amino)-	d	1	100/10,000
1563-66-2	Carbofuran		10	10/10,000
75-15-0	Carbon disulfide		100	10,000
786-19-6	Carbophenothion		500	500
57-74-9	Chlordane		1	1,000
470-90-6	Chlorfenvinfos		500	500
7782-50-5	Chlorine		10	100
24934-91-6	Chlormephos		500	500
999-81-5	Chlormequat chloride	h	100	100/10,000
79-11-8	Chloroacetic acid		100	100/10,000
107-07-3	Chloroethanol		500	500
627-11-2	Chloroethyl chloroformate		1,000	1,000
67-66-3	Chloroform		10	10,000
107-30-2	Chloromethyl methyl ether	c	10	100
542-88-1	Chloromethyl ether	h	10	100
3691-35-8	Chlorophacinone		100	100/10,000
1982-47-4	Chloroxuron		500	500/10,000
21923-23-9	Chlorthiophos	h	500	500
10025-73-7	Chromic chloride		1	1/10,000
10210-68-1	Cobalt carbonyl	h	10	10/10,000
62207-76-5	Cobalt, ((2,2′-(1,2-ethanediylbis (nitrilomethylidyne))bis (6-fluorophenylato))(2-)-N,N′,O,O′)-		100	100/10,000
64-86-8	Colchicine	h	10	10/10,000
56-72-4	Coumaphos		10	100/10,000
5836-29-3	Coumatetralyl		500	500/10,000
535-89-7	Crimidine		100	100/10,000

(Continued)

TABLE 8.1 *Continued*

CAS No.	Chemical Name	Notes	Reportable Quantity* (pounds)	Threshold Planning Quantity (pounds)
4170-30-3	Crotonaldehyde		100	1,000
123-73-9	Crotonaldehyde, (E)-		100	1,000
506-68-3	Cyanogen bromide		1,000	500/10,000
506-78-5	Cyanogen iodide		1,000	1,000/10,000
2636-26-2	Cyanophos		1,000	1,000
675-14-9	Cyanuric fluoride		100	100
66-81-9	Cycloheximide		100	100/10,000
108-91-8	Cyclohexylamine		10,000	10,000
17702-41-9	Decaborane(14)		500	500/10,000
8065-48-3	Demeton		500	500
919-86-8	Demeton-S-methyl		500	500
10311-84-9	Dialifor		100	100/10,000
19287-45-7	Diborane		100	100
111-44-4	Dichloroethyl ether		10	10,000
149-74-6	Dichloromethylphenylsilane	h	1,000	1,000
62-73-7	Dichlorvos		10	1,000
141-66-2	Dicrotophos		100	100
1464-53-5	Diepoxybutane		10	500
814-49-3	Diethyl chlorophosphate	h	500	500
71-63-6	Digitoxin	c	100	100/10,000
2238-07-5	Diglycidyl ether		1,000	1,000
20830-75-5	Digoxin	h	10	10/10,000
115-26-4	Dimefox		500	500
60-51-5	Dimethoate		10	500/10,000
2524-03-0	Dimethyl phosphorochloridothioate		500	500
77-78-1	Dimethyl sulfate		100	500
99-98-9	Dimethyl-p-phenylenediamine		10	10/10,000
75-78-5	Dimethyldichlorosilane	h	500	500
57-14-7	Dimethylhydrazine		10	1,000
644-64-4	Dimetilan	d	1	500/10,000
534-52-1	Dinitrocresol		10	10/10,000
88-85-7	Dinoseb		1,000	100/10,000
1420-07-1	Dinoterb		500	500/10,000
78-34-2	Dioxathion		500	500
82-66-6	Diphacinone		10	10/10,000
152-16-9	Diphosphoramide, octamethyl-		100	100
298-04-4	Disulfoton		1	500
514-73-8	Dithiazanine iodide		500	500/10,000
541-53-7	Dithiobiuret		100	100/10,000
316-42-7	Emetine, dihydrochloride	h	1	1/10,000
115-29-7	Endosulfan		1	10/10,000
2778-04-3	Endothion		500	500/10,000
72-20-8	Endrin		1	500/10,000
106-89-8	Epichlorohydrin		100	1,000

(Continued)

TABLE 8.1 *Continued*

CAS No.	Chemical Name	Notes	Reportable Quantity* (pounds)	Threshold Planning Quantity (pounds)
2104-64-5	EPN		100	100/10,000
50-14-6	Engocalciferol	c	1,000	1,000/10,000
379-79-3	Ergotamine tartrate		500	500/10,000
1622-32-8	Ethanesulfonyl chloride, 2-chloro-		500	500
10140-87-1	Ethanol, 1,2,-dichloro-, acetate		1,000	1,000
563-12-2	Ethion		10	1,000
13194-48-4	Ethoprophos		1,000	1,000
538-07-8	Ethylbis(2-chloroethyl)amine		500	500
371-62-0	Ethylene fluorohydrin	c, h	10	10
75-21-8	Ethylene oxide		10	1,000
107-15-3	Ethylenediamine		5,000	10,000
151-56-4	Ethyleneimine		1	500
542-90-5	Ethylthiocyanate		10,000	10,000
22224-92-6	Fenamiphos		10	10/10,000
115-90-2	Fensulfothion	h	500	500
4301-50-2	Fluenetil		100	100/10,000
7782-41-4	Fluorine	k	10	500
640-19-7	Fluoroacetamide	j	100	100/10,000
144-49-0	Fluoroacetic acid		10	10/10,000
359-06-8	Fluoroacetyl chloride	c	10	10
51-21-8	Fluorouracil		500	500/10,000
944-22-9	Fonofos		500	500
107-16-4	Formaldehyde cyanohydrin	h	1,000	1,000
50-00-0	Formaldehyde		100	500
23422-53-9	Formetanate hydrochloride	d, h	1	500/10,000
2540-82-1	Formothion		100	100
17702-57-7	Formparanate	d	1	100/10,000
21548-32-3	Fosthietan		500	500
3878-19-1	Fuberidazole		100	100/10,000
110-00-9	Furan		100	500
13450-90-3	Gallium trichloride		500	500/10,000
77-47-4	Hexachlorocyclopentadiene	h	10	100
4835-11-4	Hexamethylenediamine, N,N′-dibutyl-		500	500
302-01-2	Hydrazine		1	1,000
74-90-8	Hydrocyanic acid		10	100
7647-01-0	Hydrogen chloride (gas only)		5,000	500
7783-07-5	Hydrogen selenide		10	10
7664-39-3	Hydrogen fluoride		100	100
7722-84-1	Hydrogen peroxide (Conc. > 52%)		1,000	1,000
7783-06-4	Hydrogen sulfide		100	500
123-31-9	Hydroquinone		100	500/10,000
13463-40-6	Iron, pentacarbonyl-		100	100
297-78-9	Isobenzan		100	100/10,000
78-82-0	Isobutyronitrile	h	1,000	1,000

(Continued)

TABLE 8.1 *Continued*

CAS No.	Chemical Name	Notes	Reportable Quantity* (pounds)	Threshold Planning Quantity (pounds)
102-36-3	Isocyanic acid, 3,4-dichlorophenyl ester		500	500/10,000
465-73-6	Isodrin		1	100/10,000
55-91-4	Isofluorphate	c	100	100
4098-71-9	Isophorone diisocyanate		100	100
108-23-6	Isopropyl chloroformate		1,000	1,000
119-38-0	Isopropylmethylpyrazolyl dimethylcarbamate		1	500
78-97-7	Lactonitrile		1,000	1,000
21609-90-5	Leptophos		500	500/10,000
541-25-3	Lewisite	c, h	10	10
58-89-9	Lindane		1	1,000/10,000
7580-67-8	Lithium hydride	b	100	100
109-77-3	Malononitrile		1,000	500/10,000
12108-13-3	Manganese, tricarbonyl methylcyclopentadienyl	h	100	100
51-75-2	Mechlorethamine	c	10	10
950-10-7	Mephosfolan		500	500
1600-27-7	Mercuric acetate		500	500/10,000
21908-53-2	Mercuric oxide		500	500/10,000
7487-94-7	Mercuric chloride		500	500/10,000
10476-95-6	Methacrolein diacetate		1,000	1,000
760-93-0	Methacrylic anhydride		500	500
126-98-7	Methacrylonitrile	h	1,000	500
920-46-7	Methacryloyl chloride		100	100
30674-80-7	Methacryloyloxyethyl isocyanate	h	100	100
10265-92-6	Methamidophos		100	100/10,000
558-25-8	Methanesulfonyl fluoride		1,000	1,000
950-37-8	Methidathion		500	500/10,000
2032-65-7	Methiocarb		10	500/10,000
16752-77-5	Methomyl	h	100	500/10,000
151-38-2	Methoxyethylmercuric acetate		500	500/10,000
78-94-4	Methyl vinyl ketone		10	10
60-34-4	Methyl hydrazine		10	500
556-64-9	Methyl thiocyanate		10,000	10,000
556-61-6	Methyl isothiocyanate	h	500	500
79-22-1	Methyl chloroformate	h	1,000	500
3735-23-7	Methyl phenkapton		500	500
74-93-1	Methyl mercaptan		100	500
80-63-7	Methyl 2-chloroacrylate		500	500
676-97-1	Methyl phosphonic dichloride		100	100
74-83-9	Methyl bromide		1,000	1,000
624-83-9	Methyl isocyanate		10	500
502-39-6	Methylmercuric dicyanamide		500	500/10,000
75-79-6	Methyltrichlorosilane	h	500	500

(Continued)

TABLE 8.1 *Continued*

CAS No.	Chemical Name	Notes	Reportable Quantity* (pounds)	Threshold Planning Quantity (pounds)
1129-41-5	Metolcarb	d	1	100/10,000
7786-34-7	Mevinphos		10	500
315-18-4	Mexacarbate		1,000	500/10,000
50-07-7	Mitomycin C		10	500/10,000
6923-22-4	Monocrotophos		10	10/10,000
2763-96-4	Muscimol		1,000	500/10,000
505-60-2	Mustard gas	h	500	500
13463-39-3	Nickel carbonyl		10	1
65-30-5	Nicotine sulfate		100	100/10,000
54-11-5	Nicotine	c	100	100
7697-37-2	Nitric acid		1,000	1,000
10102-43-9	Nitric oxide	c	10	100
98-95-3	Nitrobenzene		1,000	10,000
1122-60-7	Nitrocyclohexane		500	500
10102-44-0	Nitrogen dioxide		10	100
62-75-9	Nitrosodimethylamine	h	10	1,000
991-42-4	Norbormide		100	100/10,000
95-48-7	o-Cresol		100	1,000/10,000
NONE	Organorhodium complex (PMN-82-147)		10	10/10,000
630-60-4	Ouabain	c	100	100/10,000
23135-22-0	Oxamyl	d	1	100/10,000
78-71-7	Oxetane, 3,3-bis (chloromethyl)-		500	500
2497-07-6	Oxydisulfoton	h	500	500
10028-15-6	Ozone		100	100
2074-50-2	Paraquat methosulfate		10	10/10,000
1910-42-5	Paraquat dichloride		10	10/10,000
56-38-2	Parathion	c	10	100
298-00-0	Parathion-methyl	c	100	100/10,000
12002-03-8	Paris green		1	500/10,000
19624-22-7	Pentaborane		500	500
2570-26-5	Pentadecylamine		100	100/10,000
79-21-0	Peracetic acid		500	500
594-42-3	Perchloromethyl mercaptan		100	500
108-95-2	Phenol		1,000	500/10,000
64-00-6	Phenol, 3-(1-methylethyl)-, methylcarbamate	d	1	500/10,000
4418-66-0	Phenol, 2,2′-thiobis[4-chloro-6-methyl-		100	100/10,000
58-36-6	Phenoxarsine, 10,10′-oxydi-		500	500/10,000
696-28-6	Phenyl dichloroarsine	h	1	500
59-88-1	Phenylhydrazine hydrochloride		1,000	1,000/10,000
62-38-4	Phenylmercury acetate		100	500/10,000
2097-19-0	Phenylsilatrane	h	100	100/10,000

(Continued)

TABLE 8.1 *Continued*

CAS No.	Chemical Name	Notes	Reportable Quantity* (pounds)	Threshold Planning Quantity (pounds)
103-85-5	Phenylthiourea		100	100/10,000
298-02-2	Phorate		10	10
4104-14-7	Phosacetim		100	100/10,000
947-02-4	Phosfolan		100	100/10,000
75-44-5	Phosgene		10	10
732-11-6	Phosmet		10	10/10,000
13171-21-6	Phosphamidon		100	100
7803-51-2	Phosphine		100	500
2703-13-1	Phosphonothioic acid, methyl-, O-ethyl O-(4-(methylthio)phenyl)ester		500	500
50782-69-9	Phosphonothioic acid, methyl-, S-(2-(bis(1-methylethyl) amino)ethyl) O-ethyl ester		100	100
2665-30-7	Phosphonothioic acid, methyl-, O-(4-nitrophenyl) O-phenyl ester		500	500
3254-63-5	Phosphoric acid, dimethyl 4-(methylthio) phenyl ester		500	500
2587-90-8	Phosphorothioic acid, O-dimethyl-5-(2-(methylthio) ethyl)ester	c, g	500	500
10025-87-3	Phosphorus oxychloride		1,000	500
10026-13-8	Phosphorus pentachloride	b	500	500
7719-12-2	Phosphorus trichloride		1,000	1,000
7723-14-0	Phosphorus	b, h	1	100
57-47-6	Physostigmine	d	1	100/10,000
57-64-7	Physostigmine, salicylate (1:1)	d	1	100/10,000
124-87-8	Picrotoxin		500	500/10,000
110-89-4	Piperidine		1,000	1,000
23505-41-1	Pirimifos-ethyl		1,000	1,000
151-50-8	Potassium cyanide	b	10	100
1012-450-2	Potassium arsenite		1	500/10,000
506-61-6	Potassium silver cyanide	b	1	500
2631-37-0	Promecarb	d, h	1	500/10,000
106-96-7	Propargyl bormide		10	10
107-12-0	Propionitrile		10	500
542-76-7	Propionitrile, 3-chloro-		1,000	1,000
70-69-9	Propiophenone, 4'-amino	g	100	100/10,000
109-61-5	Propyl chloroformate		500	500
75-56-9	Propylene oxide		100	10,000
75-55-8	Propylene amine		1	10,000
2275-18-5	Prothoate		100	100/10,000
129-00-0	Pyrene	c	5,000	1,000/10,000
504-24-5	Pyridine, 4-amino-	h	1,000	500/10,000
140-76-1	Pyridine, 2-methyl-5-vinyl-		500	500

(Continued)

TABLE 8.1 *Continued*

CAS No.	Chemical Name	Notes	Reportable Quantity* (pounds)	Threshold Planning Quantity (pounds)
1124-33-0	Pyridine, 4-nitro-, 1-oxide		500	500/10,000
53558-25-1	Pyriminil	h	100	100/10,000
14167-18-1	Salcomine		500	500/10,000
107-44-8	Sarin	h	10	10
7783-00-8	Selenious acid		10	1,000/10,000
7791-23-3	Selenium oxychloride		500	500
563-41-7	Semicarbazide hydrochloride		1,000	1,000/10,000
3037-72-7	Silane, (4-aminobutyl) diethoxymethyl-		1,000	1,000
13410-01-0	Sodium selenate		100	500/10,000
7784-46-5	Sodium arsenite		1	100/10,000
62-74-8	Sodium fluoroacetate		10	10/10,000
124-65-2	Sodium cacodylate		100	100/10,000
143-33-9	Sodium cyanide (Na(CN))	b	10	100
7631-89-2	Sodium arsenate		1	1,000/10,000
10102-18-8	Sodium selenite	h	100	100/10,000
26628-22-8	Sodium azide (Na(N3))	b	1,000	500
10102-20-2	Sodium tellurite		500	500/10,000
900-95-8	Stannane, acetoxytriphenyl-	g	500	500/10,000
57-24-9	Strychnine	g	10	100/10,000
60-41-3	Strychnine, sulfate		10	100/10,000
3689-24-5	Sulfote		100	500
3569-57-1	Sulfoxide, 3-chloroprophyloctyl		500	500
7446-09-5	Sulfur dioxide		500	500
7783-60-0	Sulfur tetrafluoride		100	100
7446-11-9	Sulfur trioxide	b	100	100
7664-93-9	Sulfuric acid		1,000	1,000
77-81-6	Tabun	c, k	10	10
7783-80-4	Tellurium hexafluoride	k	100	100
107-49-3	Tepp		10	100
13071-79-9	Terbufos	h	100	100
78-00-2	Tetraethyl lead	c	10	100
597-64-8	Tetraethyltin	c	100	100
75-74-1	Tetramethyllead	c	100	100
509-14-8	Tetranitromethane		10	500
10031-59-1	Thallium sulfate	h	100	100/10,000
2757-18-8	Thallous malonate	c, h	100	100/10,000
6533-73-9	Thallous carbonate	c, h	100	100/10,000
7791-12-0	Thallous chloride	c, h	100	100/10,000
7446-18-6	Thallous sulfate		100	100/10,000
2231-57-4	Thiocarbazide		1,000	1,000/10,000
39196-18-4	Thiofanox		100	100/10,000
297-97-2	Thionazin		100	500
108-98-5	Thiophenol		100	500
79-19-6	Thiosemicarbazide		100	100/10,000

(*Continued*)

TABLE 8.1 *Continued*

CAS No.	Chemical Name	Notes	Reportable Quantity* (pounds)	Threshold Planning Quantity (pounds)
5344-82-1	Thiourea, (2-chlorophenyl)-		100	100/10,000
614-78-8	Thiourea, (2-methylphenyl)-		500	500/10,000
7550-45-0	Titanium tetrachloride		1,000	100
91-08-7	Toluene-2,6-diisocyanate		100	100
584-84-9	Toluene-2,4-diisocyanate		100	500
110-57-6	trans-1,4-dichlorobutene		500	500
1031-47-6	Triamiphos		500	500/10,000
24017-47-8	Triazofos		500	500
1558-25-4	Trichloro(chloromethyl)silane		100	100
27137-85-5	Trichloro(dichlorophenyl)silane		500	500
76-02-8	Trichloroacetyl chloride		500	500
115-21-9	Trichloroethylsilane	h	500	500
327-98-0	Trichloronate	k	500	500
98-13-5	Trichlorophenylsilane	h	500	500
998-30-1	Triethoxysilane		500	500
75-77-4	Trimethylchlorosilane		1,000	1,000
824-11-3	Trimethylolpropane phosphite		100	100/10,000
1066-45-1	Trimethyltin chloride		500	500/10,000
639-58-7	Triphenyltin chloride		500	500/10,000
555-77-1	Tris(2-chloroethyl)amine	h	100	100
2001-95-8	Valinomycin	c	1,000	1,000/10,000
1314-62-1	Vanadium pentoxide		1,000	100/10,000
108-05-4	Vinyl acetate monomer		5,000	1,000
129-06-6	Warfarin sodium	h	100	100/10,000
81-81-2	Warfarin		100	500/10,000
28347-13-9	Xylylene dichloride		100	100/10,000
1314-84-7	Zinc phosphide	b	100	500
58270-08-9	Zinc, dichloro(4,4-dimethyl-5((((methylamino)carbonyl) oxy)imino) pentanenitrile),(T-4)-		100	100/10,000

*Only the statutory or final RQ is shown. For more information, see 40 CFR Table 302.4.

Notes.

a. This chemical does not meet acute toxicity criteria. Its threshold planning quantity (TPQ) is set at 10,000 pounds.

b. This material is a reactive solid. The TPQ does not default to 10,000 pounds for non-powder, non-molten, non-solution form.

c. The calculated TPQ changed after technical review as described in the technical support document.

d. Indicates that the RQ is subject to change when the assessment of potential carcinogenicity and/or other toxicity is completed.

e. Statutory reportable quantity for purposes of notification under SARA sect 304(a)(2).

f. [Reserved]

g. New chemicals added that were not part of the original list of 402 substances.

h. Revised TPQ based on new or re-evaluated toxicity data.

i. TPQ is revised to its calculated value and does not change due to technical review as in the proposed rule.

j. The TPQ was revised after proposal due to calculation error.

k. Chemicals on the original list that do not meet toxicity criteria but because of their high production volume and recognized toxicity are considered chemicals of concern ("Other chemicals").

LSUP.15   LIABILITY

The term liability is closely tied to the system of values and ethics which have been developed in this country. In concise language, define and give an example of the term. Explain how ones interpretation of liability affects waste incineration permitting, application and design.

SOLUTION: In a very real sense, this is an extension of a problem that appeared in Chapter 6, Problem LRCRA.17.

Joint and several liability is an assignment of responsibility when two or more persons fail to exercise the proper care and a division of harm is not possible. If two hunters fire their weapons and a person is killed, and there is no way to determine which projectile caused the harm, both hunters may be each held liable for the harm to the aggrieved party. In the case of joint and several strict liability, each party who managed a waste may be responsible for mitigating damages caused by the waste. For example, the generator, the transporter, the storage facility, and the incinerator operator may each individually or collectively be responsible for damages caused by mismanagement of a waste.

These provisions as described in this and in Problem LRCRA.17 provide a tremendous impetus to hazardous waste generators to dispose of their waste on site under carefully controlled conditions. This concept of liability also burdens the generator with the threat of future costs as a result of someone else's improper actions. These values follow directly from a system of government which was created to ensure that individual citizens do not suffer loss of property and freedoms (health) by the actions of others. The need for a careful choice of a contractor to carry out waste management and disposal responsibilities is also highlighted by these provisions.

LSUP.16   PREPARING FOR EMERGENCIES

Answer the following two questions.

1. List at least three specific types of chemical emergencies that might occur in a large metropolitan area.
2. Briefly describe at least five specific features of an emergency preparedness plan that would be put in place to respond to a major accidental release of a volatile hazardous chemical.

SOLUTION:
1. Examples of three types of chemical emergencies that could occur in a large metropolitan area are:
   a. A spill from a railroad tanker car containing a hazardous chemical resulting in the release of a gas cloud

b. An explosion and fire at a petrochemical plant, and

c. A corrosive acid leak from storage drums into a nearby river or lake.

These examples provide a sample of the types of industrial accidents that are common in most major and even smaller, industrialized American cities. Readers are encouraged to review local and national newspapers for current examples of such chemical accidents and emergencies.

2. Five typical features of an emergency preparedness plan are:
   a. Evacuation plans and routes for nearby dwellings, schools, offices, and industries

   b. Access routes and predicted response times for emergency teams

   c. Ability to predict the trajectory and concentration of airborne or water-borne toxic releases

   d. Ability to provide emergency services to the site, e.g., adequate municipal water flowrate and pressure to extinguish a major petrochemical fire, and

   e. Plans for mobilization of area medical personnel to treat casualties associated with major airborne accidental chemical releases.

LSUP.17 SOURCES OF LAND POLLUTION

Describe some sources of land pollution.

SOLUTION: Historically, land has been used as the dumping ground for wastes, including those removed from air and water. Improper handling, storage, and disposal of chemicals can cause serious problems. Most are familiar with the examples that follow.

In 1984, gasoline that leaked out of an underground storage tank at a service station seeped down through the soil to the water table and spread out across the surface of the groundwater. When heavy spring rains cause the water table to rise, the moisture that seeped into the gas station's basement carried gasoline with it. The fumes eventually reached explosive levels. A spark from an air compressor that controlled the lift ignited the vapors, and the gas station building was destroyed in an explosion.

In another case, a leak from an underground storage tank at a local service station in 1984 contaminated groundwater, affecting 14 wells. Property values dropped dramatically. Twelve families were forced to rely on bottled water for months. Some residents were affected by headaches, nausea, and eye and skin irritations.

In the Love Canal (Niagara Falls, New York) and in Times Beach (Missouri), improper disposal of hazardous waste resulted in contaminated land and water in surrounding communities. In addition, hundreds of drinking water wells have been contaminated by improper waste disposal throughout the United States.

Municipal wastes include household and commercial wastes, demolition materials, and sewage sludge. Solvents and other harmful household and commercial wastes are generally so intermingled with other materials that specific control of each is virtually impossible. Leachate resulting from rainwater seeping through municipal landfills may contaminate underlying groundwater. While the degree of hazard presented by this leachate is often relatively low, the volume produced is so great that it may contaminate groundwater.

Uncontrolled disposal sites containing hazardous wastes and other contaminants present some of the most serious environmental problems this nation has ever faced. These sites can contaminate groundwater, lead to explosions, and present other dangers to people and the nearby environment. In many cases, the people who disposed the waste were unaware of the problems that the sites eventually would create for public health and the environment. Most of the abandoned or inactive sites and many of the active hazardous waste facilities where hazardous wastes have escaped are linked in some fashion to the chemical and petroleum industries.

LSUP.18 HAZARD RISK ANALYSIS

Risk may be logically considered to consist of:

1. A scenario.
2. A probability of the scenario occurring.
3. A set of consequences which arise if the scenario does occur.

Consider a rotary kiln incinerator which is receiving drummed wastes from a superfund site with material that has highly variable heating values. The process train consists (as is usually the case) of the following: the feed system, the primary combustion chamber, a secondary combustion chamber designed to operate at 1200°C, an emergency vent stack, a quencher, a packed tower scrubber and two ionizing wet scrubbers in series connected to an induced draft fan and a 20 m stack.

1. Propose a scenario which may lead to a discharge of POHCs which is substantially higher than the permitted DRE.
2. Suggest a logical approach for estimating the probability of that scenario using a process flow sheet, the interdependence of equipment and/or operator intervention, and estimates of the probability of individual events occurring.
3. Describe how you would approach estimating the consequences of your scenario based upon a typical hazardous waste feed to the system.
4. For your example scenario, how could risk be reduced by dealing with one or more of the three components of risk?

SOLUTION:

1. Any number of scenarios might be proposed, such as those provided in Table 8.2 below:

TABLE 8.2   Risk Scenarios

Risk Number	Risk Scenario
1	A spill of waste from the tank or feed system.
2	A blow-back of waste upon over-charging to the kiln.
3	Excess air feed control problems.
4	Reduction or failure of flow in the fuel line.
5	Reduction or failure of flow in the quench/absorber water line.
6	Power loss.
7	Faulty emergency vent stack release, etc.

Note that all of the above scenarios are related to incinerator operations and POHC emissions.

2. Although other methods are available, the best approach for developing a probability estimate would be a fault tree analysis.

3. Given a POHC emission scenario and the probability of occurrence, the main concern would be the estimation of a maximum possible acute exposure, and determining where the exposure would take place. Dispersion modeling would be the logical choice in exposure assessment and prediction. For acute exposure, a carcinogenic risk assessment would likely be inappropriate. Rather, a comparison of maximum values with short-term exposure limits would make the most sense. Event tree analyses are often employed here.

4. For any risk scenario, the overall risk can be reduced by:
   a. Taking steps to eliminate the scenario by modifying the process
   b. Reducing the probability of the scenario by identifying key components from the fault tree analysis and modifying them to lower overall risk probability levels
   c. Taking steps to reduce the consequences, e.g., establishing a larger buffer zone, reducing acute exposure, reducing population at risk, etc.

See S. Shaefer and L. Theodore, "Probability and Statistics Application for Environmental Science," CRC Press/Taylor & Francis Group, Boca Raton, FL, 2007, for additional details.

This topic will be revisited in earnest in Chapter 9.

LSUP.19   RANKING RISK

Provide some information on the ranking risks.

SOLUTION: The public often worries about the largely publicized risks and thinks little about those that they face regularly. A study was recently performed that compared the responses of two groups—15 national risk assessment experts and 40 members of the League of Women Voters—on the risks of 30 activities and technologies. This search produced striking discrepancies, as presented in Table 8.3. The League members rated nuclear power as the number 1 risk, while experts numbered it at 20 and the League ranked X-rays at 22 while the experts gave it a rank of 7.

The rankings of perceived risks for 30 activities and technologies, based on averages in a survey of a group of experts and a group of informed lay people, and members of the League of Women Voters is provided in Table 8.3. A ranking of 1 denotes the highest level of perceived risk.

TABLE 8.3   Ranking Risks: Reality and Perception

League of Women Voters	Activity or Technology	Experts
1	Nuclear power	20
2	Motor vehicles	1
3	Handguns	4
4	Smoking	2
5	Motorcycles	6
6	Alcoholic beverages	3
7	General (private) aviation	12
8	Police work	17
9	Pesticides	8
10	Surgery	5
11	Fire fighting	18
12	Large construction	13
13	Hunting	23
14	Spray cans	26
15	Mountain climbing	29
16	Bicycles	15
17	Commercial aviation	16
18	Electric power (nonnuclear)	9
19	Swimming	10
20	Contraceptives	11
21	Skiing	30
22	X rays	7
23	High school and college football	27
24	Railroads	19
25	Food preservatives	14
26	Food coloring	21
27	Power motors	28
28	Prescription antibiotics	24
29	Home appliances	22
30	Vaccinations	25

LSUP.20 CONVINCING THE PUBLIC

As a well educated person in hazardous waste management attending a public hearing on a risk assessment study where the outcome is "no potential health hazard to the public," what are the main concerns you would have in order to be convinced that the results are reliable and acceptable?

SOLUTION: Risk assessment and conveying its results to inform the public are controversial and hard to accomplish since the public often does not comprehend the interpretation of risk assessment results in the same way as the risk assessors do. Several factors should be considered in determining the level of acceptability of the results of a risk assessment study. First, the level of acceptable risk used as the end point of the study must be agreed upon as the acceptable level of risk. This depends upon the individual's level of concern. In addition, in considering reliability and acceptability of the study, many elements leading to the conclusion of the study can be subject to criticism including the capability of the subjects used in the health studies, the method employed in assessing health effects, the nature of samples collected, and the detection limits of examined parameters. Details follow.

1. Professional experience is an extrinsic indication of the capability of the investigator to accomplish the study while qualifications surmise the principal knowledge of the individual.
2. The number of subjects being studied is statistically important in a quantitative risk assessment since the result of a risk assessment is usually reported in terms of the probability that the event might take place. It is therefore important to design into the study an adequate number of samples in order to obtain the required level of significance and reduce the chance of drawing a wrong conclusion (such as to incorrectly reject or fail to reject a hypothesis).
3. Suitability and characteristics of the subjects include their sex, age, and susceptibility to health effects from the contaminant(s) of concern.
4. Since most of the toxic substances that may affect human health are often measured at low concentrations (e.g., parts per billion or parts per trillion), methods of analysis and detection limits of the parameters will be the key decisions for determining the reliability of the data. Quality assurance and quality control (QA/QC) also play important roles in quantifying the accuracy and reproducibility of the data.

LSUP.21 SEVEN CARDINAL RULES OF RISK COMMUNICATION

List the seven cardinal rules of risk communication.

SOLUTION: There are no easy prescriptions for successful risk communication. However, those who have studied and participated in debates about risk generally agree on seven cardinal rules. These rules apply equally well to the public and private sectors. Although many of these rules may seem obvious, they are continually and consistently violated in practice. Thus, a useful way to focus on these is to focus on why they are frequently not followed.

1. *Accept and involve the public as a legitimate partner.* A basic tenet of risk communication in a democracy is that people and communities have a right to participate in decisions that affect their lives, their property, and the things they value. *Guidelines*: Demonstrate respect for the public and underscore the sincerity of the effort by involving the community early before important decisions are made. Involve all parties that have an interest or stake in the issue under consideration. If you are a government employee, remember that you work for the public. If you do not work for the government, the public still holds you accountable.
Point to consider: The goal in risk communication in a democracy should be to produce an informed public that is involved, interested, reasonable, thoughtful, solution-oriented, and collaborative; it should not be to diffuse public concerns or replace action.

2. *Plan carefully and evaluate your efforts.* Risk communication will be successful only if carefully planned.
Guidelines: Begin with clear, explicit risk communication objectives, such as providing information to the public, motivating individuals to act, stimulating response to emergencies, and contributing to the resolution of conflict. Evaluate the information available about the risks and know its strengths and weaknesses. Classify and segment the various groups in the audience. Aim communications at specific subgroups in the audience. Recruit spokespeople who are good at presentation and interaction. Train any staff, including technical staff, in communication skills; reward outstanding performance. Whenever possible, pretest your messages. Carefully evaluate efforts and learn from mistakes.
Points to Consider: There is no such entity as "the public" instead, there are many publics, each with its own interests, needs, concerns, priorities, preferences, and organizations. Different risk communication goals, audiences, and media require different risk communication strategies.

3. *Listen to the public's specific concerns.* If one does not listen to the people, one cannot expect them to listen. Communication is a two-way activity.
Guidelines: Do not make assumptions about what people know, think, or want done about risks. Take the time to find out what people are thinking: Use techniques such as interviews, focus groups, and surveys. Let all parties that have an interest or stake in the issue be heard. Identify with the audience and try to put oneself in their place. Recognize people's emotions. Let people know that you understand what they said, addressing their concerns as well as yours. Recognize the "hidden agendas," symbolic meanings, and broader economic or political considerations that often underlie and complicate the task of risk communication.
Point to Consider: People in the community are often more concerned about such issues as trust, credibility, competence, control, voluntariness, fairness, caring, and compassion than about mortality statistics and the details of quantitative risk assessment.

4. *Be honest, frank, and open.*   In communicating risk information, trust and credibility are the most precious assets.
   Guidelines: State your credentials; but do not ask or expect to be trusted by the public. If you do not know an answer or are uncertain, say so. Get back to people with answers. Admit mistakes. Disclose risk information as soon as possible (emphasizing any reservations about reliability). Do not minimize or exaggerate the level of risk. Speculate only with great caution. If in doubt, lean toward sharing more information, not less, or people may think you are hiding something. Discuss data uncertainties, strengths and weaknesses, including the ones identified by other credible sources. Identify worst-case estimates as such, and cite ranges of risk estimates when appropriate.
   Point to Consider: Trust and credibility are difficult to obtain. Once lost they are almost impossible to regain completely.

5. *Coordinate and collaborate with other credible sources.*   Allies can be effective in helping communicate risk information.
   Guidelines: Take time to coordinate all interorganizational and intraorganizational communications. Devote effort and resources to the slow, hard work of building bridges with other organizations. Use credible and authoritative intermediates. Consult with others to determine who is best able to answer questions about risk. Try to issue communications jointly with other trustworthy sources (for example, credible university scientists and/or professors, physicians, or trusted local officials).
   Point to Consider: Few things make risk communication more difficult than conflicts or public disagreements with other credible sources.

6. *Meet the needs of the media.*   The media are a prime transmitter of information on risks; they play a critical role in setting agendas and in determining outcomes.
   Guidelines: Be open and accessible to reporters. Respect their deadlines. Provide risk information tailored to the needs of each type of media (for example, graphics and other visual aids for television). Prepare in advance and provide background material on complex risk issues. Do not hesitate to follow up on stories with praise or criticism, as warranted. Try to establish long-term relationships of trust with specific editors and reporters.
   Point to Consider: The media are frequently more interested in politics than in risk; more interested in simplicity than in complexity; more interested in danger than in safety.

7. *Speak clearly and with compassion.*   Technical language and jargon are useful as professional shorthand, but they are barriers to successful communication with the public.
   Guidelines: Use simple, nontechnical language. Be sensitive to local norms, such as speech and dress. Use vivid, concrete images that communicate on a personal level. Use examples anecdotes that make technical risk data come alive. Avoid distant, abstract, unfeeling language about deaths, injuries, and illnesses. Acknowledge and respond (both in words and with action) to emotions that people express, including anxiety, fear, anger, outrage, helplessness. Acknowledge and respond to the distinctions that the public views as

important in evaluating risks, e.g., voluntariness, controllability, familiarity, dread, origin (natural or man-made), benefits, fairness, and catastrophic potential. Use risk comparisons to help put risks in perspective, but avoid comparisons that ignore distinctions which people consider important. Always try to include a discussion of actions that are under way or can be taken. Tell people what can be done. Promise only what can be done, and be sure to do what is promised.

Points to Consider: Regardless of how well one communicates risk information, some people will not be satisfied. Never let efforts to inform people about risks prevent one from acknowledging and saying that any illness, injury, or death is a tragedy. And finally, if people are sufficiently motivated, they are quite capable of understanding complex risk information, even if they may not agree.

Additional details can be found in the following excellent reference: EPA, "Seven Cardinal Rules of Risk Communication," EPA document OPA-87-020, April 1988. See also "Superfund Community Involvement Toolkit," EPA 540-K-01-004, September 2002 and "General Guidance on Risk Management Programs for Chemical Accident Prevention (40 CFR Part 68)," EPA-550-B-04-001, April 2004.

LSUP.22 BURIED DRUMS

You and your family have chosen to pursue a new lifestyle and purchased a small farm in the Midwest. While excavating the foundation for a new barn, you discover three buried drums. The drums are nearly full of a dark gray substance, but have no labels or other identifying marks to suggest origin or content.

1. What federal environmental law or laws may apply to this situation?
2. As a responsible citizen (but with limited finances), what steps should you take to assure proper disposal of this material?

SOLUTION:

1. Comprehensive Environmental Response, Compensation and Liability Act (CERCLA/Superfund), and Resource Conservation and Recovery Act (RCRA)
2. Contact the appropriate local environmental or public health agency. Have the unknown waste analyzed to determine whether it is a hazardous or non-hazardous waste. Analysis will utilize the tests for hazardous characteristics:
   a. Corrosivity
   b. Ignitability
   c. Reactivity
   d. Toxicity

Also dispose the material at a licensed facility, utilizing a Uniform Hazardous Waste Manifest if hazardous.

LSUP.23 OIL SPILL EFFECTS

There have been a number of major accidents involving oil spills in the U.S. such as those occurring near Prince William Sound, Alaska (Exxon Valdez); Galveston Bay, Texas; and, Lake Michigan, Michigan. These incidents resulted in extensive environmental damage and massive cleanup efforts by the responsible parties, and state and federal agencies. What are the possible risks of these incidents to ecological systems and to humans? Briefly describe the mechanisms of the risk (if any).

SOLUTION: Risks to marine life and to humans from accidents involving major oil spills include the following,
  Immediate effects:

1. Death of organisms through coating and asphyxiation and skin contact
2. Absorption of water soluble toxic components of oil
3. Destruction of the food sources

Long-term effects:

1. Reduce resistance to infection and stresses
2. Carcinogenic and potential mutagenic
3. Life style interruption (e.g., propagation)

Risks:

1. Risks to valuable commercial shellfish can result (e.g., oysters, scallops, soft-shell clams)
2. Risks that result from the human use of marine resources (e.g., fisheries products)
3. Risks that result from the recreational use of the marine resources from oil tar or prolonged skin contact with carcinogenic hydrocarbons that contribute to a public healthy hazard
4. Risks that result from direct water utilization

LSUP.24 EMERGENCY RESPONSE TO A TANK FAILURE

You are the manager of a 25,000 gallon tank facility storing oil and oil-based products. This facility is located on a hillside 200 ft above and 2,000 ft away from a navigable river. The river intake for a water treatment plant serving 30,000 people is located 3 miles downstream from your location. This is a relatively new facility with no history of past chemical spills. You are responsible for preparation of the Spill Prevention Control and Countermeasures (SPCC) plan for emergency response to a tank failure or other spill in this facility.

Prepare an outline listing the items or topic areas that the plan must address under the Clean Water Act Amendments, the Resource Conservation and Recovery Act (RCRA), and the Comprehensive Environmental Response, Compensation, and Liability Act (CERCLA).

SOLUTION: A Spill Prevention, Control, and Countermeasures Plan should contain the following elements as a minimum:

1. Safety measures prior to spill:
   a. Inspection and routine maintenance schedule involving material transportation areas, storage areas, safety devices/alarm systems, and emergency response equipment/facilities
   b. Containment structures/safety devices/site cleanup equipment (both on-site and community response)
   c. Emergency equipment—description, capability, and on-site location
   d. Availability and training of site personnel
   e. Identification of individuals responsible for emergency coordination and agency notifications
2. Emergency response in the event of a spill:
   a. Containment/cleanup procedures (within the facility)
   b. Agencies to be notified/parties responsible for this notification
   c. Actions required for external spill effects:
3. Immediate and short term response related to containment and health.
4. Longer term responsibilities (clean-up, environmental monitoring, population health follow-up, etc.)

LSUP.25   TANK FACILITY INFORMATION

Assume you are the manager of a 50,000 gallon tank facility storing aqueous solutions of two different pesticides. This facility is the source of one of the major hazards identified in the Local Emergency Planning Committee (LEPC) inventory. What information are you required to provide to your LEPC under the terms of EPCRA? Your answer can be in outline form, and should address the specific nature of the reporting, record keeping, and emergency response requirements.

SOLUTION: The Emergency Planning and Community Right To Know Act (EPCRA) requires that you provide the LEPC with information that contains the following elements as a minimum:

1. General Information
   a. Submission of Material Safety Data Sheets (MSDSs) for each pesticide
   b. An emergency and hazard information form for each pesticide: (the form incorporates a tiered approach)
      Program I: Yearly maximum quantity, average daily quantity, days on site, physical/health hazard information and general location for each substance
      Program II (if requested by the LEPC): Exact location(s) and storage method(s)
   c. A description of on-site emergency equipment and employee emergency response training

2. If the facility is subject to chemical release reporting requirements (non-emergency release of toxics to the environment), a toxic chemical release form must be submitted to the EPA and state/local officials (this is not part of the LEPC reporting process, but is required under EPCRA).

3. If a "reportable quantity" emergency release to the environment occurs, reporting requirements include:
   a. The identity, hazard category, health risk and precaution information appropriate to the substance
   b. Time, duration, quantity and medium of release
   c. Identification of individuals responsible for emergency coordination and agency notification, (include National Response Center and post spill reporting requirements)
   d. Known or anticipated acute/chronic health risks and, if appropriate, advice regarding medical treatment and emergency response activity

LSUP.26 CARGO EXPLOSION

In 1947, two ships docked in Texas City, Texas, with tons of ammonium nitrate fertilizer and other cargo aboard. These ships caught fire, burned and exploded over a period of more than 16 hours. The explosions were so powerful that almost 600 people were killed and more than 3,500 were injured. That area and much of the city was destroyed. One of the ship's anchors was thrown approximately 2 miles inland where it still lies today as a memorial to the incident.

As described earlier, the Emergency Planning and Community Right-to-Know Act (EPCRA) among other things, requires any facility that produces, uses or stores any chemical on a published list in excess of the "Threshold Planning

Quantity" to notify local emergency response entities (such as the fire department, police department, hospitals, etc.) of the quantity, identity, and nature of these chemicals; to cooperate with a Local Emergency Planning Committee (LEPC); and, to develop an emergency plan to be used in the event of a release.

While the regulatory definition of a "facility" includes transportation vessels and port authorities for release reporting, these entities are exempted from notification and emergency planning requirements. As a result, emergency response planning against another Texas City disaster is not a requirement of the EPCRA legislation.

Prepare a list of areas of concern that would have to be addressed if the notification and emergency planning requirements were applied to port areas. Among other things, you may wish to address matters such as the short residence time of in-transit materials and the political (as opposed to legal) ramifications of applying regulation of this kind to foreign flag carriers.

Note: This is an open ended question with many correct answers.

SOLUTION: The following areas should be addressed in the notification and emergency planning for port areas:

1. Who shall be responsible for notification and/or emergency planning—the shipper, transporter, or port operator?
2. How shall the inventory of the materials flowing in and out of the area be maintained?
3. Should there be a minimum storage time that triggers notification and emergency planning?
4. Should an emergency plan be developed for the release of every chemical that ever flowed through the port even though some of those chemicals may never be present in the area again?
5. What notification and emergency planning criteria should be adopted for large quantities of listed materials that frequency flow through the port but are present for only short periods of time?
6. Should limits be placed on quantities of some materials being stored in the port area at a given time?
7. Should ports be classified as to what materials are allowed to enter them?
8. Should segregation of cargo by compatibility groups be required for materials waiting to be loaded or trans-shipped?
9. Should port areas be rezoned to reduce the surrounding population?
10. Are evacuation plans possible for port areas in large cities?
11. Is there sufficient authority under current law to accomplish this task or is new legislation required?
12. What would be the political consequences of requiring foreign ships to adhere to these regulations?
13. What would be the cost of applying these regulations to port areas?

LSUP.27 DISPOSAL AND STORAGE METHODS

Provide information on disposal and storage methods.

SOLUTION: Unusable by-products from many chemical and metal-processing operations may often need to be contained and disposed of as hazardous wastes. Therefore, it is generally advisable to contact the Department of Environmental Protection (or equivalent at the state or local level) or the regional office of the federal EPA for specific recommendations on proper disposal. In all cases, however, disposal should be in accordance with federal, state, and local regulations.

Safe storage practices should be applied to various types of wastes. Storage procedures are intended to prevent incompatible materials from coming into direct contact with each other and to isolate such materials from sources of ignition, such as smoking and open flames. The chemical and physical properties of the materials, the quantities of the materials, and the packaging systems employed will determine the extent of isolation or separation. However, it is always advisable to seek the recommendations of the supplier to obtain more in-depth information concerning incompatible materials with which the chemical should not be stored.

LSUP.28 INCINERATOR OPERATIONAL PARAMETER

Many process and operational variables at a hazardous waste incinerator are continuously monitored. Values outside of desired ranges may signal impending problems with: a) contaminant destruction and removal efficiency, b) equipment integrity, and/or c) worker safety. For each of the following indicators, describe concerns for one or more of the problem categories listed above.

1. High CO in stack gas.
2. Low temperature in the secondary combustion chamber.
3. High combustion gas flow rate.
4. Low scrubber water flow rate.
5. Low quench water flow rate.
6. High pressure in primary or secondary combustion chamber.
7. High temperature in primary or secondary combustion chamber.
8. Loss of flame in primary or secondary combustion chamber.

SOLUTION:

1. High CO can indicate incomplete combustion. Emissions of both polycyclic organic hydrocarbons (POHCs) and products of incomplete combustion (PICs) are likely to be higher under such conditions. Immediate corrective action is necessary.

2. Low temperature in the secondary combustion chamber usually means that DREs for POHCs may not be achieved because of inadequate combustion temperature. Corrective action should occur immediately.

3. High combustion gas flow rate means that residence times within the combustion zone in the incinerator will be lower, reducing the net POHC conversion and potentially producing an exhaust gas that does not meet the required DRE.

4. Low scrubber water flow rate indicates that HCl removal will likely be below design specifications. It may also mean that a removal of 99% HCl or an HCl emission rate ≤4 lb/h required may not be continuously met. There is also a danger to downstream equipment if corrosive gases pass through the scrubber in excessive concentrations. Corrective action is usually immediate.

5. Low quench water flow rate will indicate inadequate quenching which can endanger downstream equipment (particularly plastics) because of excessive (high) fluid gas temperatures. Immediate corrective action is necessary.

6. High pressure in either combustion chamber should not occur since the entire system is operated under negative pressure from the induced draft (ID) fan. High internal incinerator pressures can develop if high Btu wastes are rapidly loaded to the incinerator. If pressure builds in a rotary kiln, emissions can occur from the seals on the kiln, and these emissions will not be thoroughly combusted nor vented from the incinerator area. Unvented seal leakage and high pressure at sufficient levels can lead to exposure and explosion hazards, with resulting danger to workers and damage to equipment.

7. High temperature in either or both combustion chambers can endanger the refractory integrity and thereby the incinerator shell, thus resulting in worker safety (explosion) concerns. High incinerator temperatures can also affect downstream equipment as indicated by low quench water flow.

8. Flame loss in either combustion chamber will allow combustible gases to accumulate within the incinerator to potentially explosive levels, resulting in increased risks to both personnel and equipment. Emissions would be unacceptably high during flame-out conditions and immediate corrective action is necessary.

LSUP.29   RISK

Define risk.

SOLUTION: There are many definitions of the word risk. It is a combination of uncertainty and damage; a ratio of hazards to safeguards; a triplet combination of event, probability, and consequences; or, even a measure of economic loss or human injury in terms of both the incident likelihood and the magnitude of the loss or injury. People face all kinds of risks everyday, some voluntarily and others involuntarily. Therefore, risk plays a very important role in today's world. Studies on cancer caused a turning point in the world of risk because it opened the eyes of risk scientists and health professionals to the world of risk assessments.

The topic of risk also falls under the domain of TSCA and the Clean Air Act. Under Section 112 (for the pre-1990 Clean Air Act), "risk" was used to identify pollutants and sources which were to be covered by NESHAP. For post 1990, CAA used it to determine whether additional controls will be necessary under the "residual risk" program which follows MACT implementation. The topic of "risk" is one of the reasons the revisions to the CAA in 1990 were implemented. For emissions of various toxic substances from certain identified source categories, the law requires controls to be put in place prior to an evaluation of "risk."

LSUP.30   RISK DEFINITIONS

Define the following terms:

1. Risk management
2. Risk communication
3. Risk estimation
4. Risk perception
5. Comparative risk assessment

SOLUTION:

1. *Risk management* is an evaluation of various options to reduce the risk to the exposed population. Risk management usually follows a risk assessment. Specific actions that may be involved in risk management include consideration of engineering constraints, regulatory issues, social issues, political issues, and economic issues.

2. *Risk communication* is the part of the risk management process that includes exchanging risk information among individuals, groups, and government agencies. The major challenge in this phase of the risk management process is transferring information from the experienced expert to the nonexperienced, but often greatly concerned, public.

3. *Risk estimation* is based on the nature and extent of the source, the chain of events, pathways, and processes that connect the cause to the effects. It is also based on the relationship between the characteristics of the impact (dose) and the type of effects (response).

4. *Risk perception* describes an individual's intuitive judgment of the risk. Risk perception is not often in agreement with the actual level of risk.

5. *Comparative risk assessment* is the comparison of potential risks associated with a variety of activities and situations so that a specific action can be placed in perspective with other risks. An attempt is often made, e.g., to compare an individual's risk of death or cancer from exposure to a hazardous

waste site with that associated with traveling in an automobile or eating a peanut butter sandwich (both of these latter events have relatively high risks but are perceived by the public to have a relatively low risk when compared to the risk of a hazardous waste site).

LSUP.31   STANDARD VALUES

To apply risk assessments to large groups of individuals, certain assumptions are usually made about an "average" person's attributes. List the average or standard values used for:

1. Body weight
2. Daily drinking water intake
3. Amount of air breathed per day
4. Expected life span
5. Dermal contact area

SOLUTION:

1. Average body weight is 70 kg for an adult and 10 kg for a child.
2. The average daily drinking water intake is 2 liters for an adult and 1 liter for a child.
3. The average amount of air breathed per day is 20 m^3 for an adult and 10 m^3 for a child.
4. The average expected life span is 70 years.
5. The average dermal contact area is 1000 cm^2 for an adult and 300 cm^2 for a child.

LSUP.32   ACUTE RISKS OR CHRONIC RISKS

In general terms, what is the difference between acute risks (AR) and chronic risks (CR).

SOLUTION: AR. These are the risks associated with short exposures, usually at high concentrations.

CR. These are the risks associated with long term exposures, usually at low concentrations.

LSUP.33   HEALTH RISK ASSESSMENT

Health risk assessment is a topic that falls under the domain of a host of regulatory acts, including OSHA. List and describe the four major steps in a health risk assessment.

SOLUTION: The four major steps in a health risk assessment are hazard identification, dose–response assessment, exposure assessment, and risk characterization (see Figure 8.1).

A *hazard* is defined as a toxic agent or a set of conditions that has the potential to cause adverse effects to human health or the environment. *Hazard identification* is a process that determines the potential human health effects that could result from exposure to a hazard. This process requires a review of the scientific literature. The literature could include information published by the EPA, federal or state agencies, and health organizations.

Dose–response, or *toxicity, assessment* is the determination of how different levels of exposure to a hazard or pollutant affect the likelihood or severity of health effects. Responses/effects can vary widely since all chemicals and contaminants vary in their capacity to cause adverse effects. The dose–response relationship can be evaluated for either carcinogenic or noncarcinogenic substances.

Exposure assessment is the determination of the magnitude of exposure, frequency of exposure, duration of exposure, and routes of exposure by contaminants to human populations and ecosystems. There are three components to this step. The first is the identification of contaminants being released. The second is

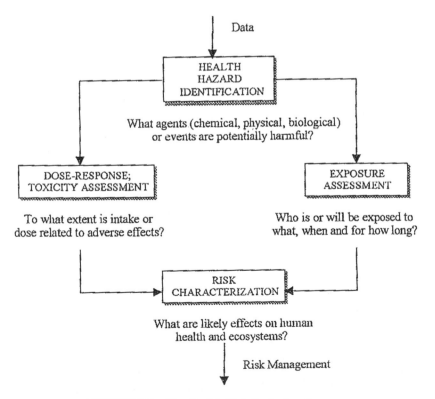

FIGURE 8.1   The Health Risk Evaluation Process.

an estimation of the amounts of contaminants released from all sources or the source of concern. Third, there is an estimation of the concentration of contaminants.

Finally, in *risk characterization*, toxicology and exposure data/information are combined to obtain a qualitative or quantitative expression of risk.

An expanded treatment of and comparative analysis with hazard risk assessment is provided in the following reference: A.M. Flynn and L. Theodore, "Health, Safety, and Accident Management in the Chemical Process Industries," Marcel Dekker (recently acquired by Taylor and Francis Group, Boca Raton, FL), New York City, 2002.

LSUP.34   DOSE RESPONSE

Describe in more detail the Dose–Response assessment process (see previous Problem).

SOLUTION: Dose–response assessment is the process of characterizing the relationship between the dose of an agent administered or received and the incidence of an adverse health effect in exposed populations, as well as estimating the incidence of the effect as a function of exposure to the agent. This process considers such important factors as intensity of exposure, age pattern of exposure, and possibly other variables that might affect response, such as sex, lifestyle, and other modifying factors. A dose–response assessment usually requires extrapolation from high to low doses and extrapolation from animals to humans, or from one laboratory animal species to a wildlife species. A dose–response assessment should describe and justify the methods of extrapolation used to predict incidence, and it should characterize the statistical and biological uncertainties in these methods. When possible, the uncertainties should be described numerically rather than qualitatively.

Toxicologists tend to focus their attention primarily on extrapolations from cancer bioassays. However, there is also a need to evaluate the risks of lower doses to see how they affect the various organs and systems in the body. Many scientific papers focused on the use of a safety factor or uncertainty factor approach since all adverse effects other than cancer and mutation-based developmental effects are believed to have a threshold, i.e., a dose below which no adverse effect should occur. Several researchers have discussed various approaches to setting acceptable daily intakes or exposure limits for developmental and reproductive toxicants. It is thought that an acceptable limit of exposure could be determined using cancer models, but today they are considered inappropriate because of thresholds.

For a variety of reasons, it is difficult to precisely evaluate toxic responses caused by acute exposure to hazardous materials. First, humans experience a wide range of acute adverse health effects, including irritation, narcosis, asphyxiation, sensitization, blindness, organ system damage, and death. In addition, the severity of many of these effects varies with intensity and duration of exposure. Second, there is a high degree of variation in response among individuals in a typical population. Third, for the overwhelming majority of substances encountered in industry, there are not enough data on toxic responses of humans to permit an accurate or precise assessment of the substance's hazard potential. Fourth, many releases involve multi-components. There are presently no rules on how these types of releases should be evaluated. Fifth, there are no toxicology testing protocols that exist for studying episodic releases on animals. In general, this has been a neglected area of toxicology research. There are many useful measures available to use as benchmarks for predicting the likelihood that a release event will result in serious injury or death. Several references review various toxic effects and discuss the use of various established toxicological criteria.

Dangers are not necessarily defined by the presence of a particular chemical, but rather by the amount of that substance one is exposed to, also known as the dose. A dose is usually expressed in milligrams of chemical received per kilogram of body weight per day. For toxic substances other than carcinogens, a threshold dose must be exceeded before a health effect will occur, and for many substances, there is a dosage below which there is no harm. A health effect will occur or at least be detected at the threshold. For carcinogens, it is assumed that there is no threshold, and, therefore, any substance that produces cancer is assumed to produce cancer at any concentration. It is vital to establish the link to cancer and to determine if that risk is acceptable. Analyses of cancer risks are much more complex than non-cancer risks.

Not all contaminants or chemicals are created equal in their capacity to cause adverse effects. Thus, cleanup standards or action levels are based in part on the compounds' toxicological properties. Toxicity data are derived largely from animal experiments in which the animals (primarily mice and rats) are exposed to increasingly higher concentrations or doses. Responses or effects can vary widely from no observable effect to temporary and reversible effects, to permanent injury to organs, to chronic functional impairment to ultimately, death.

For additional information, refer to the aforementioned text by A. M. Flynn and L. Theodore, 2002.

LSUP.35   CALCULATION OF HEALTH RISK

Outline a procedure to calculate health risk.

SOLUTION: As previously discussed, the assessment process includes hazard identification, dose-response assessment, exposure assessment, and risk characterization. The following are typical mathematical calculations involved in characterizing risk.

1. *Exposure (E)*:

$$\text{Pollutant concentration} \,(mg/m^3) \times \text{Exposure duration} \,(days) = E \,(mg \cdot days/m^3)$$

This represents the total exposure over the "days" in question. The days can often be the lifetime of an individual (25,550 days). Thus,

$$(PC)(ED) = E \qquad (8.1)$$

2. *Dose (D)*:

$$E \,(\text{mg} \cdot \text{days/m}^3) \times \text{Receptor dose factor} \,(\text{m}^3/\text{kg} \cdot \text{day}^2)$$
$$= D \,(\text{mg/kg} \cdot \text{day})$$

This represents the average mass of pollutant intake per unit mass of receptor on a daily basis. The Receptor Dose Factor, RDF, is given by

$$RDF = \text{``Contact'' rate} \,(\text{m}^3/\text{day}) \times \text{Intake fraction} \,(\%/100)/\text{Average}$$
$$\text{receptor weight} \,(\text{kg}) \times \text{Exposure duration} \,(\text{days})$$

Thus,

$$(E)(RDF) = D \qquad (8.2)$$

3. *Lifetime Individual Risk (LIR)*:

$$D \,(\text{mg/kg} \cdot \text{day})$$
$$\times \text{Dose-response relationship} \,(\text{kg} \cdot \text{day} \cdot \text{probability/mg} \cdot \text{lifetime})$$
$$= LIR \,(\text{probability/lifetime})$$

This represents an individual's risk over a lifetime. Thus,

$$(D)(DRR) = LIR \qquad (8.3)$$

4. *Risk to Exposed Population (REP)*:

$$LIR \,(\text{probability/lifetime}) \times \text{Exposed population} \,(\text{number of individuals})/$$
$$\text{Years per lifetime} \,(\text{years/lifetime}) = REP \,(\text{number of individuals/year})$$

This provides a reasonable estimate of the number of individuals or cases in an exposed population per year. Thus,

$$(LIR)(EP)/(YPL) = REP$$

The YPL is normally 70 (personal notes: L. Theodore, 1994).

QUANTITATIVE PROBLEMS (TSUP)

PROBLEMS TSUP.1–23

TSUP.1 HEAT RELEASE RATE

A rotary kiln at a Superfund site is designed to treat contaminated soil with a nominal heat release rate (HRR) of 15,000 Btu/(h · ft^3), an inside diameter of 8 ft, and a length of 30 ft. What is the design heat rate ($\dot{Q}$) in Btu/h?

SOLUTION:

$$\text{Kiln volume, } V = (\pi D^2/4)L = \frac{\pi(8)^2 30}{4}$$

$$= 1508 \text{ ft}^3$$

The design heat rate is given by

$$\dot{Q} = (V)(\text{HRR}) \tag{8.1}$$

$$= (1508 \text{ ft}^3)\left(\frac{15,000 \text{ Btu}}{\text{h} \cdot \text{ft}^3}\right) = 22,620,000 \text{ Btu/h}$$

TSUP.2 BATCH FEED TO A KILN

Occasionally a solid waste consisting of contaminated polyethylene pellets is "batch" fed to the kiln in 30-gal fiber containers. The pellets have a bulk density of 50 lb/ft^3 and a heating value (HV) of 18,350 Btu/lb. A single container is consumed in 6.5 min. Assume the kiln described in Problem TSUP.1 is employed in this operation.

1. Will the kiln operate within its design parameters?
2. Considering your answer to part 1, would you expect any combustion problems?

SOLUTION:

1. To determine if the kiln operates within its design parameters, the actual rate should be compared to the design value. The mass m in each container is

$$m = (V)(\rho)$$

$$= (30 \text{ gal})\left[\left(\frac{8.35 \text{ lb}}{\text{gal}}\right)\left(\frac{50}{62.4}\right)\right] = 201 \text{ lb}$$

The actual heat release, Q, is then

$$Q = m(\text{HV}) \tag{8.2}$$

$$= (201 \text{ lb}) \left(\frac{18,350 \text{ Btu}}{\text{lb}} \right)$$

$$= 3,688,000 \text{ Btu heat release})$$

But this will occur in 6.5 min; therefore equivalent hourly heat rate, $\dot{Q}$, is

$$\dot{Q} = \frac{60 \text{ min/h}}{6.5 \text{ min}} (3,688,000 \text{ Btu}) = 34 \text{ million Btu/h}$$

This is about 1.5 times the design heat rate of 22.6 million Btu/h so the answer to part 1 is No.

2. One problem may be that there will not be enough air in the system to burn the polyethylene waste at that rate so the kiln will most probably produce a dense black smoke.

TSUP.3   KILN OPERATOR PROBLEM

Refer to Problem TSUP.2 How could the problem be corrected if you were the operator?

SOLUTION: To correct the situation, package the polyethylene in smaller batches. The batch size will have to be reduced by a factor of 1.5. Therefore,

$$m = 201/1.5 = 134 \text{ lb/batch}$$
$$Q = (134)(18,350) = 2,459,000 \text{ Btu heat release}$$

The heat rate is then

$$\dot{Q} = (60/6.5)(2,459,000) = 22,697,500 \text{ Btu/h}$$

This is almost the same as the design value. Good practice would suggest that a safety margin of at least 10% be used so about 120 lb/batch would be a good size.

Also, the containers must be fed at a rate not greater than one every 6.5 min, e.g., 9 per hour.

TSUP.4   SIZING OF A ROTARY KILN

A rotary kiln incinerator is operating with an average energy release rate (HRR) of 28,000 Btu/h · ft³ of furnace volume. During operation, 4500 lb/h of solid waste from a Superfund site with an approximate heating value (HR) of 8000 Btu/lb is to be combusted. Assume the length to diameter ratio (L/D) of the rotary kiln to be 3.5.

1. Calculate the furnace volume required.
2. What are the dimensions of the kiln?

SOLUTION:

1. Calculate the heat released by the waste:

$$\dot{Q} = (\dot{m})(\text{HR})$$

$$= (4500)(8000) = 3.60 \times 10^7 \text{ Btu/h}$$

The volume required is

$$V = \frac{\dot{Q}}{\text{HRR}} = \frac{3.60 \times 10^7}{28,000} = 1286 \text{ ft}^3 \qquad (8.6)$$

2. The dimensions of the kiln for an L/D ration of 3.5 are obtained as follows:

$$V = \left(\frac{\pi}{4}\right)D^2 L = \left(\frac{\pi}{4}\right)D^2(3.5D) = 2.75D^3 \qquad (8.7)$$

$$D = \left(\frac{V}{2.75}\right)^{1/3} = \left(\frac{1286}{2.75}\right)^{1/3} = 7.76 \text{ ft}$$

$$L = 3.5D = 3.5(7.76) = 27.2 \text{ ft}$$

TSUP.5 CALCULATION OF RECEIVING TANK SIZE

Tests indicate that a sludge waste arriving in 55-gal drums to a treatment facility have a mean lead content of 11 ppm with a standard deviation of 10 ppm (normally distributed). The drums are unloaded into a 250-gal receiving tank. The facility is required to keep the lead concentration entering the incinerator at or below 15 ppm in order to meet the required particulate emission levels. Assume that the lead contents from one drum to the next are not correlated and that the tank is nearly full. What size should the receiving tank be to ensure, with 98% confidence, that the facility treats a waste with a mean lead concentration below 15 ppm?

SOLUTION: For this condition, the probability that the Pb concentration in the receiving tank exceeds 15 ppm is

$$1.0 - 0.98 = 0.02 \quad \text{or} \quad 2.0\%$$

The value of z_0 from the standard normal table (see Table 5.5 in Chapter 5) is then 2.05, which corresponds to the number of standard deviations above the mean tank concentration. According to the *central limit theorem*, the standard deviation of the mean is given by

$$\frac{\sigma}{\sqrt{n}} = \frac{10 \text{ ppm}}{\sqrt{n}}$$

where n is the number of drums. Therefore, for a 2% probability that the mean concentration in the tank exceeds 15 ppm, the number of drums can be found by solving

$$z_0 = \frac{\bar{x} - \mu}{\sigma/\sqrt{n}}$$

$$2.05 = \frac{15 - 11}{(10/\sqrt{n})}$$

or

$$n = 26.3$$

The tank volume V should then be

$$V = 26.3 \text{ drums } (55 \text{ gal/drum}) = 1446 \text{ gal}$$

TSUP.6   ESTIMATING INCINERATOR TEMPERATURE

One of the authors (Personal notes: L. Theodore, 1990) has developed the following equation to estimate the temperature (T,°F) in an incinerator in terms of the net heating value (NHV, Btu/lb) of any waste and the excess air (EA, fractional percent basis).

$$T = 60 + \frac{NHV}{(0.3)[1 + (1 + EA)(7.5 \times 10^{-4})(NHV)]} \tag{8.8}$$

Calculate the temperature in an incinerator burning a sludge with a NHV = 12,000 Btu/lb if the EA is

1. 25%
2. 50%
3. 75%
4. 100%
5. 0%

SOLUTION:

1. Employing the above equation given (for 25% EA):

$$T(25\% \text{ EA}) = 60 + \frac{12,000}{(0.3)[1 + (1 + 0.25)(7.5 \times 10^{-4})(12,000)]}$$

$$= 3325°F$$

2–5. Similarly,

$$T(50\% \text{ EA}) = 2819°\text{F}$$
$$T(75\% \text{ EA}) = 2448°\text{F}$$
$$T(100\% \text{ EA}) = 2165°\text{F}$$
$$T(0\% \text{ EA}) = 4060°\text{F}$$

Additional details are available in the following reference: J. Santoleri et al., "Introduction to Hazardous Waste Incineration," 2nd Edition, John Wiley & Sons, Hoboken, NJ, 2000.

TSUP.7 SUPERFUND WASTE INCINERATION TEMPERATURE SUPERFUND

Estimate the theoretical flame temperature of a waste mixture containing 25% cellulose, 35% motor oil, 15% water (vapor) and 25% inerts, by mass. Assume 5% radiant heat losses. The flue gas contains 11.8% CO_2, 13 ppm CO, and 10.4% O_2 (dry basis) by volume.

$$\text{NHV of cellulose} = 14,000 \text{ Btu/lb}$$
$$\text{NHV of motor oil} = 25,000 \text{ Btu/lb}$$
$$\text{NHV of water} = 0 \text{ Btu/lb}$$
$$\text{NHV of inerts} = -1,000 \text{ Btu/lb}$$

SOLUTION: Determine the NHV for the mixture.

$$\text{NHV} = 0.25 \, (14,000 \text{ Btu/lb}) + 0.35(25,000 \text{ Btu/lb}) + 0.15 \, (0.0 \text{ Btu/lb})$$
$$+ \, 0.25 \, (-1,000 \text{ Btu/lb})$$
$$= 12,000 \text{ Btu/lb}$$

Also determine the excess air employed.

$$\text{EA} = 0.95Y/(21 - Y); \quad Y = \text{percent oxygen} \tag{8.9}$$
$$= 0.95 \, (10.4)/(21 - 10.4)$$
$$= 0.932$$

Calculate the flame temperature using the equation provided in Problem TSUP.6.

$$T = 60 + \text{NHV}/[\{0.3\}\{1 + (1 + \text{EA})(7 \times 10^{-4})(\text{NHV})\}]$$
$$= 60 + 12,000/[\{0.3\}\{1 + (1 + 0.932)(7.5 \times 10^{-4})(12,000)\}]$$
$$= 2235°\text{F}$$

TSUP.8 CHOICE OF AN INCINERATION SYSTEM

A Superfund site has been studied and is ready for remedial action. The site contains approximately 40,000 cubic yards of soil contaminated by wood treating operations. The plant is in flat terrain in a rural location. The contaminants are from coal tar compounds and metals used to treat wood. The following data have been summarized from the remedial investigation/feasibility study (RI/FS).

The project is bound by RCRA regulations, not TSCA (Toxic Substances Control Act), as the PCP content (see Table 8.4) is low. The schedule calls for the production burn period to be less than 8 months. The required ash quality goal is <5 ppm total PAH (polyaromatic hydrocarbons) compounds.

TABLE 8.4 Incinerator Data

Item	Value
Density of soil	1.5 ton/yd^3
Moisture content	10%
Heating value	1,000 Btu/lb
Ash content	85%
Soil makeup	
Clay content	80%
Sand and gravel	20%
Major contaminants	
Creosote PAH compounds	30,000 ppm
Pentachlorophenol (PCP)	100 ppm
Chromium	1 ppm
Arsenic	0.5 ppm
Ash fusion temperature	2,100°F

Based on the above information and that provided in Table 8.4, choose an appropriate incineration system. Provide the following information:

1. Size the system in terms of throughout in tons per hour. Assume 65% capacity utilization.
2. Specify the type of incinerator (e.g., rotary kiln, fluid bed, infrared, indirect fired desorber).

SOLUTION:

1. The system size is to be based on 8 months of production burn time. This is 5840 operating hours. Dividing the tonnage (1.5 × 40,000 yards) by this value results in a feed rate of 10 tons per hour (tph). Applying a 65% capacity utilization (to account for forced and planned outages) for this project duration results in a capacity of 16 tph.
2. The predominant type of incinerator used for this type of project is the rotary kiln. A co-current (cools faster at inlet) type is preferred for feedstocks with

high heating values. A fluid bed could also be used due to the high boiling points of the PAH compounds, the stringent ash quality goal, and relatively high PAH concentrations.

Personal notes: J. Santoleri and L. Theodore, 2000.

TSUP.9 INCINERATOR EMISSION CONTROL

Refer to Problem TSUP.8 provide the following information:

1. Type of pollution control (e.g., venturi scrubber, baghouse, etc.)
2. List primary RCRA required stack tests.

SOLUTION:

1. The dust loading will be high as the feedstock is clay (which is very fine) plus metals. Therefore, the best choice for pollution control would be a baghouse to remove the particulate. A venturi scrubber would be a poor choice as it has difficulty in removing fine particulates and the scrubber water would have high suspended solids, requiring a high blowdown rate.

   Is an acid gas absorber needed? The pentachlorophenol feed rate at 16 tph of soil is 3.2 lb/h. It is not all chlorine and will produce less than 4 lb/h of HCl; hence, no acid gas absorber is required. As long as a nonchlorinated principal organic hazardous constituents (POHC) (e.g., naphthalene) is used in the destruction and removal efficiency (DRE) tests, HCl should not be a problem. Had HCl been more than the allowable limit, it could be taken care of with a wet acid gas absorber placed downstream of the baghouse or a spray dryer upstream of the baghouse (using a lime slurry as the reagent) or dry reagent injection into the baghouse if only low levels of acid were to be removed.
2. Primary RCRA regulations call for a DRE of a least 99.99% of all POHCs. The regulations also require 99% removal of HCl from the incinerator flue gas or a maximum of 4 lb/h (1.8 kg/h). RCRA performance standards (40 CFR 264.343) require a maximum particulate emission of 0.08 gr/dscf corrected to 7% oxygen in the flue gas. In addition, many states require BACT (best available control technology). Metals emissions are dealt with in a later Problem, TSUP.12.

Personal notes: J. Santoleri and L. Theodore, 2000.

TSUP.10 FUEL TANK SPILL I

On January 2, 1988, a fuel oil tank at an Ashland Oil terminal in Pennsylvania ruptured and a 35 ft high wave of 600,000 gallons of Number 2 distillate fuel oil surged out over a containment dike into the Monongahela River. In this case, the containment dike was breached by the violence of the release of oil that surged over the dike. Assume a slightly different case in which a similar tank (containing 3.9 million gal of fuel) ruptures slowly and the entire contents of the tank is to be retained by a circular 5 ft high dike.

1. If the radius of the dike is 192 ft, how far from the top of the dike will the level of the oil be?
2. The hazard in this case stems from the fact that rain begins to fall just after the rupture of the tank. The heaviest daily rainfall recorded in this region of Pennsylvania over a recent 30 yr period is 5.68 inches. The fuel is immiscible with water and has a lower density. Assume the worst foreseeable case (record breaking rainfall for a long period in the future), and calculate how long it will take for the oil level to reach the top of the dike.

SOLUTION:

1. The depth of the fuel, h, held in the dike is calculated based on the volume of the spill divided by the cross-sectional area of the dike:

$$h = V/A = /\pi r^2 \tag{8.10}$$

$$= (3.9 \times 10^6 \text{gal})(0.1337 \text{ ft}^3/\text{gal})/[\pi(192 \text{ ft})^2]$$

$$= 4.5 \text{ ft}$$

The oil will therefore rise to a level that is approximately 0.5 ft below the top of the dike.

2. The allowable height H is

$$H = 5.0 - 4.5$$

$$= 0.5 \text{ ft}$$

The time to rise to the top of the dike is

$$t = H/\text{rainfall rate}$$

$$= 0.5 \text{ ft}(12 \text{ in/ft})(1 \text{ d}/5.68 \text{ in}) = 1.05 \text{ d}$$

Because the water will sink below the oil, the oil surface will rise to the top of the dike in approximately 1 d of the heavy rains assumed in the problem statement.

TSUP.11 TANK SPILL II

Refer to Problem TSUP.10. Would this hazard be different if the tank had contained chloroform ($CHCl_3$, density $= 1.5$ g/mL, water solubility $= 1$ mL/200 mL water at 25°C)? What if it had contained ethylene glycol ($HOCH_2CH_2OH$, density $= 1.12$ g/mL, miscible with water)?

SOLUTION: Chloroform (like many other chlorinated organic compounds) is denser than water. Rainwater will collect on top of the chloroform and only that portion which dissolves in water is going to be carried over the top of the dike. Unfortunately, the solubility of chloroform in water is substantial and it is a potential human carcinogen. The hazard, in this case, is probably worse than in the case of the fuel oil.

Ethylene glycol is used as an antifreeze in automobile radiators and mixes freely with water in all proportions. Heavy rainfall will lead to serious discharges over the dike as the ethylene glycol-water mixture rises. Toxic effects from ethylene glycol ingestion resemble those of alcoholism, and a massive spill, like that assumed in this problem, could lead to health risks if it endangered nearby water supplies.

TSUP.12 MERCURY EMISSIONS

Refer to Problem TRCRA.18.

An incinerator burns mercury-contaminated waste, from a Superfund site. The waste material has an ash content of 1%. The solid waste feed rate is 1000 lb/h and the gas flowrate is 20,000 dscfm. It is reported that the average mercury content in the particulates was 2.42 μg/g when the vapor concentration was 0.3 mg/dscm. For the case where incinerator emissions meet the particulate standard of 0.08 gr/dscf (0.1832 g/dscm) with a 99.5% efficient electrostatic precipitator (ESP), calculate the amount of mercury in grams/day leaving the stack as a vapor and with the fly ash.

SOLUTION: As noted in Problem TRCRA 18, the amount of ash leaving the stack is 329 lb/day.

The amount of mercury leaving the stack with the fly ash is

$$(329 \text{ lb ash/day})(2.42 \times 10^{-6} \text{g Hg/g ash}) = 7.96 \times 10^{-4} \text{ lb Hg/day}$$
$$= 0.361 \text{ g Hg/day}$$

The amount of mercury leaving the stack as vapor is

$$\left(\frac{0.3 \times 10^{-3} \text{g Hg}}{\text{dscm}}\right)\left(\frac{20,000 \text{ dscf}}{1 \text{ min}}\right)\left(\frac{1 \text{ m}^3}{35.3 \text{ ft}^3}\right)\left(\frac{60 \text{ min}}{\text{h}}\right)\left(\frac{24 \text{ h}}{\text{day}}\right) = 244.8 \text{ g/day}$$

Total mercury leaving the stack $= 244.8 + 0.361 = 245.2$ g/day. As expected, the amount of mercury leaving the stack with the particulate is negligible relative to that leaving as vapor.

TSUP.13 WORST CASE DISCHARGE PLANNING VOLUME

In accordance with 40 CFR 112 "Oil Pollution Prevention," the owner or operator of an offshore oil production facility must determine the worst case spill discharge for that facility. These figures are then used for emergency response planning. The following problem provides an example of how the well production volume should be determined for this planning process.

A facility consists of two production wells producing under pressure, which are both less than 10,000 feet deep. The well rate of well A is 5 barrels per day, and the well rate of well B is 10 barrels per day. The facility is unattended for a maximum of 7 days. The facility operator estimates that it will take 2 days to have the response equipment and personnel on scene and responding to a blowout, and that the projected rate of recovery will be 20 barrels per day.

Calculate the production volume that should be used to calculate the worst-case discharge planning volume using the following formulas and guidelines:

Step 1:   The ratio of well rate to recovery rate (R_{well}) is first determined. If this ratio is greater than one (i.e., the well rate would overwhelm response efforts), then *Method A* is used. If this ratio is less than one, then *Method B* is used.

Step 2:   Use appropriate calculation methods below to determine production volume.
   Method A:
   (a) For wells 10,000 feet or less: Production volume = 30 days × rate of well
   (b) For wells deeper than 10,000 feet: Production volume = 45 days × rate of well

   Method B:
   Production Volume = Discharge volume #1 + discharge volume # 2
   Discharge   Volume #1 = volume released during time between blowout and arrival of response team
                = (days unattended + response days) × well rate
   Discharge Volume #2 = (<10,000-feet deep) = [30 days − (days unattended + response days)] × well rate × R_{well}
   (>10,000-feet deep) = [45 days − (days unattended + response days)] × well rate × R_{well}

SOLUTION: First apply the guidelines in step 1.
   Step 1:   10-barrels per day ÷ 20-barrels per day = 0.5
          Calculation Method = Method B
          Finally, apply the guidelines provided in Method B.
   Step 2:   Discharge Volume #1 = (7 days + 2 days) × 10 barrels per day = 90 barrels
          Discharge Volume #2 = [30 days − (7 days + 2 days)] × 10 barrels per day × 0.5 = 105 barrels
          Production Volume = 90 barrels + 105 barrels = 195 barrels

TSUP.14   LANDFILLING FEASIBILITY

The total amount of superfund contaminated soil at the LAST (Leo Stander and Theodore) company site is approximately 80,000 tons. Evaluate the cost of landfilling versus stabilization for the management of the hazardous waste at this site.

Each truck can carry 25,000 lb to the nearest suitable landfill site at Theodore Estates, at a distance of 750 miles. The trucking cost per mile is $2.50. The total stabilization cost is $62 per ton. Identify advantages and disadvantages of landfilling for this site.

SOLUTION: Stabilization cost estimate:

$$80,000 \text{ tons} \times \$62/\text{ton} = \$4,960,000$$

Trucking cost estimate:

$$80,000 \text{ tons} \left(\frac{2000 \text{ lb}}{\text{ton}}\right)\left(\frac{\text{truck}}{25,000 \text{ lb}}\right) = 6400 \text{ trucks}$$

$$6400 \text{ trucks}\left(\frac{750 \text{ miles}}{\text{trucks}}\right)\left(\frac{\$2.50}{\text{mile}}\right) = \$12,000,000$$

Because trucking costs alone are $12 million, stabilization is certainly more cost effective.

Advantages of landfilling: Landfilling permits the actual reclamation of the contaminated land at the LAST company site and eliminates one site from the national inventory of contaminated sites.

Disadvantages of landfilling: Valuable landfill space is being used by soils amenable to other treatment, precluding the use of that space by wastes better suited for landfill.

Additional details on landfilling can be found in Chapter 10.

TSUP.15 DIKE FAILURE

A chemical plant site in a low lying area has been "protected" from soil contamination via flooding by constructing a dike to hold back annual floods with a 100 yr recurrence interval.

1. What is the risk that the dike will be overtopped in a 15 yr period?
2. What recurrence interval should be used to reduce the risk to 5% that the dike will be overtopped in a 15 yr period?

SOLUTION:

1. The probability that the dike will be overtopped in a given year is $1/100 = 0.01$. The probability that the dike will not be overtopped in a given year is $1 - 0.01 = 0.99$. The probability that the dike will not be overtopped in a 15 year period is:

$$\overline{P}_{15} = (0.99)^{15} = 0.86$$

Therefore, the probability that the dike will be overtopped in a 15 yr period is:

$$P_{15} = 1 - 0.86 = 0.14$$

So the risk of failure is 14%.

2. If the risk of overtopping is to be reduced to 5% from 14% in a 15 yr period, the probability of not overtopping is:

$$\bar{P}_{15} = (1 - 0.05) = 0.95 = \bar{P}_1^{15}$$

$$\bar{P}_1 = 0.9966$$

The probability of overtopping in a year is then $1 - 0.9966 = 0.0034$, which yields a corresponding recurrence period of:

$$T = 1/0.0034 = 294 \, yr$$

TSUP.16 LANDFILL LEAK TO AQUIFER

A leak of trichloroethylene (TCE) has been detected at a superfund landfill approximately 60m from Smallville, USA. Water is moving in the aquifer at a rate of 100 m/yr, while the TCE is moving at a rate of 20 m/yr in the same direction as the ground water flow. Radial dispersion can be assumed to be insignificant relative to the axial velocity. Data indicates the ground water flow pattern in the aquifer is homogenous and isotropic. Development of a new pumping well will cost $2,000,000 and will take six months. A sampling well can be developed within one month at a cost of $50,000. Smallville has a tax base of 300 middle and lower middle class households, and has very little political clout.

1. Determine how many days it will take the TCE to reach the Smallville's aquifer.
2. Determine how soon Smallville must decide whether to relocate their wells or not in order to prevent contamination of their water supply.

SOLUTION:

1. If the contaminant stays in the ground water flowstreams and does not exhibit any significant lateral dispersion, then it may miss the Smallville aquifer completely. At the other extreme, the direct distance currently between the TCE and Smallville measures approximately 60 m. At this distance, the TCE should reach Smallville in:

$$60 \, m/(20 \, m/yr) = 3 \, yr$$

These analyses are based on available data and an assumption that the geological features are uniform in the region surrounding Smallville. The actual time to reach the aquifer could be reduced if shortcircuiting occurred through formations not evident in these preliminary survey results.

2. Assuming Smallville does not want to risk contamination, the latest day for starting a new well would be approximately 900 days from the initial detection of the plume since it takes approximately 200 days to develop a new well; however, 700 days would probably provide a reasonable margin of safety. These dates would be subsequent to obtaining the various regulatory approvals, a procedure that could add at least one year to the overall project schedule. An investment of $100,000 in sampling wells might be justified to confirm the actual contaminant movement rate and to possibly avoid avoid drilling a new well if the contaminant will not enter the Smallville aquifer.

TSUP.17 SUPERFUND AQUIFER LEAK OPTION I

Refer to Problem TSUP.16. Summarize other options Smallville might consider as alternatives to drilling new wells.

SOLUTION: Among the other alternatives that could be considered are removal of the contaminant from the ground, piping in water from another supply, and addition of a treatment process for the water withdrawn from the aquifer. This latter option is normally at a lower capital cost than the first two by at least an order of magnitude but requires a continuing expenditure for operation. The unit could probably be installed for $50,000 (assuming something as simple as an activated carbon adsorber with a TCE detector on the discharge) and would remove a variety of contaminants.

TSUP.18 SUPERFUND AQUIFER LEAK OPTION II

Refer to Problem TSUP.16. Summarize the various economic impacts associated with this incident.

SOLUTION: Funds from all operations should come first from the operator of the landfill and then from the government if the operator is unable to supply the funds. Practically speaking, it would probably take at least five years in court to obtain these funds from either source, so the township will have to arrange for interim financing for any of the options noted above.

TSUP.19 RETARDATION FACTOR CALCULATION

Refer to Problem TSDWA.13. Answer the following two questions.

1. What is the dissolved benzene retardation factor (R) assuming that the soil organic carbon fraction (f_{oc}) is 0.5%?

2. What is the rate of biodegradation expected for this benzene?

   Note: An equation that may be employed to estimate the retardation factor is:

$$R = 1 + f_{oc}(K_{oc})(\rho_b)/n \qquad (8.11)$$

where f_{oc} is the fraction of organic carbon in the soil; K_{oc} = the organic carbon normalized soil/water partition coefficient; ρ_b = the bulk density of the aquifer solids; and n = the aquifer solid total porosity. For benzene, K_{oc} is reasonably approximated by its octanol/water partition coefficient, $K_{ow} \approx 100$ (mL water/ g octanol). Typical values of n and ρ_b for aquifer solids are 0.3 and 2 g/mL, respectively.

SOLUTION:

1. The movement of benzene would be retarded by the presence of the organic matter in the aquifer solids. The retardation factor is determined using the formula, and the benzene and the aquifer input data given in the problem statement. Thus, an estimate of the retardation factor is (see Equation (8.11) above).

$$R = 1 + \frac{(0.005 \text{ g C/gsoil})(100 \text{ mL/g})(2 \text{ g/mL})}{(0.30)} = 1 + 3.33$$

$$= 4.33$$

   or the benzene will travel less then $\frac{1}{4}$ the speed of the aquifer pore water velocity.

2. Among aromatic compounds, benzene has a relatively high biodegradability. In fact, studies have shown that in many aquifer systems, benzene will degrade in the presence of excess oxygen at rates ranging from 0.1% to 10%/d at temperatures of 10°C, as long as the concentrations are below levels that would inhibit growth of microorganisms in the subsurface. This is an additional benefit in light of the significant retardation factor for benzene in these aquifer solids.

TSUP.20 TOXIC TRAVEL TIME IN SOIL

The velocity of a toxic contaminant in groundwater is slowed by the presence of organic matter in the soil into which the contaminant will partition. Under equilibrium partitioning conditions, the contaminant velocity is related to the pore water velocity by

$$v_c = v/R \qquad (8.12)$$

where v_c = contaminant velocity, length/time

v = pore water velocity, length/time

R = retardation factor, dimensionless

For naphthalene in a particular aquifer, R has been found experimentally to equal 80. If the pore water velocity from a source to a well at a distance of 2000 m is 5×10^{-3} cm/s, what is the travel time of naphthalene to the well?

SOLUTION: The pore water velocity, $v = 5 \times 10^{-3}$ cm/s, and the retardation factor, $R = 70$, are given in the problem statement. Thus, the velocity of the naphthalene is determined as follows:

$$v_c = \frac{v}{R}$$

$$= \frac{5 \times 10^{-3} \, \text{cm/s}}{80}$$

$$= 6.25 \times 10^{-5} \, \text{cm/s}$$

The naphthalene travel time is simply the distance divided by naphthalene's retardation velocity:

$$t = \frac{2000 \, \text{m}}{(6.25 \times 10^{-5} \, \text{cm/s})(1 \, \text{m}/100 \, \text{cm})}$$

$$= 3.20 \times 10^9 \, \text{s}$$

$$= 101.5 \, \text{yr}$$

This long travel time is due to naphthalene's extremely low retardation velocity.

TSUP.21   TOXIC CHEMICAL EXPOSURE

Hazardous waste sites provide an easy opportunity for individuals to be exposed to some of the most dangerous poisons known. OSHA standards (29 CFR 1910) require protection for workers involved in hazardous waste remediation. Protection must be provided through engineering controls, work practices, and/or personal protective equipment (PPE).

Hazardous waste remediation workers and industrial/manufacturing workers have a greater risk of toxic chemical overexposure and/or poisoning than the general public due to the nature of their work. Each of the chemicals listed in this problem is commonly used in industry. Toluene and methylene chloride are common solvents and act as narcotics at high concentrations. Vinyl chloride is used in the plastics industry, and Aroclor 1260 [polychlorinated biphenyl (PCB)] is used in electrical capacitors and transformers. Both are suspected carcinogens. Pentachlorophenol finds application as an insecticide and wood preservative. It causes damage to the lungs, liver, and kidneys, and also causes contact dermatitis.

New York City obtains the majority of its drinking water from several reservoirs in the Catskill region of New York State. At an abandoned site approximately $\frac{1}{4}$ mile

TABLE 8.5 Solubility and Henry's Law Constant Data

Chemical	Form	Water Solubility (mg/L)	Henry's Law Constant [atm·m^3/(gmol)]	Density (g/mL)
Methylene chloride	Liquid	6900	0.003	1.32
Vinyl chloride	Liquid	1.1	2.40	0.91
Toluene	Liquid	535.0	0.0059	0.87
Pentachlorophenol	Crystal	14.0	N/A	1.98
Aroclor 1260 (PCB)	Resin	0.0027	0.0071	>1

from one such reservoir, several dozen leaking barrels of toxic chemicals have been discovered in a buried trench.

Given the following limited information for five of these chemicals, discuss each compound in relative terms of its potential hazards to (a) hazardous waste remediation workers involved in sampling and cleanup of the site and (b) ground and surface water supplies.

Assume the barrels have corroded and have been leaking slowly over several years. The site consists of porous soil, is at a slightly higher elevation than the reservoir, and is situated over a groundwater source.

Data for the five chemicals are provided in Table 8.5.

SOLUTION: Methylene chloride, vinyl chloride, and toluene will be present in liquid form in the trench. All would also be expected to be present in the vapor state, with vinyl chloride being the most volatile due to its high Henry's law constant. The principal dangers to site remediation workers, therefore, would be from inhalation of vapors and skin absorption of contacted liquids. Pentachlorophenol and PCBs will be present in solid and semisolid forms. Minimal vapors of these compounds would be expected. The principal dangers would be from inhalation of dusts and fumes, ingestion of solids, and skin absorption due to solids contact. In the absence of engineering controls and work practices, the necessary worker PPE would include, at a minimum, respirators and chemical protective clothing.

The higher a compound's water solubility, the more easily the compound will disperse within a water source. Methylene chloride would be expected to spread rapidly upon reaching groundwater and would be most likely to find its way into the reservoir. Toluene will dissolve more readily than the remaining compounds. Thus, it too can disperse to a greater extent. Vinyl chloride, pentachlorophenol, and Aroclor 1260 have low or very low water solubilities. These compounds would not travel large distances in stagnant water.

The compounds with relatively low Henry's law constants (methylene chloride, toluene, pentachlorophenol, and Aroclor 1260) have the lowest potential volatility. They would not evaporate readily and would therefore present a higher risk to the ground and surface waters. Vinyl chloride has the highest relative volatility and would present the least risk to the ground and surface water.

Methylene chloride, pentachlorophenol, and Aroclor 1260 all have densities greater than that of water, i.e., greater than 1.0 g/mL. As such, these compounds would be expected to sink to the bottom of the ground and surface water bodies. Vinyl chloride and toluene have densities less than that of water and would be expected to float on the ground and surface water surfaces. Chemicals sinking to and settling on the bottom of a water body generally present a lesser risk to humans and animal life unless disturbed. Chemicals that float on water surfaces are more likely to contact human and animal life because they can be carried large distances by winds and currents.

TSUP.22 SOIL REMEDIATION

A Dense Non-Aqueous Phase Liquid (DNAPL) is a low-solubility compound with density greater than water; an example is trichloroethylene (TCE). DNAPLs are also called "sinkers" since they drop to the bottom of an aquifer upon contact with the water table. Once in the soil or the ground water, DNAPLs are extremely difficult to remove.

The TCE spill has contaminated an aquifer where the specific discharge (ground water flow per unit area) is $0.02 \text{ m}^3/\text{m}^2 \cdot \text{d}$.

Assume a cube of soil, 1 m on each side, where the ground water flows normal to one of the faces (see Figure 8.2 below). The porosity (η) of the soil is 0.35 (the fraction of the void space occupied by the fluid) initially and the residual saturation is 20% for TCE and 80% for water. If the proposed removal mechanism is solely by dissolution, find:

1. The number of pore volumes required to remove all the TCE. Assume that all water passing through the cube is available for dissolution of the TCE.
2. The time to remove all of the TCE.

The density of TCE is $1{,}470 \text{ kg}/\text{m}^3$ and its solubility in water is 1.1 g/L.

SOLUTION:

1. The number of pore volumes for flushing is dependent on the total TCE volume, the total TCE mass, the solubility of TCE in water, and the volume

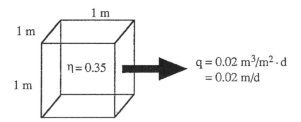

FIGURE 8.2 Remediation physical mode.

of water in a pore volume.

$$\text{initial TCE volume} = (0.35)(0.20)1\,\text{m}^3 = 0.07\,\text{m}^3$$
$$\text{initial TCE mass} = (0.07\,\text{m}^3)(1{,}470\,\text{kg/m}^3) = 103\,\text{kg}$$

The water volume needed to dissolve 103 kg of TCE is:

$$V = (103\,\text{kg})(1000\,\text{g/kg})/1.1\,\text{g/L}$$
$$= 93{,}600\,\text{L} = 93.6\,\text{m}^3$$

The water in 1 pore volume (see Figure 8.1) is:

$$0.35(1\,\text{m}^3)(0.8) = 0.28\,\text{m}^3$$

The number of pore volumes (m) to flush out the TCE is:

$$m = \text{required water volume}/\text{water volume per pore}$$

$$93.6\,\text{m}^3/0.28\,\text{m}^3 = 334.3$$

2. Removal time of TCE is related to the permeability of the aquifer and the number of pore volumes that must be moved.
   Flow through one face of the soil cube is:

$$Q = (0.02\,\text{m/d})(1\,\text{m}^2) = 0.02\,\text{m}^3/\text{d}$$

The time to dissolve all of the residual TCE is:

$$t = 93.6\,\text{m}^3/(0.02\,\text{m}^3/\text{d}) = 4{,}680\,\text{d} = 12.8\ \text{yr}$$

This required time can also be determined from the effective pore velocity and average retention time of the ground water in one pore volume.

$$\text{pore velocity} = (0.02\,\text{m/d})/[(0.35)(0.80)] = 0.071\,\text{m/d}$$

The time to travel through the cube, i.e., the pore volume retention time is:

$$\text{pore retention time} = 1\,\text{m}/(0.071\,\text{m/d}) = 14\ \text{d}$$

Therefore, the time to move 334.3 pore volumes is:

$$\text{Time to dissolve all TCE} = 334.3 \,\text{pore volumes} \, (14 \, \text{d/pore/volume})$$
$$= 4680 \,\text{d} = 12.8 \,\text{yr}$$

Note: As the TCE is flushed out, the residual saturations will change so that the water volume in the soil at the beginning of the process is less than the volume at the end. Since the water residual saturation is high, this correction is small, and can safely be neglected. Also, this assumes equilibrium conditions during "pumping"; in reality this never occurs, with reported extraction efficiency generally varying from only 10 to 50%. As this problem illustrates, it takes a long time to remediate contaminated water using pump-and-treat technology alone.

TSUP.23 CHROMIUM SLUDGE

A tannery uses a filter press to dewater its raw sludge. The dewatered sludge contains 120 mg of total chromium/kg of sludge. The filter press has a surface area of 150 ft^2 with a filtration rate of 10 gal/hr · ft^2. Assume that it is 100% efficient in separating solids. Estimate the amount of total chromium disposed/yr from this plant if the plant operates 16 hr/d, and 7 d/wk, 50 wk/yr. The following information is available.

Before dewatering:

Water content of raw sludge $= 95\%$; Solid content of raw sludge $= 5\%$

After Dewatering:

Water content $= 52\%$; Solid content $= 48\%$

SOLUTION: This is a mass balance problem. The mass of chromium disposed will be the amount that is present in the dewatered sludge. First estimate the amount of dewatered sludge from the given data.

The amount of filtrate/d is

$$q = \text{filter area (volume of filtrate/hr · ft}^2) \, (\text{h of operation/d})$$
$$= 150 \,\text{ft}^2 (10 \,\text{gal/hr} \cdot \text{ft}^2) \, (16 \,\text{h/d}) = 24{,}000 \,\text{gal/d}$$

Develop mass balance equations relating the raw sludge and the dewatered sludge as follows:

Mass Balance Equation 1:

Mass of raw sludge = mass of dewatered sludge + mass of filtrate

Mass Balance Equation 2:

$$\text{Mass of solids in the raw sludge} = \text{mass of solids in the dewatered sludge}$$
$$+ \text{mass of solid in filtrate}$$

With these equations, the following input data are available:

mass of raw sludge not known
mass of dewatered sludge not known $= m_s$
mass of filtrate $= 24{,}000 \text{ gal/d } (8.34 \text{ lb/gal}) = 200{,}000 \text{ lb/d}$
mass fraction of solids in the sludge $= 0.05$ (mass of raw sludge)
mass fraction of solids in the dewatered sludge $= 0.47$ (mass of dewatered sludge)

Dividing Equation 2 by Equation 1 yields:

$$\frac{\text{mass of solids in raw sludge}}{\text{mass of raw sludge}} = \frac{\text{mass of solids in dewatered sludge}}{\text{mass of dewatered sludge} + \text{filtrate}}$$

$$0.05 = \frac{0.47 \, m_s}{m_s + 200{,}000}$$

Rearranging and solving,

$$0.05 \, m_s + 10{,}000 = 0.47 \, m_s$$
$$10{,}000 = 0.42 \, m_s$$
$$m_s = 23{,}830 \, \text{lb/d}$$

Thus, the mass rate of dewatered sludge is 23,830 lb/d.

The annual amount of total chromium in the sludge ($\dot{m}$) is determined from the mass concentration of chromium in the sludge as:

$\dot{m} = $ mass of dewatered sludge (mass fraction of chromium in sludge)
$= (23{,}830 \, \text{lb/d})(120 \, \text{lb}/1{,}000{,}000 \, \text{ lb})$
$= 2.86 \, \text{lb/d}$
$= (2.86 \, \text{lb/d})(7 \, \text{d/wk})(50 \, \text{wk/yr}) = 1000 \, \text{lb/yr}$

CHAPTER 9

OCCUPATIONAL SAFETY AND HEALTH ACT (OSHA)

QUALITATIVE PROBLEMS (LOSHA)

PROBLEMS LOSHA.1–30

LOSHA.1 OCCUPATIONAL SAFETY AND HEALTH ACT (OSHAct)

Describe OSHAct.

SOLUTION: The Occupational Safety and Health Act (OSHAct) of 1970 was enacted "... to assure so far as possible every working man and woman in the Nation safe and healthful working conditions and to preserve our human resources." Under the Act, the Occupational Safety and Health Administration (OSHA) was created within the Department of Labor to:

1. Encourage employers and employees to reduce workplace hazards and to implement new or improve existing safety and health programs.
2. Provide for research to develop innovative ways of dealing with occupational safety and health problems.
3. Establish "separate but dependent responsibilities and rights" for employers and employees for the achievement of better safety and health conditions.

Environmental Regulatory Calculations Handbook, by Leo Stander and Louis Theodore
Copyright © 2008 John Wiley & Sons, Inc.

4. Maintain a reporting and record keeping system to monitor job-related injuries and illnesses.

5. Establish training programs to increase the number and competence of occupational safety and health personnel.

6. Develop mandatory job safety and health standards and enforce them effectively.

7. Provide for the development, analysis, evaluation and approval of state occupational safety and health programs.

While OSHA continually reviews and redefines specific standards and practices, its basic purposes remain constant. OSHA strives to implement its mandate fully and firmly with fairness to all concerned. In all its procedures, from standards development through implementation and enforcement, OSHA guarantees employers and employees the right to be fully informed, to participate actively, and to appeal actions. Because of OSHA's policy regarding the need for workers to understand occupational dangers, management has established goals in order to implement training and communication programs that form the basis of support of its regulations. One such measure for achieving these goals was to compile a list of the chemicals determined to be potential hazards in the workplace. The original list of OSHA Chemicals has varied over the years as chemicals are added or deleted form the list. At the time of the preparation of this text, 453 chemicals were so noted by OSHA. See Table 9.4, Problem LOSHA.24 later in this Problem set.

LOSHA.2   "RIGHT-TO-KNOW" LAWS

What are the "right-to-know" laws?

SOLUTION: Probably one of the most important safety and health standards ever adopted is the OSHA hazard communication standard, more popularly known as the "right-to-know" law. The hazard communication standard requires employers to communicate information to the employee on hazardous chemicals that exist within the workplace. The program requires employers to craft a written hazard communication program, keep material safety data sheets (MSDSs) for all hazardous chemicals at the workplace and provide employees with training on those hazardous chemicals, and ensure that proper warning labels are in place.

The Hazardous Waste Operations and Emergency Response Regulation enacted by OSHA addressed the safety and health of employees involved in operations at uncontrolled hazardous waste sites which are being cleaned up under government mandate, and in certain hazardous waste treatment, storage, and disposal operations conducted under RCRA. The standard provided for employee protection during initial site characterization and analysis, monitoring activities, training, and emergency response.

Four major areas are under the scope of the regulation:

1. Cleanup operations at uncontrolled hazardous waste sites that have been identified for cleanup by a government health or environmental agency.
2. Routine operations at hazardous waste TSD (Transportation, Storage, and Disposal) facilities or those portions of any facility regulated by 40 CFR parts 264 and 265.
3. Emergency response operations at sites where hazardous substances have or may be released.
4. Corrective actions at RCRA sites.

The regulations address three specific populations of workers at the above operations. First, it regulates hazardous substance response operations under CERCLA, including initial investigations at CERCLA sites before the presence or absence of a hazardous substance has been ascertained; corrective actions taken in cleanup operations under RCRA; and, those hazardous waste operations at sites that have been designated for cleanup by state or local government authorities. The second worker population to be covered are those employees engaged in operations involving hazardous waste TSD facilities. The third employee population to be covered are those employees engaged in emergency response operations for releases or substantial threat of releases of hazardous substances, and post emergency response operations to such facilities.

LOSHA.3   OSHA AND NATIONAL INSTITUTE OF OCCUPATIONAL SAFETY AND HEALTH (NIOSH)

The *Occupational Safety and Health Act* (OSHAct) identifies basic duties that must be carried out by employers. Discuss these basic duties. Also, state the major roles of the *National Institute of Occupational Safety and Health* (NIOSH) and the *Occupational Safety and Health Administration* (OSHA).

SOLUTION: Employers are bound by OSHAct to provide each employee with a working environment free of recognized hazards that cause or have the potential to cause physical harm or death. Employers must have proper instrumentation for the evaluation of test data provided by an expert in the area of toxicology and industrial hygiene. This instrumentation must be obtained because the presence of health hazards cannot be evaluated by visual inspection. This data collection effort provides the employer with substantial evidence to disprove invalid complaints by employees alleging a hazardous working situation. The law also gives employers the right to take full disciplinary action against those employees who violate safe working practices in the workplace.

OSHA's stated mission is to assure the safety and health of America's workers by setting and enforcing standards; providing training, outreach, and education;

establishing partnerships; and, encouraging continual improvement in workplace safety and health. Additional information is available at: www.osha.gov/oshinfo/mission.html.

NIOSH recommends standards for industrial exposure that OSHA uses in its regulations. OSHA has the power to enforce all safety and health regulations and standards recommended by NIOSH.

Although OSHA is in the U.S. Department of Labor, NIOSH is in the U.S. Department of Health and Human Services and is an agency established to help assure safe and healthful working conditions for working men and women by providing research, information, education, and training in the field of occupational safety and health. NIOSH provides leadership to prevent work-related illness, injury, disability, and death by gathering information, conducting scientific research, and translating the knowledge gained into products and services. NIOSH's mission is critical to the health and safety of every American worker. Each day, an average of 9,000 U.S. workers sustain disabling injuries on the job, 16 workers die from an injury suffered at work, and 137 workers die from work-related diseases. The Liberty Mutual 2005 Workplace Safety Index estimates that employers spent $50.8 billion in 2003 on wage payments and medical care for workers hurt on the job.

NIOSH objectives include:

1. Conduct research to reduce work-related illnesses and injuries.
2. Promote safe and healthy workplaces through interventions, recommendations and capacity building.
3. Enhance global workplace safety and health through international collaborations.

LOSHA.4   ASBESTOS REGULATORY CONCERNS

Discuss regulatory concerns as they apply to asbestos.

SOLUTION: For more than 30 years, the EPA and several other federal agencies have acted to prevent unnecessary exposure to asbestos by prohibiting some uses and by setting exposure standards in the workplace. The government also limits exposure to the public at large. Five agencies have major authority to regulate asbestos.

1. The Occupational Safety and Health Administration (OSHA) sets limits for worker exposure on the job.
2. The Food and Drug Administration (FDA) is responsible for preventing asbestos contamination in food, drugs, and cosmetics.
3. The Consumer Product Safety Commission (CPSC) regulates asbestos in consumer products. It has banned the use of asbestos in drywall patching compounds, ceramic logs, and clothing. The CPSC is now studying the extent of

asbestos use in consumer products generally, and is considering a ban on all nonessential product uses that can result in the release of asbestos fibers.

4. The Mine Safety and Health Administration (MSHA) regulates mining and milling of asbestos.

5. The EPA regulates the use and disposal of toxic substances in air, water, and land, and has banned all uses of sprayed asbestos materials. The effects of cumulative exposure to asbestos have been established by dozens of epidemiological studies. In addition, EPA has issued standards for handling and disposing of asbestos-containing wastes.

LOSHA.5 MSDS SHEETS

Provide a one-sentence explanation of the need for each piece of information on a typical MSDS sheet.

SOLUTION:

1. Product or chemical identity used on the label:
   This ensures that the correct chemical is being used and alerts the worker to the potential hazards of working with the chemical.

2. Manufacturer's name and address:
   Contacting the manufacturer would help clarify any uncertainties concerning the chemicals being used and could also, in the event that it has been discovered that the manufacturer has made an error in the production or delivery of a certain chemical, prevent a potential catastrophe elsewhere caused by the use of the same chemical.

3. Chemical and common names of each hazardous ingredient:
   This serves as a reference for those working with chemicals to check and see if the chemicals being used are hazardous.

4. Name, address, and phone number for hazard and emergency information:
   Their assistance may be necessary if an accident should occur or if there is any uncertainty concerning a certain chemical.

5. The hazardous chemical's physical and chemical characteristics, such as vapor pressure and flashpoint:
   This information can be used to control the environment that the hazardous chemical is going to be used in.

6. Physical hazards, including potential for fire, explosion, and reactivity:
   This aids in the analysis of a "worst-case" scenario that could result from a simple accident.

7. Known health hazards:
   This alerts workers to use special caution when working with the hazardous materials.

8. Exposure limits:
   This helps to protect those working with the material.
9. Emergency and first-aid procedures:
   In the event of an accident, these procedures could save an afflicted worker's life.
10. Toxicological information:
    Alerts workers to the potential risk of developing cancer from working with the ingredient and encourages special caution to be taken in working with the ingredient.
11. Precautions for safe handling and use:
    This helps to protect those working with the material.
12. Control measures such as engineering controls, work practices, hygienic practices or personal protective equipment required:
    Those measures are used in an attempt to minimize the risk involved in working with hazardous materials.
13. Procedures for spills, leaks, and clean-up: These procedures are used to minimize the damage caused by these accidents.

LOSHA.6 NANOTECHNOLOGY LEGISLATION

Comment on how and to what degree new legislation and rulemaking will be necessary for environmental control/concern of nanotechnology (operators) from an OSHA perspective.

SOLUTION: Workplace exposure to chemical substances (including nanoparticles) and the potential for pulmonary toxicity is subject to regulation by the OSHA under the OSHAct, including the requirement that potential hazards be disclosed on MSDS. (An interesting question arises as to whether carbon nanotubes, chemically carbon but with different properties because of their small size and structure, are indeed to be considered the same as or different from carbon black for MSDS purposes.) Both governmental and private agencies can be expected to develop the requisite threshold limit values (TLVs) for workplace exposure.

The reader is referred to the following two texts for additional information:

1. L. Theodore and R. Kunz, "Nanotechnology: Environmental Implications and Solutions," John Wiley & Sons, Hoboken, NJ, 2005.
2. L. Theodore, "Nanotechnology: Basic Calculations for Engineers and Scientists," John Wiley & Sons, Hoboken, NJ, 2006.

LOSHA.7 AMBIENT OXYGEN

When responding to a chemical emergency, ambient oxygen measurement is frequently used to assist in specifying the level of personal protective equipment required for worker safety. An oxygen deficient atmosphere, on a vol% basis, is defined by NIOSH/OSHA as (select one):

1. 25 or less.
2. 19.5 or less.
3. 15 or less.
4. 12.5 or less.
5. 10 or less.

SOLUTION: NIOSH and OSHA define oxygen deficient atmosphere as those containing 19.5% oxygen or less.

The answer is therefore (2).

LOSHA.8 CARBONYL SULFIDE

An industrial complex is leaking carbonyl sulfide, CAS463518. Emergency response personnel must work in an atmosphere containing low concentrations of carbonyl sulfide to clean up the released substance. On-site air monitoring is available and is being performed by an industrial hygienist. If full-face air purifying respirators, equipped with the appropriate organic vapor canisters are being evaluated for use by these workers, what is the maximum airborne concentration of carbonyl sulfide these respirators can be exposed to, based on NIOSH criteria, to ensure worker safety (select one)?

1. 500 ppm.
2. 1,000 ppm.
3. 5,000 ppm.
4. 10,000 ppm.
5. not recommended by NIOSH.

SOLUTION: An exposure standard has not been established for carbonyl sulfide as yet, and NIOSH recommends self-contained breathing apparatus (SCBA) for situations where no exposure limits are set.

The answer to the question is therefore (5).

LOSHA.9 PARTICULATE MATTER–2.5 (PM–2.5 μm PM$_{2.5}$)

Which factors will most affect worker mortality rates due to particulate air pollution (PM-2.5)? PM-2.5 defines particulate matter with an aerodynamic diameter equal to or less than 2.5 micrometers. Aerodynamic diameter is the equivalent diameter of a spherical particle of a density of 1,000 kg/m^3 to produce the observed settling velocity of the particle of interest.

SOLUTION: The following factors contribute to the mortality rate of human (including the working) population exposed to PM-2.5:

1. *Age.* Children and the elderly are more sensitive to particulate pollution than middle-aged adults. Currently, 63% of the children and 60% of the elderly in the U.S. live in Non-Attainment areas (areas which have not met the EPA National Ambient Air Quality Standards).

2. *Weight.* Heavier people/workers are more prone to health problems due to particulate pollution.

3. *Smoking Status.* Smokers are more likely to have lung problems, and are more sensitive to ambient particulate contaminants than non-smokers.

4. *Occupational Exposure.* Long-term exposure to particulates such as dust and metals at the workplace predisposes the workers to several health problems, including brown lung, black lung, berylliosis, and lung cancer.

5. *Chronic Health Conditions.* People with asthma, cardiac and pulmonary diseases are more sensitive to particulate air pollution due to their preexisiting impacted health conditions.

LOSHA.10 TOXICOLOGY

Discuss the science of toxicology.

SOLUTION: Toxicology is the science dealing with the effects, conditions, and detection of toxic substances or poisons. Six primary factors affect human response to toxic substances or poisons. These are detailed below:

1. The chemical itself: Some chemicals produce immediate and dramatic biological effects, whereas others produce no observable effects or produce delayed effects.

2. The type of contact: Certain chemicals appear harmless after one type of contact (e.g., skin) but may have serious effects when contacted in another manner (e.g., lungs).

3. The amount (dose) of a chemical: The dose of a chemical exposure depends upon how much of the substance is physically contacted.

4. Individual sensitivity: Humans vary in their response to chemical substance exposure. Some types of responses that different persons may experience at a certain dose are serious illness, mild symptoms, or no noticeable effect. Different responses may also occur in the same person at different exposures.

5. Interaction with other chemicals: Toxic chemicals in combination can produce different biological responses than the responses observed when exposure is to one chemical alone.

6. Duration of exposure: Some chemicals produce symptoms only after one exposure (acute), some only after exposure over a long period of time (chronic), and some may produce effects from both kinds of exposure.

LOSHA.11   ROUTES OF EXPOSURE

Briefly discuss the various routes by which a chemical can enter the body.

SOLUTION: To protect the body from hazardous chemicals, one must know the route of entry into the body. All chemical forms may be inhaled. After a chemical is inhaled into the mouth, it may be ingested, absorbed into the bloodstream, or remain in the lungs. Various types of personal protective equipment (PPE) such as dust masks and respirators prevent hazardous chemicals from entering the body through inhalation. Ingestion of chemicals can also be prevented by observing basic housekeeping rules, such as maintaining separate areas for eating and chemical use or storage, washing hands before handling food products, and removing gloves when handling food products. Wearing gloves and protective clothing prevent hazardous chemicals from entering the body through skin absorption.

After a chemical has entered the body, the body may break it down or metabolize it, the body may excrete it, or the chemical may remain deposited in the body.

The route of entry of a chemical is often determined by the physical form of the chemical. Physical chemical forms and the routes of entry are summarized in the Table 9.1.

TABLE 9.1   Routes of Exposure

Chemical Form	Principal Danger
Solids and fumes	Inhalation, ingestion, and skin absorption
Dusts and gases	Inhalation into lungs
Liquids, vapors and mists	Inhalation of vapors and skin absorption

LOSHA.12   TOXICOLOGY TERMINOLOGY

Describe the following toxicology terms:

1. Threshold limit value (TLV)
2. Immediately dangerous to life and health (IDLH)
3. Lethal dose (LD)
4. Effective dose (ED)
5. Toxic dose (TD)
6. Lethal concentration (LC)

SOLUTION: The concept of *threshold* is used to assess the toxicity of noncarcinogenic chemical substances. The dose-effect relationship is generally characterized by a threshold below which no effects can be observed. However, the threshold value for a toxic substance cannot be identified precisely. Instead, it can only be bracketed, based on the analysis of data from animal tests in which other parameters are used to evaluate the hazard.

1. The *threshold limit value* (TLV) is the maximum limit of the amount of a chemical to which a human can be exposed without experiencing toxic effects. The TLV is categorized into the TLV-TWA, TLV-STEL, and TLV-C, where -TWA, -STEL, and -C represent *time-weighted average, short-term exposure limit,* and *ceiling*, respectively. A TWA can be the average concentration over any period of time. Most often, however, a TWA is the average concentration of a chemical that workers can typically be exposed to during a 40-h week and a normal 8-h day without showing any toxic effects. A STEL is a 15 min time-weighted average exposure. Excursions to the STEL should be at least 60 min apart, no longer than 15 min in duration, and should not be repeated more than four times per day. Since the excursions are calculated into an 8 h TWA, the exposure must be limited to avoid exceeding the TWA. Ceiling values, C, exist for substances whose exposure results in a rapid and particular type of response. It is used where TWA (with its allowable Excursions) would not be appropriate. The American Conference of Governmental Industrial Hygienists (ACGIH) and the Occupational Safety and Health Administration (OSHA) state that a ceiling value should not be exceeded even instantaneously. The National Institute for Occupational Safety and Health (NIOSH) also uses ceiling values. However, its ceiling values are similar to a STEL.

Similar exposure limits employed are the *permissible exposure limit* (PEL) and the *recommended exposure limit* (REL). PELs are extracted from the TLVs and other standards including standards for benzene and 13 carcinogens. Since OSHA is a regulatory agency, its PELs are legally enforceable standards and apply to all private industries and federal agencies. RELs are used in developing OSHA standards, but there are many that have not been adopted and are in the same status as the exposure guidelines of the American Conference of Governmental Industrial Hygienists (ACGIH).

2. The IDLH (immediately dangerous to life and health) is the maximum concentration of a substance to which a human can be exposed for 30 min without experiencing irreversible health effects.

3–6. Dosages of a chemical can be described as a lethal dose (LD), effective dose (ED), or toxic dose (TD). The LD50 or LD_{50} is a common parameter used in toxicology. It represents the dose at which 50% of a test population would die when exposed to a chemical at that dose. Similarly, the lethal concentration (LC) is the concentration of a substance in air that will cause death.

The above are discussed in more detail in the next Problem.

LOSHA.13 TOXIC EXPOSURE GUIDELINES

List and describe the key toxic exposure guidelines.

SOLUTION: This solution provides additional information to that provided in the previous Problem.

1. *ACGIH TLV TWA*

    ACGIH Threshold Limit Value (TLV) and Time Weighted Average (TWA). The ACGIH, a private organization, issues its TLV TWA's annually for approximately 600 airborne substances. The TLV is the concentration below which a healthy worker could be exposed for 8 hours/day, 5 days/week, for 20 years without developing any disease. Time Weighted Average means that the TLV could be exceeded for a time if it is balanced by a time of lower exposure. For each TLV TWA, a CEILING value places an upper limit on the concentration to which the worker can be exposed regardless of how much time is spent at concentrations below the TLV TWA. A short-term exposure limit (STEL) is a fifteen minute TWA exposure that should not be exceeded at any time during a workday. For example, the TLV TWA for the HAP carbon tetrachloride is 5 parts per million (ppm) with a STEL of 30 ppm.

    Additional data (the latest) are available on the OSHA website; Tables Z1, Z2, and Z3 (found in 29 CFR 1910.1000) provide limits for air contaminants. Information on mineral dusts is also available.

2. *OSHA PEL TWA*

    Occupational Safety and Health Administration (OSHA) Permissible Exposure Limits (PEL) Time Weighted Average (TWA). This is similar to the ACGIH TLV TWA but issued as a regulation by OSHA to protect workers. PELs are published in 29 CFR 1910.1000.

3. *NIOSH REL TWA*

    National Institute of Occupational Safety and Health (NIOSH) Recommended Exposure Limits (REL) Time Weighted Average (TWA). NIOSH makes recommendations on occupational safety and health but does not issue regulations. The NIOSH REL TWA is similar to the ACGIH TLV TWA.

4. *IDLH*

    Immediately Dangerous to Life or Health. IDLH values are issued by NIOSH and reflect a concentration of a chemical that is likely to cause death or immediate or delayed permanent adverse effects. The IDLH is also the concentration of the chemical that could incapacitate a worker and thus prevent escape from a contaminated area. As a margin of safety, IDLH values are based on the effects that might occur as a consequence of a 30 minute exposure. NIOSH cautions, however, that the worker exposed to the IDLH should not continue to work for the entire 30 minute period. Every effort should be made to exit immediately.

LOSHA.14 CONCENTRATION TERMS

List and discuss various "concentration" terms employed by OSHA.

SOLUTION:

1. LC_{50}
   Lethal Concentration 50%. This is similar to LD_{50} except that the route of entry is inhalation. The concentrations of the inhaled chemicals (usually gases) are expressed as parts per million (ppm) or milligrams per cubic meter (mg/m^3).

2. LDLo
   Lethal Dose Low. The lowest dose that killed any of the animals in the study when administered by a route of entry other than inhalation.

3. LCLo
   Lethal Concentration Low. Same as LDLO except that the route of entry is inhalation.

4. TDLo
   Toxic Dose Low. The lowest dose used in the study that caused any toxic effect (not just death) when administered by a route of entry other than inhalation.

5. TCLo
   Toxic Concentration Low. Same as TDLo except that the route of entry is inhalation.

6. EC_{50}
   This is the median effective concentration calculated to affect 50% of a test population during continuous exposure over a specified period of time.

LOSHA.15 TOXICITY FACTORS

Define and compare the following pairs of parameters used in toxicology.

1. NOEL and NOAEL
2. LOEL and LOAEL
3. ADI and RfD

SOLUTION: The parameters NOEL and NOAEL, LOEL and LOEAL, and ADI and RfD are used to establish thresholds.

The NOEL (*no observed effect level*) is the highest dose of the toxic substance that will not cause an effect. The NOAEL (*no observed adverse effect level*) is the highest dose of the toxic substance that will not cause an adverse effect.

The LOEL (*lowest observed effect level*) is the lowest dose of the toxic substance tested that shows effects. The LOAEL (*lowest observed adverse effect level*) is the lowest dose of the toxic substance tested that shows adverse effects. The LOEL and LOAEL give no indication of individual variation in susceptibility.

The ADI (*acceptable daily intake*) is the level of daily intake of a particular substance that will not produce an adverse effect. The RfD (*reference dose*) is an estimate of the daily exposure level for the human population. The RfD development follows a stricter procedure than that followed for the ADI. This sometimes results in a lower value for the ADI. The ADI approach is used extensively by the Food and Drug Administration (FDA) and the World Health Organization (WHO), while the RfD is a contemporary replacement for the ADI used by the EPA.

When using data for LOELs, LOAELs, NOELs, or NOAELs, it is important to be aware of their limitations. Statistical uncertainty exists in the determination of these parameters due to the limited number of animals used in the studies to determine the values. In addition, any toxic effect might be used for the NOAEL and LOAEL so long as it is the most sensitive toxic effect and considered likely to occur in humans.

LOSHA.16 TOXICITY DATA

What are toxicity data.

SOLUTION: This data usually reflects the results of animal testing. The table of relative acute toxicity criteria given in Table 9.2 was published by the NIOSH in the Registry of the Toxic Effects of Chemical Substances (RTECS) in 1967. It is still widely used to interpret animal toxicity data. As the table below indicates, for animal toxicity data, the lower the number, the greater the toxicity.

Data on animal toxicity usually lists the route of entry into the body (oral ingestion, inhalation, adsorption through the skin, etc.) first, followed by the test animal (mouse, rat, human, etc.), followed by the measure of toxicity.

TABLE 9.2 Toxicity Ratings

Rating	Keywords	LD50 Single Oral Dose* (mg/kg)	LC50 Inhalation Vapor Exposure* (ppm)	LD50 Skin** (mg/kg)
4	Extremely hazardous	#1	#10	#5
3	Highly hazardous	50	100	43
2	Moderately hazardous	500	1000	340
1	Slightly hazardous	5,000	10,000	2,800
0	No significant hazard	>5,000	>10,000	>2,800

*Rats, **Rabbits.

LOSHA.17 REFERENCE DOSE (RfD)

RfD has occasionally been used in worker health studies. Describe and illustrate the process of correlating a RfD with a schematic of a dose–response curve. Label both axes and the critical points on the curve.

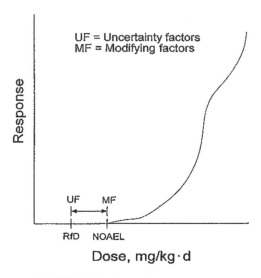

FIGURE 9.1   Dose–response curve.

SOLUTION: A RfD estimates the lifetime dose that does not pose a significant risk to the human population. This estimate may have an uncertainty of one order of magnitude or more. The RfD is determined by dividing the *no observed adverse effect level* (NOAEL) dose of a substance by the product of the uncertainly and modifying factors as shown in the following equation

$$RfD = \frac{NOAEL}{(UF)(MF)} \tag{9.1}$$

The uncertainty factor (UF) is usually represented as a multiple of 10 to account for variation in the exposed population (to protect sensitive subpopulations), uncertainties in extrapolating from animals to humans, uncertainties resulting from the use of subchronic data instead of data obtained from chronic studies, and uncertainties resulting from the use of the lowest observable adverse effect level (LOAEL) instead of the NOAEL. The modifying factor (MF) reflects qualitative professional judgment of additional uncertainties in the data.

The schematic of the dose–response curve shown in Figure 9.1 illustrates that the value of the reference dose is less than the value of the NOAEL by a safety factor.

LOSHA.18   HAZARD OPERABILITY STUDY (HAZOP)

List and discuss the key guide words that are employed in a HAZOP study.

SOLUTION: A HAZOP study—as demonstrated below—is a very useful technique that may lead to a more reliable and safer process. Whether it is applied at preliminary design stages or to the detailed layout of an existing plant, its benefits can be

TABLE 9.3 HAZOP Terms

Guide Words	Meaning	Examples
NO or NOT	No part of the intention is achieved but nothing else happens.	No flow, no agitation, no reaction.
MORE and LESS	Quantitative increase(s) or decrease(s) to the intended activity.	More flow, higher pressure, lower temperature, less time.
AS WELL AS	All of the intention is achieved but some additional activity occurs.	Additional component contaminants, extra phase.
PART OF	Only part of the intention is achieved, part is not.	Component omitted, part of multiple destinations omitted.
REVERSE	The opposite of the intention occurs.	Reverse flow, reverse order of addition.
OTHER THAN	No part of the intention is achieved; something different happens.	Wrong component, startup and shutdown problems, utility failure.

invaluable. It reduces the possibility of accidents for the process involved, improves on-stream availability of the process, can lead to a better understanding of the process and possible malfunctions, and provides training for the evaluation of any process. Finally, it is also a way of optimizing a process and providing a reliable and cost-effective system.

Generally, HAZOP focuses on a major piece of equipment, although a lesser piece of equipment such as a pump or a valve may be chosen depending upon the nature of the materials being handled and the operating conditions. Once an intended operation is defined, a list of possible deviations from the intended operation is developed. The degrees of deviation from normal operation are conveyed by a set of guide words, some of which are listed in Table 9.3.

The purpose of these guide words is to develop the thought process and encourage discussion that is related to any potential deviation(s) in the system. Upon recognizing a possible deviation, the possible cause(s) and consequence(s) can be determined.

Additional details can be found in the aforementioned Flynn and Theodore text, 2002.

LOSHA.19 THE FAR CONCEPT

The acronym FAR is the number of fatal accidents per 1000 workers in a working lifetime (10^8h). A responsible chemical company typically employs a FAR equal to 2 for chemical process risks such as fires, toxic releases, or spillage of corrosive chemicals.

Identify potential problem areas that may develop for a company if acceptable FAR numbers are exceeded.

SOLUTION: Potential problems that may develop for a company within the community if acceptable FAR numbers are exceeded include:

1. Adverse publicity by media
2. Adverse community relations
3. Decreased public trust in company

Potential problems that may develop for a company concerning legal and regulatory issues if acceptable FAR numbers are exceeded include:

1. Legal action against the company by those affected
2. Potential notices of violations by appropriate regulatory agencies, e.g., NIOSH, OSHA, EPA, etc.

LOSHA.20 RESPIRATORS

Briefly describe the role respirators can play in worker health risk management.

SOLUTION: Respirators provide protection against inhaling harmful materials. Different types of respirators may be used depending on the level of protection desired. For example, supplied-air respirators (e.g., a self-contained breathing apparatus) may be required in situations where the presence of highly toxic substances is known or suspected and/or in confined spaces where it is likely that toxic vapors may accumulate. On the other hand, a full-face or half-face air-purifying respirator may be used in situations where measured air concentrations of identified substances will be reduced by the respirator below the substance's threshold limit value (TLV) and the concentration is within the service limit of the respirator (i.e., that provided by the canister).

Air-purifying respirators contain cartridges (or canisters) that contain an adsorbent, such as charcoal, to adsorb the toxic vapor and thus purify the breathing air. Different cartridges can be attached to the respirator depending on the nature of the contaminant. For example, a cartridge for particulates will contain a filter rather than charcoal. The charcoal in a cartridge acts like a fixed-bed adsorber. The performance of any charcoal cartridge may be evaluated by treating it as a fixed-bed adsorber.

OSHA covers worker safety issues other than respiratory, including ladders, confined space entry, clothing, respirators, eye protection, ear protection, and safe working conditions.

Additional details on characteristics and factors used for respirator selection is provided in Problem TOSHA.11.

LOSHA.21 ACCIDENT DEFINITION

Define an accident.

SOLUTION: An accident is an unexpected event that has undesirable consequences. The causes of accidents have to be identified in order to help prevent accidents from occurring. Any situation or characteristic of a system, plant, or process that has the potential to cause damage to life, property, or the environment is considered a hazard. A hazard can also be defined as any characteristic that has the potential to cause an accident. The severity of a hazard plays a large part in the potential amount of damage a hazard can cause if it occurs. Risk is the probability that human injury, damage to property, damage to the environment, or financial loss will occur. An acceptable risk is a risk whose probability is unlikely to occur during the lifetime of the plant or process or event. An acceptable risk can also be defined as an accident that has a high probability of occurring, with negligible consequences. Risks can be ranked qualitatively in categories of high, medium, and low. Risk can also be ranked quantitatively as an annual number of fatalities per million affected individuals. This is normally denoted as a number times one millionth, e.g., 3×10^{-6} can indicate that on average 3 workers will die every year out of one million individuals.

LOSHA.22 HAZARD RISK ASSESSMENT

Describe the hazard risk assessment process.

SOLUTION: Risk evaluation of accidents serves a dual purpose. It estimates the probability that an accident will occur and also assesses the severity of the consequences of an accident. Consequences may include damage to the surrounding environment, financial loss, or injury to life. The problem is primarily concerned with the methods used to identify hazards and the causes and consequences of accidents. Issues dealing with health risks have been explored previously. Risk assessment of accidents provides an effective way to help ensure either that a mishap does not occur or reduces the likelihood of an accident. The result of the risk assessment allows concerned parties to take precautions to prevent an accident before it happens.

There are several steps in evaluating the risk of an accident (see Figure 9.2). These are detailed below if the system in question is a chemical plant.

1. A brief description of the equipment and chemicals used in the plant is needed.
2. Any hazard in the system has to be identified. Hazards that may occur in a chemical plant include:

Fire	Explosions
Toxic vapor release	Rupture of a pressurized vessel
Slippage	Runaway reactions
Corrosion	

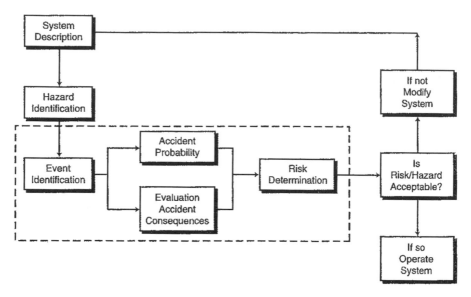

FIGURE 9.2   Hazard Risk Assessment Flowchart.

3. The event or series of events that will initiate an accident has to be identified. An event could be a failure to follow correct safety procedures, improperly repaired equipment, or failure of a safety mechanism.

4. The probability that the accident will occur has to be determined. For example, if a chemical plant has a 10-year life, what is the probability that the temperature in a reactor will exceed the specified temperature range? The probability can be qualitatively ranked from low to high. A low probability means that it is unlikely for the event to occur in the life of the plant. A medium probability suggests that there is a possibility that the event will occur. A high probability means that the event will probably occur during the life of the plant.

5. The severity of the consequences of the accident must be determined. This will be described later in more detail.

6. If the probability of the accident and the severity of its consequences are low, then the risk is usually deemed acceptable and the plant should be allowed to operate. If the probability of occurrence is too high or the damage to the surroundings is too great, then the risk is usually unacceptable and the system needs to be modified to minimize these effects.

The heart of the hazard risk assessment algorithm provided is enclosed in the dashed box (Figure 9.2). For example, an example of a hazard identification is/could be an accidental emission from a nuclear power plant. The algorithm allows for reevaluation of the process if the risk is deemed unacceptable (the process is repeated starting with either step one or two).

LOSHA.23 ACCIDENTAL SPILL OF BATTERY ACID

Outline accident prevention efforts that could address the accidental spill of battery acid in an auto repair shop.

SOLUTION:

1. Obtain data on the chemical and toxicological properties of sulfuric acid.
2. Train personnel in the proper procedures and precautions for handling automobile batteries.
3. Train personnel in the proper methods for clean up of an acid spill.
4. Have protective gear available for personnel who are designated to clean up any spill.
5. Have spill clean-up equipment available in a prominent and unobstructed location in the shop.

LOSHA.24 OSHA CHEMICALS

Provide a list of regulated OSHA chemicals.

SOLUTION: The list below (see Table 9.4) was adapted from J. Spero, B. Devito, and L. Theodore, "Regulatory Chemicals Handbook," Marcel Dekker (recently acquired by CRC Press/Taylor & Francis Group, Boca Raton, FL), New York City, 2000.

In addition to Spero et al., the reader may refer to following references for specific details on each of the chemicals.

TABLE 9.4 Occupational Safety and Health Administration Chemicals (OSHA)

Chemical or Trade Name	CAS No.
Acetaldehyde	75-07-0
Acetic acid	64-19-7
Acetic anhydride	108-24-7
Acetone	67-64-1
Acetonitrile	75-05-8
2-Acetylaminofluorine	53-96-3
Acetylene tetrabromide	79-27-6
Acrolein	107-02-8
Acrylamide	79-06-1
Acrylonitrile	107-13-1

(Continued)

TABLE 9.4 *Continued*

Chemical or Trade Name	CAS No.
Aldrin	309-00-2
Allyl alcohol	107-18-6
Allyl chloride	107-05-1
Allyl glycidyl ether	106-92-3
Allyl propyl disulfide	2179-59-1
alpha-Alumina	1334-28-1
Aluminum, metal (as Al)	7429-90-5
4-Aminodiphenyl	92-67-1
2-Aminopyridine	504-29-0
Ammonia	7664-41-7
Ammonium sulfamate	7773-06-0
n-Amyl acetate	628-63-7
sec-Amyl acetate	626-38-0
Aniline and homologs	62-53-3
Anisidine (o-, p-isomers)	29191-52-4
Antinomy and compounds (as Sb)	7440-36-0
ANTU (alpha Naphthythiourea)	86-88-4
Arsenic, inorganic compounds (as As)	7440-38-2
Arsenic, organic compounds (as As)	7440-38-2
Arsine	7784-42-1
Asbestos	1332-21-4
Azinphos-methyl	86-50-0
Barium, soluble compounds (as Ba)	7440-39-3
Barium sulfate	7727-43-7
Benomyl	17804-35-2
Benzene	71-43-2
Benzidine	92-87-5
Benzoyl peroxide	94-36-0
Benzyl chloride	100-44-7
Beryllium and beryllium compounds (as Be)	7440-41-7
Bismuth telluride, undoped	1304-83-1
Boron oxide	1303-86-2
Boron trifluoride	7637-07-2
Bromine	7726-95-6
Bromoform	75-25-2
Butadiene	106-99-0
2-Butanone	78-93-3
2-Butoxyethanol	111-76-2
n-Butyl-acetate	123-86-4
sec-Butyl acetate	105-46-4
tert-Butyl acetate	540-88-5
n-Butyl alcohol	71-36-3
sec-Butyl alcohol	78-92-2

(Continued)

TABLE 9.4 *Continued*

Chemical or Trade Name	CAS No.
tert-Butyl alcohol	75-65-0
Butylamine	109-73-9
tert-Butyl chromate (as CrO_2)	1189-85-1
n-Butyl glycidyl ether (BGE)	2426-08-6
Butyl mercaptan	109-79-5
p-tert-Butyltoluene	98-51-1
Cadmium (as Cd)	7440-43-9
Calcium carbonate	1317-65-3
Calcium hydroxide	1305-62-0
Calcium oxide	1305-78-8
Calcium silicate	1334-95-2
Calcium sulfate	7778-18-9
Camphor, synthetic	76-22-2
Carbaryl (Sevin)	63-25-2
Carbon black	1333-86-4
Carbon dioxide	124-38-9
Carbon disulfide	75-15-0
Carbon monoxide	630-08-0
Carbon tetrachloride	56-23-5
Cellulose	9004-34-6
Chlordane	57-74-9
Chlorinated camphene	8001-35-2
Chlorinated diphenyl oxide	55720-99-5
Chlorine	7782-50-5
Chlorine dioxide	10049-04-4
Chlorine trifluoride	7790-91-2
Chloroacetaldehyde	107-20-0
a-Chloroacetophenone	532-27-4
Chlorobenzene	108-90-7
o-Chlorobenzylidene malononitrile	2698-41-1
Chlorobromomethane	74-97-5
Chlorodiphenyl (42% Chlorine) (PCB)	53469-21-9
Chlorodiphenyl (54% Chlorine) (PCB)	11097-69-1
Chloroform	67-66-3
bis(Chloromethyl)ether	542-88-1
Chloromethyl methyl ether	107-30-2
1-Chloro-1-nitropropene	600-25-9
Chloropicrin	76-06-2
beta-chloroprene	129-99-8
2-Chloro-6-(trichloromethyl)pyridine	1929-82-4
Chromic acid and chromates (as CrO_3)	not available
Chromium compounds (as Cr)	7440-47-3
Clopidol	2971-90-6

(Continued)

TABLE 9.4 *Continued*

Chemical or Trade Name	CAS No.
Coal dust (less than 5% SiO$_2$), respirable fraction	not available
Coal dust (grater than or equal to 5% SiO$_2$), respirable fraction	
Coal tar pitch volatiles	65966-93-2
Cobalt metal, dust, and furne (as Co)	7440-48-4
Coke oven emissions	not available
Copper	7440-50-8
Cotton dust	not available
Crag herbicide	136-78-7
Cresol, all isomers	1319-77-3
Crotonaldehyde	123-73-9
Cumene	98-82-8
Cyanides (as CN)	not available
Cyclohexane	110-82-7
Cyclohexanol	108-93-0
Cyclohexanone	108-94-1
Cyclohexene	110-83-8
Cyclopentadiene	542-92-7
2,4-D (Dichlorophenoxyacetic acid)	94-75-7
Decaborane	17702-41-9
Demeton (Systox)	8065-48-3
Diacetone alcohol	123-42-2
Diazomethane	334-88-3
Diborane	19287-45-7
1,2-Dibromo-3-chloropropane	96-12-8
Dibutyl phosphate	107-66-4
Dibutyl phthalate	84-74-2
o-Dichlorobenzene	95-50-1
p-Dichlorobenzene	106-46-7
3,3'-Dichlorobenzidine	91-94-1
Dichlorodifluoromethane	75-71-8
1,3-Dichloro-5,5-dimethyl hydantoin	118-52-5
Dichlorodiphenyltrichloroethane (DDT)	50-29-3
1,1-Dichloroethane	75-34-3
1,2-Dichloroethylene	540-59-0
Dichloroethyl ether	111-44-4
Dichloromonofluoromethane	75-43-4
1,1-Dichloro-1-nitroethane	594-72-9
Dichlorotetrafluoroethane	76-14-2
Dichlorovos	62-73-7
Dichloropentadienyl iron	102-54-5
Dieldrin	60-57-1
Diethylamine	109-89-7
2-Diethylaminoethanol	100-37-8

(*Continued*)

TABLE 9.4 *Continued*

Chemical or Trade Name	CAS No.
Difluorodibromomethane	75-61-6
Diglycidyl ether (DGE)	2238-07-5
Diisobutyl ketone	108-83-8
Diisopropylamine	108-18-9
4-Dimethylaminoazobenzene	60-11-7
Dimethyl acetamide	127-19-5
Dimethylamine	124-40-3
Dimethylaniliine	121-69-7
Dimethyl-1,2-dibromo-2,2-dichloroethyl phosphate	300-76-5
Dimethylformamide	68-12-2
1,1-Dimethylhydrazine	57-14-7
Dimethylphthalate	131-11-3
Dimethyl sulfate	77-78-1
ortho-dinitrobenzene	528-29-0
meta-dintrobenzene	99-65-0
para-dinitrobenzene	100-25-4
Dinitro-o-cresol	534-52-1
Dinitrotoluene	25321-14-6
Dioxane	123-91-1
Diphenyl	92-52-4
Dipropylene glycol methyl ether	34590-94-8
di-sec octyl phthalate	117-81-7
Emery	12415-34-8
Endosulfan	115-29-7
Endrin	72-20-8
Epichlorohydrin	106-89-8
EPN	2104-64-5
Ethanolamine	141-43-5
2-Ethoxyethanol	110-80-5
2-Ethoxyethyl acetate	111-15-9
Ethyl acetate	141-78-6
Ethyl acrylate	140-88-5
Ethyl alcohol	64-17-5
Ethylamine	75-04-7
Ethyl amyl ketone	541-85-6
Ethyl benzene	100-41-4
Ethyl bromide	74-96-4
Ethyl butyl ketone	106-35-4
Ethyl chloride	75-00-3
Ethyl ether	60-29-7
Ethyl formate	109-94-4
Ethyl mercaptan	75-08-1
Ethyl silicate	78-10-4

(Continued)

TABLE 9.4 *Continued*

Chemical or Trade Name	CAS No.
Ethylene chlorohydrin	107-07-3
Ethylenediamine	107-15-3
Ethylene dibromide	106-93-4
Ethylene dichloride	107-06-2
Ethylene glycol dinitrate	628-96-6
Ethyleneamine	151-56-4
Ethylene oxide	75-21-8
N-Ethylmorpholine	100-74-3
Ferbam	14484-64-1
Ferrovanadium dust	12604-58-9
Fluorides (as F)	not available
Fluorine	7782-41-4
Fluorotrichloromethane	75-69-4
Formaldehyde	50-00-0
Formic acid	64-18-6
Furfural	98-01-1
Furfuryl alcohol	98-00-0
Grain dust	not available
Glycerin	56-81-5
Glycidol	556-52-5
Graphite, natural, respirable dust	7782-42-5
Graphite, synthetic	not available
Gypsum	13397-24-5
Hafnium	7440-58-6
Heptachlor	76-44-8
Heptane	142-82-5
Hexachloroethane	67-72-1
Hexachloronaphthalene	1335-87-1
n-Hexane	110-54-3
2-Hexanone	591-78-6
Hexone	108-10-1
sec-Hexyl acetate	108-84-9
Hydrazine	302-01-2
Hydrogen bromide	10035-10-6
Hydrogen chloride	7647-01-0
Hydrogen cyanide	74-90-8
Hydrogen fluoride (as F)	7664-39-3
Hydrogen peroxide	7722-84-1
Hydrogen selenide (as Se)	7783-07-5
Hydrogen sulfide	7783-06-4
Hydroquinone	123-31-9
Iodine	7553-56-2
Iron oxide fume	1309-37-1
Isoamyl acetate	123-92-2

(Continued)

TABLE 9.4 *Continued*

Chemical or Trade Name	CAS No.
Isoamyl alcohol (primary and secondary)	123-51-3
Isobutyl acetate	110-19-0
Isobutyl alcohol	78-83-1
Isophorone	78-59-1
Isopropyl acetate	108-21-4
Isopropyl alcohol	67-63-0
Isopropylamine	75-31-0
Isopropyl ether	108-20-3
Isopropyl glycidyl ether (IGE)	4016-14-2
Kaolin	1332-58-7
Ketene	463-51-4
Lead, inorganic (as Pb)	7439-92-1
Limestone	1317-65-3
Lindane	58-89-9
Lithium hydride	7580-67-8
L.P.G. (Liquified petroleum gas)	68476-85-7
Magnesite	546-93-0
Magnesium oxide fume	1309-48-4
Malathion	121-75-5
Maleic anhydride	108-31-6
Manganese compounds (as Mn)	7439-96-5
Manganese fume (as Mn)	7439-96-5
Marble	1317-65-3
Mercury (aryl and inorganic) (as Hg)	7439-97-6
Mercury (organo) alkyl compounds (as Hg)	7439-97-6
Mercury (vapor) (as Hg)	7439-97-6
Mesityl oxide	141-79-7
Mcthoxychloɪ	72-43-5
2-Methoxyethanol	109-86-4
2-Methoxyethyl acetate	110-49-6
Methyl acetate	79-20-9
Methyl acetylene	74-99-7
Methyl acetylene-propadiene mixture (MAPP)	not available
Methyl acrylate	96-33-3
Methyial (Dimethoxy-methane)	109-87-5
Methyl alcohol	67-56-1
Methylamine	74-89-5
Methyl n-amyl ketone	110-43-0
Methyl bromide	74-83-9
Methyl chloride	74-87-3
Methyl chloroform	71-55-6
Methylcyclohexane	108-87-2
Methylcyclohexanol	25639-42-3
o-Methylcyclohexanone	583-60-8

(Continued)

TABLE 9.4 *Continued*

Chemical or Trade Name	CAS No.
Methylene chloride	75-09-2
Methyl formate	107-31-3
Methyl hydrazine	60-34-4
Methyl iodide	74-88-4
Methyl isoamy ketone	110-12-3
Methyl isobutyl carbinol	108-11-2
Methyl isocyanate	624-83-9
Methyl mercaptan	74-93-1
Methyl methacrylate	80-62-6
alpha-Methyl styrene	95-83-9
Methylene bisphenyl isocyanate (MDI)	101-68-8
Methylenedianiline	101-77-9
Molybdenum	7439-98-7
Monomethyl aniline	100-61-8
Morpholine	110-91-8
Naphtha (Coal tar)	8030-30-6
Naphtalene	91-20-3
alpha-Naphthylamine	134-32-7
beta-Naphthylamine	91-59-8
Nickel carbonyl (as Ni)	13463-39-3
Nickel, metal and insoluble compounds (as Ni)	13463-39-3
Nickel, soluble compounds (as Ni)	13463-39-3
Nicotine	54-11-5
Nitric acid	7697-37-2
Nitric oxide	10102-43-9
p-Nitroaniline	100-01-6
Nitrobenzene	98-95-3
p-Nitrochlorobenzene	100-00-5
4-Nitrodiphenyl	92-93-3
Nitroethane	79-24-3
Nitrogen dioxide	10102-44-0
Nitrogen trifluoride	7783-54-2
Nitroglycerin	55-63-0
Nitromethane	75-52-5
1-Nitropropane	108-03-2
2-Nitropropane	79-46-9
N-Nitrosodimethylamine	not available
o-Nitrosodimethylamine	88-72-2
m-Nitrosodimethylamine	99-08-1
p-Nitrosodimethylamine	99-99-0
Octachloronaphthalene	2234-13-1
Octane	1111-65-9
Oil mist, mineral	8012-95-1
Osmium tetroxide (as Os)	20816-12-0

(*Continued*)

TABLE 9.4 *Continued*

Chemical or Trade Name	CAS No.
Oxalic acid	144-62-7
Oxygen difluoride	7783-41-7
Ozone	10028-15-6
Paraquat, respirable dust	4685-14-7
Parathion	56-38-2
Particulates not otherwise regulated (PNOR)	not available
Pentaborane	19624-22-7
Pentachloronaphthalene	1321-64-8
Pentachlorophenol	87-86-5
Pentaerythritol	115-77-5
Pentane	109-66-0
2-Pentanone	107-87-9
Perechloroethylene	127-18-4
Perchloromethyl mercaptan	594-42-3
Perchloryl fluoride	7616-94-6
Perlite	93763-70-3
Petroleum distillates	8002-70-3
Phenol	108-95-2
p-Phenylene diamine	106-50-3
Phenyl ether, vapor	101-84-8
Phenyl ether-biphenyl mixture, vapor	not available
Phenyl glycidyl ether (PGE)	122-60-1
Phenylthydrazine	100-63-0
Phosdrin	7786-34-7
Phosgene	75-44-5
Phosphine	7803-51-2
Phosphoric acid	7664-38-2
Phosphorus (yellow)	7723-14-0
Phorphorus pentachloride	10026-13-8
Phosophorus pentasulfide	1314-80-3
Phosphorus trichloride	7719-12-2
Phthalic anhydride	85-44-9
Picioram	1918-02-1
Picric acid	88-89-1
Pindone	83-26-1
Plaster of Paris	26499-65-0
Platinum (as Pt)	7440-06-4
Portland cement	65997-15-1
Propane	74-98-6
beta-Propriolactone	57-57-8
n-Propyl acetate	109-60-4
n-Propyl alcohol	71-23-8
n-Propyl nitrate	627-13-4
Propylene dichloride	78-87-5

(Continued)

TABLE 9.4 *Continued*

Chemical or Trade Name	CAS No.
Propylene imine	75-55-8
Propylene oxide	75-56-9
Pyrethrum	8003-34-7
Pyridine	110-86-1
Quinone	106-51-4
Rhodium (as Rh), metal, fume, and insoluble compounds	7440-16-6
Rhodium (as Rh), soluble compounds	7440-16-6
Ronnel	299-84-3
Rotenone	83-79-4
Rouge	not available
Selenium compounds (as Se)	7782-49-2
Selenium hexafluoride (as Se)	7783-79-1
Silica, amorphous, precipitated and gel	112926-00-8
Silica, amorphous, diatomaceous earth, containing less than 1% crystalline silica	61790-53-2
Silica, crystalline cristobalite, respirable dust	14464-46-1
Silica, crystalline quartz, respirable dust	14808-60-7
Silica, crystalline tripoli (as quartz), respirable dust	1317-95-9
Silica, crystalline tridymite, respirable dust	15468-32-3
Silica, fused, respirable dust	60676-86-0
Mica (respirable dust)	12001-26-2
Soapstone, total dust	not available
Soapstone, respirable dust	not available
Talc (containing asbestos)	not available
Talc (containing no asbestos), respirable dust	14807-96-6
Tremolite, asbestiform	not available
Silicon	7440-21-3
Silicon carbide	409-21-2
Silver, metal and soluble compounds (as Ag)	7440-22-4
Sodium fluoracetate	62-74-8
Sodium hydroxide	1310-73-2
Starch	9005-25-8
Stibine	7803-52-3
Stoddard solvent	8052-41-3
Strychnine	57-24-9
Styrene	100-42-5
Sucrose	57-50-1
Sulfur dioxide	7446-09-5
Sulfur hexafluoride	2551-62-4
sulfuric acid	7664-93-9
Sulfur monochloride	10025-67-9
Sulfur pentafluoride	5714-22-7

(Continued)

TABLE 9.4 *Continued*

Chemical or Trade Name	CAS No.
Sulfuryl fluoride	2699-79-8
2,4,5-T (2,4,5-Tetrachlorophenoxyacetic acid)	93-76-5
Tantalum, metal and oxide dust	7440-25-7
TEDP	3689-24-5
Tellurium and compounds (as Te)	13494-80-9
Tellurium hexafluoride (as Te)	7783-80-4
Temephos	3383-96-8
TEPP	107-49-3
Terphenyls	26140-60-3
1,1,1,2-Tetrachloro-2,2,-difluoroethane	76-11-9
1,1,2,2-Tetrachloro-2,2-difluoroethane	76-12-0
1,1,2,2-Tetrachloroethane	79-34-5
Tetrachloronaphthalene	1335-88-2
Tetraethyl lead (as Pb)	78-00-2
Tetrahydrofuran	109-99-9
Tetramethyl lead (as Pb)	75-74-1
Tetramethyl succinonitrile	3333-52-6
Tetranitromethane	509-14-8
Tetryl	479-45-8
Thallium, soluble compounds (as Ti)	7440-28 0
4,4'-Thiobis	96-69-5
Thiram	137-26-8
Tin, inorganic compounds (except oxides) (as Sn)	7440-31-5
Tin, organic compounds (as Sn)	7440-31-5
Titanium oxide	13463-67-7
Toluene	108-88-3
Toluene-2,4-diisocyanate	584-84 9
o-Toluidine	95-53-4
Tributyl phosphate	126-73-8
1,1,2-Trichloroethane	79-00-5
Trichloroethylene	79-01-5
Trichloronaphthalene	1321-65-9
1,2,3-Trichloropropane	96-18-4
1,1,2-Trichloro-1,2,2-trifluoroethane	121-44-8
Triethylamine	121-44-8
Trifluorobromomethane	75-63-8
2,4,6-trinitrotoluene	118-96-7
Triorthocresyl phosphate	78-30-8
Triphenyl phosphate	115-86-6
Turpentine	8006-64-2
Uranium (as U)	7440-61-1
Vanadium	1314-62-1
Vegetable oil mist	not available
Vinyl chloride	75-01-4

(Continued)

TABLE 9.4 *Continued*

Chemical or Trade Name	CAS No.
Vinyl toluene	25013-15-4
Warfarin	81-81-2
Xylenes (o-, m-, p-isomers)	1330-20-7
Xylidine	1300-73-8
Yttrium	7440-65-5
Zinc chloride fume	7646-85-7
Zinc oxide fume	1314-13-2
Zinc oxide	1314-13-2
Zinc stearate	557-05-1
Zirconium compounds	7440-67-7

1. R. Lewis, "Sax's Dangerous Properties of Industrial Materials," 9th edition, Van Nostrand Reinhold, New York City, 1996.
2. "Suspect Chemicals Sourcebook," Roytech Publications, Bethesda, Maryland, 1996.

LOSHA.25 MERCURY

An OSHA chemical of major concern to the government, industry, and public is mercury, provide key information on this chemical.

SOLUTION: Mercury, aryl and inorganic compounds (Hg, 200.6)

CAS/DOT IDENTIFICATION #: 7439-97-6/UN2809

Synonyms: Synonyms may vary depending upon specific compound.

Physical Properties: Most inorganic mercury compounds are white powders or crystals; mercuric sulfide (cinnabar) is red and turns black when exposed to light; insoluble in hydrochloric or similar acids; soluble in nitric acid and hot concentrated sulfuric acid; water soluble salts include mercuric chlorate, cyanide, chloride and acetate; oxides, sulfates and most other common salts, including mercurous chloride, are sparingly soluble or decomposed in water; appearances vary from colorless crystals to yellow, red (oxide, sulfide, iodide), and brown or black (sulfide); MP ($-39°C$, $-38°F$); BP ($357°C$, $674°F$); DN (13.534 g/cm^3 at $25°C$); SG (13.5); VD (not applicable); VP (0.0012 mmHg at $20°C$).

Chemical Properties: Mercury salts yield metallic mercury when heated with sodium carbonate; mercury salts may be reduced to metal by hydrogen peroxide in presence of alkali hydroxide; soluble ionized mercuric salts give a yellow precipitate or mercuric oxide with sodium hydroxide and a red precipitate of mercury diiodide with alkali iodide; mercurous salts give a black precipitate with alkali hydroxides and a white precipitate of mercurous chloride (calomel) with hydrogen chloride or soluble chlorides; decomposes slowly on exposure to sunlight.

Explosion and Fire Concerns: Not combustible; NFPA rating (not rated); mercurous chloride is incompatible with bromides, iodides, alkali chlorides, sulfates, sulfites, carbonates, hydroxides, ammonia, silver salts, copper salts, hydrogen peroxide, iodine, and iodoform; mercuric oxide reacts explosively with acetyl nitrate, chlorine and hydrocarbons, butadiene and ethanol and iodine (at 35°C), and hydrogen peroxide and traces of nitric acid; forms heat or shock-sensitive explosive mixtures with metals and non-metals; contact with acetylene, acetylene products, or ammonia gases may from solid products that are sensitive to shock and which can initiate fires of combustible materials; decomposition emits highly toxic fumes of Hg; use water spray, fog, or foam for firefighting purposes.

Health Symptoms: Inhalation (irritates eyes, skin and respiratory system); skin absorption (central nervous system damage, kidney damage, weight loss).

First Aid: wash eyes immediately with large amounts of water; wash skin immediately with soap and water; provide oxygen and respiratory support.

Human Toxicity Data: Inhalation-man TDLo 44,300 $\mu g/m^3/8H$; toxic effect: central nervous system, liver, MET; inhalation-woman TCLo 150 $\mu g/m^3/$46D; toxic effect: central nervous system, gastrointestinal tract; skin-man TDLo 129 $mg/kg/5H$; toxic effect: ear, central nervous system, skin.

Acute Health Risks: Irritation of eyes, skin, and mucous membranes; severe nausea; vomiting; abdominal pain; renal damage; prostration; chest pain; dyspnea; bronchitis; pneumonitis; insomnia; headache; fatigue; weakness; irritability; gastrointestinal disturbances; anorexia; low weight; proteinuria.

Chronic Health Risks: Tremors; trouble remembering and concentrating; increased salivation; gum problems; loss of appetite and weight; changes in mood and personality; hallucinations; psychosis; clouding of the eyes; skin allergies; grayish skin color; kidney damage; decreased sex drive.

Exposure Guidelines: ACGIH TLV TWA 0.1 $\mu g(Hg)/m^3$ (skin); OSHA PEL CL 0.1 mg (Hg)$/m^3$ (skin); NIOSH REL CL 0.1 mg$/m^3$ (skin); IDLH 10 mg (Hg)$/m^3$.

Personal Protection: Wear full protective clothing (suits, gloves, footwear, headgear, etc.); wear chemical safety goggles and face shield; full facepiece respiratory protection is recommended; eye wash fountains should be provided in the immediate work area.

Spill Clean-Up: Ventilate area of spill; use a specialized charcoal-filtered vacuum or suction pump to collect all visible material; sprinkle the entire area of the spill with elemental zinc powder; use a 5–10% sulfuric acid solution to dampen the zinc powder to create a paste-like consistency; after paste dries to a light gray color, it may be swept up and disposed of properly; residual material is removed with soap and water.

Disposal And Storage Methods: Contain and dispose of mercury as a hazardous waste; contact a Department of Environmental Protection (or the equivalent) or the regional office of EPA for specific recommendations; store in tightly closed containers in a cool, well-ventilated area; keep away from acetylene, ammonia and nickel; store in secure poison area.

Regulatory Information: A1; DOT hazard class/division (6.1); labels (poison).

Other Comments: Inorganic salts of mercury, such as ammoniated mercuric chloride or mercuric iodide have been used in skin lightening creams; mercuric chloride has been used as a topical antiseptic or disinfectant agent; mercuric sulfide and mercuric oxide are used as pigments in paints; mercuric sulfide is also used as a pigment for tattoos; some inorganic mercury compounds are also used in fungicides.

LOSHA.26   DRUM LABELING

Someone at the back of a parking lot found an old drum containing some unknown chemical. Corrosion and weathering have obliterated all markings on the drum except one. The remaining label looks as follows:

1. Interpret this National Fire Prevention Association (NFPA) Hazard Identifi-

FIGURE 9.3   Drum label.

    cation System label as fully as possible and suggest at least one chemical substance whose chemical characteristics are similar to those of the contents of this drum.

2. Who should be contacted about this container?

SOLUTION: The last label on the drum is a National Fire Protection Association (NFPA) Hazard Identification System label (see NFPA 49 and 325M).

The upper code number provides an indication of flammability hazard. The number on the left is an indication of the health hazard of the material, while the number on the right is a measure of its chemical reactivity hazard. The lower symbol indicates special hazards, i.e., oxidizers, water reactivity, radioactivity, etc. The number codes vary from 0 (no hazard) to 4 (extreme hazard).

The number 1 on the left means that there is a minor health hazard associated with this material.

The number 3 on the top means that the contents of the drum can be ignited under almost all ambient temperature conditions.

The number 3 on the right means that the contents of the drum are capable of detonation or explosive reaction, but that they require a strong initiating source (or must be heated under confinement, or react explosively with water).

The "W" with the line through it means that the contents of the drum react with water.

This label is unusual in that most substances that react strongly with water will present a more than minor health hazard. While NFPA does not list any chemical whose classification codes exactly match those given in the problem, Figures 9.4–9.7 are examples of water reactive substances illustrating the variety of materials that have significant health, fire or reactivity hazards.

Among others, the local EPA and/or local fire department could be contacted about such a find.

FIGURE 9.4 NFPA label for sodium metal, a reactive metal that produces hydrogen gas and heat when it reacts with water.

FIGURE 9.5 NFPA label for lithium hydride, a strong reducing agent.

FIGURE 9.6 NFPA label for diethyl zinc, an organometallic compound.

FIGURE 9.7   NFPA label for thionyl chloride, an acid chloride that liberates HCl when it reacts with water.

LOSHA.27   NOISE DEFINITION AND HEARING PROTECTION

Describe the measures of sound intensity, and concerns for hearing protection.

SOLUTION: An estimated 20 million Americans—many of whom are workers—are exposed to noise that poses a threat to their hearing. Everyone at some time or another has experienced the effects of noise pollution. Many people are unaware that the sounds that cause them so much annoyance may also be affecting their hearing. Hearing loss is one of the most serious health threats that is a result of noise pollution.

Noise pollution is traditionally not placed among the top environmental problems facing the nation; however, it is one of the more frequently encountered sources of pollution in everyday life. Noise pollution can be defined simply by combining the meaning of both environmental terms, *noise* and *pollution*. Noise is typically defined as unwanted sound, and pollution is generally defined as the presence of matter or energy whose nature, location, or quantity produces undesired environmental effects. Noise is typically thought of as a nuisance rather than a source of pollution. This is due in part because noise does not leave a visible impact on the environment as do other sources of pollution.

The damage done by the traditional pollution of air and water is widely recognized. The evidence is right before one's eyes—in contaminated water, oil spills, and dying fish, as well as in smog that burns the eyes and sears the lungs. However, noise is a more subtle pollutant. Aside from sonic booms that can break windows, noise usually leaves no visible evidence, although it also can pose a hazard to health and well-being. Approximately 15 million Americans are exposed to noise that poses a threat to their hearing on the job. Another 15 million are exposed to dangerous noise levels without knowing it from trucks, airplanes, motorcycles, stereos, lawnmowers, and kitchen appliances.

Sound travels in waves through the air like waves through water. The higher the wave, the greater its power. The greater the number of waves a sound has, the greater is its frequency or pitch.

The strength of sound or sound level is measured in decibels (dB). The decibel scale ranges from 0, which is regarded as the threshold of hearing for normal, healthy ears, to 94, which is regarded as the theoretical maximum for pure tones. Since the decibel scale, like the pH scale, is logarithmic, 20 dB is 100 times louder than 0; 30 dB is 1000 time louder; 40 dB is 10,000 time louder, etc. Thus at high levels, even a small reduction in level values can make a significant difference in noise intensity.

The frequency is measured in hertz (Hz) (cycles per second) and can be described as the rate of vibration. The faster the movement, the higher the frequency of the sound pressure waves created. The human ear does not hear all frequencies. Normal hearing ranges from 20 to 20,000 Hz, or, roughly, from the lowest note to the highest note on a violin.

The human ear also does not hear all sounds equally. Very low and very high notes sound more faint to one's ear than do 1000-Hz sounds of equal strength. This is the way ears function. The human voice in conversation covers a median range of 300–4000 Hz. The musical scale ranges from 30 to 4000 Hz. Noise in these ranges sound much louder than do very low or very high pitched noises of equal strength.

Since hearing also varies widely among individuals, what may seem loud to one person may not be to another. Although loudness is a personal judgment, precise measurement of sound is made possible by use of the decibel scale. This scale measures sound pressure or energy according to international standards.

The reader should also note that EPA has noise requirements that address environmental noise issues. OSHA has worker safety requirements for noise. Some of these requirements are discussed in the next two Problems.

LOSHA.28 OCCUPATIONAL NOISE EXPOSURE

What is occupational noise exposure?

SOLUTION: Noise, or unwanted sound, is one of the most pervasive occupational health problems. It is a by-product of many industrial processes. Sound consists of pressure changes in a medium (usually air), caused by vibration or turbulence. These pressure changes produce waves emanating away from the turbulent or vibrating source. As discussed previously, exposure to high levels of noise causes hearing loss and may cause other harmful health effects as well. The extent of damage depends primarily on the intensity of the noise and the duration of the exposure.

Noise-induced hearing loss can be temporary or permanent. Temporary hearing loss results from short-term exposures to noise, with normal hearing returning after a period of rest. Generally, prolonged exposure to high noise levels over a period of time gradually causes permanent damage.

OSHA's hearing conservation program is designed to protect workers with significant occupational noise exposures from hearing impairment even if they are subject to such noise exposures over their entire working lifetimes.

LOSHA.29 HEARING PROTECTION REQUIREMENTS

When is an employer required to provide hearing protectors?

SOLUTION: Employers must provide hearing protectors to all workers exposed to 8-hour TWA noise levels of 85 dB or above. This requirement ensures that employees have access to protectors before they experience any hearing loss.

Employees must wear hearing protectors:

1. For any period exceeding 6 months from the time they are first exposed to 8-hour TWA noise levels of 85 dB or above until they receive their baseline audiograms if these tests are delayed due to mobile test van scheduling;
2. If they have incurred standard threshold shifts that demonstrate they are susceptible to noise; and
3. If they are exposed to noise over the permissible exposure limit of 90 dB over an 8-hour TWA.

Employers must provide employees with a selection of at least one variety of hearing plug and one variety of hearing muff. Employees should decide, with the help of a person trained to fit hearing protectors, which size and type protector is most suitable for the working environment. The protector selected should be comfortable to wear and offer sufficient protection to prevent hearing loss.

Hearing protectors must adequately reduce the noise level for each employee's work environment. Most employers use the Noise Reduction Rating (NRR) that represents the protector's ability to reduce noise under ideal laboratory conditions. The employer then adjusts the NRR to reflect noise reduction in the actual working environment.

The employer must reevaluate the suitability of the employee's hearing protector whenever a change in working conditions may make it inadequate. If workplace noise levels increase, employers must give employees more effective protectors. The protector must reduce employee exposures to at least 90 dB and to 85 dB when an STS already has occurred in the worker's hearing [an STS is an average shift in either ear of 10 dB or more at 2,000, 3,000, and 4,000 Hz]. Employers must show employees how to use and care for their protectors and supervise them on the job to ensure that they continue to wear them correctly.

(OSHA Informational Booklet, OSHA 3074, 2002—Revised).

LOSHA.30 BUILDING COLLAPSE

A large office building has collapsed to a pile of rubble and debris after an explosion and a subsequent fire. The building was originally constructed in 1966 was fully occupied since. Search, rescue, and clean up has been ordered.

1. What materials can be expected in the ash, rubble, and debris?
2. Will respiratory protection be required for workers?
3. If so, what types of respiratory protection are available?
4. Will training be necessary?

SOLUTION:

1. As the building was constructed before the 1970s, asbestos may have been used around beams, pipes, etc. Concrete dust, insulation material, glass shards, and ash will make up a large portion of the material. Some of the dust particles may be respirable. As the building housed office workers, one can expect to find electronic debris, i.e., e-waste, which consists of crushed computers, monitors, batteries, copiers, microwaves, etc. E-waste contains substances that are considered to be hazardous, including lead, mercury, and cadmium.

2. Employees need to wear respirators whenever engineering and work practice control measures are not adequate to prevent atmospheric contamination at the worksite. As the work site cannot be adequately controlled, respirators will be required.

3. When selecting respirators, employers must consider the chemical and physical properties of the contaminant, as well as the toxicity and concentration of the hazardous material and the amount of oxygen present. Other selection factors are the nature and extent of the hazard, work rate, area to be covered, mobility, work requirements and conditions, as well as the limitations and characteristics of the available respirators.

    Air-purifying respirators use filters or sorbents to remove harmful substances from the air. They range from simple disposable masks to sophisticated devices. They do not supply oxygen and must not be used in oxygen-deficient atmospheres or in other atmospheres that are immediately dangerous to life or health (IDLH).

    Atmosphere-supplying respirators are designed to provide breathable air from a clean air source other than the surrounding contaminated work atmosphere. They include supplied-air respirators (SARs) and self-contained breathing apparatus (SCBA) units. The time needed to perform a given task, including the time necessary to enter and leave a contaminated area, is an important factor in determining the type of respiratory protection needed. For example, SCBAs, gas masks, or air-purifying chemical-cartridge respirators provide respiratory protection for relatively short periods. On the other hand, an atmosphere-supplying respirator that supplies breathable air from an air compressor through an air line can provide protection for extended periods.

    If the total concentration of atmospheric particulates is low, particulate filter air-purifying respirators can provide protection for long periods without the need to replace the filter.

4. Training is essential for correct respirator use. Employers must teach supervisors and workers how to properly select, use, and maintain respirators. All employees required to use respiratory protective equipment must receive instruction in the proper use of the equipment and its limitations. Employers should develop training programs based on the employee's education level and language background.

    Training must be comprehensive enough for the employee to demonstrate knowledge of the limitations and capabilities of the respirator, why the respirator is necessary, and how improper fit, usage, or maintenance can compromise the respirator.

Training must include an explanation of the following:

1. Why respirator use is necessary;
2. Nature of the respiratory hazard and consequences of not fitting, using, and maintaining the respirator properly;
3. Reason(s) for selecting a particular type of respirator;
4. Capabilities and limitations of the selected respirator;
5. How to inspect, put on and remove, and check the seals of the respirator;
6. Respirator maintenance and storage requirements;
7. How to use the respirator effectively in emergency situations, including when the respirator malfunctions; and
8. How to recognize medical signs and symptoms that may limit or prevent the effective use of the respirator.

Users should know that improper respirator use or maintenance may cause overexposure. They also should understand that continued use of poorly fitted and maintained respirators can cause chronic disease or death from overexposure to air contaminants.

"Respiratory Protection," OSHA Publication 3079 (2002, Revised).

QUANTITATIVE PROBLEMS (TOSHA)

PROBLEMS TOSHA.1–16

TOSHA.1 IDLH AND LETHAL LEVEL

As described earlier, *the immediately dangerous to life and health* (IDLH) level is the maximum concentration of a substance to which one can be exposed for 30 min without irreversible health effects or death. A lethal level is the concentration at which death is almost certain to occur. The IDLH values were determined by the *National Institute for Occupational Safety and Health* (NIOSH) for the purpose of respirator selection. Respirators provide protection against the inhalation of toxic or harmful materials and may be necessary in certain hazardous situations.

Carbon dioxide is not normally considered to be a threat to human health. It is exhaled by humans and is found in the atmosphere at about 3000 parts per million (ppm). However, at high concentrations it can be a hazard and may cause headaches, dizziness, increased heart rate, asphyxiation, convulsions, or coma.

   Two large bottles of flammable solvent were ignited by an undetermined ignition source after being knocked over and broken by a janitor while cleaning a 10 ft × 10 ft × 10 ft research laboratory. The laboratory ventilator was shut off and the fire was fought with a 10-1b CO_2 fire extinguisher. As the burning solvent had covered much of the floor area, the fire extinguisher was completely emptied in extinguishing the fire.

   The IDLH level for CO_2 set by NIOSH is 50,000 ppm. At that level, vomiting, dizziness, disorientation, and breathing difficulties occur after a 30 min exposure. At a 10% level (100,000 ppm), death can occur after a few minutes even if the oxygen in the atmosphere would otherwise support life.

   Calculate the concentration of CO_2 in the room after the fire extinguisher emptied. Does it exceed the IDLH value? Assume that the gas mixture in the room is uniformly mixed, that the temperature in the room is 30°C (warmed by the fire above normal room temperature of 20°C), and that the ambient pressure is 1 atm.

SOLUTION: First, calculate the number of moles of CO_2, n_{CO_2}, discharged by the fire extinguisher.

$$n_{CO_2} = (10\,\text{lb}\,CO_2)(454\,\text{g/lb})/(44\,\text{g/gmol}\,CO_2)$$
$$= 103\,\text{gmol of}\,CO_2$$

The volume of the room, V, is

$$V = (H)(W)(L)$$
$$= (10\,\text{ft})(10\,\text{ft})(10\,\text{ft})(28.3\,\text{L/ft}^3)$$
$$= 28{,}300\,\text{liters}$$

The ideal gas law is used to calculate the total number of moles of gas in the room, n:

$$n = \frac{PV}{RT}; T = 273° + °C$$
$$= \frac{(1\,\text{atm})(28{,}300\,\text{L})}{\left(0.08206\,\dfrac{\text{atm}\cdot\text{L}}{\text{gmol}\cdot\text{K}}\right)(303\,\text{K})} \qquad (9.2)$$
$$= 1138\,\text{gmol gas}$$

The concentration or mole fraction of CO_2 in the room, y_{CO_2}, may now be calculated:

$$y_{CO_2} = (\text{gmol } CO_2)/(\text{gmol gas})$$
$$= (103 \text{ gmol } CO_2)/(1138 \text{ gmol gas})$$
$$= 0.0905$$
$$= 9.05\%$$

The IDLH level is 5.0% and the level is 10.0%. Therefore, the level in the room of 9.05% does exceed the IDLH level for CO_2. It is also dangerously close to the lethal level. The person extinguishing the fire is in great danger and should take appropriate safety measures.

If a dangerous level is present, consideration must be given to using protective equipment such as a respirator. As described in the qualitative section of this Chapter (LOSHA problem) respirators protect the individual from harmful materials in the air. Air-purifying respirators will clean the air but will not protect users against an oxygen-deficient atmosphere. Thus, air-purifying respirators are generally not used in IDLH applications. The only respirators that are recommended for fighting fires are self-contained breathing apparatuses with full facepieces. Recommendations for the selection of the proper respirator are based on the most restrictive of the occupational exposure limits.

TOSHA.2   REACTOR SEAL RUPTURE

One of the first basic steps in an OSHA study is to estimate the concentration of any pollutants that are being generated in the process. A large laboratory with a volume of 1100 m^3, at 22°C and 1 atm contains a reactor which may emit as much as 1.50 gmol of a hydrocarbon (HC) into the room if a seal ruptures. If the hydrocarbon mole fraction in the room air becomes greater than 850 ppb it constitutes a health hazard.

1. Calculate the total number of gmol of air in the room.
2. Suppose the reactor seal ruptures and the maximum amount of HC is emitted almost instantaneously. Assume that the air flow in the room is sufficient to make the room behave like a continuous stirred tank reactor (CSTR), i.e., the air composition is spatially uniform. Calculate the ppb of hydrocarbon in the room. Is there a health risk?

SOLUTION: First calculate the gmol of air in the room employing Charles' law, noting that at 0°C, 1.0 gmol of an ideal gas occupies 22.4 L (1,000 gmol occupies 22.4 m^3)

$$n_{air} = (1100 \text{ m}^3)(1000 \text{ gmol}/22.4 \text{ STP m}^3)[273 \text{ K}/(22 + 273) \text{ K}]$$
$$= 45,440 \text{ gmol air}$$

Calculate the mole fraction of hydrocarbon in ppm and ppb

$$y_{HC} = 1.5\,\text{gmol HC}/45,440\,\text{gmol air}$$
$$= 3.3\ \text{ppm}$$
$$= 3,300\ \text{ppb}$$

Since 3,300 ppb $\gg$ 850 ppb, there is a definite health risk.

TOSHA.3 WORKPLACE EXPOSURE

Workers in a certain plant are exposed for 8 h periods simultaneously to acetone (300 ppm), sec-butyl acetate (100 ppm), and methyl ethyl ketone (150 ppm). The TLV standard for acetone is 750 ppm, for sec-butyl acetate is 200 ppm, and for methyl ethyl ketone is 200 ppm. Note that these 1989 "Threshold Limit Values" (TLV) are the maximum exposures permitted for one person for a 40 hour week. Exposure to multiple hazards exceed the standard when Σ(Concentration/TLV) $>$ 1.0. Is the atmosphere to which the workers are exposed in excess of the standards?

SOLUTION: Exposure to a mixture of three substances exceeds the TLV standard when the following equation holds:

$$(C_1/TLV_1) + (C_2/TLV_2) + (C_3/TLV_3) > 1 \qquad (9.3)$$

In this case,

$$(300/750) + (100/200) + (150/200) = 1.65 > 1$$

Therefore, the exposure is greater than the acceptable TLV level. Note that this method of estimating combined exposure is valid only when the various substances all have similar toxic effects.

TOSHA.4 TETRACHLOROETHYLENE (TCE) EXPOSURE

Normally, a certain employee spends all morning (9 am to 12 N) at her desk, 1 h (12 N to 1 pm) in the cafeteria, all afternoon in a production facility (1 pm to 4:45 pm) where the TCE concentration is 100 ppm, and one quarter of an hour in the quality control laboratory just before she goes home. In the quality control laboratory she spends most of her time looking closely at beakers of samples taken from the production line. The TCE concentration directly above the beakers is 600 ppm. The 1989 TLV-TWA level for TCE is 100 ppm. Calculate the time weighted average (TWA) exposure for this worker. Is this level of exposure in excess of the standard?

SOLUTION: TWA exposure is calculated using the following equation:

$$\text{TWA exposure} = \frac{t_1\,(\text{Concentration 1}) + t_2\,(\text{Concentration 2}) + \cdots}{\text{total time}} \qquad (9.4)$$

For this exposure example, the TWA is:

$$\text{TWA} = \frac{4\,\text{h}\,(0\ \text{ppm}) + 3.75\,\text{h}\,(100\ \text{ppm}) + 0.25\,\text{h}\,(600\ \text{ppm})}{8\,\text{h}} = 66\ \text{ppm}$$

Thus the worker's exposure is less than the TLV-TWA standard. But it should not be forgotten that the relevant short term exposure levels should also be checked (TLV-STEL). For TCE, this limit is 300 ppm for any 5 min during a 3 h period. The time spent in the quality control laboratory therefore exceeds this limit.

TOSHA.5 TIME WEIGHTED AVERAGE (TWA) OVERTIME EXPOSURE

Refer to Problem TOSHA.4. This particular employee has been working up to 15 h of overtime in the production area each week. Recalculate her TWA exposure based on this new information. Is this level safe?

SOLUTION: When the worker works 3 h overtime in a day, the TWA becomes

$$\text{TWA} = 66\ \text{ppm} + 3\,\text{h}\,(100\ \text{ppm})/8\,\text{h}$$
$$= 66 + 37 = 103\ \text{ppm}$$

Thus, her exposure exceeds the TLV standard. However, TLV-TWA standards are based on an 8 h exposure and on a 40 h work week. Thus, applying this TLV-TWA standard to this employee working overtime may allow her to be exposed to a much greater hazard than was contemplated when the standard was set. When exposure exceeds the TLV standards, additional personal protective equipment must be used to reduce the worker's overall exposure to within acceptable limits.

TOSHA.6 ACCIDENTAL VAPOR EMISSION

Many industrial chemicals are toxic or flammable, or sometimes both. Regardless of whether the chemical is toxic or flammable, it can present a danger to plant operators and the public if it is released from its container. Substantial efforts should be taken to assure that toxic or flammable materials are not spilled or released from containment. There is always a chance, however, that such materials might be released. Therefore, provisions must be made to protect the plant operators and anyone who lives or works in the vicinity.

A certain poorly ventilated chemical storage room (10 ft × 20 ft × 8 ft) has a ceiling fan but no air conditioner. The air in the room is at 51°F and 1.0 atm pressure. Inside this room, a 1 lb bottle of iron(III) sulfide (Fe_2S_3) sits next to a bottle of sulfuric acid containing 1 lb H_2SO_4 in water. An earthquake (or perhaps the elbow of a passing technician) sends the bottles on the shelf crashing to the floor where the bottles break, and their contents mix and react to form iron(III) sulfate [$Fe_2(SO_3)$] and hydrogen sulfide (H_2S).

Calculate the maximum H_2S concentration that could be reached in the room assuming rapid mixing by the ceiling fan with no addition of outside air (poor ventilation). Compare the result with the TLV (10 ppm) and IDLH (300 ppm) levels for H_2S.

SOLUTION:

1. Balance the chemical equation:

amount	Fe_2S_3	+	$3H_2SO_4$	→	$Fe_2(SO_4)_3$	+	$3H_2S$
before	1 lb		1 lb		0		0
reaction	0.0048 lbmol		0.010 lbmol		0		0

The molecular weights of Fe_2S_3 and H_2SO_4 are 208 and 98, respectively.

The terms *limiting reactant* and *excess reactant* refer to the actual number of moles present in relation to the stoichiometric proportion required for the reaction to proceed to completion. From the stoichiometry of the reaction, 3 lbmol of H_2SO_4 are required to react with each lbmol of Fe_2S_3. The sulfuric acid is the limiting reactant and the iron(III) sulfide is the excess reactant. In other words, 0.0144 lbmol of H_2SO_4 is required to react with each 0.0048 lbmol of Fe_2S_3, or 0.030 lbmol of Fe_2S_3 is required to react with 0.010 lbmol of H_2SO_4.

Calculate the moles of H_2S generated, n_{H_2S}:

$$n_{H_2S} = (0.010 \text{ lbmol } H_2SO_4)(3H_2S/3H_2SO_4)$$
$$= 0.010 \text{ lbmol}$$

Next, convert the moles to mass:

$$m_{H_2S} = (0.010 \text{ lbmol } H_2S)(34 \text{ lb/lbmol } H_2S)$$
$$= 0.34 \text{ lb}$$

The final H_2S concentration in the room in ppm, C_{H_2S}, can now be calculated. At 32°F and 1 atm, one lbmol of an ideal gas occupies 359 ft^3; at 51°F, one lbmol occupies

$$V = 359\left(\frac{460 + 51}{460 + 32}\right) = 373 \text{ ft}^3$$

Therefore

$$C_{H_2S} = \frac{(0.34\,\text{lb})\left(\dfrac{373\,\text{ft}^3}{\text{lbmol air}}\right)\left(\dfrac{\text{lbmol air}}{29\,\text{lb}}\right)(10^6)}{1600\,\text{ft}^3}$$

$$= 2733\,\text{ppm}$$

This concentration of H_2S far exceeds the TLV (10 ppm) as well as the IDLH (300 ppm).

This result explains why metal sulfides should never be stored near strong mineral acids. Note that the obnoxious odor of H_2S is overpowering at much lower levels than these. Without self-contained breathing apparatus, it is not likely that anything could be done to minimize the result of such a spill.

TOSHA.7 H$_2$S CLOUD

Refer to Problem TOSHA.6. Later, when exhausted from the room, the H_2S mixes with outside air. What will be the final volume of the H_2S cloud when the concentration finally reaches the TLV?

SOLUTION: The volume of the H_2S cloud after dilution is obtained as follows:

$$10\,\text{ppm} = \frac{10\,\text{ft}^3\,H_2S}{10\,\text{ft}^3\,\text{air}} = \left(\frac{0.34\,\text{lb}\,H_2S}{34\,\text{lb}\,H_2S/\text{lbmol}\,H_2S}\right)\left(\frac{373\,\text{ft}^3 H_2S}{\text{lbmol}\,H_2S}\right)\bigg/V$$

following for V,

$$V = (10^5)(0.01)(373)$$

$$= 3.73 \times 10^5\,\text{ft}^3$$

TOSHA.8 STORAGE ROOM EVAPORATION

It is not unusual to find reaction flasks containing volatile solvents stored in research laboratory refrigerators.

Consider the following: a 500 mL flask of diethyl ether (MW = 74.14 g/gmol, specific gravity = 0.713) is stored in a 15 ft^3, unventilated refrigerator at 41°F and 1 atm. (*Note*: the vapor pressure of diethyl ether at 41°F is 200 mm Hg).

1. How many grams of the diethyl ether must evaporate to achieve the minimum percent of ether vapors producing a flammable mixture? This minimum

percentage of vapors is defined as the lower flammable limit (LFL) which for diethyl ether is 1.9% by volume (v/v).

2. How many grams of the ether must evaporate to reach the maximum percent of ether vapors that would still be flammable? This maximum percentage of vapors is the upper flammable limit (UFL) which for diethyl ether is 36% v/v.

SOLUTION:

1. The quantity of diethyl ether needed to reach a concentration of 1.9 vol% in the refrigerator may be found by first calculating the number of gmol of air in the refrigerator.

To find the gmol air in the refrigerator, use the Ideal Gas Law, $PV = nRT$, using $P = 1.00$ atm, and $T = 41°F = 5°C = 278$ K.

Volume of the refrigerator $= (15.0\ ft^3)(0.0283\ m^3/ft^3) = 0.425\ m^3 = 425\ L$

$$n = \frac{(1\ atm)(425\ L)}{(0.0821\ atm\text{-}L/gmol\text{-}K)(278\ K)} = 18.6\ gmol\ air\ in\ the\ refrigerator$$

To find the number of gmol of diethyl ether needed to reach the 1.9% LFL, the gmol of air in the refrigerator is multiplied by 0.019:

$$(18.6\ gmol\ air)(0.019) = 0.353\ gmol\ diethyl\ ether\ at\ the\ LFL$$

To convert to g diethyl ether, the following calculation is made:

$$(0.353\ gmol\ diethyl\ ether)(74.1\ g/gmol) = 26.1\ g\ diethyl\ ether\ at\ the\ LFL$$

The LFL is reached in the refrigerator when 26.1 g of diethyl ether in the flask have evaporated.

To better evaluate the chance of reaching the diethyl ether LFL in the refrigerator it would be useful to find what percent of the diethyl ether in the flask would have to evaporate in order to produce this situation:

$$\frac{26.1\ g\ diethyl\ ether\ evaporating}{357\ g\ diethyl\ ether\ in\ flask} \times 100 = 7.3\%$$

Since a vapor pressure of 200 mm Hg is considerable, the chance of as little as 7.3% of the diethyl ether evaporating to reach the LFL is quite high. Those periods of time, such as weekends, when the refrigerator might not be opened, would result in the situations of greatest risk.

2. The quantity of diethyl ether needed to reach a concentration of 36 vol% in the refrigerator, its UFL, may be found by multiplying the gmol of air in the refrigerator by 0.36, and finally, converting gmol to g of diethyl ether.

$$\text{The gmol air in the refrigerator} = 18.6 \, \text{gmol air}$$

The number of gmol of diethyl ether needed to reach the 36% UFL is:

$$(18.6 \, \text{gmol air})(0.36) = 6.70 \, \text{gmol diethyl ether at the UFL}$$

To convert to g diethyl ether, the following calculation is made:

$$(67.0 \, \text{gmol diethyl ether})(74.1 \, \text{g/gmol}) = 4964 \, \text{g diethyl ether at the LFL}$$

The mass of diethyl ether in the flask is determined as:

$$(500 \, \text{mL diethyl ether})(0.713 \, \text{g/mL}) = 357 \, \text{g diethyl ether in the flask}$$

Since 4964 g of diethyl ether represents more diethyl ether than is in the flask, the diethyl ether UFL cannot be reached in the refrigerator.

TOSHA.9 STORAGE PRACTICE HAZARD

Refer Problem TOSHA.8. Comment on the hazards of such a storage practice and suggest safer alternatives.

SOLUTION: As there is no possibility of exceeding the UFL, the hazard is not reduced by further evaporation. A storage practice such as this, which results in a significant chance of producing a flammable mixture of vapors within an enclosed space, is not advisable. An explosion proof ventilated refrigerator should be used, or the solvent should be changed/eliminated.

TOSHA.10 DILUTION VENTILATION

Estimate the dilution ventilation required in an indoor work area where a toluene-containing adhesive is used at a rate of 3 gal/8-h workday. Assume that the specific gravity of toluene (C_7H_8) is 0.87, that the adhesive contains 40 vol% toluene, and that 100% of the toluene is evaporated into the room air at 20°C. The plant manager has specified that the concentration of toluene must not exceed 80% of its threshold limit value (TLV) of 100 ppm.

The following equation can be used to estimate the dilution air requirement:

$$q = K(q_c/C_a) \tag{9.5}$$

where q = dilution air flowrate

K = dimensionless mixing factor that accounts for less than complete mixing characteristics of the contaminant in the room, the contaminant toxicity level, and the number of potentially exposed workers. Usually, the value of K varies from 3 to 10, where 10 is used under poor mixing conditions and when the contaminant is relatively toxic (TLV < 100 ppm).

q_c = volumetric flowrate of pure contaminant vapor, c

C_a = acceptable contaminant concentration in the room, volume or mole fraction (ppm × 10^{-6}).

SOLUTION: The dilution air is estimated from

$$q = K(q_c/C_a)$$

Since the TLV for toluene is 100 ppm and C_a is 80% of the TLV,

$$C_a = [0.80(100)] \times 10^{-6} = 80 \times 10^{-6} \text{ (volume fraction)}$$

The mass flowrate of toluene is

$$\dot{m}_{tol} = \left(\frac{3 \text{ gal}_{adhesive}}{8 \text{ h}}\right)\left(0.4 \frac{\text{gal}_{toluene}}{1 \text{ gal}_{adhesive}}\right)\left[\frac{(0.87)(8.34 \text{ lb})}{1 \text{ gal}_{toluene}}\right]$$

$$= 1.09 \text{ lb/h}$$

$$= \left(\frac{1.09 \text{ lb}}{1 \text{ h}}\right)\left(\frac{454 \text{ g}}{1 \text{ lb}}\right)\left(\frac{1 \text{ h}}{60 \text{ min}}\right)$$

$$= 8.24 \text{ g/min}$$

Since the molecular weight of toluene is 92,

$$\dot{n}_{tol} = 8.24/92$$

$$= 0.0896 \text{ gmol/min}$$

The resultant toluene vapor volumetric flowrate, q_{tol}, is calculated directly from the ideal gas law:

$$q_c = q_{tol} = \frac{(0.0896 \text{ gmol/min})[0.08206 \text{ atm} \cdot \text{L/(gmol} \cdot \text{K)}](293 \text{ K})}{1 \text{ atm}}$$

$$= 2.15 \text{ L/ min}$$

Therefore, the required diluent volumetric flowrate is (assuming a K value of 5)

$$q = (K)(q_c/C_a)$$
$$= \frac{(5)(2.15 \, L/min)}{80 \times 10^{-6}}$$
$$= 134{,}375 \, L/min$$
$$= \left(134{,}375 \frac{L}{min}\right)\left(\frac{1 \, ft^3}{28.36 \, L}\right)$$
$$= 4748 \, ft^3/min$$

TOSHA.11   RESPIRATORS

Discuss the general subject of respirators.

SOLUTION: Where there are high concentrations of contaminants, an atmosphere-supplying respirator such as the positive-pressure SAR offers protection for a long period. SARs eliminate the need for concern about filter breakthrough times, change schedules, or using end-of-service-life indicators (ESLI) for airborne toxic materials, factors that must be considered when using air-purifying respirators.

Respirators must not impair the worker's ability to see, hear, communicate, and move as necessary to perform the job safely. For example, atmosphere-supplying respirators may restrict movement and present other potential hazards.

SARs with their trailing hoses can limit the area the wearer can cover and may present a hazard if the hose comes into contact with machinery. Similarly, a SCBA that includes a back-mounted, compressed-air cylinder is both large and heavy. This may restrict climbing and movement in tight places and the added weight of the air cylinder presents an additional burden to the wearer.

Another factor to consider when using respirators is the air-supply rate. The wearer's work rate determines the volume of air breathed per minute. The volume of air supplied to meet the breathing requirements is very significant when using atmosphere-supplying respirators such as self-contained and airline respirators that use cylinders because this volume determines their operating life.

The peak airflow rate also is important in the use of a constant-flow SAR. The air-supply rate should always be greater than the maximum amount of air being inhaled in order to maintain the respiratory enclosure under positive pressure.

Higher breathing resistance of air-purifying respirators under conditions of heavy work may cause the user breathing difficulty, particularly in hot, humid conditions. To avoid placing additional stress on the wearer, use the lightest respirator possible that presents the least breathing resistance.

TABLE 9.5   Hazard–Respirator Information

Hazard	Respirator
Oxygen deficiency Gas, vapor contaminants and other highly toxic air contaminants	Full-facepiece, pressure-demand SCBA certified for a minimum service life of 30 minutes. A combination full- facepiece, pressure-demand SAR with an auxiliary self-contained air supply.
Contaminated atmospheres—for escape	Positive-pressure SCBA. Gas mask. Combination positive-pressure SAR with escape SCBA.
Not immediately dangerous to life or health:	
Gas and vapor contaminants	Positive-pressure SAR. Gas mask. Chemical-cartridge or canister respirator.
Particulate contaminants	Positive-pressure SAR including abrasive blasting respirator. Powered air-purifying respirator equipped with high-efficiency filters. Any air-purifying respirator with a specific particulate filter.
Gaseous and particulate contaminants	Positive-pressure supplied-respirator. Gas mask. Chemical-cartridge respirator with mechanical filters.
Smoke and other fire-related contaminants	Positive-pressure SCBA.

SCBAs and some chemical canister respirators provide a warning of remaining service time. This may be a pressure gauge or timer with an audible alarm for SCBAs or a color ESLI on the cartridge or canister. The user should understand the operation and limitations of each type of warning device.

For the many gas masks and chemical-cartridge respirators with no ESLI devices, the employer must establish and enforce a cartridge or canister change schedule. In addition, employees should begin each work shift with new canisters and cartridges.

Table 9.5 list presents a simplified version of characteristics and factors used for respirator selection. It does not specify the contaminant concentrations or particle size. Some OSHA substance-specific standards include more detailed information on respirator selection as it applies to IDLH.

TOSHA.12   CARCINOGENS

Determine the action level in $\mu g/m^3$ for an 80-kg person with a life expectancy of 70 years exposed to benzene over a 15-year period. The acceptable risk is one incident of cancer per one million persons, or 10^{-6}. Assume a breathing rate of $15 \, m^3/day$ and an absorption factor of 75% (0.75). The potency factor for benzene is 1.80 mg/(kg · day).

The following equation has been used in health risk assessment studies for carcinogens:

$$C_m = \frac{(R)(W)(L)}{(P)(I)(A)(ED)} \tag{9.6}$$

where $C_m =$ action level, i.e., the concentration of carcinogen above which remedial action should be taken

$R =$ acceptable risk or probability of contracting cancer

$W =$ body weight

$L =$ assumed lifetime

$P =$ potency factor

$I =$ intake rate

$A =$ absorption factor, the fraction of carcinogen absorbed by the human body

$ED =$ exposure duration

SOLUTION: Using the equation provided in the Problem statement, the action level for the carcinogen benzene for an 80-kg worker with a life expectancy of 70 years and an exposure duration of 15 years is:

$$C_m = \frac{(R)(W)(L)}{(P)(I)(A)(ED)}$$

$$= \frac{(10^{-6})(80\,\text{kg})(70\,\text{yr})}{\left(1.80\frac{\text{kg}\cdot\text{d}}{\text{mg}}\right)\left(15\frac{\text{m}^3}{\text{d}}\right)(0.75)(15\,\text{yr})\left(\frac{\text{mg}}{1000\,\mu\text{g}}\right)}$$

$$= 0.0184\,\mu\text{g}/\text{m}^3$$

The risk for this person would be classified as unacceptable if the exposure exceeds $0.0184\,\mu\text{g}/\text{m}^3$ in a 15-year period.

The above calculation was based on an individual exposed to the substance for 24 hours per day and 7 days per week. This calculation can easily be used to estimate the amount of risk due to worker exposure if the hours, days, and years are adjusted accordingly. This comment also applies to Problem TTSCA.5.

TOSHA.13 TANK CAR ACCIDENT

A train has collided with a truck at an intersection in the industrial section of a major city. A tank car and a flatbed car filled with containers have derailed. The tank car is lying in a ditch alongside the tracks, surrounded by some containers that have broken loose from the flatbed car. The following information is known.

1. The tank car is labeled "hydrogen fluoride".
2. The UNNA number shown on all the containers is 1806. (The UNNA number is the United Nations North America number. This numbering system was

developed by the U.S. Department of Transportation, and has since become the UN standard system for classifying hazardous materials.)

If you were responding to this incident, what additional information would you want to know?

SOLUTION:

1. Obviously, information (particularly health related) on hydrogen fluoride.
2. Determine the chemical from the UNNA number. The number 1806 represents phosphorus pentachloride.
3. Obtain information on phosphorus pentachloride.
4. Determine what problems could occur if these chemicals were to be mixed together when crews right the tank car and collect the containers.

TOSHA.14 TOBACCO SMOKE VS. GASOLINE VAPORS

Environmental tobacco smoke and gasoline vapors both contain mixtures of trace amounts of many of the individual compounds regulated as Air Toxics under Section 112 of the Clean Air Act. Many gasoline attendants are more likely to be exposed to these mixtures during the course of their lives than to specific compounds on the air toxics list. Hence, estimating cancer risk resulting from worker exposure to these mixtures is a useful and relevant exercise.

Which air toxic mixture imposes the greatest cancer risk to worker breathing either of the two gases below at an average concentration of 5 $\mu g/m^3$? The following data is provided.

Environmental tobacco smoke [unit cancer risk 2.80×10^{-5} $(\mu g/m^3)^{-1}$]

Gasoline vapors [unit cancer risk 1.60×10^{-6} $(\mu g/m^3)^{-1}$].

SOLUTION: To solve this problem, compute the cancer risk for each air toxics mixture by multiplying the average concentration by the unit cancer risk.

1. For environmental tobacco smoke,

$$\text{Risk} = (5\,\mu g/m^3)\,(2.80 \times 10^{-5}) = 1.40 \times 10^{-4} = 140 \text{ in } 10^6$$

2. For gasoline vapors

$$\text{Risk} = (5\,\mu g/m^3)(1.60 \times 10^{-5}) = 8.00 \times 10^{-5} = 80 \text{ in } 10^6$$

Since 140 in 10^6 exceeds 80 in 10^6, the conclusion is, that for identical durations of exposure to identical concentrations in air, environmental tobacco smoke poses a greater cancer risk than gasoline vapors. However, both mixtures exhibit high cancer risks at trace levels.

TOSHA.15   INSTANTANEOUS "PUFF" MODEL

Refer to Problem TCAA.38. As noted, the factory workers will be adversely affected if the concentration of the gas is greater than $1.0 \, \mu g/L$. Is there any impact on the workers in the factory? Assume that stability category D applies so that

$$\sigma_y = 16\text{mm at 200 mm downstream}$$
$$\sigma_z = 8.5\text{mm at 200 mm downstream}$$

SOLUTION: The maximum concentration at the factory can now be calculated using Equation (3.33) given in the Problem statement TCAA.38:

$$C(x, y, \text{ø}, H) = \left[\frac{2\,m_T}{(2\pi)^{1.5}\sigma_x\sigma_y\sigma_z}\right]\left\{\exp\left[-0.5\left(\frac{x - ut}{\sigma_x}\right)^2\right]\right\}$$

$$\left\{\exp\left[-0.5\left(\frac{H}{\sigma_z}\right)^2\right]\right\}\left\{\exp\left[-0.5\left(\frac{y}{\sigma_y}\right)^2\right]\right\} \tag{3.33}$$

Note that for the maximum concentration, $x = ut$ and the first exponential term at the maximum concentration becomes 1.0 (unity).

Substituting yields

$$C(200, 50, 0, 20) = \left[\frac{2(72{,}000)}{(2\pi)^{1.5}(16)(16)(8.5)}\right][1]\left\{\exp\left[-0.5\left(\frac{20}{8.5}\right)^2\right]\right\}$$

$$\times\left\{\exp\left[-0.5\left(\frac{50}{16}\right)^2\right]\right\}$$

$$= 2.00 \times 10^{-3} \, \text{g/m}^3$$

The concentration can also be expressed in units of micrograms/liter:

$$C = (2.00 \times 10^{-3} \, \text{g/m}^3)(1 \times 10^6 \, \mu g/g)/(1000 \, \text{L/m}^3)$$
$$= 2.00 \, \mu g/L$$

Since the calculated maximum concentration, $2.00 \, \mu g/L$ is greater than $1.0 \, \mu g/L$, there is an impact on workers in the factory.

TOSHA.16 PERFORMANCE OF A CARBON CARTRIDGE RESPIRATOR

A respirator cartridge contains 80 g of a blend of charcoal, and tests have shown that breakthrough (when toxic A starts to be emitted from the cartridge) will occur when 80% of the charcoal is saturated. How long will this cartridge be effective if the ambient concentration of toxic A is 700 ppm and the temperature is 30°C? Assume that the breathing rate of a normal person is 45 L/min (45,000 cm^3/min).

For a particular blend of carbon cartridge (for organic vapor), the adsorption potential (equilibrium concentration) for A can be expressed by the following equation:

$$\log_{10}(C_A) = -0.11\left[\left(\frac{T}{V}\right)\log_{10}\left(\frac{p'_A}{p_A}\right)\right] + 2.076 \qquad (9.7)$$

where C_A = amount of A adsorbed in charcoal, cm^3 liq/100 g charcoal at partial pressure p_A.

T = temperature, K

V = molar volume of A (as liquid) at the normal boiling point, 100 cm^3/gmol for substance A

p'_A = saturation (vapor pressure) of A

p_A = partial pressure of A

The vapor pressure of A at 30°C is 60 mm Hg and the liquid density is 1.12 g/cm^3. The molecular weight is 110.

SOLUTION: First, calculate the partial pressure of toxic A, p_A, in units of mm Hg:

$$p_A = (y_A)(p)$$
$$= (700 \times 10^{-6})(760\,\text{mm Hg}) \qquad (9.8)$$
$$= 0.532\,\text{mm Hg}$$

The ratio of vapor pressure to the partial pressure, p'_A/p_A, is then

$$p'_A/p_A = (60\,\text{mm Hg})/(0.532\,\text{mm Hg})$$
$$= 112.8$$

The amount of A adsorbed per 100 g of charcoal in $cm^3/100$ g charcoal is

$$\log_{10}(C_A) = -0.11\left[\left(\frac{T}{V}\right)\log_{10}\left(\frac{p'_A}{p_A}\right)\right] + 2.076$$

$$= -0.11\left[\left(\frac{30 + 273\,K}{100\,cm^3/gmol}\right)\log_{10}(112.8)\right] + 2.076$$

$$= 1.392$$

$$C_A = 24.65\,cm^3/100\,g$$

The mass of A adsorbed in the cartridge in grams at 80% saturation (i.e., at breakthrough) is

$$m_A = (\text{breakthrough fraction})(m_c)(p_A)$$

$$= (0.8)\left(\frac{24.65\,cm^3\text{of A}}{100\,g\,\text{charcoal}}\right)(80\,g\,\text{charcoal})(1.12\,g\,A/cm^3)$$

$$= 17.67\,g\,\text{of A}$$

Next, calculate the volumetric flowrate of A inhaled through the cartridge from the vapor concentration and breathing rate:

$$q_A = y_A\,q_{air}$$

$$= (700 \times 10^{-6})\,(45{,}000\,cm^3/min)$$

$$= 31.5\,cm^3/min$$

The intake mass flowrate of A is (applying the ideal gas rate)

$$\dot{m}_A = \frac{pq_A(MW)}{RT}$$

$$= \frac{(1\,atm)\,(31.5\,cm^3/min)\,(110\,g/gmol)}{\left(82.06\,\frac{cm^3\cdot atm}{mol\cdot K}\right)(303\,K)}$$

$$= 0.139\,g/min$$

Finally, the time for breakthrough in minutes is

$$t = (17.67\,g)/(0.139\,g/min)$$

$$= 127\,min$$

The typical economic concentration limit of an organic vapor cartridge is approximately 1000 ppm. If the ambient or local concentration is above 1000 ppm, other methods of personal protection are recommended. It is also important to select the appropriate type of adsorbent since some toxic chemicals are not readily adsorbed by charcoal. For example, hydrogen cyanide is not well adsorbed by charcoal. Since the *immediately dangerous to life and health* (IDLH) (see Chapter 9 for more details) level of hydrogen cyanide is 50 ppm and the odor threshold is greater than 50 ppm, by the time the worker smells the vapor, it will be too late to avoid death. Therefore, it is crucial to select a cartridge adsorbent that is specifically designed for hydrogen cyanide. Examples of such special cartridges are the cartridge for chlorine gas and pesticide vapors. Typically, cartridges are color coded so that they are easily distinguished (e.g., black for common organic vapors).

CHAPTER 10

POLLUTION PREVENTION ACT (PPA)

QUALITATIVE PROBLEMS (LPPA)

PROBLEMS LPPA.1–29

LPPA.1   WHY POLLUTION PREVENTION?

Describe the background and objectives of pollution prevention.

SOLUTION: When EPA was established in the early 1970's, it had to focus first on controlling and cleaning up the most immediate problems. Those efforts yielded major reductions in pollution. Over time, however, the Agency learned that traditional "end-of-pipe" approaches not only can be expensive and less than fully effective but also can sometimes transfer pollution from one medium to another. Additional improvements to environmental quality were required to move "upstream" to prevent pollution from occurring in the first place.

As will be shown later, preventing pollution also offers important economic benefits, as pollution never created avoids the need for expensive investments in waste management or cleanup. Pollution prevention has the potential for both protecting the environment and strengthening economic growth through more efficient manufacturing and raw material use.

Pollution prevention is influenced by a number of factors, including EPA regulations and state programs, collaborative efforts that offer recognition and technical

Environmental Regulatory Calculations Handbook, by Leo Stander and Louis Theodore
Copyright © 2008 John Wiley & Sons, Inc.

assistance, public data, the availability of clean technologies, and the practices and policies of large public agencies. To be effective, pollution prevention programs must establish the following objectives for each of these areas:

1. Regulations and Compliance: The mainstream activities at EPA such as regulatory development permitting, inspections, and enforcement, must reflect commitment to reduce pollution at the source and minimize the cross-media transfer of waste.

2. State and Local Partnerships: Increasingly, state and local agencies are the "face of government" for the general public. The national network of state and local prevention programs needs to be strengthened and must seek to integrate prevention into state and local regulatory, permitting, and inspection programs supported with federal funds.

3. Private Partnerships: The program must identify and pioneer new cooperative efforts that emphasize multi-media prevention strategies, reinforce the mutual goals of economic and environmental well-being, and represent new models for government/private sector interaction.

4. Federal Partnerships: The program must work closely with counterparts in other agencies to ensure that pollution prevention guides management and procurement decisions, and pursues opportunities for reducing waste at the source in the non-industrial sector.

5. Public Information/The Right-to-Know: Programs must collect and share useful information that helps identify pollution prevention opportunities, measure progress, and recognizes success.

6. Technological Innovation: Programs must attempt to meet high priority needs for new pollution prevention technologies that increase competitiveness and enhance environmental stewardship through partnerships with other federal agencies, universities, states, and the private sector.

7. New Legislation: Where justified, pollution prevention programs will not hesitate to seek changes in federal environmental law that will encourage investment in source reduction (defined in next problem).

LPPA.2   DESCRIBE THE POLLUTION PREVENTION ACT

SOLUTION: The most important piece of pollution prevention legislation enacted to date is the Pollution Prevention Act. The Pollution Prevention Act, established pollution prevention as a "national objective." The act notes that (see also Chapter 2):

"There are significant opportunities for industry to reduce or prevent pollution at the source through cost-effective changes in production, operation, and raw materials use.... The opportunities for source reduction are often not realized because existing regulations, and the industrial resources they require for compliance, focus upon treatment and disposal rather than source reduction .... Source reduction is fundamentally different and more desirable than waste management and pollution control.

The Act establishes the pollution prevention hierarchy as national policy, declaring that pollution should be prevented or reduced at the source wherever feasible, while pollution that cannot be prevented should be recycled in an environmentally safe manner. In the absence of feasible prevention or recycling opportunities, pollution should be treated; disposal or other releases into the environment should be used as a last resort.

Source reduction is defined in the law to mean any practice which reduces the amount of any hazardous substance, pollutant, or contaminant entering any waste stream or otherwise released into the environment (including fugitive emissions) prior to recycling, treatment, or disposal; and, any practice which reduces the hazards to public health and the environment associated with the release of such substances, pollutants, or contaminants."

The definition of source reduction is fairly clear: simply produce less pollution at the source by any combination of strategies. Likewise, recycling (in its many forms) is a well understood principle in today's society, although the true value to the environment seems to be lost on a sizable portion of the population. Any reuse of a substance without chemical change is considered to be recycling. Distillation, melting, filtration and physical changes are permitted. Treatment can include a number of processes, including physical, biological, and chemical methods. Incineration is included in this category as a chemical method of treatment (oxidation). Disposal includes landfilling, landfarming, deep well injection and ocean dumping. (These four options are treated in more detail in four later Problems).

In the current working definition used by EPA, source reduction and recycling are considered to be the most viable pollution prevention techniques, preceding treatment and disposal. In its original "Pollution Prevention Policy Statement" published in the January 26, 1989 *Federal Register* (this was later superceded by the Pollution Prevention Act), EPA encouraged organizations, facilities, and individuals to fully utilize source reduction and recycling practices and procedures to reduce risk to public health, safety, and the environment. Additional details on both the Pollution Prevention Act and the Pollution Prevention Policy Statement can be found in the CRC text/reference book *Pollution Prevention*, authored by R. Dupont, L. Theodore, and K. Ganesan.

Two additional points need to be made

1. The Pollution Prevention Act has no provisions related to enforcement and contains no penalties for failure to comply.
2. The effectiveness of implementation of the Pollution Prevention Act relies heavily on voluntary compliance as indicated below:
    a. The Pollution Prevention Act calls on companies to disclose and report a great deal about their operations. Widespread inspections to determine compliance would be very expensive. It would also severely strain the government's manpower.
    b. The law aims at creating a more cooperative relationship between the environmental agencies and industry. Strict enforcement provisions with

penalties for incomplete compliance could do the opposite and actually create a disincentive to critical self-auditing, self-policing and voluntary disclosure.

c. Penalties for willful non-compliance, however, could be effective in reducing any open flaunting of the law.

d. Companies have an incentive to voluntarily comply with the law because having smaller quantities of chemicals to dispose of could actually save money while giving the company a public relations edge. This is discussed in more detail in later Problems.

LPPA.3 WASTE MINIMIZATION/WASTE MANAGEMENT HIERARCHY

Describe waste minimization/waste management hierarchy.

SOLUTION: Describe EPA: Figure 10.1 depicts the EPA hierarchy of preferred approaches to integrated waste management. As one proceeds down the hierarchy, more and more waste is potentially generated; and more energy is expended in the management of this waste. As illustrated by Figure 10.1, integrated waste management may be broken down into the following four major components.

Additional details on these four components are provided in an expanded version in the Figure 10.2.

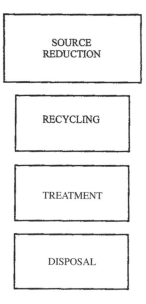

FIGURE 10.1 U.S. EPA's Integrated Waste Management and Pollution Prevention Hierarchy.

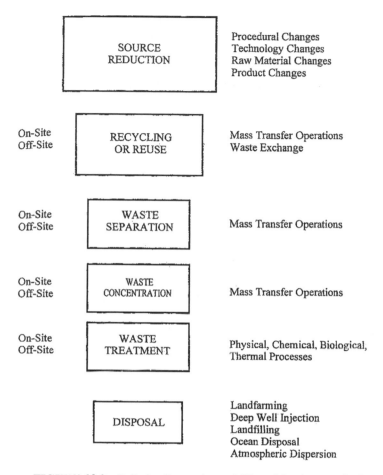

FIGURE 10.2   Pollution Prevention and Waste Management Options.

Waste minimization is defined by the EPA as: "The reduction, to the extent feasible, of hazardous waste that is generated or subsequently treated, stored, or disposed of. It includes any source reduction or recycling activity undertaken by a generator that results in either: (1) the reduction of total volume or quantity of hazardous waste, or (2) the reduction of toxicity of hazardous waste, or both, so long as such reduction is consistent with the goal of minimizing present and future threats to human health and the environment."

There is also need to differentiate between pollution control and pollution prevention. These are briefly defined below:

1. Pollution control is an end-of-pipe process. It is performed after the waste is generated. It often transfers the pollutant from one environmental medium to another.

2. Pollution prevention (P2 or P^2) focuses on the reduction of pollution at the source. This implies "up-front" limiting of waste.

LPPA.4   EPA'S POLLUTION PREVENTION STRATEGY: MULTIMEDIA STRATEGY FOR PRIORITY PERSISTENT, BIOACCUMULATIVE AND TOXIC (PBT) POLLUTANTS

Describe EPA's *Pollution Prevention Strategy*.

SOLUTION: EPA's *Pollution Prevention Strategy*, released in February 1991, was developed by the Agency in consultation with all program and regional offices. The strategy provides guidance on incorporating pollution prevention principles into EPA's ongoing environmental protection efforts and includes a plan for achieving substantial voluntary reductions of targeted high risk industrial chemicals. The strategy was aimed at maximizing private sector initiatives while challenging industry to achieve ambitious prevention goals. A major component of that strategy was the *Industrial Toxics Project*, known more generally as the *33/50 Program*. The overall goals of the 33/50 program were to reduce national pollution release and off-site transfers of priority toxic chemicals and to encourage pollution prevention. The specific program goal was to voluntarily reduce releases and transfers of 17 toxic chemicals by 33 percent by the end of 1992 and 50 percent by the end of 1995. Both goals were reached 1 year ahead of schedule.

The 17 pollutants identified as targets of the industrial toxics project presented both significant risks to human health and the environment and significant opportunities to reduce such risks through prevention. The list (Table 10.1) was drawn from recommendations submitted by program offices, taking into account such criteria as health and ecological risk, potential for multiple exposures or cross-media contamination, technical or economic opportunities for pollution prevention, and limitations for their effective treatment using conventional technologies.

All of the targeted chemicals are included in EPA's Toxic Release Inventory (TRI); thus, reductions in their releases can be measured in each year's TRI reports. Despite company participation being strictly voluntary, almost 1,300

TABLE 10.1   The 33/50 Program Target Chemicals

Benzene	Methyl Ethyl Ketone (MEK)
Cadmium	Methyl Isobutyl Ketone (MIBK)
Carbon Tetrachloride	Nickel
Chloroform	Tetrachloroethylene (PCE)
Chromium	Toluene
Cyanide	1,1,1-Trichloroethane (TCA)
Dichloromethane	Trichloroethylene (TCE)
Lead	Xylene
Mercury	

parent companies, operating more than 6,000 facilities in the U.S., participated in the 33/50 program. Companies examined their own industrial process and identified and implemented cost-effective pollution prevention practices for toxic chemicals. Information on a variety of "success stories" is available on EPA's web site at: http://www.epa.gov/opptintr/3350/. These success stories detail not only technical innovations, but new managerial and cost-accounting approaches that have led to significant reductions in hazardous materials releases by participating companies. Some key results of the Program are listed below.

1. The 33/50 Program achieved its goal in 1994, one year ahead of schedule, primarily through program participants' efforts.
2. Although the largest reductions in 33/50 Program chemicals reflected U.S. action to phase out ozone-depleting chemicals under the Montreal Protocol, facilities also reduced releases and transfers of the other 33/50 chemicals by 50% from 1988 to 1995. The most recent data show a 1988 baseline total for the 17 chemicals of 1.496 billion pounds of on-site releases and transfers off-site to treatment and disposal and a 1994 total of 748 million pounds. In 1995, releases and transfers of the 33/50 chemicals totaled 672 million pounds, and by 1996, releases and transfers had dropped nearly 60% from the 1988 baseline, to 601 million pounds.
3. Facilities reported more source reduction activity (pollution prevention) for 33/50 chemicals than for other TRI chemicals and this activity covered a greater percentage of production-related waste for 33/50 chemicals than for other TRI chemicals.
4. Reductions continued at a higher rate for 33/50 chemicals than for other TRI chemicals in the year after the 33/50 Program ended.

The 33/50 program has been replaced by the multimedia strategy for priority persistent, bioaccumulative, and toxic (PBT) pollutants. The goal of this strategy is to further reduce risks to human health and the environment from existing and future exposure to priority persistent, bioaccumulative, and toxic (PBT) pollutants. The EPA has developed this strategy to overcome the remaining challenges in addressing priority PBT pollutants. These pollutants pose risks because they are toxic, persist in ecosystems, and accumulate in fish and in the food chain. The PBT challenges remaining stem from the pollutants' ability to travel long distances, to transfer rather easily among air, water, and land, and to linger for generations, making EPA's traditional single-statute approaches less than the full solution to reducing risks from PBTs. Due to a number of adverse health and ecological affects linked to PBT pollutants—especially mercury, PCBs, and dioxins—it is key for EPA to aim for further reductions in PBT risks. The fetus and child are especially vulnerable. EPA is committing, through this strategy, to create an enduring cross-office system that will address the cross-media issues associated with priority PBT pollutants.

This multimedia strategy reinforces and builds on existing commitments related to priority PBTs, such as the 1997 Canada—U.S. Binational Toxics Strategy (BNS), the North American Agreement on Environmental Cooperation, and the Clean

Water Action Plan. EPA is forging a new approach to reduce risks from and exposures to priority PBT pollutants through increased coordination among EPA national and regional programs. This approach also requires the significant involvement of stakeholders, including international, state, local, and tribal organizations, the regulated community, environmental groups, and private citizens.

LPPA.5 PROCESS AND EQUIPMENT MODIFICATIONS

Explain how process and equipment modifications are related to pollution prevention.

SOLUTION: One of the primary procedures available for source reduction involves technology changes. This procedure involves process and equipment modifications to reduce waste, primarily in a production setting. Technology changes can range from minor changes that can be implemented quickly and at low cost to major changes involving replacement of equipment process or processes at a very high cost. Since technology modifications usually require greater personnel and capital cost than procedural changes, they are generally investigated after all possible procedural changes have been instituted. Categories of technology modifications include:

1. Process changes
2. Equipment, piping, or layout changes
3. Changes to operational settings
4. Additional automation

These four modifications are discussed below.

Process changes often utilize innovative technology to develop a new process to achieve the same end product while reducing waste originally produced. The pollution prevention program should encompasses process development activities. Research and development efforts are encouraged in order to bring about new processes resulting in a reduced volume of waste generation. Process redesign can include alteration of an existing process by addition of new unit operations or by implementation of new technology to replace out-of-date operations. For example, a metal manufacturer modified a process by using a two-stage abrasive cleaner and eliminated the need for a chemical cleaning bath.

Equipment changes can reduce waste generation by reducing equipment-related inefficiencies. The required capital involved in using more efficient equipment can be justified by higher productivity, reduced raw material costs, and reduced waste management costs. Modifications to certain types of equipment can require a detailed evaluation of process characteristics. In this case, equipment vendors should be consulted for information regarding the applicability of particular types of equipment for a given process. Many equipment changes can be very simple and inexpensive. Examples include installing better seals on equipment to eliminate leakage or simply putting drip pans under equipment to collect leaking material for reuse.

Another minor modification is to increase agitation and alter temperatures to prevent the formation of fouling deposits resulting from crystallization, sedimentation, corrosion, and chemical reactions during formulating and blending procedures.

Changes to operational settings involve changes in the way process equipment operate, including adjustments to the temperature, pressure, flow rate, and residence time parameters. These changes often represent the easiest and least expensive ways to reduce waste generation. Process equipment is designed to operate most efficiently at the optimum parameter settings. Therefore, less waste will be generated when equipment operates efficiently. Trial runs can be used to determine the actual optimum settings. For example, a plating company can change the flow rate of chromium in the plating bath to the optimum setting and reduce the chromium concentration used, thereby resulting in less chromium waste requiring treatment.

Additional controls and/or automation can result in improved monitoring and adjustment of operating parameters to ensure the greatest level of efficiency. Simple steps involving one-stream set point controls or advanced statistical process control systems can be utilized. Automation can reduce the likelihood of human errors resulting in spills and costly downtime. The resulting increase in efficiency owing to automation can increase product yields.

LPPA.6 RECYCLING

Briefly discuss recycling.

SOLUTION: Recycling or reuse can take two forms: preconsumer and postconsumer applications. Preconsumer recycling involves raw materials, products, and by-products that have not reached a consumer for an intended end-use, but are typically reused within an original process. Postconsumer recycled materials are those that have served their intended end-use by a business, consumer, or institutional source and have been separated from municipal solid waste for the purpose of recycling.

Recycling techniques allow waste materials to be used for a beneficial purpose. A material is recycled if it is used, reused, or reclaimed. Recycling through use and/or reuse involves returning waste material either to the original process as a substitute for an input material or to another process as an input material. Recycling through reclamation is the processing of waste for recovery of a valuable material or for regeneration. Recycling of a waste can provide a very cost-effective waste management alternative. This option can help eliminate waste disposal costs, reduce raw material costs, and provide income from saleable waste.

Recycling is the second most preferred option in the pollution prevention hierarchy and as such should be considered only when all source reduction options (see earlier Problem) have been investigated and implemented. Reducing the amount of waste generated at the source will often be more cost-effective than recycling, since waste is primarily lost raw material or product which requires time and money to recover. It is important to note that recycling can increase a generator's risk or liability as a result of the associated handling and management of the materials involved. The measure of effectiveness in recycling is dependent upon the ability to separate any recoverable waste from other process waste that is not recoverable.

LPPA.7 ON-SITE AND OFF-SITE RECYCLING

To prevent pollution and minimize waste, industry has to choose between on-site recovery and off-site recovery. Briefly discuss the advantages and the disadvantages of each technique.

SOLUTION:

1. Some advantages of on-site recovery include:

   Less (or no) waste leaves the facility

   Owner control of purity of reclaimed solvent

   Reduced reporting

   Lower liability

   Possible lower unit cost of reclaimed solvent.

2. Some disadvantages of on-site recovery include:

   Capital investment for equipment

   Workers liability: health, fires, explosion, leaks, spills and other risks

   Operator training

   Additional-operating costs

3. Some advantages of off-site recovery are:

   Since several wastes are received at the off-site facility, this will improve the economics for processing

   Outlets for reclaimed chemicals

   Fractionation capabilities

   Technical support

   Ability to blend by-products to obtain most cost-effective method

4. Some disadvantages of off-site recovery are:

   No control of the fate or what is to be done with the waste

   Considerable waste leaves the facility.

Reuse involves finding a beneficial purpose for a recovered waste in a different process. Three factors to consider when determining the potential for reuse are:

1. The chemical composition of the waste and its effect on the reuse process.
2. Whether the economic value of the reused waste justifies modifying a process in order to accommodate it.
3. The extent of availability and consistency of the waste to be reused.

LPPA.8 TREATMENT METHODS

Discuss treatment method from a pollution prevention perspective.

SOLUTION: Recognizing that not all wastes can be eliminated through source reduction or recycling methods, viable treatment processes must then be looked to for managing remaining wastes. Aside from incineration, there are a host of additional treatment alternatives for reducing wastes. According to the EPA definition, treatment is "any practice, other than recycling, designed to alter the physical, chemical, or biological character or composition of a hazardous substance, pollutant, or contaminant, so as to neutralize said substance, pollutant, or contaminant or to render it nonhazardous through a process or activity separate from the production of a product or the providing of a service." Treatment can generally be divided into three categories:

1. Chemical treatment
2. Biological treatment
3. Physical treatment

One should note that the treatment methods are generally not geared for resource recovery, reuse, or recycling. These are basically methods that are available for treating waste once it has been generated.

Numerous chemical processes are used in the treatment of waste. Most treatment schemes are used in conjunction with other methods to achieve an end result. Some treatment methods are directly applicable to recycling efforts. Examples of chemical processes (treatment methods) include calcination, catalysis, chlorinolysis, electrolysis, hydrolysis, neutralization, oxidation, photolysis, reduction, and (chemical) precipitation.

Biological processes also involve chemical reactions, but are differentiated from the chemical category in that these reactions take place in or around microorganisms. The most common use of biological processes in waste treatment is for the decomposition of organic compounds. Examples of biological processes are activated sludge, aerated lagoons, anaerobic digestion, composting, enzyme treatment, trickling filters, and waste stabilization ponds.

There are more than 20 types of physical treatment processes known to be used in the handling of wastes; however, very few of these are fully developed or commonly used in industry. Some techniques have been found to have little potential use, so that further research in these areas is unlikely. Zone refining, freeze drying, electrophoresis, and dialysis all fall into this category. The difficulty of the operation and/or the high cost of these processes overshadows any possible future use. The most common processes used today are sedimentation, filtration, flocculation, and solar evaporation. Most other processes fall in between these two extremes; i.e., they show some potential for future use but are not presently used to any great extent.

Physical treatment may be separated into two categories: phase separation and component separation processes. In the latter, a particular species is separated from a single-phase, multicomponent system. The various physical treatments may fall into one or both of these categories. Sedimentation and centrifugation are used in phase separation; liquid ion exchange and freeze crystallization are used for component separation; and, distillation and ultrafiltration are used in both. Phase separation processes are employed to reduce waste volume and to concentrate the waste into one phase before further treatment and material recovery are performed. Such waste streams as slurries, sludges, and emulsions, which contain more than one phase, are the usual candidates for this category.

For example, one type of physical process is *adsorption*. Gas adsorption in a mass transfer process in which gas molecules are removed from an air stream because they adhere to the surface of a solid. In an adsorption system, the contaminated air stream is passed through a layer of solid particles referred to as the adsorbent bed. As the contaminated air stream passes through the adsorbent bed, the pollutant molecules absorb or "stick" to the surface of the solid adsorbent particles. Eventually the adsorbent bed becomes "filled" or saturated with the pollutant. The adsorbent bed must then be disposed of and replaced, or the pollutant vapors must be desorbed before the adsorbent bed can be reused.

LPPA.9   PETROLEUM STORAGE TANKS

Describe the various types of petroleum storage tanks, particularly as they apply to pollution prevention.

SOLUTION: A fixed-roof petroleum storage tank consists of a cylindrical, steel shell covered by a cone shaped roof. Fixed-roof storage tanks are used for storage of gasoline, crude oils, fuel oils, and other types of petroleum liquids. The tanks are located at refineries, at terminals, at bulk plants, at tank farms, along pipelines, in oil production fields, and at other industrial facilities. Fixed-roof storage tanks which have capacities less than 1,600,000 liters (420,000 gal) are used to store only produced crude oil and condensate prior to lease custody transfer are exempt. Normally, a tank is equipped with pressure/vacuum relief valves set to operate at low pressure variations. When pressure inside the tank exceeds the relief pressure, VOC vapors are released to the atmosphere. The expansion of vapors in the tank due to changes in ambient temperature and pressure result in VOC emissions termed "breathing losses." Additional VOC emissions termed "working losses" occur during filling and emptying operations.

External and internal floating roof tanks are the preferred storage tank from a pollution prevention point of view. Total emissions from floating roof tanks are the sum of standing storage losses and withdrawal losses. Standing storage losses from internal floating roof tanks include rim seal, deck fitting, and deck seam losses. Standing storage losses from external floating roof tanks, as discussed here, include only rim seal losses since deck fitting loss equations have not been developed. There is no deck seam loss, because the decks have welded sections. A typical internal floating roof generally incorporates two types of primary seals: resilient foam filled seals and wipers. Similar in design to those in external floating roof tanks, these seals close the annular vapor space between the edge of the floating roof and the tank wall. Secondary seals are not commonly used with internal floating roof tanks.

External floating roof tanks consist of a cylindrical steel shell equipped with a roof which floats on the surface of the stored liquid, rising and falling with the liquid level. The liquid surface is completely covered by the floating roof, except at the small annular space between the roof and the tank wall. A seal (or seal system) attached to the roof contacts the tank wall (with small gaps, in some cases) and covers the annular space. The seal slides against the tank wall as the roof is raised or lowered. The purpose of the floating roof and the seal (or seal system) is to reduce the evaporation loss of the stored liquid.

An internal floating roof tank has both a permanent fixed roof and a deck inside. The deck rises and falls with the liquid level and either floats directly on the liquid surface (contact deck) or rests on pontoons several inches above the liquid surface (non-contact deck). The terms "deck" and "floating roof" can be used interchangeably in reference to the structure floating on the liquid inside the tank. There are two basic types of internal floating roof tanks: tanks in which the fixed roof is supported by vertical columns within the tank and tanks with a self-supporting fixed roof and no internal support columns. Fixed roof tanks that have been retrofitted to employ a floating deck are typically of the first type, while external floating roof tanks typically have a self-supporting roof when converted to an internal floating roof tank. Tanks initially constructed with both a fixed roof and a floating deck may be of either type.

LPPA.10 BULK PLANTS

Describe bulk plants, particularly as they apply to pollution prevention.

SOLUTION: A "bulk plant" is defined as a gasoline distribution facility having a daily gasoline throughput of 76,000 liter (20,000 gal) or less per day. The daily gasoline throughput at a typical size bulk plant is 14,000 to 17,000 liter/day (4,000 to 5,000 gal/day).

The VOC losses occur from the storage tanks when temperature induced pressure expels vapor laden air (breathing loss). Losses also occur during draining and filling of the tanks (working loss). Fugitive losses can occur from valves, truck hatches, piping, and pump seals. Account tank trucks are used to deliver gasoline to bulk plant customers. The tank trucks are filled at loading docks similar to the types described for loading tank trucks at terminals. A typical bulk plant has one loading rack.

Three levels of increasingly more effective control alternatives are applicable to bulk plants. The alternatives are:

1. Submerged filling of tank trucks.
2. Alternative 1 plus a vapor balance (displacement) system to control VOC emissions from filling of bulk plan storage tanks. The vapors displaced from the storage tank are transferred to the tank truck being unloaded. Ultimately, the vapors are recovered when the tank truck returns to the terminal.

3. Alternative 2 plus a vapor balance system to control VOC emissions from the filling of account tank trucks. The vapors displaced from the account tank truck are transferred to the storage tank.

EPA guidance recommends application of Alternative 2 to all bulk plants. It may not be reasonable to apply Alternative 3 as the costs of installing a vapor balance system on existing loading racks and tank trucks may be too high for small, independent bulk plants, i.e., for bulk plants operating at less than 15,000 liter/day (4,000 gal/day) of gasoline throughput.

LPPA.11   RECOVERY OF GASOLINE VAPORS FROM TANK TRUCKS

Describe the recovery and control of gasoline vapors from tank trucks.

SOLUTION: A "terminal" is a gasoline distribution facility which has a gasoline throughput greater than 76,000 liter/day (20,000 gal/day). The gasoline throughput at a typical size terminal is 950,000 liter/day (250,000 gal/day).

It is necessary to meter and to deliver the gasoline from the storage tanks to the tank trucks. The equipment includes pumps, piping, valves, fittings, meters, and loading arm assemblies. The loading rack may also be a ground level facility for bottom loading of tank trucks.

As a pollution prevention measure, an active vapor control system must be installed to condense, absorb, adsorb, or incinerate the VOC Vapors. Control equipment to fulfill this requirement is presently commercially available. However, leakage from the vapor control system and tank trucks must be minimized by a good maintenance and inspection program. For more information on control equipment, consult the Theodore Tutorial entitled "Air Pollution Control Equipment," by L. Theodore and R. Allen, East Williston, NY, 1995.

With these control measures VOC emissions can be reduced significantly. Also note that VOC emissions are a function of the concentration of gasoline vapor in an empty tank truck.

LPPA.12   CUTBACK ASPHALT

Describe the cutback asphalt process.

SOLUTION: Asphalt surface and pavements are composed or compacted aggregate and an asphalt binder. Aggregate materials are produced from rock quarries as manufactured stone or are obtained from natural gravel or soil deposits. Metal ore refining processes produce artificial aggregates as a byproduct. The aggregate performs three functions in asphalt. It transmit the load from the surface to the base course, takes the abrasive wear of traffic, and provides a nonskid surface.

The asphalt binder holds the aggregate together, preventing displacement and loss of aggregate, and provides a waterproof cover for the base.

Asphalt binders take the form of asphalt cement (the residue of the distillation of crude oils) and liquified asphalts. To be used for pavement, asphalt cement, which is semisolid, must be heated prior to mixing with the aggregate. The resulting hot mix asphalt concrete is generally applied in thicknesses of from two to six inches.

Liquified asphalts are:

1. asphalt cutbacks (asphalt cement thinned or "cutback" with volatile petroleum distillates such as naphtha, kerosene, etc.) and

2. asphalt emulsions (nonflammable liquids produced by combining asphalt and water with an emulsifying agent, such as soap).

Liquified asphalts are used in tack and seal operations, in priming roadbeds for hot mix application, and for paving operations up to several inches thick.

Cutback asphalts fall into three broad categories: rapid cure (RC), medium cure (MC), and slow cure (SC) road oils. SC, MC, and RC cutbacks are prepared by blending asphalt cement with heavy residual oils, kerosene-type solvents, or naphtha and gasoline solvents, respectively. Depending on the viscosity desired, the proportion of solvent added generally ranges from 25 to 45 percent by volume.

Emulsified asphalts are of two basic types. One type relies on water evaporation to cure. The other type (cationic emulsions) relies on ionic bonding of the emulsion and the aggregate surface. Emulsified asphalt can substitute for cutback in almost any application. Emulsified asphalts are gaining in popularity because of the energy and environmental problems associated with the use of cutback asphalts.

In most applications, cutback asphalt is sprayed directly on the road surface as a bonding agent or as part of a new surface layer. It is also mixed with an aggregate for pothole repair and shoulder work.

Another use of asphalt is as a hot mix which is used for paving new roads or resurfacing existing roads. In this application the asphalt is liquefied by melting at high temperatures prior to mixing with an as aggregate to produce the hot mix. Such a use of asphalt does not require the addition of distillates and therefore is not a source of VOCs.

Note: The EPA guidance document recommends the sole use of emulsified asphalt. Because true emulsified asphalt contains no volatile organic compounds, substitution of emulsified asphalt for cutback asphalt reduces VOC emissions by 100%. Concerns have been raised that emulsified asphalt may be unsuitable for use as a penetrating prime coat, for long term stockpiling, and for use in cold weather. However, emulsified asphalt is successfully being used in the states of New York, Pennsylvania, and Indiana during winter conditions.

LPPA.13 DEGREASING OPERATIONS

Solvent metal cleaning or "degreasing" involves using organic solvents to clean and remove oils, greases, and other soils from metal surfaces. There are three basic types of organic solvent degreasers: cold cleaners; open-top vapor degreasers, and conveyorized degreasers. A cold cleaner consists of a tank containing a nonboiling organic solvent into which parts are batch loaded for cleaning. An open-top vapor degreaser consists of a tank containing an organic solvent heated to its boiling point. Open-top vapor degreasers are also batch loaded. Additional details are provided in the next Problem. A conveyorized degreaser allows continuous degreasing operations using either nonboiling or boiling organic solvents. Metal parts may be loaded continuously into the degreaser by various types of conveyor systems. Degreasing operations have been identified as significant sources of volatile organic compounds and halogenated solvents which are listed hazardous air pollutants.

A variety of control options and pollution prevention techniques are available for each type and size of degreaser. Control options can include equipment as simple as a manually operated tank cover for small units or as complex as a carbon adsorption system to reduce the amounts of solvents vented to the atmosphere. Pollution prevention techniques such as use of the least amount of heat needed to keep the solvent at a low boil and spill reduction measures also reduce emissions.

Consider a degreasing operation in a metal finishing process. Give an example of process modifications that might be made to this part of the process that represent:

1. Source reduction.
2. Recycling.
3. Waste treatment.
4. Waste disposal.

SOLUTION:

1. Substitution of a different, nontoxic or non-polluting solvent (e.g., citric acid based) for a toxic solvent such as TCE.
2. Filtering and reuse of solvent for noncritical cleaning.
3. Distillation of spent solvent to recover high purity solvent that can be reused in the process.
4. Disposal of still bottoms from a solvent distillation process by shipping to an off-site landfill.

LPPA.14 OPEN TOP VAPOR DEGREASERS

Describe open top degreasers and provide some information on pollution prevention measures.

SOLUTION: Open top vapor systems, often referred to as open top vapor degreasers, are batch loaded boiling degreasers that clean with condensation of hot solvent vapor on colder metal parts. Vapor degreasing uses halogenated solvents (usually perchloroethylene, trichloroethylene, or 1,1,1-trichloroethane) because they are not flammable and their vapors are much heavier than air.

A described in the previous Problem, a typical vapor degreaser is a sump containing a heater that boils the solvent to generate vapors. The height of these pure vapors is controlled by condenser coils and/or a water jacket encircling the device. Solvent and moisture condensed on the coils are directed to a water separator, where the heavier solvent is drawn off the bottom and is returned to the vapor degreaser. A "free-board" extends above the top of the vapor zone to minimize vapor escape. Parts to be cleaned are immersed in the vapor zone, and condensation continues until they are heated to the vapor temperature. Residual liquid solvent on the parts rapidly evaporates as they are slowly removed from the vapor zone. Lip mounted exhaust systems carry solvent vapors away from operating personnel. Cleaning action is often increased by spraying the parts with solvent below the vapor level or by immersing them in the liquid solvent bath. Nearly all vapor degreasers are equipped with a water separator which allows the solvent to flow back into the degreaser.

An example of requirements which appear in State and Federal regulations for some of the open top vapor degreasers is provided below:

Open top vapor degreasers which have a degreaser opening which is greater than ten square feet (0.93 square meters) shall:

1. Be equipped with:
   a. a cover that can be opened or closed easily without disturbing the vapor zone;
   b. a safety switch which shuts off the sump heat if condenser coolant is either not circulating or too warm (condenser flow switch and thermostat);
   c. a safety switch which shuts off the spray pump if the vapor level drops more than four inches (10 cm); and
   d. a permanent, conspicuous label summarizing the operating requirements found in paragraph (3) of this subsection.
2. Be equipped with one or more of the following control devices:
   a. freeboard ratio greater than or equal to 0.75 and, if the degreaser opening is greater than 10 square feet (0.93 square meters), the cover must be powered;
   b. refrigerated chiller;
   c. enclosed design in which the cover or door opens only when the dry part is actually entering or exiting the degreaser; and
   d. carbon adsorption system, with ventilation greater than 50 cfm/ft^2 (15 m^3/min/m^2) of air vapor area when the cover is open, and exhausting less than 25 parts per million of solvent averaged over one complete adsorption cycle.
3. Be operated in accordance with the following requirements:
   a. keep cover closed at all times except when processing work loads through the degreaser;
   b. minimize solvent carry-out by: racking all parts to allow full drainage; moving parts in and out of the degreaser at less than 11 feet per minute (3.3 meters per minute); degreasing the workload in the vapor zone at least 30 seconds or until condensation ceases; tipping out any pools of

solvent on the cleaned parts before removal; and, allowing parts to dry within the degreaser for at least 15 seconds or until visually dry;

c. do not degrease porous or absorbent materials such as cloth, leather, wood or rope;

d. work loads should not occupy more than half of the degreaser's open top area;

e. never spray above the vapor level;

f. repair solvent leaks immediately or shut down the degreaser;

g. do not dispose of waste solvent or transfer it to another party such that greater than 20% of the waste by weight will evaporate into the atmosphere; store waste only in closed containers;

h. exhaust ventilation should not exceed $65 \, \text{cfm/ft}^2$ $(20 \, \text{m}^3/\text{min}/\text{m}^2)$ of degreaser open area, unless necessary to meet OSHA requirements; ventilation fans should not be used near the degreaser opening; and,

i. water should not be visually detectable in solvent exiting the water separator.

LPPA.15 WASTE MINIMIZATION ASSESSMENT PROCEDURE

Discuss in general terms waste minimization procedures.

SOLUTION: Once opportunities for waste reduction have been identified by a pollution prevention assessment, source reduction techniques should be implemented wherever feasible. As discussed earlier, source reduction involves the reduction of pollutant wastes at their source, usually within a process, and is the most desirable option in the pollution prevention hierarchy. By avoiding the generation of wastes, sources reduction eliminates the problems associated with the handling and disposal of wastes. A wide variety of facilities can adopt procedures to minimize the quantity of waste generated.

Many source reduction options involve a change in procedural or organizational activities rather than in technology. For this reason, these options tend to affect the managerial aspect of production, and usually do not demand large capital and time investments. This makes implementation of many source reduction options affordable to companies of any size. Source reduction includes changes (as noted earlier) in procedure, technology, and input material. Each of these three options are further described below.

Once all pollution prevention options have been identified, priorities should be assigned to those of greater importance, i.e., which should be concentrated on first. Ideally, assessments should be performed on all options. However, this is not always practical when considering the availability of (limited) resources. Options of lesser importance can be assessed as time and budgets permit. Prioritization of options should take into consideration the following:

1. compliance with current and future regulations
2. costs of (waste) disposal and treatment

3. potential safety and environmental liability
4. quantity of waste and its hazardous makeup
5. potential for minimization
6. available budget for the PPA

A practical approach in selecting an option first is to choose one with a high probability for successful reduction of wastes. A successful implementation will help secure commitment for further pollution prevention efforts.

Additional details are available for EPA; Training manual, Chapter 2, "A Methodolgy for Pollution Prevention." The material was derived from the United Nations Industrial Development Organizations Manual "Audit and Reduction Manual for Industrial Emissions and Wastes."

LPPA.16 WASTE REDUCTION PROGRAMS

The major components of a waste reduction program must include the following: management commitment, communication of the program to the rest of the company, waste audits, cost/benefit analysis, implementation of pollution prevention program, and follow-up. Describe each of these general principles necessary for a successful pollution prevention program.

SOLUTION: A waste reduction plan must be specific to the industry. However, some general principles that must be incorporated for the success of the pollution prevention plan are discussed below.

1. *Management Commitment*: Gaining the approval and support of top management is vital to the success of the waste reduction plan. It will be necessary to educate management about the pollution prevention program and its benefits through seminars and meetings. A short video tape and a brochure, such as the one used by 3M, may help. 3M has successfully used a 12-minute video on "Pollution Prevention Pays" to communicate the program to management and employees.
2. *Communicating the Program to the Rest of the Company*: Middle managers and employees with direct process line experiences are in the best position to make suggestions as to where process improvements can be made. In addition, a monetary incentive and corporate recognition of employees for their practical pollution prevention ideas will also be very effective.
3. *Performing a Waste Audit*: The company must identify the processes, the product, and the waste streams in which hazardous chemicals are used. Mass balances of specific hazardous chemicals will help to identify waste reduction opportunities. Engineering interns could be very valuable in conducting such audits. Since managers and employees are often busy performing their assigned duties, an outside person may be more able to focus on pollution prevention opportunities, cutting through some of the management and personnel barriers of the industry, and may achieve significant progress.

4. *Cost/Benefit Analysis*: Any change or modification in the process requires additional capital plus operation and maintenance costs. A cost analysis must be included to help management make informed decisions. Factors including cost avoidance, enhanced productivity and decreased liability risks from the pollution prevention effort, should be factored into the study. Federal and State agencies provide matching grants to small industries to implement pollution prevention programs.

5. *Implementation of Pollution Prevention Programs*: Resistance to change by management and employees will still be an impediment for the implementation of the P2 program. People known to resist new ideas and changes in the company must be included in the planning stages of the program. The CEO must be convinced of the merits of the program and must fully support the implementation of it.

6. *Follow Up*: Reduced energy costs, reduced materials, and reduced waste disposal fees must be tracked and communicated to the company personnel. The tracking information will be very useful in the filing of the company's waste manifest and biennial reports on its waste reduction efforts that may be required by regulations. In addition, this information will demonstrate to employees and management that pollution prevention programs not only make sense environmentally, but economically as well.

LPPA.17   POLLUTION PREVENTION CLASSIFICATION

Identify the following sources as pollutants, waste, or neither and classify them as primary, secondary, fugitive, or process emissions. Suggest ways to eliminate or reduce each source and classify that suggestion as source reduction, recycle/reuse, treatment, or ultimate disposal.

1. A truck delivers a load of coal to a plant. The coal is dumped at the site into a storage bin generating coal dust.

2. A solvent bath has a lid that is closed during down-time. When the bath is in use the liquid is exposed to the air.

3. Spent solvent from a bath must be replaced periodically. The used solvent is stored in a drum until it is removed from the site.

4. Solvent is emitted from the stack of a chemical plant. In the atmosphere it undergoes chemical reaction(s) and contributes to aerosol (ozone) formation. Classify both the solvent and the aerosol.

5. In a plating operation, a part is dipped in a chemical bath, and the dipped piece is transferred to a rinse bath several feet away. During the transfer, the dipped piece drips onto the plant floor (dragout). Most of the resulting chemical spill evaporates and the rest is washed down the drain, which flows into the city sewer system.

6. A pump through which an organic solvent is flowing has a small, undetected leak.

7. A sulfuric acid solution which is used to regenerate an ion exchange column.

8. A sulfuric acid stream is neutralized by mixing with a caustic stream from another part of the process. Classify both of the reagent streams and the neutralized stream from the process.

SOLUTION:

1. The coal dust is a *fugitive emission*. An estimate of the amount of dust generated is very difficult to make, as are suggestions for reducing this source. However, proper enclosure (source reduction) of the transfer point and exhausting the dust to a control system (treatment) may minimize the fugitive emissions and may help to reduce/eliminate and explosion hazard. This dust contributes to the ambient particulates loading and as such is considered a pollutant. Emissions factors may exist for this source.

2. The evaporating solvent is classified as a *fugitive emission*. An estimate of the magnitude of this emission can be made by comparing volumes of fresh and spent solvent in the bath. One possibility is for the development of an automated system (source reduction) that removes the lid only for purposes of inserting or removing parts. It may also be possible to reduce evaporation by temperature control of the bath (providing the bath is effective at reduced temperatures) or of the air space above the solvent liquid surface, i.e., a vapor condenser (source reduction). Another option might be to install an exhaust hood (treatment) to remove fumes, which can then be treated (e.g., adsorbed before the air is vented to the atmosphere).

3. The spent solvent is a *primary emission* and may be classified as a hazardous waste, depending on its chemical composition. The amount of solvent waste may be reduced by recovering (recycle/reuse) the solvent via distillation or other appropriate methods and reusing it. It may be possible to replace the solvent with one that is not environmentally hazardous (source reduction).

4. The solvent emitted from the stack (as a gas) is a *primary emission*. The amount of solvent lost in this way may be reduced by condensation (treatment) of the solvent and separation from the stack gas. The aerosol is classified as a *secondary pollutant*. This term refers to pollutants that result from further reaction of primary emissions. Frequently, secondary pollution occurs at different locations (e.g., downwind) from the source of the primary emissions. The production of secondary pollutants depends on a number of ambient conditions, primarily ambient volatile organic levels, NO_x, temperature, and solar radiation. In general, ozone levels can be reduced by a concerted program of emissions abatement (treatment) from automobiles and industry, leading to reduced ambient concentrations.

5. The "spill" occurs as a part of the processing of the piece and can be classified as a *process emission*. Depending upon the chemicals used in the bath, the water used to wash the spill down the drain may become a waste or hazardous waste stream. It may be possible to reduce the problem by changing the location (source reduction) of the rinse bath so that the path of the piece remains over collection tanks, and by suspending the piece for a longer

amount of time over the bath (source reduction) to allow most of the bath solution to drip back into the tank.

6. The leaking solvent is a *fugitive emission*. An estimate of the amount of solvent which becomes a fugitive emission can be made by performing a material balance for the solvent over an entire process. Checking the integrity of the pump (source reduction) may reveal a repairable leak, or it may be necessary to switch to a different type of pump that is less likely to leak. It may also be possible to reduce this type of loss in process piping (at gaskets, for example) by shrink-wrapping the joint or otherwise containing the vapor (source reduction). If possible, another solvent (source reduction) might be considered that has a lower vapor pressure or that is less corrosive to the pump interior.

7. The sulfuric acid solution is a necessary *process emission*. If the acidic stream has a pH < 2, it is classified as a hazardous waste and must undergo further treatment before disposal. Similarly, if a caustic effluent has a pH > 12, it is classified as a hazardous waste. It may be possible to replace (source reduction) the acidic cleaner with one that is not classified as hazardous. It may also be possible to recover spent solution (treatment) and recycle (recycling/reuse) it for cleaning purposes, reducing the amount of waste generated.

8. As with step (7) the sulfuric acid solution is a *primary emission*. If either the acidic or caustic stream is classified as hazardous (see step 7), the neutralized stream is also considered hazardous even if 2 < pH < 12. This is known as the "derived-from" rule, which assigns the hazardous classification to derivatives of hazardous wastes. It may also be possible to reduce the amount of neutralized solution to be processed by concentrating (treatment) it. The water thus recovered can by recycled (recycling/reuse) for use in other parts of the process.

LPPA.18   INDUSTRIAL WASTE REDUCTION TECHNIQUES

Explain the following waste reduction techniques that are used in industry:

1. Good housekeeping
2. Material substitution
3. Equipment design modification
4. Waste exchange
5. Detoxification

SOLUTION:

1. *Good housekeeping*. Improper labeling, storage and dumping of hazardous chemicals often increase the risk of spillage, groundwater contamination and waste treatment plant upsets. Educating the employees on the proper handling and disposal of hazardous chemicals reduces such events tremendously.

2. *Material substitution*. This method involves replacing the polluting ingredients of a product with a less toxic one. Material substitution is most

economical when a product is being developed for the first time. With an existing product, new chemicals often require an additional investment in new equipment. However, the additional investment may be less expensive than the cost of additional control systems necessary to meet new emission standards. For example, replacement of an organic based solvent as a carrier for tablet coating with a water-based coating cost a pharmaceutical company $60,000. Alternatively the air pollution control system that would be needed to meet emission standards for the organic solvent had an initial capital cost of $180,000 and recurring $30,000 annual operating costs. Clearly, this pharmaceutical company will replace the organic based coating with a water-based one to realize the economic and pollution prevention benefit this material substitution provides.

3. *Equipment design modification.* Old equipment may contribute to the production of harmful by-products and excessive emission of pollutants. Replacing such process equipment might eliminate or minimize the generation of harmful by-products. Low interest federal loans may be available to assist in these changes.

4. *Waste exchange.* (One of the authors of this text visited a waste exchange facility in Phoenix, AZ.) Waste from one industry may actually be a valuable resource as a raw material in another industry. Avoided disposal costs for the waste generator and inexpensive raw material for the receiving industry are the inherent advantages of this waste reduction technique. Most of the chromium used in U.S. industries is imported. Recovery and reuse of chromium will not only minimize the amount of waste but also make the U.S. less dependent on imports.

5. *Detoxification.* After considering all other options for waste reduction, detoxification of waste by neutralization or other techniques may be considered. Thermal, biological, and chemical treatments are some of the detoxification processes commonly used. However, energy costs, along with the potential of generating new compounds and product streams, must be evaluated thoroughly to justify the perceived environmental benefits of proposed detoxification processes.

LPPA.19 APPLICABLE AND LIMITATIONS OF POLLUTION PREVENTION OPTIONS

Consider the list of pollution problems below and briefly discuss which of the pollution prevention options discussed earlier would be most applicable and the limitations of the others.

1. Soil at railroad station contaminated with PCBs.
2. The Greenhouse effect.
3. Volatile organic (VOCs), e.g. xylene, escaping from a stack.
4. Ozone layer depletion.

5. Radioactive wastes from a nuclear power plant.
6. Chromium salts from the printing/photographic industries.
7. Platinum catalyst in an industrial process.
8. Sulfuric acid waste from the pharmaceutical industry.

SOLUTION:

1. Source reduction has already occurred since PCBs have been banned. Recycling is not applicable. The best solution to this problem is probably incineration, followed by landfilling.
2. The best solution, if there is one, is source reduction. There are a variety of sources of carbon dioxide, the leading cause of the Greenhouse effect, but the major source is the combustion of fossil fuels—oil, coal, gasoline, and wood. The destruction of the world's forests is also a contributing cause since plants tie up carbon dioxide in the form of cellulose, generating oxygen. Recycling is not an option (except through photosynthesis as mentioned above) and treatment is impossible since there are such vast quantities of the gas. Disposal is not an option.
3. The best solution is first to reduce the amount of xylene used and lost by the process. Recovery and recycling should then be considered. However, this solution is dependent on a number of variables, including the concentration. The third best option would be treatment by incineration. However, if the VOC is adsorbed onto a bed of activated carbon, both of the options above (recycling and treatment) are viable, as well as disposal in a landfill, the last choice.
4. Again, the best hope lies in source reduction and to find alternatives to CFCs. Treatment, recycling, and disposal are not reasonable options for the CFCs at the altitudes where they are destroying the ozone layer. At ground level, recycling of refrigerants and disposal are activities which are being conducted. The major effort is to find substitutes for the CFCs which are more friendly to the ozone layer. Unfortunately, the HCFCs which are proposed are still capable of damaging the ozone layer.
5. There is only one solution for wastes already generated and scientists are working to find the best means of disposal. Leading the list of candidates is "glassification" whereby the radioactive wastes are imbedded in a stable glass form and placed in a geologically stable storage area. Since the half-life of many radioactive isotopes may be thousands of years, the best solution to future wastes is source reduction. Radioactivity is not affected by any type of chemical change, so treatment is not an option. It has been proposed to recycle wastes to recover the plutonium and to use this highly toxic substance in breeder reactors. This would generate power and more plutonium, but is not safe at the present time.
6. The current solution is to reduce chromium (VI) to chromium (III) and then to precipitate it with a base. The resulting solid is then sent to a landfill. A better long term solution is to reduce its use (source reduction). This can be accomplished in some instances by finding alternatives. For example, lab glassware

was once routinely cleaned with a solution of chromium trioxide in sulfuric acid. The rinsing went down the sink. Today, alternative cleaning agents are used. Alternatives have been found to chromium salts in oxidation reactions. Treatment would not alter its toxicity and recycling is probably too expensive relative to finding on alternative.

7. Recovery and recycling is the best solution for this precious heavy metal. Economically, this is preferable to landfilling. Since it is used as a catalyst, source reduction is usually not possible. Alternative catalysts are also heavy metals. That which is not recycled should be landfilled.

8. Usually the best option for these wastes is treatment with a strong base to form a salt and water at a pH of about 7. At this point the waste water treatment plant can easily deal with the stream. In some favorable instances, the sulfuric acid can be cleaned up, concentrated to its original molarity, and reused. Depending on the application, source reduction may alleviate the problem, but the above options will be needed for whatever waste remains. Disposal is not an option.

LPPA.20 MATERIAL BALANCES

Explain material balances.

SOLUTION: The conservation law for mass can be applied to any process or system. The general form of this law is given in the equation below:

$$\text{mass in} - \text{mass out} + \text{mass generated} = \text{mass accumulated} \tag{10.1}$$

This equation may be applied to the total mass involved or to a particular species, on either a mole or mass basis. The conservation law for mass can be applied to steady-state or unsteady-state processes and to batch or continuous systems. To isolate a system for study, it is separated from the surroundings by a boundary or envelope. This boundary may be real (e.g., the walls of a vessel) or imaginary. Mass crossing the boundary and entering the system is part of the mass-in term above, whereas that leaving the system is part of the mass-out term. The equation may be written for any compound whose quantity is not changed by chemical reaction, or any chemical element, regardless of whether it has participated in a chemical reaction. It may be written for one piece of equipment, around several pieces of equipment, or around an entire process. It may be used to calculate an unknown quantity directly, to check the validity of experimental data, or to express one or more of the independent relationships among the unknown quantities in a particular problem.

A steady-state process is one in which there is no change in conditions (temperature, pressure, etc.) or rates of flow with time at any given point in the system. The accumulation term is then zero. If there is no chemical reaction, the generation term is also zero. All other processes are classified as unsteady-state. In a batch process, the container holds the product or products. In a continuous process, reactants are fed in an unending flow to a piece of equipment or to several pieces in series, and products are continuously removed from one or more points. A continuous process may or may not be steady-state. As indicated previously, the equation may be applied to the total mass of each stream (referred to as an overall or total material balance) or to the individual components of the streams (referred to as a componential or component material balance). Often the primary task in preparing a material balance is to develop the quantitative relationships among the streams. For more information on the conservation of mass, refer to the Theodore Tutorial entitled "*Material and Energy Balances*" by J. Reynolds, East Williston, NY, 1994.

The conservation law for mass finds its major application in performing pollution prevention assessments. A pollution prevention assessment is a systematic, planned procedure with the objective of identifying ways to reduce or eliminate waste. The procedure consists of a careful review of a plant's operation and waste streams, and the selection of specific areas to assess. After a specific waste stream or area is established as the focus, a number of options with the potential to minimize waste are developed and screened. The selected options are then evaluated for technical and economic feasibility. Finally, the most promising options are selected for implementation. Additional details are provided in the next problem.

LPPA.21   POLLUTION PREVENTION ASSESSMENT PROCEDURE

Describe in detail the pollution prevention assessment procedure.

SOLUTION: As discussed in previous Problems, pollution prevention assessments are procedures for identifying and assessing pollution prevention opportunities within a plant site. This process consists of a careful review of the plant's operations and waste streams and the selection of specific streams and/or operations to assess. It is an extremely useful tool in diagnosing how the facility can reduce or eliminate wastes. A pollution prevention assessment can also focus on how to reduce wastes destined for treatment and/or disposal. This is unlike an environmental audit, which focuses on regulatory compliance and environmental protection. Whereas audits are often thought of in terms of legal compliance, a pollution prevention assessment is commonly referred to technically as a search for pollution prevention opportunities, generally without concern for legal implications.

The assessment process should characterize the selected waste streams and processes—an ongoing part of this type of analysis. Some of the data may be collected during the first review of site-specific data; however, some of the information may not be collected until the actual (later) audit inspection. Regardless, as much information as possible should be collected at an early date. The information-gathering process can be helpful in establishing specific goals for the program. Information about the facility's waste stream can come from a variety of sources such as personnel involved with the process and the waste stream, hazardous waste manifests, biennial reports, environmental audits, emission inventories, engineering studies, waste assays, and permits.

Simplified mass balances (application of the aforementioned conservation law for mass) should be developed for each of the important waste-generating operations noted and gain a better understanding of the origins of each waste stream. As noted in the previous Problem, a mass balance is essentially a check to make sure that what goes into the process (i.e., the total mass of all raw materials, water, etc.) leaves the process (i.e., total mass of the product, waste, and by-products). The material balance should be made individually for all components that enter and leave the process. When chemical reactions take place in a system, there is an advantage to doing "elemental balances" for specific chemical elements in a system.

Material balances can assist in determining concentrations of waste constituents where analytical test data are limited. They are particularly useful when there are points in the production process where it is difficult (because of inaccessibility) or uneconomical to collect analytical data. If this is the case, one is often confronted with situations where there are more unknowns than equations available for solution (or the reverse).

The aforementioned mass balance calculations are particularly useful for quantifying fugitive emissions such as evaporative losses. Waste stream data and mass balances will enable one to track the flow and characteristics of the waste streams over time. Since in most cases the accumulation equals zero, it can then be assumed that any buildup is actually leaving the process through fugitive emissions or other means. This will be useful in identifying trends in waste/pollutant generation and will also be critical in the task of measuring the performances of the pollution prevention options that are implemented.

LPPA.22 REACTOR SEAL CORRECTIVE STEPS

Refer to Problem TOSHA.2: What might be done to decrease the environmentally hazardous nature of the reactor? What pollution prevention measures could be implemented?

SOLUTION: Have the seal piped to discharge into a hood or a duct in case of rupture. Also consider replacing the material in the reactor with material that has a lower vapor pressure or is less toxic (input substitution).

LPPA.23 BARRIERS AND INCENTIVES TO POLLUTION PREVENTION

List some of the barriers and incentives to pollution prevention.

SOLUTION: There are numerous reasons why industry does not reduce wastes generated. The following "dirty dozen" are common disincentives.

1. Technical limitations
2. Lack of information
3. Consumer preference obstacles
4. Concern over product quality decline
5. Economic concerns

6. Resistance to change
7. Regulatory barriers
8. Lack of markets
9. Management apathy
10. Institutional barriers
11. Lack of awareness of pollution prevention advantages
12. Concern over the dissemination of confidental product information

Various means exist to encourage pollution prevention through regulatory measures, economic incentives, and technical assistance programs. Since the benefits of pollution prevention can surpass preventive barriers, a "baker's dozen" incentives is presented below:

1. Economic benefits
2. Regulatory compliance
3. Liability reduction
4. Enhanced public image
5. Federal and state grants
6. Market incentives
7. Reduced waste-treatment costs
8. Potential incentives
9. Decreased worker exposure
10. Decreased energy consumption
11. Increased operating efficiencies
12. Competitive advantages
13. Reduced negative environmental impacts

More details on both the incentives and barriers are available in the literature cited below

1. R. Perry and D. Green, "Perry's Chemical Engineers Handbook," 7th Edition, McGraw-Hill Book Co., New York City, 1997.
2. R. Dupont, L. Theodore, and K. Ganesan, "Pollution Prevention. The Waste Management Approach for the 21st Century," Lewis-CRC Publishers, Boca Raton, FL, 2000.

LPPA.24 FROM POLLUTION CONTROL TO POLLUTION PREVENTION

Provide information on the evolution of waste management approaches.

SOLUTION: The solution to the problem is provided in Table 10.2 below.

TABLE 10.2   Evolution of Waste Management Approaches in the United States

Prior to 1945	No control
1945–1960	Little control
1960–1970	Some control
1970–1975	Greater control, EPA established
1975–1980	More sophisticated control
1980–1985	Beginning of waste-reduction management
1985–1990	Waste-reduction management
1990–1995	Formal pollution prevention programs (Pollution Prevention Act of 1990)
1995–2000	Widespread acceptance of pollution prevention
>2000	Green manufacturing, sustainable development, designs for the environment,...

LPPA.25   FINANCIAL CONSIDERATIONS

Discuss financial considerations as they apply to pollution prevention.

Note: Much of this material was derived from the EPA document "A Primer for Financial Analysis of Pollution Prevention Projects."

SOLUTION: As companies incorporate pollution prevention approaches in their strategic planning, capital investment priorities, and process design decisions, it is vital that they understand both the quantitative and qualitative dimensions of assessing pollution prevention projects. These projects tend to reduce or eliminate costs that may not be captured in cursory financial analyses due to the way costs are categorized and allocated by conventional management accounting systems. Additionally, pollution prevention projects often have impacts on a broad range of issues, such as market share and public impact that are difficult to quantify but that can be of strategic importance. Identifying and analyzing all costs and less tangible items is an important step in an evaluation of the potential benefits of pollution prevention projects. This is discussed in more detail in the next two problems.

The process for assessing pollution prevention projects, particularly the financial analysis component, fits within the *framework* of a standard capital budgeting model. However, the process described below, referred to as *capital budgeting for pollution projects prevention* expands on and broadens the way capital budgeting is often practice. The approach described here attempts to address this tendency of financial analyses to omit environmentally-related costs, which typically are lumped into overhead accounts, allocated to products, or overlooked in the cost identification process. This problem also focuses relatively greater attention on the more qualitative impacts of projects. While this is not the only way to evaluate a project, it does provide an accurate method for ensuring that important benefits and potential impacts are included in the analysis.

Pollution prevention can take many forms—from simple "housekeeping" improvements, which cost little to carry out, to installation of expensive capital equipment. Although many pollution prevention projects, such as material substitution or process redesign, do not require large outlays for the purchase of equipment, they may require significant engineering expense, create incremental costs or savings, or may require extensive qualitative assessments related to such issues as product quality or employee health and safety. The analytical tools described in this Problem are applicable to the assessment of most pollution prevention initiatives that fit under the umbrella of the capital budgeting process.

Pollution prevention projects involving capital budgeting generally include:

1. New manufacturing equipment
2. Replacement equipment
3. Plant expansion and construction

Capital budgeting is a process of evaluating capital investment options based on a company's needs and analyzing the impact of an investment on a company's cash flow over time. Pollution prevention and other capital projects are justified by showing how the project will increase revenue and how the added revenue will not only recover costs, but substantially increase the company's earnings as well. Financial tools demonstrate the importance of the pollution prevention investment on a life cycle or total cost basis in terms of revenue, expenses, and profits. Key concepts and factors used in capital budgeting are described below.

1. *Life Cycle Costing (LCC)*: Also referred to as Total Cost Accounting by some, this method analyzes the costs and benefits associated with a piece of equipment or a procedure over the entire time the equipment or procedure is to be used. The concept was first applied to the purchase of weapons systems for the U.S. military. Experience showed initial purchase price was a poor indicator of the total cost: costs such as those associated with maintainability, reliability, disposal/salvage value, and training/education needed to be considered in the financial decision making process. Similarly, in justifying pollution prevention, all benefits and costs must be spelled out in the most concrete terms possible over the life of each option. More details follow.

2. *Present Worth (PW)*: The importance of present worth, or present value, lies in the fact that time is money. The preference between a dollar now or a dollar a year from now is driven by the fact that the dollar in-hand can earn interest. Mathematically, this relationship is as follows:

$$\text{Present Value} = \frac{\text{Future Value}}{(1 + \text{interest})^{\text{number of years}}} \qquad P = \frac{F}{(1 + r)^n} \qquad (10.2)$$

where P is the present worth or present value, F is the future value, r is the interest or discount rate (fractional basic), and n is number of periods. For example, $1,000,000 in one year at 5% interest compounded annually would have a computed present value of:

$$P = \frac{1,000,00}{(1 + .05)^1} = \$950,000$$

Because money can "work," at 5% interest, there is no difference between $950,000 now and $1,000,000 in one year because they both have the same value at the present time. Similarly, if the $1,000,000 was to be received in 3 years, the present value would be:

$$P = \frac{1,000,000}{(1 + .05)^2} = \$860,000$$

In considering either multiple payments or case into and out of a firm, the present values are additive. For example, at 5% interest the present value of receiving both $1,000,000 in one year and 1,000,000 in 3 years would be $950,000 + $860,000 = $1,810,000. Similarly, if one was to receive $1,000,000 in one year, and pay $1,000,000 in 3 years, the present value would be $950,000 − $860,000 = $90,000. As a result, present worth calculations allow both costs and benefits which are expended or earned in the future to be expressed as a single lump sum at their current or present value.

3. *Comparative Factors for Financial Analysis*: The more common methods for comparing investment options all utilize the present value equation presented in Equation (10.2). Generally, one of the following four factors is used:

1. *Payback Period (PP)*: This factor measures how long it takes to return the initial investment capital. Conceptually, the project with the quickest return is the best investment.

2. *Internal Rate of Return (ROI)*: This factor is also called "return on investment" or ROI. It is the interest rate that would produce a return on the invested capital equivalent to the project's return. For example, a pollution prevention project with an internal rate of return of 23% would indicate that pursing the project would be equivalent to investing the money in a bank and receiving 23% interest.

3. *Benefits Cost Ratio*: This factor is a ratio determined by taking the total present value of all financial benefits of a pollution prevention project and dividing by the total present valve of all costs of the project. If the ratio is greater than 1.0, the benefits outweigh the costs and the project is economically worthwhile to undertake.

4. *Present Value of Net Benefits*: This factor shows the worth of a pollution prevention project as a present value sum. It is determined by calculating the present values of all benefits, doing the same for all costs and subtracting the two totals. The new result would be an amount of money that would represent the tangible value of undertaking the project.

While firms can use any of these factors, the importance of life cycle costing (or total cost analysis) makes the present value of net benefits the preferred method. Additional details of LCC and TCA follow.

Life Cycle Costing (LCC) tool and the Total Cost Assessment (TCA) tool are introduced as concept overviews. Both tools can be used to establish economic criteria to justify pollution prevention projects. TCA is often used to describe *internal* costs and savings, including environmental criteria. LCC includes all *internal* costs plus *external* costs incurred throughout the entire life cycle of a product, process, or activity.

LCC has been used for many years by both the public and private sector. It associates economic criteria and societal (external) costs with pollution prevention opportunities. The purpose of LCC is to quantity a series of time-varying costs for a given opportunity over an extended time horizon, and to represent these costs as a single value. These time varying cost usually include the following.

1. *Capital Expenditures.* Costs for large, infrequent investment with long economic lives (e.g., new structures, major renovations and equipment replacements).
2. *Non-recurring Operations and Maintenance (O&M).* Costs reflecting items that occur on a less frequent than annual basis that are not capital expenditures (e.g., repair or replacement of parts in a solvent distillation unit).
3. *Recurring O&M.* Costs for items that occur on an annual or more frequent basis (e.g., oil and hydraulic fluid changes).
4. *Energy.* All energy or power generation related costs. Although energy costs can be included as a recurring O&M cost, they are usually itemized because of their economic magnitude and sensitivity to both market prices and building utilization.
5. *Residual Value.* Costs reflecting the value of equipment at the end of the LCC analysis period. This considers the effects of depreciation and service improvements.

By considering all costs, a LCC analysis can quantify relationships that exist between cost categories. For example, certain types of capital improvements will reduce operations, maintenance, and energy costs while increasing the equipment's residual value at the end of the analysis period.

Societal (external) costs include those resulting from health and ecological damages, such as those related to unregulated air emissions, wetland loss, or deforestation, can also be reflected in a LCC analysis either in a qualitative or quantitative manner. LCC includes the following cost components.

1. *Extraction of Natural Resources.* The costs of extracting the material for use and any direct or indirect environmental cost for the process.
2. *Production of Raw Materials.* All of the costs of processing the raw materials.
3. *Making the Basic Components and Product.* The total cost of material fabrication and product manufacturing.
4. *Internal Storage.* The cost of storage of the product before it is shipped to distributors and/or retail stores.
5. *Distribution and Retail Storage.* The cost of distributing the products to retail stores including transportation costs, and the cost of retail storage before purchase by the consumer.

6. *Product Use.* The cost of consumer use of the product. This could include any fuels, oils, maintenance, and repairs which must be made to the equipment.

7. *Product Disposal or Recycling.* The costs of disposal or recycling of the product.

The TCA tool is especially interesting because it usually employs both economic and environmental criteria. As with the LCC analysis, the TCA study is usually focused on a particular process as it affects the bottom-line costs to the user. Environmental criteria are not explicit, i.e., success is not measured by waste reduction or resource conservation, but by cost savings. However, since the purpose of TCA is to change accounting practices by including environmental costs, environmental goals are met through cost reductions.

Because of its focus on cost and cost effectiveness, TCA shares many of the features of LCC analysis by tracking direct costs, such as capital expenditures and O&M expenses/revenues. However, TCA also includes indirect costs, liability costs and less tangible benefits—subjects that are not customarily included in LCC analysis. By factoring in these indirect environmental costs TCA achieves both economic and environmental goals. Because of its private sector orientation, TCA most often uses Net Present Value (NPV) and Internal Rate of Return (IRR) as well as other economic comparison methods.

LPPA.26 TRADITIONAL ECONOMIC ANALYSIS

Discuss some of the problems associated with the tradition type of economic analysis.

SOLUTION: The main problem with the traditional type of economic analysis is that it is difficult—nay, in some cases impossible—to quantify some of the not-so-obvious economic merits of a pollution prevention program. Several considerations discussed in the previous Problems surfaced as factors that need to be taken into account in any meaningful economic analysis of a pollution-prevention effort. What follows is a summary listing of these considerations, most of which have been detailed earlier.

1. Decreased long-term liabilities etc.
2. Regulatory compliance.
3. Regulatory recordkeeping.
4. Dealings with the EPA.
5. Dealings with state and local regulatory bodies.
6. Elimination or reduction of fines and penalties.
7. Potential tax benefits.
8. Customer relations.
9. Stockholder support (corporate image).

10. Improved public image.
11. Reduced technical support.
12. Potential insurance costs and claims.
13. Effect on borrowing power.
14. Improved mental and physical well-being of employees.
15. Reduced health-maintenance costs.
16. Employee morale.
17. Other process benefits.
18. Improved worker safety.
19. Avoidance of rising costs of waste treatment and/or disposal.
20. Reduced training costs.
21. Reduced emergency response planning etc.

Many proposed pollution-prevention programs have been squelched in their early stages because a comprehensive economic analysis was not performed. Until the effects described above are included, the true merits of a pollution-prevention program may be clouded by incorrect and/or incomplete economic data. Can something be done by industry to remedy this problem? One approach is to use a modified version of the standard Delphi panel. In order to estimate these other economic benefits of pollution prevention, several knowledgeable individuals within and perhaps outside the organization are asked to independently provide estimates, with explanatory details, on these economic benefits. Each individual in the panel is then allowed to independently review all responses. The cycle is then repeated until the group's responses approach convergence.

More details or quantifying the economic merits of a pollution prevention project are available in the literature cited below.

1. R. Perry and D. Green, "Perry's Chemical Engineers' Handbook," 7th Edition, McGraw-Hill Book Co., New York City, 1997.
2. R. Dupont, L. Theodore, and K. Ganesan, "Pollution Prevention. The Waste Management Approach for the 21st Century," Lewis-CRC Publishers, Boca Raton, FL, 2000.

LPPA.27   ECONOMIC CONSIDERATIONS

Summarize some of the economic considerations associated with pollution prevention programs.

SOLUTION: The greatest driving force behind any pollution prevention plan is the promise of economic opportunities and cost savings over the long term. Pollution prevention is now recognized as one of the lowest-cost options for waste/pollutant management. Hence, an understanding of the economics involved in pollution prevention programs/options is quite important in making decisions at both the engineering and management levels. Every engineer should be able to execute an economic evaluation of a proposed project. If the project cannot be justified economically after *all* factors—include those discussed in Problem LPPA.26—have been taken into account, it should obviously not be pursued. The earlier such a project is identified, the fewer resources will be wasted.

Before the true cost or profit of a pollution-prevention program can be evaluated, the factors contributing to the economics must be recognized. As discussed earlier there are two traditional contributing factors (capital costs and operating costs), but there are also other important costs and benefits associated with pollution prevention that need to be quantified if a meaningful economic analysis is going to be performed. Table 10.3 demonstrates the evolution of various cost-accounting methods.

The TCA aims to quantify not only the economic aspects of pollution prevention but also the social costs associated with the production of a product or service from cradle to grave (i.e., life cycle). The TCA attempts to quantify less tangible benefits such as the reduced risk derived from not using a hazardous substance. The future is certain to see more emphasis placed on the TCA approach in any pollution-prevention program. For example, a utility considering the option of converting from a gas-fired boiler to coal-firing is usually not concerned with the environmental effects and implications associated with such activities as mining, transporting and storing the coal prior to its usage as an energy feedstock. Pollution prevention approaches in the future will become more aware of this need.

The economic evaluation referred to above is usually carried out using standard measures of profitability. As noted, each company and organization has its own economic criteria for selecting projects for implementation. (For example, a project can be judged on its payback period. For some companies, if the payback period is more than 3 years, it is a dead issue). The economic analysis presented above represents a preliminary, rather than a detailed, analysis. For smaller facilities with only a few (and perhaps simple) processes, the entire pollution-prevention

TABLE 10.3   Economic Analysis Timetable

Prior to 1945	Capital costs only
1945–1960	Capital and some operating costs
1960–1970	Capital and operating costs
1970–1975	Capital, operating, and some environmental control costs
1975–1980	Capital, operating, and environmental control costs
1980–1985	Capital, operating, and more sophisticated environmental control costs
1985–1990	Capital, operating, and environmental controls, and some life-cycle analysis (Total Systems Approach)
1990–1995	Capital, operating, and environmental control costs and life-cycle analysis (Total Systems Approach)
1995–2008	Widespread acceptance of Total Systems Approach
>2008	???

assessment procedure will tend to be much less formal. In this situation, several obvious pollution-prevention options such as the installation of flow controls and good operating practices may be implemented with little or no economic evaluation. In these instances, no complicated analyses are necessary to demonstrate the advantages of adopting the selected pollution-prevention option. A proper perspective must also be maintained between the magnitude of savings that a potential option may offer and the amount of man power required to do the technical and economic feasibility analyses. More details or economic considerations are available in the literature cited below

1. R. Perry and D. Green, "Perry's Chemical Engineers' Handbook," 7th Edition, McGraw-Hill Book Co., New York City, 1997.
2. R. Dupont, L. Theodore, and K. Ganesan, "Pollution Prevention. The Waste Management Approach for the 21st Century," Lewis-CRC Publishers, Boca Raton, FL, 2000.

LPPA.28 POLLUTION PREVENTION LITERATURE

List some of the key EPA literature on pollution prevention.

SOLUTION: Information from the following websites should be useful.

1. The Pollution Prevention Information Clearinghouse is located at http://www.epa.gov/opptintr/ppic
2. Introduction to Pollution Prevention Training Manual is located at http://www.epa.gov/opptintr/ppic/pubs/intropollutionprevention.pdf.
3. The Guide to Industrial Assessments for Pollution Prevention and Energy Efficiency located at http://www.epa.gov/Pubs/2001/energy/complete.pdf.
4. The final discussion of the EPA program concerning the 33-50 program can be found at http://www.epa.gov/opptintr/3350/ with the final report at http://www.epa.gov/opptintr/3350/3350-fnl.pdf.
5. A discussion of the EPA program concerning "Persistent Bioaccumulative and Toxic Chemical Program" (PBT) is located at http://www.epa.gov/pbt/ with the Waste Minimization Program discussed at http://www.epa.gov/epaoswer/hazwaste/minimize with priority chemicals listed at http://www.epa.gov/epaoswer/hazwaste/minimize/chemlist.htm.
6. Compliance assistance for a variety of industrial sectors can be viewed at the EPA Website concerning compliance center notebooks at http://www.epa.gov/compliance/resources/publications/assistance/sectors/notebooks/
7. General information
   a. http://www.epa.gov/glossary
   b. http://www.epa.gov/P2/pubs/basic.htm.

The following publications are available from, "Guide to Industrial Assessment for Pollution Prevention and Energy Efficiency," (Table 10.4).

TABLE 10.4 Pollution Prevention Publications

Title	EPA Document Number
The Automotive Refinishing Industry	EPA 625/791/016
The Automotive Repair Industry	EPA 625/791/013
The Commercial Printing Industry	EPA 625/790/008
The Fabricated Metal Products Industry	EPA 625/790/006
The Fiberglass-Reinforced and Composite Plastics Industry	EPA 625/791/014
The Marine Maintenance and Repair Industry	EPA 625/791/015
The Mechanical Equipment Repair Industry	EPA 625/R92/008
Metal Casting and Heat Treating Industry	EPA 625/R-92-009
The Metal Finishing Industry	EPA 625/R92/011
Municipal Pretreatment Programs	EPA 625/R93/006
Non-Agricultural Pesticide Users	EPA 625/R93/009
The Paint Manufacturing Industry	EPA 625/790/005
The Pesticide Formulating Industry	EPA 625/790/004
The Pharmaceutical Industry	EPA 625/791/017
The Photoprocessing Industry	EPA 625/791/012
The Printed Circuit Board Manufacturing Industry	EPA 625/790/007
Research and Educational Institutions	EPA 625/790/010
Selected Hospital Waste Streams	EPA 625/790/009
Wood Preserving Industry	EPA 625/R93/014
OTHER MANUALS:	
Facility Pollution Prevention Guide	EPA 625/R92/088
Opportunities for Pollution Prevention Research to Support the 33/50 Program	EPA/600/R92/175
Life Cycle Design Guidance Manual	EPA/600/R92/226
Life Cycle Assessment: Inventory Guidelines and Principles	EPA/600/R92/245
Pollution Prevention Case Studies Compendium	EPA/600/R92/046
Industrial Pollution Prevention Opportunities for the 1990s	EPA/600/891/052
Achievements in Source Reduction and Recycling for Ten Industries in the United States	EPA/600/291/051
Background Document on Clean Products Research and Implementation	EPA/600/290/048
Waste Minimization Practices at Two CCA Wood Treatment Plants	EPA/600/R93/168
WMOA Report and Summary—Fort Riley, Kansas	EPA/600/S2-90/031
WMOA Report and Summary—Philadelphia Naval Shipyard/Governors Island	EPA/600/S2-90/062

(Continued)

TABLE 10.4 *Continued*

Title	EPA Document Number
Management of Household and Small-Quantity-Generator Hazardous Waste in the United States	EPA/600/S2-89/064
WMOA Report and Summary—Naval Undersea Warfare Engineering Station, Keport, WA	EPA/600/S2-91/030
WMOA Report and Summary—Optical Fabrication Laboratory, Fitzsimmons Army Medical Center, Denver, Colorado	EPA/600/S2-91/031
WMOA Report and Summary—A Truck Assembly Plant	EPA/600/S2-91/038
WMOA Report and Summary—A Photofinishing Facility	EPA/600/S2-91-039
WMOA Report and Summary—Scott Air Force Base	EPA/600/S2-91/054
Guidance Document for the WRITE Pilot Program with State and Local Governments	EPA/600/S8-89/070
Machine Coolant Waste Reduction by Optimizing Coolant Life	EPA/600/S2-90/033
Recovery of Metals Using Aluminum Displacement	EPA/600/S2-90/032
Metal Recovery/Removal Using Non-Electrolytic Metal Recovery	EPA/600/S2-90/033
Evaluation of Five Waste Minimization Technologies at the General Dynamics Pomona Division Plant	EPA/600/S2-91/067
An Automated Aqueous Rotary Washer for the Metal Fabrication Industry	EPA/600/Sr-92/188
Automotive and Heavy Duty Engine Coolant Recycling by Filtration	EPA/600/S2-91/066
Automotive and Heavy Duty Engine Coolant Recycling by Distillation	EPA/600/Sr-92/024
Onsite Waste Ink Recycling	EPA/600/Sr-92/251
Diaper Industry Workshop Report	EPA/600/S2-92/251
Hospital Pollution Prevention Case Study	EPA/600/S2-91/024
Waste Minimization Audit Report: Case Studies of Minimization of Cyanide Waste From Electroplating Operations	EPA/600/S2-87/055
Waste Minimization Audits at Generators of Corrosive and Heavy Metal Wastes	EPA/600/S2-87/056
Waste Minimization Audit Report: Case Studies of Minimization of Solvent Wastes From Parts Cleaning and From Electronic Capacitor Manufacturing Operations	EPA/600/S2-87/057
Waste Minimization in the Printed Circuit Board Industry—Case Studies	EPA/600/S2-88/008
Waste Minimization Audit Report: Case Studies of Minimization of Solvent Wastes and Electroplating Wastes at a DOD Installation	EPA/600/S2-88/010

(Continued)

TABLE 10.4 *Continued*

Title	EPA Document Number
Waste Minimization Audit Report: Case Studies of Minimization of Mercury-Bearing Wastes at a Mercury Cell Chloroalkali Plant	EPA/600/S2-88/011
Pollution Prevention Opportunity Assessment: USDA Beltsville Agricultural Research Center, Beltsville, Maryland	EPA/600/Sr-93/008
Pollution Prevention Opportunity Assessment for Two Laboratories at Sandia National Laboratories	EPA/600/Sr-93/015
Ink and Cleaner Waste Reduction Evaluation for Flexographic Printers	EPA/600/Sr-93/086
Mobile Onsite Recycling of Metalworking Fluids	EPA/600/Sr-93/114
Evaluation of Ultrafiltration to Recover Aqueous Iron Phosphating/Degreasing Bath	EPA/600/Sr-93/144
Recycling Nickel Electroplating Rinse Waters by Low Temperature Evaporation and Reverse Osmosis	EPA/600/Sr-93/160
WASTE MINIMIZATION ASSESSMENT FOR:	
Acrial Lifts	EPA 600/S-94-011
Aluminum and Steel Parts	EPA 600/S-94-010
Aluminum Cans	EPA 600/M91/025
Aluminum Extrusions	EPA 600/S-92-010
Automotive Air Conditioning Condensers and Evaporators	EPA 600/S-92-007
Baseball Bats and Golf Clubs	EPA 600/S-93-007
Caulk	EPA 600/S-94-017
Can-Manufacturing Equipment	EPA 600/S-92-014
Chemicals	EPA 600/S-92-004
Commercial Ice Machines and Ice Storage Bins	EPA 600/S-92-012
Components for Automobile Air Conditioners	EPA 600/S-92-009
Compressed Air Equipment Components	EPA 600/M91/024
Custom Molded Plastic Products	EPA 600/S-92-034
Cutting and Welding Equipment	EPA 600/S-92-029
Electrical Rotating Devices	EPA 600/S-94-018
Felt Tip Markets, Stamp Pads, and Rubber Cement	EPA 600/S-94-013
Fine Chemicals Using Batch Process	EPA 600/S-92-055
Finished Metal and Plastic Parts	EPA 600/S-94-005
Finished Metal Components	EPA 600/S-92-030
Gravure-Coated Metalized Paper and Metalized Film	EPA 600/S-94-008
Heating, Ventilating, and Air Conditioning Equipment	EPA 600/M91/019
Industrial Coatings	EPA 600/S-92-028
Injection-Molded Car and Truck Mirrors	EPA 600/S-92-032
Iron Castings and Fabricated Sheet Metal Parts	EPA 600/S-95-008
Labels and Flexible Packaging	EPA 600/S-95-004
Machined Parts	EPA 600/S-92-031

(Continued)

TABLE 10.4 *Continued*

Title	EPA Document Number
Metal Bands, Clamps, Retainers, and Tooling	EPA 600/S-92-015
Metal-Plated Display Racks	EPA 600/S-92-019
Microelectronic Components	EPA 600/S-94-015
Military Furniture	EPA 600/S-92-017
Motor Vehicle Exterior Mirrors	EPA 600/S-92-020
New and Reworked Rotogravure Printing Cylinders	EPA 600/S-95-005
Orthopedic Implants	EPA 600/S-92-064
Outdoor Illuminated Signs	EPA 600/M91/016
Paper Rolls, Ink Rolls, Ink Ribbons, and Magnetic and Thermal Transfer Ribbons	EPA 600/S-95-003
Parts for Truck Engines	EPA 600/S-94-019
Penny Blanks and Zinc Products	EPA 600/S-92-037
Permanent-Magnet DC Electric Motors	EPA 600/S-92-016
Pliers and Wrenches	EPA 600/S-94-004
Prewashed Jeans	EPA 600/S-94-006
Printed Circuit Boards	EPA 600/M91/022
Printed Circuit Boards	EPA 600/S-92-033
Printed Labels	EPA 600/M91/047
Printed Plastic Bags	EPA 600/M90/017
Product Carriers and Printed Labels	EPA 600/S-93-008
Prototype Printed Circuit Boards	EPA 600/M91/045
Rebuilt Railway Cars and Components	EPA 600/M91/017
Refurbished Railcar Bearing Assemblies	EPA 600/M91/044
Rotogravure Printing Cylinders	EPA 600/S-94-009
Screwdrivers	EPA 600/S-92-003
Sheet Metal Cabinets and Precision Metal Parts	EPA 600/S-92-021
Sheet Metal Components	EPA 600/S-92-035
Silicon-Controlled Rectifiers and Schottky Rectifiers	EPA 600/S-94-036
Surgical Implants	EPA 600/S-92-009
Treated Wood Products	EPA 600/S-92-022
Water Analysis Instrumentation	EPA/600/S-92/013
WASTE REDUCTION ACTIVIES AND OPTIONS FOR:	
Printer of Forms and Supplies for the Legal Profession	EPA/600/S-92/003
Nuclear Powered Electrical Generating Station	EPA/600/S-92/025
State DOT Maintenance Facility	EPA/600/S-92/026
Local Board of Education in New Jersey	EPA/600/S-92/027
Manufacturer of Finished Leather	EPA/600/S-92/039
Manufacturer of Paints Primarily for Metal Finishing	EPA/600/S-92/040
Manufacturer of Writing Instruments	EPA/600/S-92/041
Manufacturer of Room Air Conditioner Units and Humidifiers	EPA/600/S-92/042
Autobody Repair Facility	EPA/600/S-92/043

(Continued)

TABLE 10.4 *Continued*

Title	EPA Document Number
Fabricator and Finisher of Steel Computer Cabinets	EPA/600/S-92/044
Manufacturer of Artists' Supply Paints	EPA/600/S-92/045
Manufacturer of Wire Stock Used for Production of Metal Items	EPA/600/S-92/046
Manufacturer of Commercial Refrigeration Units	EPA/600/S-92/047
Waste Reduction: Pollution Prevention Publications Transporter of Bulk Plastic Pellets	EPA/600/S-92/048
Manufacturer of Electroplated Wire	EPA/600/S-92/049
Manufacturer of Systems to Produce Semiconductors	EPA/600/S-92/050
Remanufacture of Automobile Radiators	EPA/600/S-92/051
Manufacturer of Fire Retardant Plastic Pellets and Hot Melt Adhesives	EPA/600/S-92/052
Printing Plate Preparation Section of a Newspaper	EPA/600/S-92/053
Manufacturer of General Purpose Paints and Painting Supplies	EPA/600/S-92/054
Manufacturer of Fine Chemicals Using Batch Processes	EPA/600/S-92/055
Laminator of Cardboard Packages	EPA/600/S-92/056
Manufacturer of Hardened Steel Gears	EPA/600/S-92/057
Scrap Metal Recovery Facility	EPA/600/S-92/058
Manufacturer of Electroplating Chemical Products	EPA/600/S-92/059
Manufacturer of Plastic Containers by Injection Molding	EPA/600/S-92/060
Fossil Fuel-Fired Electrical Generating Stations	EPA/600/S-92/061
Manufacturer of Commercial Dry Cleaning Equipment	EPA/600/S-92/062
Electrical Utility Transmission Systems Monitoring and Maintenance Facility	EPA/600/S-92/063
Manufacturer of Orthopedic Implants	EPA/600/S-92/064

Pollution prevention opportunity case studies are also available in the Appendix of the above Guide. The case studies referred to in the Guide, include:

1. Construction and Demolition Waste Recycling etc.
2. Packaging Reuse.
3. Oil Analysis Program.
4. Maintenance Fluid Recycling.
5. Metal Working Fluid Substitution.
6. Use of Automated Aqueous Cleaner.
7. Recycling of Cleaner Through Filtration.
8. Proper Rinsing Set-Up for Chemical Etching.
9. Waste Reduction in the Chromate Conversion Process.
10. Plating Process Bath Maintenance.
11. Closed-Loop Plating Bath Recycling Process.

12. Water-Borne Paint as a Substitute for Solvent-Based Coatings.
13. High Velocity Low Pressure (HVLP) Paint System.
14. Replacing Chemical Stripping with Plastic Media Blasting.
15. White Water and Fiber Reuse in Pulp and Paper Manufacturing.
16. Chemical Substitution in Pulp and Paper Manufacturing.
17. On-Site Ink Recycling.
18. Solvent Reduction in Commercial Printing Industry etc.

The reader should not underestimate the value of both Guide and Table 10.4. These documents provide a near infinite amount of invaluable information to those involved with P2 activities. In many instances, one need only review the appropriate document to (immediately in some cases) implement a P2 program. States such as California and New York are another source of valuable P2 literature.

LPPA.29 A FAMOUS QUOTE

Who was it that once said 'an ounce of prevention is worth a pound of cure?'

SOLUTION: (The answer can be found at the end of the Index.)

QUANTITATIVE PROBLEMS (TPPA)

PROBLEMS TPPA.1–18

TPPA.1 COAL DISPOSAL

A 6% ash coal is the source of fossil fuel at a local utility. The average coal feed rate to the boiler is approximately 28,000 lb/hr. Estimate the amount of ash that must be disposed of for the following percentages of ash in the coal that "flies," i.e., leaves the boiler with the gas.

  a. 0%
  b. 25%
  c. 50%
  d. 75%
  e. 100%

The ash that does not "fly" exits the bottom of the boiler (as a solid waste that must be disposed of in a controlled landfill). Assume steady state round-the-clock operation. Comment on pollution prevention measures that can be instituted to reduce the emission of fly ash from the boiler.

SOLUTION:

1. Calculate the rate of ash feed to the boiler in lb/hr.

$$F = (0.06)(28000)$$
$$= 1680\,\text{lb/hr}$$

2. Calculate the rate of ash feed to the boiler in lb/week.

$$F = (1680\,\text{lb/hr})(24\,\text{hr/day})(7\,\text{days/wk})$$
$$F = 282,240\,\text{lb/wk}$$

3. Determine the ash collected in the boiler bottoms in lb/wk if 0% of the ash "flies."

$$m_0 = (1 - 0)(282,240)$$
$$= 282,240\ \text{lb/wk}$$

4. Determine the ash collected in the boiler bottoms in lb/wk if 25% of the ash "flies."

$$m_{25} = (1 - 0.25)(282,240)$$
$$= 211,680\,\text{lb/wk}$$

5. Determine the ash collected in the boiler bottoms in lb/wk if 50% of the ash "flies."

$$m_{50} = (1 - 0.5)(282,240)$$
$$= 141,120\,\text{lb/wk}$$

6. Determine the ash collected in the boiler bottoms in lb/wk if 75% of the ash "flies."

$$m_{75} = (1 - 0.75)(282,240)$$
$$= 70,560\,\text{lb/wk}$$

7. Determine the ash collected in the boiler bottoms in lb/wk if 100% of the ash "flies."

$$m_{100} = (1 - 1)(282,240)$$
$$= 0\,\text{lb/wk}$$

8. Comment on pollution prevention measures.

Source reduction: Use an alternative fuel rather than coal.

Treatment: Air pollution control equipment can be utilized. See the Theodore Tutorial by L. Theodore and R. Allen entitled "Air Pollution Control Equipment," East Williston, NY, 1995.

The reader is left an exercise of calculating the rate of ash normally discharged (from the boiler) in the flue gas. An electrostatic precipitator (ESP) or baghouse (fabric filter) is normally used to treat the flue gas to reduce the quantity of ash discharged to the atmosphere. Particulate and/or ash collection efficiencies rarely are below 99% by mass; most operate in excess of 99.5% efficiency.

TPPA.2   CUTBACK ASPHALT APPLICATION

A local asphalt company has provided the state Air Pollution Agency with their records, indicating that 40,000 lb of rapid curing cutback asphalt (containing 45% diluent by volume) is annually employed. Assume the density of naphtha (the diluent) and asphalt cement to be 5.9 lb/gal and 9.3 lb/gal, respectively. As part of the state's pollution prevention program, the company has been asked to determine the annual VOC emissions. Field data suggests that 95% of the diluent evaporates.

SOLUTION:

1. Write a conservation of mass equation relating the total asphalt mass with the volume of naphtha (x) and volume of asphalt (y).

$$40{,}000\,\text{lb} = (x\ \text{gal})(5.9\ \text{lb/gal}) + (y\ \text{gal})(9.3\,\text{lb/gal})$$

2. Using a volume balance, express the volume fraction of diluent in terms of x and y.

$$0.45 = x/(x+y)$$

3. Solve the two equations obtained in steps (1) and (2) for x and y in gal.
First, solve for y in terms of x using the equation from step (2):

$$0.45(x+y) = x$$
$$0.45y = 0.55x$$
$$y = 1.22x$$

Next, substitute this value of y into the equation in step (1):

$$40{,}000 = 5.9x + (9.3)(1.22x)$$
$$= 17.3x$$
$$x = 2312\,\text{gal}$$

and

$$y = 1.22x$$
$$= (1.22)(2312)$$
$$= 2821\,\text{gal}$$

4. Calculate the amount of diluent employed by the company in lb/yr.

$$x = (2312)(5.9)$$
$$= 13{,}640\,\text{lb/yr}$$

5. Calculate the VOC emissions in lb/yr.

$$VOC = (0.95)(13{,}640\,lb/yr)/(365\,days/yr)/(24\,hr/day)$$

$$= 1.48\,lb/hr$$

$$= 12{,}958\,lb/yr$$

TPPA.3 COLD CLEANER APPLICATION

A cold cleaner degreaser presently operates at a temperature of 20°C. At this temperature the vapor pressure of the solvent is 19 mm Hg.

1. Calculate the partial pressure of the solvent at the air-solvent interface.
2. Calculate the ppm of the solvent in the gas at the interface.
3. A vent is employed to control emissions into the workplace. If the vent air flow rate is 220 ft³/min (measured at 20°C), calculate the *maximum* emission rate of the solvent in lb/hr.
4. Comment on the results of (3).
5. Based on the above information and the degreaser regulations provided in the comment section below, design a cold cleaner degreaser that has an opening of 20 ft² that contains some simple pollution prevention measures.

SOLUTION:

1. Obtain the partial pressure of the solvent at the interface.

$$p_s = p'_s$$
$$= 19\,mm\,Hg$$

2. Calculate the ppm of the solvent.

$$ppm_s = (p_s/P) \times 10^6$$

$$= 19/760 \times 10^6$$

$$= 25{,}000\,ppm$$

This corresponds to a mole (or volume) fraction y_s of 0.025.

3. Calculate the maximum emission rate of the solvent.

$$m_s = (q)(y_s)(e)$$

$$= (220\,ft^3/min)\,(60\,min/hr)(0.025)(0.075\,lb/ft^3)$$

$$= 24.75\,lb/hr$$

4. Comment on the result of step (3). This is a fairly large emission rate.
5. Offer design suggestions for the degreaser.

The design must include (at a minimum):

a. A cover to prevent evaporation when the unit is not in use.

b. Equipment for draining cleaned parts.

c. A permanent, conspicuous label summarizing the operating requirements.

TPPA.4 SOLID WASTE VOLUME REDUCTION DUE TO HOME COMPACTORS

Household garbage compactors are useful in reducing the volume of waste collected by garbage collectors. Garbage that is only discarded in containers and then collected can, if applicable, shorten the life of the community landfill and make more work for the collectors.

The following table provides a weekly volume estimate for some common household solid wastes (see Table 10.5).

Estimate the percent volume reduction that would be achieved in the solid waste collected at the community if all households installed garbage compactors which compacted the solid wastes to a density of 20 lb/ft^3. Note that garden trimmings, wood, ferrous metals, and dirt, ash, and brick are usually not placed in trash compactors.

SOLUTION:

1. Calculate the total weight of all solid wastes.

$$W_{tot} = 10 + 50 + 10 + 2 + 1 + 2 + 10 + 5 + 2 + 4 + 1$$
$$= 97 \text{ lb}$$

2. Calculate the total weight, excluding items not placed in the compactor.

$$W_{tcom} = 10 + 50 + 10 + 2 + 1 + 2 + 2$$
$$= 77 \text{ lb}$$

TABLE 10.5 Household Waste Volume

Waste	Weight (lb)	Density (lb/ft^3)	Volume (ft^3)
Food	10	18.0	0.6
Paper	50	5.0	10.0
Cardboard	10	6.5	1.5
Plastics	2	4.0	0.5
Textiles	1	4.0	0.3
Leather	2	10.0	0.2
Garden Trimmings	10	6.5	1.5
Wood	5	15.0	0.3
Nonferrous Metals	2	10.0	0.2
Ferrous Metals	4	20.0	0.2
Dirt, Ash, Brick, etc.	1	30.0	0.03

Note: It is assumed that glass and aluminum cans are recycled by the household.

3. Calculate the total volume of all solid wastes.

$$V_{tot} = 0.6 + 10.0 + 1.5 + 0.5 + 0.3 + 0.2 + 1.5 + 0.3 + 0.2$$
$$+ 0.2 + 0.03$$
$$= 15.33 \ ft^3$$

4. Calculate the total volume, excluding items not placed in the compactor.

$$V_{tcom} = 0.6 + 10.0 + 1.5 + 0.5 + 0.3 + 0.2 + 0.2$$
$$= 13.30 \ ft^3$$

5. Calculate the volume of the compacted waste with a bulk density of 20 lb/ft^3.

$$V_{com} = 77/20$$
$$= 3.85 \ ft^3$$

6. Determine the percent volume reduction for the solid waste deposited into the compactor.

$$VR = (100)(13.30 - 3.85)/(13.30)$$
$$= 71.1\%$$

7. Determine the overall percent volume reduction achieved with the household compactor (include garden trimmings, wood, ferrous metals, dirt, ashes, and brick).

$$V_{t, fin} = (15.33 - 13.30) + 3.85$$
$$= 5.88$$
$$VR - (100)(15.33 - 5.88)/(15.33)$$
$$= 61.6\%$$

TPPA.5 COMMUNITY RECYCLING

As described earlier, recycling is an excellent way of reducing the quantity of solid waste that must be burned, landfilled, or otherwise disposed of in this country. Key factors which can contribute to the success of a recycling program are listed below:

1. The quantity of recyclables in the waste stream.
2. The efficiency of the separation at the source. In other words, does the recycler actually put 100% of any particular item in the recycle bin or does the recycler sometimes get lazy or make a mistake. For whatever reason, it is a virtual certainty that 100% of the items do not make it to the recycler bin.
3. The percentage of the people that actually try to comply with the recycling effort. Not everyone is civic minded.

Thus, the quantity actually recycled (RE) can be defined as follows:

$$RE = Q \times P \times E \tag{10.3}$$

where Q = quantity of recyclable material

  P = people participating on a fractional basis

  E = efficiency of separation on a fractional basis

Suppose the Town of East Williston would like to begin a curbside recycling program. The town would like to start with a modest program that picks up some paper goods, plastics, glass and metals. The town's target recyclables are as follows:

  Paper—newspapers, magazines, and corrugated material

  Plastic—all PET (polyethylene terephthalate) and HDPE

   (high density polyethylene) container plastics

  Glass—all green, amber and brown container glass

  Metal—all ferrous and nonferrous metal containers

See Table 10.6 for the quantities of these recyclables.

A supposed statistically valid study was performed in order to estimate the efficiency of this program. The study concluded that 80% of the people in town would be willing to participate in a recycling program that involved only the separation of paper. Only 60% (P = 0.6) thought they could ever get used to separating out paper, plastics, glass, and metal.

Based on the waste stream characteristics given in Table 10.6, estimate the reduction in tons per day (TPD) of solid waste sent to the town's landfill if it (a) implements only paper recycling, and (b) implements the complete program.

The separation efficiency (E) within each household is estimated at 85% for the paper-only program. Separation efficiencies for the complete program are estimated at 85%, 80%, 90% and 85% for the paper, plastic, glass and metal, respectively.

Suppose the town's existing landfill has an estimated remaining capacity of 1,273,750 tons. How long will the landfill last if no recycling is implemented? How long will it last if the full recycling program is implemented? Assume solid waste generation grows 3% annually. Indicate how the town can extend the landfill site even further? Use the data provided in Table 10.6.

SOLUTION:

1. Determine the individual quantities of recyclables, Q, in the waste stream for paper, plastics, glass, and metals.

   Paper:

   $$Q = 8.4 + 3.5 + 6.3$$
   $$= 18.2 \text{ TPD}$$

   Plastics:

   $$Q = 2.1 + 1.6 + 4.1$$
   $$= 7.8 \text{ TPD}$$

TABLE 10.6   Characteristics of Target Recyclable Waste Stream

Component	Quantity (TPD)
Paper	
Newspaper	8.4
Magazines	3.5
Corrugated/brown paper	6.3
Books	0.2
Office paper	0.4
Other	5.0
Plastics	
PET (1 liter)	2.1
PET (2 liter)	1.6
HDPE	4.1
Other	10.2
Glass	
Green	2.6
Amber	1.2
Brown	7.1
Other	0.4
Ferrous metals	
Food cans	3.6
Auto parts	0.8
Other	4.7
Nonferrous metals	
Aluminum cans	1.7
Aluminum foil	1.3
Other	3.2
Total of all other nonrecyclable items	81.6
Total of all items	150

Glass:

$$Q = 2.6 + 1.2 + 7.1$$
$$= 10.9 \text{ TPD}$$

Metal:

$$Q = 3.6 + 1.7$$
$$= 5.3 \text{ TPD}$$

2. Determine the total quantity of recyclables, Q_T, in the waste stream.

$$Q = 18.2 + 7.8 + 10.9 + 5.3$$
$$= 42.2 \text{ TPD}$$

3. Determine the reduction in solid waste for the paper only recycling program, $RED_{(paper)}$, in tons/day.

$$RED_{(paper)} = QPE$$
$$= (18.2)(0.6)(0.85)$$
$$= 9.3 \text{ TPD}$$

4. Determine the reduction in solid waste for the complete recycling program, $RED_{(total)}$, in tons/days.

$$RED_{(paper)} = (18.2)(0.6)(0.85) = 9.3 \text{ TPD}$$
$$RED_{(plast)} = (7.8)(0.6)(0.80) = 3.7 \text{ TPD}$$
$$RED_{(glass)} = (10.9)(0.6)(0.90) = 5.9 \text{ TPD}$$
$$RED_{(metal)} = (5.3)(0.6)(0.85) = 2.7 \text{ TPD}$$
$$RED_{(total)} \qquad\qquad\quad = 21.6 \text{ TPD}$$

5. Calculate the solid waste sent to the landfill in year one, W_1, with no recycling. Note that the total of all items in Table 10.6 is 150 TPD.

$$W_1 = (150 \text{ TPD})(365 \text{ days}) = 54{,}750 \text{ tons}$$

6. Determine the waste generated in tons after 10 years with 3% growth per year.

$$w_n = (150)(1.03)^n(365)(n)$$
$$w_{10} = (150)(1.03)^{10}(365)(10)$$
$$= 735{,}794 \text{ tons}$$

7. Obtain the present capacity of the landfill in tons.

$$W_{total} = 1{,}273{,}750 \text{ tons}$$

8. Calculate the life expectancy in years of the landfill.

$$(150)(1.03)^n(365)(n) = 1{,}273{,}750$$
$$n(1.03)^n = 23.26$$

Trial and error solution gives:

$$n = 14.95 \text{ yrs}$$
$$= 15 \text{ yrs}$$

9. Determine the life of the landfill with a full recycling program.

$$(150 - 21.6)(1.03)^n(365)(n) = 1{,}273{,}750$$
$$n(1.03)^n = 27.18$$

Trial and error solution gives:

$$n = 16.63 \text{ yrs}$$

10. Determine how long the landfill life can be extended by recycling.

$$16.63 - 14.95 = 1.68 \text{ yrs}$$

11. How can the landfill life be extended even further? If P, the percentage of people taking part in recycling, can be increased, the life of the landfill can be extended.

Recycling is not the only answer to all of the nation's solid waste problems. Any plan to deal effectively with any solid waste problem should also include source reduction and waste-to-energy programs.

TPPA.6 INDUSTRIAL PAPER WASTE REDUCTION

As described earlier the two most important aspects of any waste minimization program are source reduction and recycling. Although source reduction is often the most effective way of preventing pollution, it does not always achieve the best results by itself. In some situations, a combination of source reduction and recycling can be the most effective method of minimizing waste. Due to the fact that landfills are closing fast and that pollution is becoming a major concern to society, reducing the generation of wastes by any means can help preserve the environment.

This problem pertains to the use of measures to improve a company's method of handling paper waste. At present, all of the company's paper waste is combined with the rest of their waste and is hauled off to landfills. They are seeking a more environmentally sound method of dealing with the waste paper produced by their 1850 employees who, on average, produce 1.5 pounds of paper waste per day. Approximately one third of the paper waste is recyclable. The average number of days per year that the employees are at work is 250.

It has been estimated that for each ton of paper which is recycled there is a savings of about 60 lb of air pollution and 3 cubic yards of landfill space. Also, since it takes 60% less energy to manufacture paper from recycled stock than from virgin sources, there exists an estimated energy savings of 4200 kW per ton recycled. In addition, 7000 gallons of water per ton of paper recycled are saved—enough to meet the needs of 30 households for an entire day.

Apply source reduction and recycling strategies to this situation assuming that 75% of the employees will be actively participating in a source reduction and recycling program. Assume an average decrease in waste paper generation of 0.3 lb per day per participating person as well as an estimated savings by recycling 0.5 lb per day per participating person. Discuss the program's usefulness and calculate the change in waste production, energy use, and pollutants generated as a result of this program.

SOLUTION:

1. Calculate the annual rate of waste paper initially being generated at the company.

$$A_1 = (1850 \text{ people})(1.5 \text{ lb/day})(250 \text{ day/yr})$$

$$= 693{,}750 \text{ lb/yr}$$

2. Calculate the annual waste paper generated after a source reduction of 0.3 lb/day.

$$A_s = (0.75)(1850 \text{ people})(1.5 - 0.3 \text{ lb/day})(250 \text{ day/yr})$$

$$+ (0.25)(1850 \text{ people})(1.5 \text{ lb/day})(250 \text{ day/yr})$$

$$= 416{,}250 + 173{,}438$$

$$= 589{,}688 \text{ lb/yr}$$

3. Calculate the annual waste paper generated after implementing both reduction and recycling (0.5 lb/day) measures.

$$A_r = (0.75)(1850 \text{ people})(1.5 - 0.3 - 0.5 \text{ lb/day})(250 \text{ day/yr})$$

$$+ (0.25)(1850 \text{ people})(1.5 \text{ lb/day})(250 \text{ day/yr})$$

$$= 242{,}813 + 173{,}438$$

$$= 416{,}250 \text{ lb/yr}$$

4. Determine the decrease in the amount of waste paper generated after the pollution prevention program is put into effect.

$$D = 693{,}750 - 416{,}250$$

$$= 277{,}500 \text{ lb/yr}$$

5. Calculate the annual amount of paper recycled in tons per year.

$$R = (0.75)(1850 \text{ people})(0.5 \text{ lb/day})(250 \text{ day/yr})$$

$$= 173{,}438 \text{ lb/yr}$$

$$= 86.7 \text{ ton/yr}$$

6. Estimate the annual reduction in pollutants generated due to the recycling effort.

$$P_r = (86.72 \text{ ton/yr})(60 \text{ lb/ton rec})$$

$$= 5{,}202 \text{ lb/yr}$$

7. Calculate the annual energy savings due to the pollution prevention program.

$$E_r = (86.7 \text{ ton/yr})(4{,}200 \text{ kW/ton})$$

$$= 364{,}140 \text{ kW/yr}$$

8. Determine the annual water savings in gal/yr.

$$W_r = (86.72 \text{ ton/yr})(7000 \text{ gal } H_2O \text{ saved/ton})$$
$$= 606,900 \text{ gal/yr}$$

9. Estimate the number of households that the water savings calculated in step (8) can accommodate on both a daily and an annual basis.

$$H = (606,900 \text{ gal})(30 \text{ households})/(7000 \text{ gal})$$
$$= 2600 \text{ households for one day}$$

or

$$H = (606,900 \text{ gal})(30 \text{ households})/(7000 \text{ gal})/(365 \text{ days})$$
$$= 7 \text{ households for one year}$$

It seems that the above combination of source reduction and recycling proved to be effective in this program.

TPPA.7 ALTERNATE ENERGY SOURCES

By combusting fossil fuels, power plants today are huge polluters. In addition, the much vaunted nuclear plants are not immune as they are responsible for thermal pollution and the problems associated with both spent fuel and radioactive wastes.

Alternate methods can be used to provide power with minimal amounts of any pollution generated. These methods are not used largely for economic reasons. As pollution becomes more and more of a problem, as witnessed by ozone depletion and global warming, society can no longer ignore the pollution generated from traditional energy sources. Alternate energy sources should be reevaluated and their feasibility determined. One such method is investigated in this problem.

A lake which can be natural or man-made is located at the top of a mountain. A power plant has been constructed at the bottom of the mountain. The potential energy of the water traveling downhill can be used to spin turbines and generate electricity. This is the operating mode in the daytime at peak electrical demand. At night, when demand is reduced, the water is pumped back up the mountain. The operation is shown below in Figure 10.3.

Using the method of power "production" described above, determine how much power (watts) is generated by a lake located at an elevation of 3000 ft above the power plant. The flow rate of water is 500,000 gpm. Turbine efficiency is 30%. Neglect friction effects.

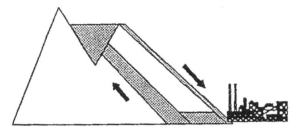

FIGURE 10.3   Reservoir Application.

SOLUTION:

1. Convert height and flow rate to SI units in order to solve for the power in watts.

$$(3000 \text{ ft})(0.3048 \text{ m/ft}) \quad = 914.4 \text{ m}$$

$$(500{,}000 \text{ gal/min})(0.00378 \text{ m}^3/\text{gal}) = 1890 \text{ m}^3/\text{min}$$

2. Determine the mass flow rate of the water in kg/s.

$$(1890 \text{ m}^3/\text{min})(1000 \text{ kg/m}^3)/(60 \text{ s/min}) = 31{,}500 \text{ kg/s}$$

3. Determine the potential energy, U, of the water flow.

$$U = mgh$$

$$= (31{,}500 \text{ kg/s})(9.8 \text{ m/s}^2)(914.4 \text{ m})$$

$$= 2.82 \times 10^8 \text{ kg-m/s}^3$$

$$= 2.82 \times 10^8 \text{ N/s}$$

$$= 282 \text{ MW}$$

4. Determine the actual power output.

$$P = (0.30)(282) = 84.7 \text{ MW}$$

This is enough power for a small town. No pollution is generated because no fossil fuel is required. The initial construction expense would be quite high but the long term cost of producing electricity would probably be very economical. An economic analysis is suggested.

TPPA.8   RETROFITTING OFFICE LIGHTING SYSTEMS

In the United States, lighting accounts for the consumption of about 500 billion kilowatt hours of energy each year-roughly the output of 100 standard power plants. Lighting is a very prevalent component of the total energy used in most commercial buildings, accounting for 30–50% of the consumed energy. Since lighting constitutes such a major part of a building's energy consumption, economics would seem to be the first concern in regards to the effect that inefficient lighting systems have. However, another issue relating to the excessive use of energy is the effect that such abuse has on the environment. This arises due to the production of energy by non-nuclear methods that utilize the combustion of fossil fuels. Regarding lighting, the amount of pollutant emissions from power plants can be decreased simply by reducing the amount of energy used for lighting, especially in large commercial structures where annual energy consumption (on lighting alone) can often easily exceed one million kilowatt hours. Since most commercial buildings were originally equipped with a lighting system that is up to 30% over the necessary size, the amount of energy used can be reduced by removing a percentage of the lights. Lights can also be replaced by more efficient and higher quality lights. These two methods can be used separately or in unison.

The Aldo Leone corporation's 100,000 square foot office center contains 1253 standard four lamp fluorescent fixtures consuming an average of 174 watts per fixture or about 43.5 watts per lamp. These lamps operate for an average of 16 hours a day, 6 days a week, thus accounting for a yearly operational time of 4,992 hours. Also, it is estimated that the local coal-fired electric power plant emits 0.0175 pounds of sulfur dioxide, 0.00824 pounds of nitrous oxides, and 2.25 pounds of carbon dioxide per kilowatt-hour generated.

Apply the most effective method of pollution prevention, i.e., source reduction, to effectively decrease the amount of pollutants resulting from the consumption of electricity in the building. It has been suggested that the number of lights to be reduced by 20% and the watts per fixture should be reduced to 106 watts per fixture. An effective guideline for properly retrofitting a new lighting system is to keep the watts per square foot to a maximum of 1.5.

1. Calculate the present watts per square foot, the new watts per square foot, and the number and new wattage of lamps to be installed.

2. Determine the total present lighting load, the new lighting load, and the load reduction (in kW) as well as the present annual load, the new annual load, and the annual load reduction (in kW · hr).

3. Calculate the effects of this energy consumption reduction on the amount of pollutants emitted from the local power plant.

SOLUTION:

1. Regarding the current present load, calculate the total number of lights.

$$N = (1253 \text{ units})(4 \text{ lights/unit})$$
$$= 5012 \text{ lights}$$

Calculate the present lighting load in W and kW.

$$P = (1253 \text{ units})(174 \text{ W/unit})$$
$$= 218,000 \text{ W}$$
$$= 218 \text{ kW}$$

Calculate the present annual load.

$$P = (218 \text{ kW})(16 \text{ h/day})(6 \text{ day/wk})(52 \text{ wk/yr})$$
$$= 1.09 \times 10^6 \text{ kW} \cdot \text{h}$$

Calculate the annual pollution contribution of SO_2, CO_2, and NO_x.

SO_2: $(0.0175 \text{ lb/kW} \cdot \text{h})(1.09 \times 10^6 \text{ kW} \cdot \text{h}) = 19,075 \text{ lb/yr}$

CO_2: $(2.25 \text{ lb/kW} \cdot \text{h})(1.09 \times 10^6 \text{ kW} \cdot \text{h}) = 2.45 \times 10^6 \text{ lb/yr}$

NO_x: $(0.00824 \text{ lb/kW} \cdot \text{h})(1.09 \times 10^6 \text{ kW} \cdot \text{h}) = 8982 \text{ lb/yr}$

Calculate the present watts per square foot.

$$(218,000 \text{ W})/(100,000 \text{ ft}^2) = 2.18 \text{ W/ft}^2$$

2. For retrofitting the lighting system, calculate the number of lights removed.

$$(5012 \text{ lights}) (0.2) = 1002 \text{ lights} = 250 \text{ units}$$

3. Calculate the new lighting load in W and kW.

$$P = (1253 - 250)(106)$$
$$= 106,000 \text{ W}$$
$$= 106 \text{ kW}$$

Calculate the new annual load in kW-h.

$$P_s = (106 \text{ kW})(16 \text{ h/day})(6 \text{ day/wk})(52 \text{ wk/yr})$$
$$= 529,000 \text{ kW} \cdot \text{h}$$

Calculate the new watts per square foot.

$$(106,000 \text{ W})/(100,000 \text{ ft}^2) = 1.06 \text{ W/ft}^2$$

This is below the guideline.
Calculate the load reduction in kW.

$$P_{red} = 218 - 106$$
$$= 112 \text{ kW}$$

Calculate the annual load reduction in kW · h/yr.

$$(112 \text{ kW})(16 \text{ h/day})(6 \text{ day/wk})(52 \text{ wk/yr}) = 559{,}000 \text{ kW} \cdot \text{h/yr}.$$

4. Calculate the new amounts of pollutants emitted in lb/yr.

SO_2: $(0.0175 \text{ lb/kW} \cdot \text{h})(529{,}000 \text{ kW} \cdot \text{h}) = 9{,}257 \text{ lb/yr}$

CO_2: $(2.25 \text{ lb/kW} \cdot \text{h})(529{,}000 \text{ kW} \cdot \text{h}) = 1.19 \times 16^6 \text{ lb/yr}$

NO_x: $(0.00824 \text{ lb/kW} \cdot \text{h})(529{,}000 \text{ kW} \cdot \text{h}) = 4360 \text{ lb/yr}$

5. Calculate the amount of pollutants reduced by the new lighting system.

SO_2: $19{,}075 - 9{,}257 = 9800 \text{ lb/yr}$

CO_2: $2.45 \times 10^6 - 1.19 \times 10^6 = 1.26 \times 10^6 \text{ lb/yr}$

NO_x: $8982 - 4360 = 4622 \text{ lb/yr}$

TPPA.9 HEAT EXCHANGER NETWORK

A plant has three streams to be heated and three streams to be cooled. Cooling water (90°F supply, 155°F return) and steam (saturated at 250 psia) are available. As part of a pollution prevention project, you have been asked to devise a network of heat exchangers that will make full use of heating and cooling streams against each other, using utilities only if necessary.

The three streams to be heated are listed in Table 10.7.
The three streams to be cooled are listed in Table 10.8.
Saturated steam at 250 psia has a temperature of 401°F.

SOLUTION: The heating duties for all streams are first calculated. The results are tabulated below in Table 10.9.

TABLE 10.7 Hot Fluid Data

Stream	Flowrate (lb/h)	C_P [Btu/lb. °F]	T_{in} (°F)	T_{out} (°F)	Duty, Btu/h
4	60,000	0.70	420	120	12,600,000
5	40,000	0.52	300	100	4,160,000
6	35000	0.60	240	90	3,150,000

TABLE 10.8   Cold Fluid Data

Stream	Flowrate (lb/h)	C_P [Btu/lb. °F]	T_{in} (°F)	T_{out} (°F)	Duty, Btu/h
1	50,000	0.65	70	300	7,475,000
2	60,000	0.58	120	310	6,612,000
3	80,000	0.78	90	250	9,984,000

The total heating and cooling duties are next compared.

$$\text{Heating:} \quad 7,475,000 + 6,612,000 + 9,984,000 = 24,071,000 \, \text{Btu/h}$$
$$\text{Cooling:} \quad 12,600,000 + 4,160,000 + 3,150,000 = 19,910,000 \, \text{Btu/h}$$

As minimum 4,161,000 Btu/h (heating–cooling) will have to be supplied by steam.

Figure 10.4 represents a system of heat exchangers that will transfer heat from the hot streams to the cold ones in the amounts desired. It is important to note that this is but one of many possible schemes. The optimum system would require a trial-and-error procedure that would examine a host of different schemes. Obviously, the economics would come into play.

It should also be noted that in many chemical and petrochemical plants there are cold streams that must be heated and hot steams that must be cooled. Rather than use steam to do all the heating and cooling water to do all the cooling, it is often advantageous, as demonstrated in this problem, to have some of the hot streams heat the cold ones. The problem of optimum heat exchanger networks has been extensively studied and is available in the literature. This problem gives one simple illustration.

Finally, highly interconnected networks of exchangers can save a great deal of "usable" energy in a chemical plant. The more interconnected they are, however, the harder the plant is to control, start-up, and shut down. Often auxiliary heat sources and cooling sources must be included in the plant design in order to ensure that the plant can operate smoothly.

TABLE 10.9   Heat Exchanger Duties

	Stream	Duty, Btu/h
1	Heating	7,475,000
2	Heating	6,612,000
3	Heating	9,984,000
4	Cooling	12,600,000
5	Cooling	4,160,000
6	Cooling	3,150,000

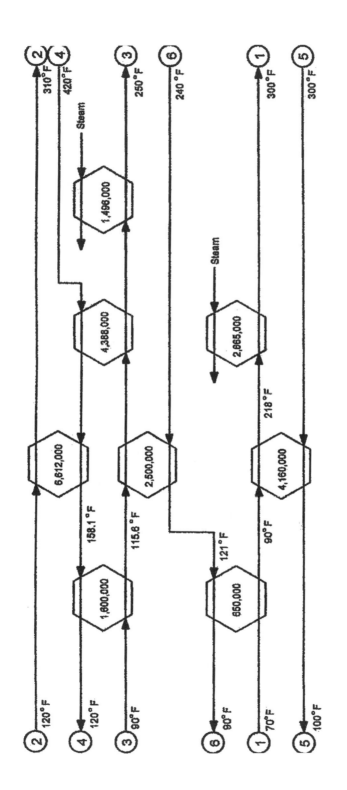

Numbers in heat exchanger boxes have units of Btu/hr

FIGURE 10.4   Flow diagram for Problem TPPA.9.

TPPA.10 ENHANCED RECOVERY OF POTASSIUM NITRATE BY RECYCLING

The processing industry has given operations involving heat transfer to boiling liquid the general name *evaporation*. The most common application is the removal of water from a processing stream. Evaporation is used in the food, chemical, and petrochemical industries, and it usually results in an increase in the concentration of a certain species. The factors that affect the evaporation process are concentration in the liquid, solubility, pressure, temperature, scaling, and materials of construction. An evaporator is a type of heat transfer equipment designed to induce boiling and evaporation of a liquid. The major types of evaporators are open kettle or pan, horizontal-tube natural convection, vertical-tube natural convection, and forced-convection evaporators.

Crystallization is the process of forming a solid phase from solution. It is employed heavily as a separation process in the inorganic chemistry industry, particularly where salts are recovered from aqueous media. In the production of organic chemicals, crystallization is also used to recover product, to refine intermediate chemicals, and to remove undesired salts. The feed to a crystallization system consists of a solution from which solute is crystallized through one or more of a variety of processes. The solids are normally separated from the crystallizer liquid, washed, and discharged to downstream equipment for further processing.

Consider the following Problem. Potassium nitrate is obtained from an aqueous solution of 20% KNO_3. During the processes, the aqueous solution of potassium nitrate is evaporated, leaving an outlet stream with a concentration of 50% KNO_3, which then enters a crystallization unit where the outlet product is 96% KNO_3 (anhydrous crystals) and 4% water. A residual aqueous solution which contains 0.55 gram of KNO_3 per gram of water also leaves the crystallization unit and is mixed with the fresh solution of KNO_3 at the evaporator inlet. Such recycling provides a means of preventing pollution by minimizing the loss of valuable raw materials. The flow diagram in Figure 10.5 depicts the process.

What is the mass flowrate, M, in kg/h of the recycled material (R) when the feedstock flowrate is 5000 kg/h?

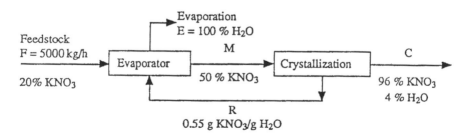

FIGURE 10.5 Crystallization Process.

SOLUTION:

1. Using Figure 10.5, perform an overall material balance on KNO_3 and solve it for the rate of product leaving the crystallizer.

$$0.96C = 0.2(5000)$$
$$C = 1000/0.96$$
$$= 1042 \, kg/hr$$

2. Write the equation for an overall mass balance around the crystallizer.

$$M = R + C$$
$$= R + 1042$$

3. Write the equation for the KNO_3 mass balance around the crystallizer.

$$0.5M = 0.96(1042) + (0.55/1.55)R$$
$$= 1000 + 0.355R$$

4. Solve for the amount of KNO_3 leaving the evaporator in terms of the amount of recycle.

$$(5000)(0.2) + 0.355R = 0.5M$$
$$M = 2000 + 0.71R$$

5. Calculate the mass flow rate of the recycle stream.
   Use the equation from step (2).

$$2000 + 0.710R = R + 1042$$
$$0.29R = 958$$
$$R = 3301 \, kg/h$$

If an insufficient amount of waste is generated on-site to make an in-plant recovery system cost-effective, or if the recovered material cannot be reused on-site, off-site recovery is preferable. Some materials commonly reprocessed off-site are oils, solvents, electroplating sludges, and process bath, scrap metal, and lead-acid batteries. As noted earlier the cost of off-site recycling is dependent upon the purity of the waste and the market for the recovered material.

TPPA.11   NICKEL ELECTROPLATING-OPERATION

A nickel electroplating line uses a dip-rinse tank to remove excess plating metals from the parts. Currently, a single tank is used which required R gal/h of fresh rinsewater to clean F parts/h (see Figure 10.6 below). Assume the cleaning is governed by the following equilibrium relation:

$$\lambda = f_1/r_1 = (\text{ounces of metal residue/part})/(\text{ounces of metal residue/gal bath})$$

1. Calculate the reduction in rinsewater flow rate (a pollution prevention measure) if a two-stage countercurrent rinse tank is used (as compared to the single-stage unit), and 99% of the residue must be removed. Assume the drag-out volume is negligible.
2. Has the total metal content of the exit rinsewater been altered? Discuss implications for further wastewater treatment/reuse.

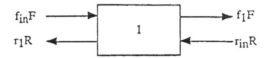

FIGURE 10.6   Flow Diagram for One Stage Operation.

SOLUTION:

1. Using the flow diagram provided in Figure 10.6, write a material balance for the residue.

$$f_{in}F + r_{in}R = f_1F + r_1R$$

2. Rearrange the equation in step (1) in terms of λ. For the single stage operation, set $i = 1$.

$$R/F = (f_{in} - f_1)/r_1 = \lambda(f_{in} - f_1)/f_1$$

3. Determine the fraction of residue removed, x.

$$X = (f_{in} - f_1)/f_{1n}$$

4. Express R/F in terms of λ and x.

$$R/F = \lambda[x/(1 - x)]$$

5. Draw a flow diagram for a two-stage (stage 1, stage 2) countercurrent operation. See Figure 10.7.

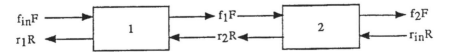

FIGURE 10.7   Flow Diagram for Two-Stage Operation.

6. Write material balances on the residue for each stage.

$$\text{Stage 1:} \quad f_{in}F + r_2R = f_1F + r_1R$$
$$\text{Stage 2:} \quad f_2F + r_{in}R = f_2F + r_2R$$

7. Solve each equation from step (6) for R/F in terms of λ and f_1 using the defining equation for λ.

$$\text{Stage 1:} \quad R/F = \lambda(f_{in} - f_1)/(f_1 - f_2)$$
$$\text{Stage 2:} \quad R/F = \lambda(f_1 - f_2)/f_2$$

8. Substitute x into the mass balance expressions.

$$\text{Stage 1:} \quad R/F = (f_{in} - f_1)/(f_1 - f_{in}(1 - x))$$
$$\text{Stage 2:} \quad R/F = [f_1 - f_{in}(1 - x)]/[f_{in}(1 - x)]$$

9. Set the right hand side (RHS) of the stage 1 and stage 2 equations equal to each other.

$$(f_{in} - f_1)/[f_1 - f_{in}(1 - x)] = [f_1 - f_{in}(1 - x)]/[f_{in}(1 - x)]$$

10. Rearrange the equation in step (9) to obtain a quadratic equation in f_1.

$$f_1^2 - [(1 - x)f_{in}]f_1 - [x(1 - x)f_{in}^2] = 0$$

11. Solve the equation in step (10) for f_1.

$$f_1 = [f_{in}(1 - x) \pm [(-f_{in}(1 - x))^2 - 4(1)x(1 - x)f_{in}^2]^{0.5}]/2$$

Through algebra, this can be reduced to:

$$f_1 = (f_{in}/2)(1 - x \pm [(3x + 1)(1 - x)]^{0.5})$$

If $(1 - x)$ is factored out, the following expression results:

$$f_1 = [f_{in}(1 - x)/2][1 \pm [(3x + 1)/(1 - x)]^{0.5}]$$

Only the "+" term is physically reasonable, so:

$$f_1 = [f_{in}(1 - x)/2][1 + [(3x + 1)/(1 - x)]^{0.5}]$$

12. Substitute f_1 into the stage 2 mass balance equation.

$$R/F = \lambda\{[f_{in}(1 - x)/2][1 + [(3x + 1)/(1 - x)]^{0.5}] - f_2\}/f_2$$

However, f_2 can be expressed in terms of x and f_{in}:

$$f_2 = f_{in}(1 - x)$$

This can be substituted into the above equation:

$$R/F = \lambda\{[f_{in}(1-x)/2][1 + [(3x+1)/(1-x)]^{0.5}] - f_{in}(1-x)\}/[f_{in}(1-x)]$$

This reduces to:

$$R/F = (\lambda/2)\{[(3x+1)/(1-x)]^{0.5} - 1\}$$

13. Calculate the rinsewater requirements for both a single and two stage countercurrent unit.

$$\text{Single-Stage } R = \lambda[0.99/(1-0.99)]F$$

$$= 99.0\lambda F$$

$$\text{Two-Stage } R = (\lambda/2)\{[(3(0.99)+1)/(1-0.99)]^{0.5} - 1\}F$$

$$= 9.46\lambda F$$

14. Calculate the rinsewater flow rate reduction

$$(99.0\lambda F - 9.46\lambda F)/(99.0\lambda F) = 0.904 = 90.4\% \text{ reduction}$$

15. Comment on the results.

For fixed residue removal, the total mass of metals in the rinsewater will be the same; with reduced water duty, the metals concentration increases. A smaller water volume now needs to be processed (or sewered). Alternatively, if the metals concentration is sufficiently high, it can be returned to the plating bath for reuse or can be recovered.

The reader should consider the merits of employing a three (or more) stage unit to accomplish the necessary cleaning. What happens to cost as the number of stage approaches infinity?

TPPA.12   OPEN TOP VAPOR DEGREASER MODIFICATION

As part of a P2 program, the LAST company has decided to include a lip hood on an open top vapor degreaser to reduce fugitive VOC vapor losses. The degreaser is 5 ft long and 3 ft wide. The company has hired Dr. Louis Theodore, a purported "grease" expert to calculate the ventilation rate through the slots (which is exhausted to an existing adsorber) and the required width of the slots.

EPA has provided data in the form of two graphs to estimate an appropriate ventilation rate and the slot (or lip) width. This procedure is based on a velocity through the slots of approxiamately 3 ft/s. The two graphical approaches have been expressed in equation form (L. Theodore: Personal notes, 1991) as:

$$VR = 10.0 \text{ (TW)}^{1.15} \tag{10.4}$$

$$SW = 0.03 \text{ (TW)}^{1.15} \tag{10.5}$$

where

VR = minimum ventilation rate, acfm/ft of degreaser length (on one side)

TW = tank width, in

SW = slot width, in

The above equations apply if the degreaser is equipped with lip hoods along both lengths of the unit. If a lip hood is located along only one length of the degreaser and the opposite side of the unit is bounded by a vertical wall, EPA suggests doubling the tank width in determining the calculated ventilation rate. The actual ventilation rate should then be set equal to one half the calculated rate from the graph (or equation).

SOLUTION:

1. Estimate the minimum ventilation rate in acfm/ft of degreaser length.

$$VR = 10.0(TW)^{1.15}$$

$$= (10.0)(36)^{1.15}$$

$$= 616.2 \text{ acfm/ft of degreaser length}$$

2. Calculate the exhaust flowrate, q, to the adsorber.

$$q = (VR)(\text{degreaser length})$$

$$= (616.2)(5)$$

$$= 3081.2 \text{ acfm}$$

3. Estimate the slot width in inches for the proposed lip hood.

$$SW = 0.03(TW)^{1.15}$$

$$= 0.03)(36)^{1.15}$$

$$= 1.85 \text{ in}$$

TPPA.13 CAPITAL RECOVERY FACTOR

The equipment (adsorber) costs for a proposed pollution prevention project were estimated to be $712,700 in October 1999. Your company has now proposed a similar type pollution prevention project for your facility in October 2006. What is the cost of the "new" project? If the total installation cost is 60% of the total equipment cost, what is the annualized capital cost of this project? Assume an annual rate of return of 10% and a operating period of 10 years. The Equipment Cost Index for 1999 and 2006 are 436.5 and 610.4, respectively.

TABLE 10.10 FECI Index

Year	Index
1994	361.3 (estimated)
1993	360.8
1992	358.2
1991	361.3
1990	357.6
1989	355.4
1988	342.5
1987	323.8
1986	318.4
1985	325.3
1984	322.7
1982	309.1
1980	289.3
1975	192.2
1970	122.7
1960	101.2

SOLUTION: Detailed cost estimates are beyond the scope of this Handbook. Such procedures are capable of producing accuracies in the neighborhood of $\pm 5\%$; however, such estimates often require many hours of engineering work. This problem is designed to give the reader a basis for a *preliminary* cost analysis only.

A simple procedure is available to estimate equipment cost from past cost data. The method consists of adjusting the earlier cost data to present values using factors that correct for inflation. A number of such indices are available; one of the most commonly used is the *Chemical Fabricated Equipment Cost Index* (FECI), past values of which are provided in Table 10.10. More recently, these values are drawn from an Equipment Index which can found in Chemical Engineering Magazine. The following formula should be employed when using the index:

$$\text{Cost}_{\text{year B}} = \text{Cost}_{\text{year A}} [\text{FECI}_{\text{year B}}/\text{FECI}_{\text{year A}}] \qquad (10.6)$$

1. Obtain the Equipment Cost Index (ECI) for 1999 and 2006.

$$\text{ECI}_{1999} = 436.5$$
$$\text{ECI}_{2006} = 610.4$$

2. Estimate the capital cost of the adsorber in 2006.

$$\begin{aligned}
\text{Cost}_{2006} &= \text{Cost}_{1999}[\text{ECI}_{2006}/\text{ECI}_{1999}] \\
&= (712{,}700)(610.4/436.5) \\
&= \$996{,}600
\end{aligned}$$

3. Calculate the installation cost, IC.

$$IC = (0.6)(\$966,600)$$
$$= \$598,000$$

4. Calculate the total capital cost, TCC.

$$TCC = CC + IC$$
$$= 996,600 + 598,000$$
$$= \$1,594,600$$

5. Calculate the capital recovery factor, CRF.

$$CRF = i(1 + i)^n / [(1 + i)^n - 1]$$

For $i = 0.1, n = 10$,

$$CRF = (0.1)(1.1)10 / [(1.1)10 - 1]$$
$$= 0.1627$$

6. Calculate the annualized capital cost, ACC, of the equipment in $/yr.

$$ACC = (TCC)(CRF)$$
$$= (1,594,600)(0.1627)$$
$$= \$259,400$$

TPPA.14 TOTAL CAPITAL COST

As part of a company's ongoing pollution prevention program, an existing appliance coating operation is to be replaced by a minibell automatic electrostatic spray system. The company estimates that the coating operation is presently operating at a 50% transfer efficiency; it has received a guarantee of 90% transfer efficiency with the new unit. The company operates 250 days/yr and processes 200 appliance units per day. An average of 2 gal of paint are presently required to coat each unit. The coating material has a VOC content of 5.0 lb VOC/gal solids. Information on the minibell system is given below. If the paint is worth $5/gal, show that an annual profit will be realized by replacing the exiting spray system with the more efficient minibell unit. Cost information on the proposed minibell system is given below.

Installed cost = $1,125,000
Operation and maintenance cost = $70,000/yr
Lifetime guarantee = 15 yrs
Interest rate for purchase cost = 8%

SOLUTION:

1. Obtain the annual operating and maintenance cost. From the Problem Statement,

$$\text{Annual O\&M cost } 70,000/\text{yr}$$

2. Calculate the annualized capital cost, ACC.

$$
\begin{aligned}
\text{ACC} &= (\text{IC})(\text{CRF}) \\
&= (\text{IC})(i)(1+i)^n/[(1+i)^n - 1] \\
&= (1,125,000)(0.08)(1.08)^{15}/[(1.08)^{15} - 1] \\
&= \$131,436
\end{aligned}
$$

3. Calculate the total annualized cost, TAC, if the minibell system is installed.

$$
\begin{aligned}
\text{TAC} &= \text{O\&M} + \text{ACC} \\
&= 70,000 + 131,436 \\
&= \$201,500
\end{aligned}
$$

4. Calculate the total annual operating savings, TAS, if the minibell system is installed. Since the new spray system operates more efficiently, a savings will be realized because of the reduced paint usage. At 50% transfer efficiency only one gallon of the two gallons of paint used per unit is actually applied. The paint usage for the minibell system, operating at 90% transfer efficiency, is $(1.0/0.9)$ or 1.11 gal. Thus, the gallons saved per unit is $2.0 - 1.11 = 0.89$ gal. The total annual savings, TAS, becomes

$$
\begin{aligned}
\text{TAS} &= (\text{gas used/number units} \cdot \text{day})(\text{operating days})(\text{cost}) \\
&= (0.89 \text{ gal})(200 \text{ units/day})(250 \text{ days/yr})(5.0 \text{ lb VOC/gal}) \\
&= \$222,500
\end{aligned}
$$

5. Calculate the annual savings.

The savings, S, can be calculated by subtracting the result of step (3) from that of step (4).

$$
\begin{aligned}
S &= 222,500 - 201,500 \\
&= \$21,000
\end{aligned}
$$

Once a particular process scheme has been selected, it is common practice to optimize the process from a capital cost and O&M (operation and maintenance) standpoint. There are many optimization procedures available, most of them too detailed for meaningful application to simple pollution prevention projects. These sophisticated optimization techniques, some of which are routinely used in the design of conventional chemical and petrochemical plants, invariably involve computer calculations. However, use of these techniques in simple pollution prevention projects and analyses is usually not warranted.

One simple optimization procedure that is recommended is the *perturbation study*. This involves a systematic change (or *perturbation*) of variables, one by one, in an attempt to locate the optimum design from a cost and operation viewpoint. To be practical, this often means that the engineer must limit the number of variables by assigning constant values to those process variables that are known beforehand to play an insignificant role. Reasonable guesses and simple or short-cut mathematical methods can further simplify the procedure. Much information can be gathered from this type of study since it usually identifies those variables that significantly impact on the overall performance of the process and also helps identify the major contributors to the total annualized cost.

TPPA.15 POLLUTION PREVENTION CREDITS

An engineer has compiled the data below for two different project options—one that includes a comprehensive pollution prevention program/option, and one that does not. From an economic point of view, which project should the engineer select? The lifetime of the equipment is 10 yrs; the interest rate is 10%.

SOLUTION:

1. For the project with pollution prevention, calculate the total capital cost, TCC (W), in $.

$$TCC(W) = 1,294,000 + 786,000$$
$$= \$2,080,000$$

TABLE 10.11   Pollution Prevention Cost Data

	Project with Pollution Prevention (W)	Project w/o Pollution Prevention (WO)
Equipment cost	$1,294,000	$1,081,000
Installation cost	$786,000	$659,000
Operating labor	$39,900/yr	$8,500/yr
Maintenance	$43,000/yr	$17,000/yr
Utilities	$958,000/yr	$821,000/yr
Overhead	$51,300/yr	$13,900/yr
Taxes, insurance, and administration	$86,200/yr	$72,600/yr
Credits	$380,000/yr	$0

2. Calculate the capital recovery factor, CRF (W).

$$CRF(W) = i(1 + i)^n/[(1 + i)^n - 1]$$
$$= (0.1)(1.1)^{10}/[(1.1)^{10} - 1]$$
$$= 0.1627$$

3. Calculate the annualized capital cost, ACC (W).

$$ACC(W) = (2,080,000)(0.1627)$$
$$= \$338,500$$

4. For the project without pollution prevention, calculate the total capital cost, TCC (WO), in \$.

$$TCC(WO) = 1,081,00 + 659,000$$
$$= \$1,740,000$$

5. Calculate the capital recovery factor, CRF (WO).

$$CRF(WO) = i(1 + i)^n/[(1 + i)^n - 1]$$
$$= CRF(W)$$
$$= 0.1627$$

6. Calculate the annualized capital cost, ACC (WO).

$$ACC(WO) = (1,740,000)(0.1627)$$
$$= \$283,200$$

7. Which project is the better choice? Based on the results calculated and presented in Table 10.12, the project with pollution prevention is the better choice.

TABLE 10.12   Pollution Prevention Cost Results

	Project with Pollution Prevention	Project w/o Pollution Prevention
Annualized capital cost	\$338,500/yr	\$282,300/yr
Operating labor	\$39,900/yr	\$8,500/yr
Maintenance	\$43,000/yr	\$17,000/yr
Utilities	\$958,000/yr	\$821,000/yr
Overhead	\$51,300/yr	\$13,900/yr
Taxes, insurance, and administration	\$86,200/yr	\$72,600/yr
Credits (see Table 10.11)	− \$380,000/yr	\$0
Total annual cost	\$1,136,900/yr	\$1,215,300/yr

TPPA.16   CLEANING METAL PARTS: TIME VALUE OF MONEY

A firm cleans metal parts with a chlorinated solvent that is hazardous to workers (see Figure 10.8). In addition, the wastewater that results from rinsing the parts must be treated before it can be discharged to the environment. Because of regulatory concerns the company is considering ways to reduce the volume of wastewater generated. The firm's current costs for parts cleaning are provided in Table 10.13.

Calculate the total cost of the current cleaning system over the next 10 year based on a 10% interest rate.

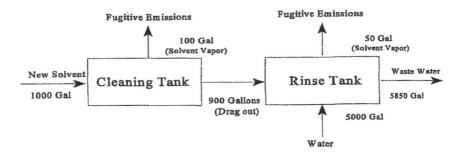

FIGURE 10.8   Annual Material Balance for Cleaning Metal Parts Process.

TABLE 10.13   Current Costs for Parts Cleaning

Item	Cost/Unit	Units	Cost/Year
Solvent	$3.25/gal	1,000 gal	$3,250.00
Water	$2.10/1,000 gal	5,000 gal	$10.50
Waste treatment	$2.50/gal	5,850	$14,625.00
		Total Annual Cost	$17,885.50

SOLUTION: Although the first step would be to examine expected business changes such as business expansions, new accounts, rising prices, etc. For simplicity, the costs and volumes from Table 10.13 are assumed to be constant. This means that the current annual costs will be same in the other years except for one very important aspect, the time value of money.

Due to the assumptions made regarding constant cost, the $17,885 total annual cost shown in Table 10.13 can be assumed to repeat each year. The present value calculations shown earlier enable this annual expenditure to be expressed as a single sum which includes the effects to interest. The first year's cost, assuming the bills are paid at the end of the year, would be the amount of money that would have to be banked starting today, to pay a $17,885 bill in one year. Using a 10% interest rate, the calculation is as follows.

$$P = \frac{\$17,885}{(1 + 10)^1} = \$16,260$$

TABLE 10.14 Present Value Calculations

Year	Expenditure	Present Value
1	$17,885	$16,260
2	$17,885	$14,781
3	$17,885	$13,437
4	$17,885	$12,216
5	$17,885	$11,105
6	$17,885	$10,096
7	$17,885	$9,178
8	$17,885	$8,343
9	$17,885	$7,585
10	$17,885	$6,895
Total		$109,896

This means that if $16,260 is banked at 10% interest, it would provide enough money to pay the $17,885 bill at the end of the year. Similarly, the second, third, fourth, etc., years expenditures can also be expressed in present value. The results of this calculation are provided in Table 10.14.

The bottom line to the analysis is that the total cost of the current cleaning system over the next 10 years, given a 10% interest rate, is $109,896 in present value terms. In other words, $110,000 invested today at 10% interest would be sufficient to pay the entire material and disposal costs for the parts cleaning operation for the next 10 years. Hence, any changes to the operation of the firm can now be compared to this $110,000 baseline. Any change which would result in a lower 10 year cost would be a benefit because it would save money; any option with a higher cost would be more expensive and should not be adopted from a financial or economic standpoint.

TPPA.17 RECHARGEABLE BATTERIES

One non-point source of environmental pollution has been the introduction of heavy metals, particularly mercury and cadmium, into landfills from the casual disposal of household nonrechargeable batteries. The major use of such batteries was once the type D size found in flashlights, but with the popularization of portable radios and tape players, Type AA batteries have far surpassed the Type D's. However, battery manufactures will now take back the spent batteries for recycle and/or disposal in a safe manner. Most communities have one or more collection sites—the town library, schools, etc.—where the public can turn in the hazardous waste.

A better solution (the authors believe) to the battery problem is to switch to rechargeable batteries which can be reused over and over, lasting for years. These do require a charging device at some initial cost and a convenient 120 volt outlet for recharging. An even better system, one which is sustainable, employs the sun

TABLE 10.15   Rechargeable Battery Data

	Conventional AA	Rechargeable Ni-Cd AA
Number required:	4	8
Cost (each):	$0.89	$2.75
Rotation frequency:	Monthly	Monthly
Solar Recharge Cost:	0	$14.00
Lifetime:	One month[1]	5 yr (1,000 charges)[2]

to do the recharging to eliminate the middle person, the power company, Again, a special solar charger is a necessity as is, of course, a sunny day or two.

Calculate the annual cost for the use of the Rechargeable Nickel-Cadmium AA battery and compare it to the annual cost for conventional AA batteries. They are to be used in a portable radio which uses 4 such batteries. The rechargeable batteries are recharged every month with a solar battery charger, which is not designed to get wet, and which requires two to four days to recharge the batteries. What is the payback time, if any? The payback time as defined earlier is the time required to realize (recover) the principal of an initial investment.

Data is provided in Table 10.15.

SOLUTION:

1. Calculate the annual cost for Conventional AA batteries.

$$(4\,\text{batteries}/\text{mo})(12\,\text{mo}/\text{yr})(\$0.89/\text{battery}) = \$42.72/\text{yr}$$

2. Calculate the cost for Rechargeable Ni-Cd AA batteries.

$$(8\,\text{batteries})(\$2.75/\text{battery}) + (1\,\text{charger})(\$14/\text{charger}) = \$36.00$$

3. Calculate the annual cost for Rechargeable Ni-Cd batteries for the first five years.

$$(\$36.00)/(5\,\text{yr}) = \$7.20/\text{yr}$$

4. Calculate the payback period (in months) of the original capital cost of the Ni-Cd batteries.

$$\$36/[(4\,\text{batteries}/\text{mo})(\$0.89/\text{battery})] = 10.1\,\text{months}$$

After 10 months, the solar-rechargeable system is free, at least until replacement is required. If the claim of 1,000 charges is accurate, then 5 yr is a conservative estimate. Obviously, if they last more than 5 yr, then the economics improve. Assuming the batteries last for 1,000 charges and the charger lasts indefinitely (not unreasonable), then the cost for four "new" batteries becomes less than $0.04, or a penny or two each. The cost of conventional batteries will not change (assuming no fluctuation in price).

As this problem illustrates, it makes good economic and environmental sense to switch to solar rechargeable batteries. There will be some inconvenience due to the necessity of charging the spent batteries, but not unlike having to run to the store (more wasted energy) to buy new conventional batteries. Of course, if the battery charger is left out in the rain, one may need a new one. A way to enhance the economic picture is to recognize that the acts of not polluting and of not using resources is worth money. However, it is difficult to place a dollar value on this. Finally, it is important to remember that eventually even the rechargeable batteries will need to be replaced and the spent batteries should not be disposed of in the household trash; rechargeable batteries should be "disposed" of in an environmentally safe manner.

TPPA.18 WEIGHTED SUM METHOD

The Weighted Sum Method is a semi-quantitative method for screening and ranking process and design options. This method can and has been applied in quantifying the important criteria that affect waste management at a particular facility. This method involves three steps.

1. Determine what the important criteria are in terms of the program goals and constraints, and the overall corporate goals and constraints, Example criteria as applied to pollution prevention are (see earlier Problems):
   a. reduction in waste quantity
   b. reduction in waste hazard (e.g., toxicity, flammability, reactivity)
   c. reduction in waste treatment/disposal costs
   d. reduction in raw material costs
   e. reduction in liability and insurance costs
   f. previous successful use within the company
   g. previous successful use in industry
   h. not detrimental to product quality
   i. low capital cost
   j. low operating and maintenance costs
   k. short implementation period with minimal disruption of plant operations
   l. improved public relations
   m. reduced workman's compensation
   n. improved employee morale
   o. reduction or elimination of liability
   p. reduction or elimination of regulatory concerns

A weight factor for each criteria is assigned. This is defined as the *weight of the criteria*. The weights (on a scale of 0 to 10, for example) are determined for each of the criteria in relation to their importance. For example, if reduction in waste treatment and disposal costs are very important, while previous success-ful use within the company is of minor importance, then the reduction in waste costs is given a high weight of 9 or 10 and the previous use within the company is given a low weight of either 1 or 2, Criteria that are not important are not included or are given a weight of 0.

2. Each criteria is then rated on effectiveness for the various options to be inves-tigated. These are defined as *effectiveness factors*. Again a scale of 0 to 10 can be used (0 for low and 10 for high).

3. Finally, the *effectiveness factor* (step 2) for a particular criterion is multiplied by the *weight of the criterion* (step 1). An option's *overall rating* is the sum of the products of the *effectiveness factor* and the *weight of the criterion*.

The option(s) with the best overall rating(s) is (are) then selected for further tech-nical and/or economic feasibility analyses.

The LAST (Leo and Stander Theodore) Corporation has instituted a new pollution prevention program that is geared to reduction and/or eliminating problems for a known element associated with the operation of hazardous/toxic pollutants at one of their facilities. Three process options are under consideration:

X: Purchase sophisticated control equipment and provide modest and supply operator training.

Y: Purchase control equipment and provide in-house operator and full training.

Z: Retrofit existing control equipment and provide advanced incineration operator training.

Based on the data provided below, you have been asked—as the environmental engineer assigned to this project—to use the weighted sum method to determine which option is most attractive.

LAST Corporation has determined that reduction in waste treatment costs is the most important criterion, with a weight factor of 10. Other significant criteria include reduction in safety hazard (weight of 8), reduction in liability (weight of 7), and ease of implementation (weight of 5). Options X, Y, and Z have also been assigned effectiveness factors. Option X is expected to reduce waste by nearly 80%, and is given a rating of 8. It is given a rating of 6 for safety hazards, 4 for reducing liability, and 2 for ease of implementation. The corresponding effectiveness factors for Y and Z are 6, 3, 4, 2, and 3, 8, 5, 8, respectively.

SOLUTION:

1. Generate the overall rating for option X. See step (3) above and proceed to calculate OR_x.

$$OR_x = (10)(8) + (8)(6) + (7)(4) + (5)(2)$$
$$= 166$$

TABLE 10.16   Weighted Sum Method Result

Criteria	Weight	Effectiveness Factors X	Y	Z
Reduce treatment costs	10	8	6	3
Reduce safety hazards	8	6	3	8
Reduce liability	7	4	4	5
Ease of implementation	5	2	2	8
Sum of Weight times Effectiveness Factors		166	122	169

2. Generate the overall rating for option Y.

$$OR_y = (10)(6) + (8)(3) + (7)(4) + (5)(2)$$
$$= 122$$

3. Generate the overall rating for option Z.

$$OR_z = (10)(3) + (8)(8) + (7)(5) + (5)(8)$$
$$= 169$$

Generate a table illustrating the overall ratings for the three options. See Table 10.16. Compare the results for options X, Y, and Z. From this screening, option Z rates the highest with a score of 169. Option X's score is 166 and option Y's score 122. In this case, both option Z and option X should be selected for further evaluation because their scores are high and close to each other. Consideration should also be given to combining the "best" features of options X and Z.

This is a procedure that has been recommended by the EPA for screening and ranking pollution prevention options. It may also be used in the analyses of other environmental projects. In 1979, one of the authors of this Handbook (as part of a USDOE project) developed a similar weighing method to analyze options of fine particulate control (air pollution) for coal-fired boilers.

APPENDIX

There are three sections to the Appendix: International Regulations, ISO 14000, and Miscellaneous Topics. The first two sections contain an Introduction that is followed by several problems with solutions. The last (third) section addresses seven topics: State Regulatory Agency Names, Federalism and Preemption, Hybrid Systems, EMFs, Life Cycle Analysis, Sustainable Development, and Environmental Justice. There are no problems in this last section.

Environmental Regulatory Calculations Handbook, by Leo Stander and Louis Theodore
Copyright © 2008 John Wiley & Sons, Inc.

APPENDIX A

INTERNATIONAL REGULATIONS

In many parts of Eastern Europe, the former Soviet Union, and the developing countries, in Asia, Africa and South America, pollution conditions persist today. Even in Western Europe, where air pollution is now far less visible than it was in Charles Dickens' time, poor air quality contributes to hundreds of thousands of deaths and millions of ailments each year. Similarly it endangers forests and lakes, and corrodes structures. Industrial activities are increasingly emitting pollutants of worrisome toxicity. Millions of tons of carcinogens, mutagens, and poisons are released into the environment each year, damaging health and habitats near their sources and sometimes thousands of miles away by way of wind currents.

The environment also affects the physical health of people all over the world. In greater Athens, the number of deaths rises sixfold on heavily polluted days. In Hungary, the National Institute of Public Health concluded that every 24th disability and every 17th death is caused by smoking 10 cigarettes a day. In Mexico, the capital has been declared a hardship post for diplomats because of its air pollution levels, some governments advise women not to have children while being posted there.

In an effort to resolve these problems, countries are trying to focus on the source of these pollutants. The focus has primarily been on the greenhouse effect, the depletion of the ozone layer, the increase of acid rain, health hazards, and a country's perspectives on the problem. Each of these topics is discussed below.

Additional and more current information on international regulations is available from the World Health Organization (WHO) and other international organizations.

Environmental Regulatory Calculations Handbook, by Leo Stander and Louis Theodore
Copyright © 2008 John Wiley & Sons, Inc.

GREENHOUSE EFFECT

The sources of the world's energy include solar, nuclear and geothermal power, and the burning of fossil fuels. Some fossil fuels are oil, gasoline natural gas, coal, and wood. Burning any of these substances uses up oxygen and produces carbon dioxide gas. Humans also produce and exhale carbon dioxide. Plants, algae, and plankton, on the other hand, take in carbon dioxide and produce oxygen. Unfortunately, modern industrial society and its need for electric power produces far more carbon dioxide than the planet's vegetation can consume. As this excess carbon dioxide rises into the atmosphere, it acts as a one-way mirror, trapping the heat reflected from the earth's surface. Many leading scientists expect that this "greenhouse" effect from increased levels of carbon dioxide and other heat-trapping gases will cause an increase in global temperatures of 5°C by the middle of this century. This rise in temperature should result in increased cloudiness, for warmer air is capable of absorbing more moisture. During the day, the additional clouds could offset the greenhouse effect by shielding the surface of the earth from the sun's rays. But at night, those same clouds could serve as a kind of atmospheric blanket, preventing heat from being radiated away.

The rise in temperature can also endanger human settlements in low-lying coastal areas in all parts of the world as well as destroy coastal wetlands. Some climatic models suggest that temperatures would increase and rainfall would decrease simultaneously in the central plains of the U.S., with significant soil moisture decreases and agricultural implications. Global warming could also result in an expansion of the earth's arid zones and increased desertification. Overall, the projected climatic changes would be expected to have major impacts on earth's ecosystem, water resources, air quality, vegetation, and biological systems, as well as on human beings themselves.

Infrared radiation from carbon dioxide (CO_2) escapes from the stratosphere rather than being trapped in the lower atmosphere. Increased CO_2 thus leads to lower stratospheric temperatures, which can alter chemical reaction rates and atmospheric dynamics. Similar remarks apply to changes in concentration of all other "greenhouse" gases such as CO, N_2O, CH_4, and halocarbons that may be present.

Additional and more current information on the Greenhouse Effect is available in a series of 4 reports developed by the International Panel on Climate Change entitled: "IPCC Third Assessment Report – Climate Change 2001". This report is available on line at: http://www.grida.no/climate/ipcc_tar/index.htm. Copies of diagrams or pictures can be perused and reproduced from this same website.

OZONE

An discussed in Chapter 3, ozone plays several extremely important roles in the atmosphere. First, ozone absorbs virtually all solar ultraviolet radiation between wavelengths of about 240 and 290 nm which would otherwise be transmitted to the surface. Such radiation is lethal to simple unicellular organisms and to the surface cells of higher plants and animals. Ultraviolet radiation in the wavelength range 290 to 320 nm (so-called UV-B) is also biologically active and prolonged exposure to it may cause skin cancer in susceptible individuals. It should, however, be pointed out that the kinds of cancer that can definitely be attributed to sunlight (basal cell and squamous cell cancers) are not terribly dangerous since, if caught in time, they may be successfully treated. However, there are also much more dangerous cancers of the skin, e.g., the melanomas. These are relatively rare and are usually found on parts of the body that are not exposed to sunlight. Ozone absorption also provides a significant energy source for driving the circulation of the mesosphere and forcing tides in the upper mesosphere and thermosphere. Second, upper atmospheric meteorology is greatly influenced by the heating that follows absorption by ozone of UV, and visible and thermal IR radiation. The ozone layer, however, is being destroyed by chlorofluorocarbons.

Ozone is formed in the stratosphere when ultraviolet radiation splits diatomic molecules of oxygen (O_2) into two atoms. Oxygen atoms combine with diatomic molecules of oxygen to produce ozone (O_3). The reaction is as follows:

$$2O + 2O_2 \rightarrow 2O_3$$

The ozone molecule is then broken down by ultraviolet radiation to form the original diatomic molecule of oxygen and an atom of oxygen (O). Under natural conditions there is an equilibrium between the creation and the destruction of ozone.

This equilibrium was recognized as shifting when it became apparent that stratospheric ozone concentrations have been declining over the past decades. The dramatic depletions of ozone over the Antarctic each year (noted as "Antarctic ozone hole") could only be explained in terms of chemistry perturbed by the release of halogen-containing compounds. The discovery and the reality of the Antarctic ozone hole came as a complete surprise to atmospheric scientists. Their models had failed to predict it and its magnitude was initially inexplicable. A number of contending theories were proposed to explain its formation. Some suggested a chemical and anthropogenic cause, while some suggested natural changes in polar atmospheric circulation.

These changes were known but not implemented until the Rowland-Molina theory emerged. This theory of ozone depletion was responsible for the banning of CFC use as aerosol propellants by the U.S., Canada, and Sweden in 1978. Most other developed countries that used CFCs industrially and commercially viewed the Rowland-Molina hypothesis as an unverified theory and continued their use without any limits.

In the late 1970s, Rowland put forth another proposal that the O_3 layer may also be threatened by the increasing release of N_2O to the atmosphere from denitrification processes of ammonium and nitrate-based fertilizers. In this new theory, N_2O would migrate to the stratosphere where it would be photolytically destroyed to produce NO. Nitrogen oxide is an O_3-destroying chemical associated with previous concerns about the O_3-depleting potential of high-speed, high-altitude supersonic transport (SSTs) and atmospheric nuclear testing.

Up until 1985 everything was theory or experimentally based. No "real life" situations were discovered. In 1985 ozone depletion moved to the stark reality of the "Antarctic Ozone Hole". British scientists working in Hally Bay (76°S latitude) had observed from ground-based measurements significant declines of stratospheric ozone during October. Such declines began in the late 1970s and accelerated sharply thereafter. An examination of satellite data confirmed these observations revealing that the area of ozone depletion extended over several millions of square miles. Ozone declines over the Antarctic were particularly large during the austral spring of 1987. O_3 levels declined to 50% of their 1979 levels, with depletions of as much as 95% at altitudes of 9 to 12 miles. The ozone hole is a seasonal phenomenon usually lasting 5 to 6 weeks or so at the beginning or the austral spring. These changes in O_3 levels over the Antarctic were associated with the ozone hole.

Continued studies that were conducted in 1986 and 1987 demonstrated the presence of high concentrations of chlorine ion (CI) and chlorine oxide (CIO) in the region of the ozone hole. These studies verified that the ozone hole was caused by the chemistry of C1 atoms in conjunction with the unique meterology of the Antarctic. Ozone destruction in this region occurs as a consequence of complex heterogeneous chemistry involving polar stratospheric clouds, hydrochloric acid, chlorine nitrate, and chlorine.

After the experiments on CFC, the Western nations attempted to control ozone depletion in 1974 by controlling the amount of CFC produced. They went from producing 5 to 10 times as much CFCs as the rest of the world to a production that has been steady and then falling. Legislation was passed in the 1970s forbidding certain uses of the CFCs in the U.S. As a result, production in 1981 was more than 20% less than it had been in 1974. A much more wide-ranging control was embodied in the "Montreal Protocol on Substances that Deplete the Ozone Layer" that was agreed upon in September 1987, and entered into force in January 1989. Each party to the Protocol was to freeze and then reduce, according to an agreed timetable, its production and consumption of five CFCs: $CFCl_3$ (trichlorofluorocarbon), CF_2Cl_2 (dichlorofluorocarbon), CF_2CIFCl_2, CF_2CICF_2Cl and CF_2CICF_3 (1,2-dichlorfluoro carbon). It would also freeze consumption and production of the halons.

There is also a commitment by industry to develop products and processes that do not use CFCs and to share these substitutes with other countries. Many nations now evaluate all possible substitutes to make sure they do not present new health or environmental problems. This threat to the ozone layer illustrates an important principle: one cannot simply outlaw an environmental problem, one must work toward an acceptable solution that is both comprehensive and economical.

Additional and more current information on ozone is available. A report prepared under the auspices of World Meterological Organization (WMO) and the United Nations Environment Programme (UNEP) entitled "Scientific Assessment of Ozone Depletion: 2002" is available on the web at: http://esrl.noaa.gov/csd/assessments/2002/ Information on the Montreal Protocol and subsequent amendments can viewed at http://www.unep.ch/ozone/index.shtml.

ACID RAIN

All rainfall is by nature somewhat acidic. Decomposing organic matter, the movement of the sea, and volcanic eruptions all contribute to the accumulation of acidic chemicals in the atmosphere. However, the principal factor that contributes to acid rain is atmospheric carbon dioxide, which causes a slightly acidic rainfall (pH of 5.6) even in the most pristine of environments. In some parts of the world, the acidity of rainfall has fallen well below 5.6.

It is estimated that 100 million tons of SO_2 are emitted worldwide every year from coal- and oil-fired power stations, industries that consume fossil fuels, and smelters. The geographical concentration of SO_2 emissions varies, with the highest levels originating in the industrial centers of Europe, North America, and the Far East. However, according to the U.S. National Research Council, this is about one-half the sulfur dioxide (SO_2) in the atmosphere.

It is estimated that Athenian monuments have deteriorated more in the past 50-years from pollution than in the previous 2400 years. Damage to historical artifacts and edifices also is evident throughout Italy. In the Katowice region of southern Poland, trains must slow down in certain places because the railway tracks have corroded, apparently from acid rain.

The Third World is following the example of the Free World. The Taj Mahal appears to be endangered by emissions from an upwind oil refinery that are eroding its marble and sandstone surfaces. Recent research found that acid rain falling on the Yucatan Peninsula and much of southern Mexico is similarly destroying the temples, murals, and megaliths of the Mayans.

Since the initial alarm in West Germany where signs of widespread forest damage from acid deposition rose from 8% in 1982 to 34% in 1983, concern about forest damage has spread throughout the world. In Europe, several countries have initiated annual surveys. The results are now brought together in a yearly assessment by the U.N. Economic Commission for Europe.

Damage in China's southwest forests is being increasingly linked by scientists to acid rain caused by a heavy reliance on high-sulfur coal. In Sichuan's Maocaoba pine forest, more than 90% of the trees have died. On Nanshan Hill in Chongqing (Chungking), the largest city in southwest China, a forest of dense masson pine has been reduced by almost half. Both these regions have highly acidic rain and elevated levels of sulfur dioxide. China reported in May 1989 that acid rain is causing serious damage in Hunan Province as well, including crop losses worth about $260 million. China has increased coal output more than 20-fold between 1949 and 1982 and planned to double consumption by the end of the 20th century. In India, SO_2 emissions from coal and oil nearly tripled between the early 1960s and the late 1970s. Growing urbanization in much of the Third World means that 'ever-increasing' numbers of people are being exposed to polluted city air.

With problems come possible solutions. One of the possible solutions is to use lower-sulfur fuels. This is often the cheapest way to reduce sulfur dioxide emissions as the purchase of technological controls is a great deal more expensive. The net effect of fuel conversion by many consumers could lead to shifts in production and employment away from the traditional higher-sulfur coal regions.

Although the material discussed above was published in 1981, much of this is still applicable. The reader is referred to Chapters 2 and 3 for additional material on acid rain.

HEALTH HAZARDS

Pollutants have also been known to endanger human beings within days of exposure. These pollutants include toxic metals and chemical emissions. Evidence suggests that toxic chemicals emitted into the air can be carried great distances before falling to the ground, a phenomenon already well known in the case of acid rain. These toxic chemicals are the cause for a variety of human ailments and property damage. Other studies have found high cancer rates in communities near certain types of factories. Measurements of lead and cadmium in the soil of the upper Silesian towns of Olkosz and Slawkow in Poland are among the highest ever recorded anywhere in the world. Their government is considering a ban on growing vegetables in several Silesian towns due to soil concentrations of cadmium, lead, mercury, and zinc that are 30 to 70% higher than the World Health Organization (WHO) norms.

There have been many local air pollution situations that have been devastating. One of these occurred December 4 to 10, 1952, in London, England. On December 4, a high-pressure area began to center on the city, shrouding it in several layers of clouds. The unburned particles of coal floated into the sky from thousands of chimneys. Visibility was only a few feet and the humidity had risen to nearly 100%. As the smoke accumulated, coughing was heard everywhere in the city. Conditions worsened on December 7. Patients with respiratory diseases crowded the London hospitals, and many died. On December 9 the high-pressure area finally began to move and, with the wind blowing fairly steadily, the skies began to clear. By December 10 a cold front had passed over the area bringing fresh, clean air from the North Atlantic. The emergency was over, but during those few days it was estimated that 4000 Londoners died.

On July 1976, when a stable air mass had developed over Milan, Italy, a chemical plant at nearby Seveso accidentally released a cloud of highly toxic dioxin (tetrachlorodibenzo-p-dioxin) into the atmosphere. The pollutant remained in the area for about 3 weeks, forcing the evacuation of 700 people, at least 500 of whom exhibited symptoms of poisoning. Pregnant women who were affected were advised to have abortions because the poison causes malformations in fetuses. About 600 animals were poisoned and had to be destroyed. All contaminated crops had to be burned. Medical experts recommended that all residents of the area have periodic medical examinations for the rest of their lives.

Toxic chemical emissions are likely to rise rapidly in developing countries as industrialization continues. The Third World's share of global iron and steel production continues to rise. In India, pesticide production has significantly increased. Production of dyes and pigments has grown at a comparable pace. Most of these countries have few pollution controls and token environmental regulations. The reader is referred to the literature — A. M. Flynn and L. Theodore, "Health and Safety in the Chemical Process Industries," Marcel-Dekker (recently acquired by CRC/Taylor & Francis Group, Boca Raton, FL), New York City, 2000.

INTERNATIONAL PERSPECTIVES

Among the solutions to control pollutants are efforts to provide incentives to countries decreasing the use of pollutants. One of these incentives is to develop a more environmentally benign energy system depending critically on correcting market imperfections. The first step is to ensure that subsidies are removed so that the cost of energy reflects its true value, providing an incentive to use less. Centrally planned or controlled economies in eastern Europe, the former Soviet Union, and the developing world have the potential to make particularly rapid strides in this direction as they move toward market-based economies. In China, for example, energy efficiency has improved since the country's economic reform program began in 1979. As energy use per unit of gross national product declines, so does the amount of pollution.

Environmental pollution and its damaging health and ecological effects have proliferated around the globe, crossing borders with the wind, international cooperation is critical. Nations that receive the bulk of their pollution from other countries have an obvious interest in sharing and financing the technical means to reduce that pollution. International negotiations within the framework of both the Economic Commission for Europe and the European Community, as well as the ongoing negotiations on global warming, offer invaluable opportunities to trade insights and experience.

More recently, the European Union (EU) has established directives which are to be adopted by the various countries. Based on a review of some of the various country regulations, that appears to have been accomplished.

PROBLEMS

A.1
Which is the biosphere and what are biogeochemical cycles?

SOLUTION: The biosphere is defined as that part of the planet that sustains life. It encompasses the lower part of the atmosphere, the hydrosphere (oceans, lakes, rivers and streams), and the lithosphere (the earth's crust) down to a depth of approximately 2 km.

Biogeochemical cycles are transport pathways, and the chemical and physical interactions of the elements within and among these regions of the biosphere.

A.2

What specific international regulations exist on the production of ozone-depleting gases?

SOLUTION: The production of ozone depleting gases is regulated under a 1987 international agreement known as the "Montreal Protocol on Substances that Deplete the Ozone Layer". The protocol established legally binding controls on the national production and consumption of ozone-depleting gases. Since 1987, as scientific basis of ozone depletion became more certain and substitute gases became available, the Montreal Protocol was strengthened with Amendments and Adjustments. The following is a list of the major protocol adjustments:

> London (1990)
> Copenhagen (1992)
> Vienna (1995)
> Montreal (1997)
> Beijing (1999)

A.3

Discuss hazardous waste incineration emission regulations in Europe.

SOLUTION: On March 23, 1992, the European Economic Community (EEC) (today the European Union) issued a proposal for a Council Directive on the Incineration of Hazardous Waste. The directive required all its member states to establish laws, regulations, and administrative procedures to comply with the directive. The directive stipulated that any new incinerator must comply immediately and existing facilities were to comply by June 30, 1997.

The regulatory approach adopted in the 1992 EEC directive established a wide array of continuous emission monitoring requirements, including continuous monitors for carbon monoxide and total dust emission levels, and monthly measurements for metals, dioxins, and furans. A summary of European guidelines and limits for PM and dioxins/furans is presented in Table A.1.

Waste incineration has been in use in Europe longer than in North America. The air pollution control device (APCD) systems are similar. However, because the majority of the European facilities have undergone retrofits and have faced more stringent emission standards, design differences exist. Incinerators in Europe currently incorporate some sort of particulate control device, such as wet or dry electrostatic precipitators (ESP), or fabric filters (FF). Most facilities have added multistage wet and dry scrubbers or spray drying plus dry absorption processes for controlling acid gas and heavy-metal emissions. The future trend is expected to be toward wet scrubbers, even though all APCD systems must be zero liquid discharge systems. Some new technologies that are emerging include adding selective catalytic reduction DeNO$_x$ reactors, activated carbon filters, and gas suspension absorbers. As new options arise, it appears that the general practice in Europe is to continue to retrofit facilities with new APCDs in series with existing equipment.

TABLE A.1   Emission Guidelines/Limits for Waste Incineration in Europe

Pollutant (daily average)	EEC guideline	Netherlands limit	Germany limit
Total dust (mg/m^3)	5	3	5
Dioxin/furan $(ngTEQ/m^3)^{[1]}$	0.1	0.1	0.1

[1]It is important to note that the European guideline or limit of 0.1 TEQ (toxicity equation quantity) is corrected to 11% oxygen, and compliance is based on daily averaging. EPA requires that dioxin/furan emissions be corrected to a stack gas oxygen level of 7%. A 0.1 limit at a 11% oxygen correction factor is equivalent to a 0.14 limit at a 7% correction factor. Further, EPA requires hazardous waste burning devices operating under RCRA regulations to comply with emissions standards generally on a hourly rolling averaging period. The European guidelines/limits are based on daily averaging, a less stringent approach in terms of operation variability. Finally, RCRA regulations require a facility to comply with the emissions standard for each of three triplicate runs during a Trial Burn or Compliance Test. Compliance with the European guideline/limits is based on the average of test runs.

The latest EU directive for Hazardous Waste Incinerators is Directive 94/67/EC and was adopted December 16, 1994. The web site with the standards can be reviewed at http://eur-lex.europa.eu/LexUriServ/LexUriServ.do?uri = CELEX: 31994L0067:EN:HTML. The new standards differ slightly from those noted above.

A.4

A risk assessment tool commonly used in Great Britain is the Fatal Accident Rate (FAR), a term that represents the accidents per 1,000 workers in a working lifetime. A responsible chemical company typically assigns a FAR = 2 for chemical process risks such as fires, toxic releases or spillage of corrosive chemicals. Identify potential problem areas that may develop for a company if acceptable FAR numbers are exceeded.

SOLUTION: If acceptable FAR numbers are not maintained within a company, a lack of concern for worker health and safety becomes apparent. From this lack of concern for health and well-being of its employees, a number of items will likely occur. These include:

1. Legal action against the company by those affected.
2. Adverse publicity by the media against the company.
3. Adverse community relations.
4. Potential notices of violations by appropriate regulatory agencies.
5. Advisory actions regarding permit compliance.
6. Employee safety concerns and discontent.
7. Increased employee turnover.
8. Increased insurance costs.
9. Decreased product quality.
10. Decreased profits.

The reader is referred to Chapter 9 for additional details.

A.5

You are advising an administrator who has a limited budget for the mitigation of hazards in a certain chemical plant in Great Britain. The plant employs two kinds of workers: day employees who work one 8 h shift daily, and shift employees who rotate through three 8 h shifts each day. A report reveals two kinds of accidents are possible during plant operation. Accidents of the first kind result in the death of one day employee per incident and occur with a frequency of 2.92×10^{-5} accidents/yr. Accidents of the second type result in the deaths of 100 shift workers per incident and occur with a frequency of 8.76×10^{-7} accidents/yr.

1. Calculate the Fatal Accident Rate (FAR) for the two kinds of accidents in units of deaths/1000 worker lifetimes (10^8 h).
2. What considerations would influence your advice on the allocation of funds to reduce these hazards?

SOLUTION: For accidents of the first type:

$$FAR = (1 \text{ fatality/accident}) (2.92 \times 10^{-5} \text{ accidents/yr}) (1 \text{ yr}/365 \text{ d})$$

$$(1 \text{ d}/8 \text{ h}) (10^8 \text{ h}/1000 \text{ worker} \cdot \text{lifetimes})$$

$$= 1 \text{ fatality}/1000 \text{ worker} \cdot \text{lifetimes}$$

For accidents of the second type:

$$FAR = (100 \text{ fatalities/accidents}) (8.76 \times 10^{-7} \text{accidents/yr}) (1 \text{ yr}/365 \text{ d})$$

$$(1 \text{ d}/24 \text{ h}) (10^8 \text{ h}/1000 \text{ worker} \cdot \text{lifetimes})$$

$$= 1 \text{ fatality}/1000 \text{ worker} \cdot \text{lifetimes}$$

While recognizing that any answer proposed for this question is likely to be incomplete, the following is suggested as thought provoking:

FARs for the two kinds of accidents are equal and over a long period of time, it is likely that an equal number of deaths may be expected from both types of accidents. This does not mean that equal consequences will result to the company from the two kinds of accidents. Accidents of the first kind involve a low (but, perhaps, steady) loss of life. Public opinion and the press are likely to grow tired and apathetic to this loss of life. Accidents of the second type, however, are sure to attract a great deal of attention in the media; adverse public reaction is nearly certain, as well as unfavorable attention from public officials.

Secondly, accidents of the second kind have a catastrophic effect on production. Not only will the entire facility be demolished (in all likelihood), but a large fraction of the pool of trained personnel will be lost all at once. Who will train the replacement personnel if everyone is lost in the disaster? Similar concerns make disruption in the local community much greater for accidents of the second kind. Some of these considerations can be factored into decision making as direct economic losses that increase the burden to the company and to the community for accidents of the second kind.

One view, to many the only acceptable one, is to give highest priority to prevention of both kinds of accidents.

APPENDIX B

ISO 14000

The acronym ISO stands for *International Organization for Standardization*. It is a worldwide program that was founded in 1947 to promote the development of international manufacturing, trade, and communication standards. ISO membership includes over 100 countries. The American National Standards Institute (ANSI) is the U.S. counterpart to ISO and is the U.S. representative to ISO.

ISO essentially receives input from government, industry, and other interested parties before developing a standard. All standards developed by ISO are voluntary; thus, there are no legal requirements to force countries to adopt them. However, countries and industries often adopt ISO standards as requirements for doing and maintaining business.

ISO develops standard in all industries except those related to electrical and electronic engineering. Standards in these areas are developed by the Geneva-based International Electrotechnical Commission (IEC), which has more than 40 member countries, including the United States.

The purpose and goal of ISO is to improve the climate for international trade by "leveling the playing field." The concept is that by encouraging uniform practices around the world, barriers to trade will be reduced. If the management processes of companies in any country could be compared more readily with the management processes of companies in any other country, then international trade would be made simpler.

ISO 14000 is a voluntary standard for environmental management systems. It does NOT require compliance with the regulations of the country in which the company is located. In some countries, it is possible that regulations may be

Environmental Regulatory Calculations Handbook, by Leo Stander and Louis Theodore
Copyright © 2008 John Wiley & Sons, Inc.

more stringent than the standard. It seems likely, however, that in some countries achieving certification of adherence to the standard would improve the quality of environmental practices in that country. If, as expected, many countries adopt laws that require imported products to have been produced by companies certified to be adhering to ISO 14000, then environmental practices will almost certainly be improved worldwide.

ISO 14000 describes in considerable detail what a company must do without prescribing how it must or can be accomplished. When completed, ISO 14000 will be comprised of approximately 20 components. It will be sufficiently specific so that it will be possible to audit companies for their conformance with the standard.

Examples of the components of the ISO 14000 environmental management systems are:

1. Environmental management principles
2. Environmental labeling
3. Environmental performance evaluation
4. Life cycle assessment
5. Principles of environmental auditing
6. Terms and definitions

The environmental management system (EMS) of ISO 14001 is part of the general management system that includes organizational structure, planning activities, responsibilities, practices, procedures, processes, and resources for developing, implementing, achieving, reviewing, and maintaining the environmental policy of an organization. It is a structured process for the achievement of continual improvement related to environmental matters. The facility has the flexibility to define its boundaries and many choose to carry out this standard with respect to the entire organization or to focus the EMS on specific operating units or activities of the organization.

The EMS enables an organization to identify the significant environmental impacts that may have arisen or that may arise from the organization's past, existing, or planned activities, products, or services. It helps the organization to identify relevant environmental, legislative, and regulatory requirements that may be imposed on it. Finally, the EMS helps in planning, monitoring, auditing, corrective action, and review activities to assure compliance with established policy and allows a company to be proactive in terms of meeting anticipated new standards and compliance objectives.

Advantages and disadvantages of the ISO 14000 series of standards are listed below.

Advantages

1. The ISO 14000 standards provide industry with a structure for managing their environmental problems, which presumably will lead to better environmental performance.

2. It facilitates trade and minimizes trade barriers by harmonization of different national standards. As a consequence, multiple inspections, certifications, and other conflicting requirements could be reduced.

3. It expands possible market opportunities.

4. In developing countries, ISO 14000 can be used as a way to enhance regulatory systems that are either nonexistent or weak in their environmental performance requirements.

5. A number of potential cost savings can be expected, including:

   Increased overall operating efficiency
   Minimized liability claims and risk
   Improved compliance record (avoided fines and penalties)
   Lower insurance rates

Disadvantages

1. Implementation of ISO 14000 standards can be a tedious and expensive process.

2. ISO 14000 standards can indirectly create a technical trade barrier to both small businesses and developing countries due to limited knowledge and resources (e.g., complexity of the process and high cost of implementation, lack of registration and accreditation infrastructure, etc.)

3. ISO 14000 standards are voluntary. However, some countries may make ISO 14000 standards a regulatory requirement that can potentially lead to a trade barrier for foreign countries who cannot comply with the standards.

4. Certification/registration issues, including

   The role of self-declaration versus third-party auditing
   Accreditation of the registrars
   Competence of ISO 14000 auditors
   Harmonization and worldwide recognition of ISO 14000 registration

Auditing a facility for certification involves several steps. Proper planning and management are very essential for effective auditing. The (lead) auditor must prepare an audit plan to ensure a smooth audit process. The audit plan must, in general, remain flexible so that any changes to the audit that are found necessary during the actual audit process can be made without compromising the audit.

An audit plan must include the following 10 items:

1. A stated scope and objective(s) for the audit. This includes the reason for conducting the audit, the information required, and the expectation of the audit.

2. Specification of the place, the facility, the date of the audit, and the number of days required to perform the audit.

3. Identification of high-priority items of the facility's and/or organization's EMS.

4. Identification of key personnel who will be involved in the auditing process.
5. Identification of standards and procedures (ISO 14001) that will be used to determine the conformance of various EMS elements.
6. Identification of audit team members including their special skills, experience, and audit background.
7. Specification of opening and closing meeting times.
8. Specification of confidentiality requirements during the audit process.
9. Specification of the format of the audit report, the language, distribution requirement, and the expected date of issue of the final report.
10. Identification of safety and related issues associated with entry and inspection of various portions of the facility, along with other equipment required to conduct an effective and efficient audit.

PROBLEMS

B.1

Read the following 10 statements regarding the ISO 14000 series of standards. Carefully justify your answers as TRUE (T) or FALSE(F).

1. ISO 14000 standards are based on a principle assumption that better environmental management will lead to better environmental performance, increased efficiency, and a greater return on investment. The standards do not explicitly indicate how to achieve these goals, nor prescribe what environmental performance standards an industry must achieve.
2. ISO 14000 standards are regulatory standards developed by the International Organization for Standardization (ISO).
3. ISO 14000 standards are market-driven and therefore are based on voluntary involvement of all interests in the market place.
4. The adoption of ISO 14000 is a one-time commitment. The company, however, needs to renew the certificate yearly.
5. A main driving force of ISO 14000 standards is the need of the EPA for the replacement of an obsolete regulatory system.
6. A minimum education requirement for an auditor with 5 years "appropriate work experience" is a high school diploma or equivalent.
7. Companies can only demonstrate compliance through third-party registration.
8. A single ISO certificate can cover several sites or facilities, or portions of sites or facilities within a single company.

SOLUTION:

1. True. ISO 14000 standards are process standards, not performance standards. They do not prescribe to a company what environmental performance they

must achieve. They provide a building block for a system to achieve environmental goals. As a consequence, these standards will lead a company to cost savings through better performance of the environmental aspects of an organization's operations.

2. False. ISO 14000 standards are international, voluntary standards developed by Technical Committee 207 (TC 207) of the International Organization for Standardization (ISO).

3. True. ISO 14000 standards are market-driven and therefore are based on voluntary involvement of all interests in the marketplace.

4. False. The adoption of ISO 14000 standards is a continual commitment. Top management must establish the company's environmental policy and make a commitment to continual improvement and prevention of pollution and to comply with relevant environmental legislation and regulations. Once the company is certified; the certificate is normally valid for 3 years. This may vary depending upon the certification body. The certification body must conduct surveillance audits no less frequently than once a year and carry out a full audit after 3 years.

5. False. U.S. EPA has been participating actively in the standards development process. At present there is no indication of adoption of the ISO 14000 standards as a possible regulatory requirement. The driving force of the ISO 14000 series of standards is mainly from the private sector.

6. True. General qualification criteria for environmental management system auditors include education and work experience. Auditors should have completed at least a secondary education or equivalent with 5 years "... appropriate working experience or a college degree with 2 years appropriate work experience."

7. False. Companies can demonstrate compliance through either a self-declaration or third-party registration.

8. True. Under ISO 14001 certification, a single certificate can cover a specific site of a company, a specific facility, several facilities or portions of sites or facilities. For example, one ISO 14001 certificate might encompass four different sites of a company in four different states and a portion of a site in a fifth state if these sites are audited at the same time against the same standard.

B.2

A third-party audit is conducted in a pulp and paper mill facility. You are one of the auditors assigned to verify written procedures that are in place and thus to verify the effectiveness of the facility's environmental management system. One of the personnel you are interviewing is the facility's environmental manager. You asked the manager to verify that she has access to all of the regulations and laws that are applicable to the plant. The manager signals you to follow her to a library in the next room. She stops in the library and proudly points toward the four bookcases occupied with texts on environmental laws.

You are impressed but you continue to ask the environmental manager another question. "Can you show me the procedure you use to evaluate regulatory compliance within your facility?" The manager immediately responds by saying that she does not need a procedure for compliance evaluation because she is intimately familiar with what documentation needs to be reviewed when a compliance issue arises. Discuss the above situation with regard to conformance and non-conformance.

SOLUTION: This is a major non-conformance because the facility lacks a procedure to evaluate its compliance with applicable regulations. The ISO 14001 standard, Section 4.2.2, states that "That organization shall establish and maintain a procedure to identify and have access to legal and other requirements to which the organization subscribes directly applicable to the environmental aspects of its activities products or services." Thus, the facility is in non-conformance with Section 4.2 and Clause 4.2.2.

The issuance of non-conformance does not mean that the facility cannot be certified. Since this is a major non-conformance, however, the facility must take corrective action and show that the procedure is in place and effective in order to come into conformance with ISO 14001 standards. Having all this in place, the facility may request that the registrar reconsider its certification.

APPENDIX C

MISCELLANEOUS TOPICS

C.1 STATE REGULATORY AGENCY NAMES

What's in a name?

Fourteen states "*protect*" their environments. Two (California and Illinois) are the only ones that, just like their Federal counterpart, are named an **Environmental Protection Agency**. Nine (Connecticut, Florida, Kentucky, Maine, Massachusetts, New Jersey, Ohio, Pennsylvania, and West Virginia) are called **Department of Environmental Protection**. One overall agency name is a **Division of Environmental Protection** (Georgia) and two (Iowa and Nevada) have an **Environmental Protection Division** or **Division of Environmental Protection** under their overall environmental agency.

Four states "*manage*" their environments. Three (Alabama, Indiana, and Rhode Island) do it under their **Department of Environmental Management**, and Hawaii has an **Environmental Management Division**.

Three states (Alaska, New York and Tennessee) "*conserve*" their environment through a **Department of Environmental Conservation**. Texas conserves its natural resources through a **Natural Resources Conservation Commission**, and Nevada has a **Department of Conservation and Natural Resources**.

Four states (Colorado, Hawaii, Kansas, and North Dakota) link the environment with "*health*" under **Departments of Public Health & Environment, Health,** or **Health and Environment**. South Carolina has a **Department of Health and Environment Control**. Delaware and Minnesota are also "*control*" orientated—with a

Environmental Regulatory Calculations Handbook, by Leo Stander and Louis Theodore
Copyright © 2008 John Wiley & Sons, Inc.

Department of Natural Resources and Environmental Control, and **Pollution Control Agency**, respectively.

Nine states (Delaware, Iowa, Missouri, Nevada, North Carolina, South Dakota, Texas, Vermont, and Wisconsin) get back to nature with "**Natural Resources**" somewhere in the title of their environmental agency.

Perhaps the more goal orientated states (Arizona, Arkansas, Idaho, Louisiana, Michigan, Mississippi, Missouri, Montana, Nebraska, Oregon, Utah, Virginia, and Wyoming)—a baker's dozen in all—label theirs as **Departments** or **Divisions of Environmental Quality**. New Mexico wins the brevity and to the point award with its **Environment Department** just beating out Maryland's **Department of the Environment**. Nevada covers all bets with its **Nevada Department of Conservation and Natural Resources—Division of Environmental Protection**.

New Hampshire, Vermont and Washington are unique with their **Department of Environmental Services, Agency of Natural Resources, and Department of Ecology**, respectively, the latter two joining a list that includes three others (Minnesota, Texas and Wisconsin) that don't have the "environment" search string somewhere in their agency or division names. Minnesota tells it like it is— its environmental agency is a **Pollution Control Agency**.

In addition to its name, the website of each state agency varies significantly. For example, Table C.1 describes Alabama's approach to pollution prevention.

C.2 FEDERALISM AND PREEMPTION

Federalism is basically a sharing of power between the states and national (federal) government. In 1787, the new federal government only exercised limited or enumerated powers granted to it by the Constitution, such as making treaties and printing money. The Tenth Amendment to the Constitution clarified that all other powers belonged to the states—"The powers not delegated to the United States by the Constitution, nor prohibited by it to the states, are reserved to the states respectively, or to the people."

The Supremacy Clause of the U.S. Constitution (U.S. Constitution, article, VI, cl. 2.) declares that all laws made pursuant to the Constitution shall be the supreme law of the land. An earlier Chief Justice recognized the statutory hierarchy of the federal-state system by observing that state laws that interfere with, or are contrary to, federal law must yield to federal statutes.

Congress may preempt state law by expressly stating so. Absent explicit preemption, Congress's intent to supercede state law may be inferred in the following situations:

1. When Congress comprehensively has regulated the field, leaving no room for states or localities to supplement federal law.
2. When Congress's interest is so dominant that the federal statutory scheme must be assumed to preclude enforcement of state laws.
3. When the object that the federal law in question seeks to obtain is the same as that which the state law seeks.

TABLE C.1 Pollution Prevention

Division Home Page: N/A

DESCRIPTION

Alabama takes a voluntary approach to pollution prevention as opposed to a regulatory approach. It encourages businesses to utilize P^2 by providing information when requested, referrals to a nonprofit foundation that will perform free waste reduction opportunity assessments at the facility in a confidential environment, and by offering seminars about pollution prevention. The pollution prevention program at the Alabama Department of Environmental Management is funded solely by the federal Pollution Prevention Incentives for States grants. Matching funds for the grant come from donations to a private foundation that offers free waste reduction opportunity assessments to businesses upon request. Pollution prevention activities at the ADEM fall into one of six categories: Regulatory Integration, Infrastructure Development and Support, Technical Assistance, Education and Outreach, Awards and Recognition, and Data Collection and Analysis.

CONTACT	**ADDRESS**
oeomail@adem.state.al.us	Alabama Department of
Phone (334) 394-4360	Environmental Management
Fax (334) 394-4383	P.O. Box 301463
	Montgomery, AL 36130-1463

Pollution Prevention Programs: Alabama Department of Environmental Management stresses pollution prevention in the form of awards, on-site assessments, P^2 fact sheets, training, presentations, workshops, and seminars.

Waste Reduction and Technology Transfer (WRATT) Foundation: provides free, confidential, non-regulatory, and non-binding waste reduction opportunity assessments to businesses or other entities upon request.

The above has been drawn from W. K. Matystik, L. Theodore and R. Diaz, "State Environmental Agencies on the Internet," Government Institute, Rockville, MD, 2000.

Lastly, state law is preempted to the extent that it actually conflicts with federal law.

When Congress makes its intentions to preempt state law clear—through explicit language in the statute or in the statute's legislative history, or when a properly issued regulation promulgated pursuant to the federal statute explicitly preempts state law, there is said to be *explicit preemption*. *Implicit preemption*, on the other hand, involves inferring preemption through those less than crystal clear tests cited by the Supreme Court.

C.3 HYBRID SYSTEMS

Traditional environmental control equipment may not be capable of achieving a standard set by regulation. For certain applications, hybrid systems may be employed.

Hybrid systems are defined as those types of control devices that involve combinations of control mechanisms, for example, fabric filtration combined with electrostatic precipitation. Unfortunately, the term *hybrid system* has come to mean different things to different people. The two most prevalent definitions employed today for hybrid systems are:

1. Two or more pieces of different air pollution control equipment connected in series, e.g., a baghouse followed by an absorber.
2. An air pollution control system that utilizes two or more collection mechanisms simultaneously to enhance pollution capture, e.g., an ionizing wet scrubber (IWS).

C.4 ELECTROMAGNETIC FIELDS (EMFS)

Most individuals are surrounded by low-level electric and magnetic fields (EMFs) from electric power lines, and appliances and electronic devices. During the 1980s, the public became concerned about such fields because of media reports of cancer clusters in residences and schools near electric substations and transmission lines. In addition, a series of epidemiological studies showed a weak association between exposure to power-frequency electromagnetic fields and childhood leukemia or other forms of cancer.

The high standard of living in the United States is due large measure to the use of electricity. Technological society developed electric power generation, distribution, and utilization with little expectation that exposure to the resultant electric and magnetic fields might possibly be harmful beyond the obvious hazards of electric shocks and burns, for which protective measures were instituted. Today, the widespread use of electric energy is clearly evident by the number of electric power lines and electrically energized devices. Because of the extensive use of electric power, most individuals in the United States are today exposed to a wide range of EMF. It is estimated that at least 100,000 people have been exposed throughout their lives to technology-generated electric and magnetic fields.

EMFs are invisible lines of force surrounding any electrical device. Power lines, electrical wiring and appliances all produce EMFs. Electric fields are produced by voltage, measured in volts per meter (v/m) or kilovolt per meter (kv/m), and are easily shielded by conducting objects like trees and buildings. Electric fields decrease in strength with increasing distance from the source. Magnetic fields are produced by current, measured in gauss (G) or tesla (T), and are not easily shielded by most materials. Magnetic fields decrease in strength, as electric fields do, with increasing distance from the source.

There are no federal health standards and regulations related to EMF exposure. Six states have set standards for transmission line electric fields (Florida, Minnesota, Montana, New Jersey, New York and Oregon). The states of New York and Florida have also set standards for transmission line magnetic fields. To date, 14 studies have analyzed a possible association between proximity to power lines and various types of childhood cancer. Four of the 14 studies showed a statistically significant association with acute lymphocytic leukemia, the most common form of leukemia.

The most frequently reported health effect of EMFs is cancer, particularly elevated risks of leukemia, lymphoma, and nervous systems cancer in children. Also, birth defects, behavioral changes, slowed reflexes, and spontaneous abortions have been noted.

If scientists eventually reach a consensus that low-level electromagnetic fields do cause cancer or some other adverse health effect, then regulations defining some safe exposure level will have to be written at some later date. But so far the data are not complete enough for regulators.

Future research is also questionable at this time because much of the research into the effects of electromagnetic fields, especially that on mechanisms that could cause health effects, is cross-disciplinary, highly complicated, and has raised more questions than it was answered. It may take more than a decade to elucidate the mechanisms. It appears that the technical profession presently does not know if EMF exposure is harmful (aside from the concern for electric shocks and burns for extreme exposure). It does not know if certain levels of EMFs are safer or less safe than other levels. With most chemicals, one assumes exposure at higher levels is worse than less exposure at lower levels. This may or may not be true for EMFs. More research is required to identify dose-response relationships. There is some evidence from laboratory studies that suggest that there may be "windows" for effects. This means that biological effects are observed at some frequencies and intensities but not at others. Also, it is not known if continuous exposure to a given field intensity causes a biological effect, or if repeatedly entering and exiting of the field causes effects. There is no number to which one can point and say "that is a safe or hazardous level of EMF." Many years may pass before scientists have clear answers on cancer or on any other possible health problems that could be caused by electromagnetic fields. But over the long run, avoiding research probably will not be acceptable. It appears that the public will continue to demand research funding and answers to these questions.

C.5 SUSTAINABLE DEVELOPMENT

Sustainability involves simultaneous progress in four major areas: human, economic, technological, and environmental. The United Nations (World Commissions on Environment and Development 1987 Report, *Our Common Future* defined "sustainable development" as:

> Development that meets the need of the present without compromising the ability of future generations to meet their own needs.

Sustainability requires conservation of resources, minimizing depletion of non-renewable resources, and using sustainable practices for managing renewable resources. There can be no product development or economic activity of any kind without available resources. Except for solar energy, the supply of resources is finite. Efficient designs conserve resources while also reducing impacts caused by material extraction and related activities. Depletion of nonrenewable resources and overuse of otherwise renewable resources limits their availability to future generations.

Another principal element of sustainability is the maintenance of ecosystem structure and function. Because the health of human populations is connected to the health of the natural world, the issue of ecosystem health is a fundamental concern to sustainable development. Sustainability requires that the health of all diverse species as well as their interrelated ecological functions be maintained. As only one species in a complex web of ecological interactions, humans cannot separate their survivability from that of the total system.

Sustainable development demands change. Consumption of energy, natural resources, and products must eliminate waste. The manufacturing industry can develop green products that can meet the sustainability requirements. Life cycle analysis, design for environment and toxic use reduction are elements that help sustainability. Sustainable manufacturing, for example, extends the responsibility of industry into material selection, facility and process design, marketing, cost accounting, and waste disposal. Extending the life of a manufactured product is likely to minimize waste generation. Design engineers must consider many aspects of the product including its *durability, reliability, remanufacturability* and *adaptability*.

Designing a product that can withstand wear, stress, and degradation extends its useful life. This, in many cases, reduces the cost and impact on the environment. Reliability is the ability of a product or system to perform its function for the length of an expected period under the intended environment. Reducing the number of components in a system and simplifying the design can enhance the reliability. Screening out potentially unreliable parts and replacing with more reliable parts helps to increase the system reliability.

Adaptable designs rely on interchangeable parts. For example, consumers can upgrade components as needed to maintain state-of-the-art performance. In remanufacturing, used worn products are restored to "like-new" condition. Thus remanufacturing minimizes the generation of waste. Products that are expensive but not subject to rapid change are the best candidates for remanufacturing. Design continuity between models in the same product line increases interchangeable parts. The parts must be designed for easy disassembly to encourage remanufacturing.

As discussed in Chapter 10, design of products that emphasizes efficient use of energy and materials reuse and recycling reduces waste and supports sustainability. By effective recycling, material life can be extended. Materials can be recycled through open loop or closed loop pathways. For example, post-consumer material is recycled in an open loop one or more times before disposal. However, in a closed-loop pathway, such as with solvents, materials within a process are recovered and used as substitutes for virgin material. Minimizing the use of virgin materials supports sustainability. Thus, resource conservation can reduce waste and directly lower environmental impact. Manufacturing a less material-intensive product not only saves materials and energy but may also be lighter, thus reducing energy and costs related to product transportation. Process modifications and alterations specifically focused on replacing toxic materials with more benign ones minimize the health risk and the environmental impact of material used and product manufacturing. It also improves the health and safety of employees. Process redesign may also yield "zero discharge" by completely eliminating waste discharges. Thus, sustainability can be accomplished through several different approaches. Evaluating these options up-front will aid in developing truly sustainable processes and products.

C.7 ENVIRONMENTAL JUSTICE

The issue *environmental justice* has come to mean different things to different people. EPA indicates that environmental justice is the fair treatment and meaningful involvement of all people, regardless of race, color, national origin, or income, with respect to the development, implementation, and enforcement of environmental laws, regulations, programs, and policies. Fair treatment means that no racial, ethnic, or socioeconomic group should bear a disproportionate share of the negative environmental consequences resulting form industrial, municipal, and commercial operations, or from the execution of federal, state, local, or tribal programs and policies. Thus, EPA has established this goal for all communities and persons across this nation. The goal will be achieved when everyone enjoys the same degree of protection from environmental and health hazards, and equal access to the decision-making process to have a healthy environment in which to live, learn, and work.

Additional details are available at: http://www.epa.gov/compliance/environmental justice/index.html.

INDEX

Environmental Regulatory Calculations Handbook, by Leo Stander and Louis Theodore
Copyright © 2008 John Wiley & Sons, Inc.